Contents

EXCEL MANUAL
for Moore and McCabe's

Introduction to the Practice of Statistics

Third Edition

Revised Edition
for Excel 97 (Windows) and Excel 98 (Macintosh)

Fred M. Hoppe
McMaster University

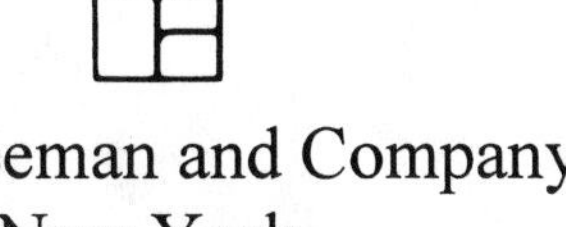

W. H. Freeman and Company
New York

Excel screen shots reprinted with permission from the Microsoft Corporation.

Cover image: Andy Warhol, 200 Campbell's Soup Cans, 1962. 72 × 100 inches. © 1998 Andy Warhol Foundation for the Visual Arts/ARS, New York, NY. Photo courtesy Leo Castelli Photo Archives.

Printed in the United States of America

ISBN: 0-7167-4003-6

First printing 1999

Preface

In recent years there has been a paradigm shift in the teaching of statistics away from a mathematical presentation and toward an emphasis on the exploratory and data analytic nature of the subject, often with reference to substantive fields.

A major theme in this new statistical symphony is that results of any analysis be supplemented with a graph. Another theme is that students who complete a first course should be able to practically, not merely theoretically, do a statistical analysis. Given a data set and a set of questions to be addressed, students should know how to input the data using the appropriate tool (from the list of tools they have been taught), examine the data graphically for evidence that confirms or dispels their assumptions, draw justifiable conclusions, and finally export the results for presentation in report form.

Central then to the success of such an approach to teaching statistics is reliable and easy-to-learn software. Many excellent software packages do exist but contain encyclopedic user manuals unsuitable for a first course at the undergraduate level. Even massive packages are not suited for all purposes. For instance, although SAS can do logistic regression, the best tool for the task is S-plus.

For a number of years at McMaster University I have taught a statistics course taken by first-year science students who pine for a career in the life sciences. Statistics is often a prerequisite for upper-level courses. Computing is a central and tested aspect of this course because students are expected to carry out statistical analyses in laboratory courses involving real data. To this end, small groups of students meet in computer labs to learn at least one statistics package.

The software we have traditionally used in first-year courses at McMaster is Minitab (in higher years, students are exposed to SAS and S-plus, as well as Matlab) because it is relatively easy to learn and McMaster has a site license for Minitab within its computer network. Unfortunately, the cost of legitimate individual copies of Minitab, even the student version, makes purchase of the software prohibitive for most students. This means that the only time many of them have for using Minitab is in the lab.

I have noticed that the reports submitted by some students, especially for assignments that require graphical output, contain more sophisticated graphs than are available with Minitab. I have learned that many choose to work in Excel through Microsoft Office instead of Minitab.

Excel is a spreadsheet, a tool for organizing data contained in columns and rows. Operations on data mimic those described by mathematical functions, and the formulas required for data analysis can therefore be expressed as spreadsheet operations. Although originally developed as a business application to display numbers in a tabular format and to automatically recalculate values in response to changes in numbers in the table, current spreadsheets enjoy built-in functions and display capabilities that can be used for statistical analysis.

Spreadsheets are therefore an alternative to specialized statistical software. Because Excel is part of an integrated word processing/graphics/database package, data can be easily input from other applications and exported into report form. For instance, I regularly download stock prices into Excel from the Internet to plot the course of investments, and my students enjoy this particular application as an incentive for learning statistics.

For the last four years I have used both *Introduction to the Practice of Statistics* by David Moore and George McCabe and *The Basic Practice of Statistics* by David Moore. Before, I had used another book by David Moore called *Statistics, Concepts and Controversies* when I was at the University of Michigan, Ann Arbor. These books are leading expositions of the new approach to the teaching of statistics.

So when Christopher Spavins, the W.H. Freeman representative at McMaster, with whom I've had many interesting conversations on the teaching of statistics, asked if I was interested in developing an Excel manual to accompany the third edition of *Introduction to the Practice of Statistics (IPS)*, I agreed. This guide, produced over a span of two months, is the result.

Some key features of this guide:

- Introductory chapter on Excel for those with no prior knowledge of Excel.
- Written for Excel 97/98 with an Appendix for Excel 5/95. Explanations are given in parallel for Excel 97/98 and Excel 5/95 when the interface differs.
- Figures from Excel 98 for Macintosh, Excel 97 for Windows, and Excel 5 for Macintosh showing differences and similarities in the user interface.
- Presentation follows *IPS* with fully worked and cross-referenced examples and exercises from *IPS*.
- Detailed exposition of the ChartWizard for graphical displays.
- Extensive use of Named Ranges to make formulas more transparent.
- Nearly 200 figures accompanied by step-by-step descriptions.
- Detailed use of simulation to explain randomness by simulating the Central Limit Theorem and the Law of Large Numbers. General distributions are simulated from the uniform (0,1) distribution by using critical values obtained in Excel to find the inverse probability transformation.
- Construction of a normal table and graphs of the normal and the Student t densities by a general method applicable to other densities.
- Development of weighted least-squares for logistic regression.

- Templates provided for one-sample procedures, chi-square tests, and nonparametric statistics.
- Development of side-by-side boxplots to supplement Excel.

I believe that Excel can satisfy all of a student's needs in a course based on a book such as *IPS*. In fact, every technique discussed in *IPS* can be developed within Excel. Students who learn statistics using Excel will find it extremely valuable in future courses (a point brought home by one of my third-year students who informed me that she used Excel in her physics course for plotting data because she "didn't feel like doing it by hand").

In some of my own consulting I have found researchers, self-taught in statistics, using Excel. Much of experimental research is exploratory in nature and Excel's excellent graphical capabilities and easy-to-use interface make it the statistics tool of choice for many in the life sciences, engineering, and business. Finally, given that Excel is produced by Microsoft, one can predict without fear a long, useful, and upgradeable life for Excel. All these considerations lead me to believe that Excel will grow in popularity in the teaching of statistics.

I will be putting up a Web site at http://www.mathematics.net/ where updates, macros, and other useful information will appear. If you have any material you think may be useful for this site, contact me. It is also my intention to install a bulletin board to facilitate communication by faculty and students who use this manual with *IPS*.

In addition to Chris Spavins, I thank Patrick Farace, statistics editor at W.H. Freeman, for his assistance in bringing this book to market and for showing me life in the fast lane of publishing, and I am grateful to Erica Seifert and Jodi Isman for their editorial and copyediting skills. The book was typeset in LaTeX by Debbie Iscoe, who, as with my earlier book, did a superb job with both care and good cheer. Brian Golding kindly showed me how to include eps files as figures.

Finally, thanks to my terrific wife, Marla, and super kids, Daniel and Tamara, for putting up with a kitchen table that became the extension of my desk and for lovingly and unquestioningly giving me the time (though not necessarily quiet time) to complete this project.

FRED M. HOPPE
DUNDAS, ONTARIO
JUNE 11, 1998
E-mail: *hoppe@mcmaster.ca*

Introduction

This book is a supplement to *Introduction to the Practice of Statistics*, Third Edition, by David S. Moore and George P. McCabe, referred to as *IPS*. Its purpose is to show how to use Excel in performing the common statistical procedures in *IPS*.

I.1 What Is Excel?

Microsoft Excel is a spreadsheet application whose capabilities include graphics and database applications. A spreadsheet is a tool for organizing data. Originally developed as a business application for displaying numbers in a table, numbers that were linked by formulas and updated whenever any part of the data in the spreadsheet changed, Excel now has built-in functions, tools, and graphical features that allow it to be used for sophisticated statistical analyses.

Windows or Macintosh?

It doesn't matter which you use. This book is designed equally for Macintosh or Windows operating systems. The Macintosh and Windows versions of Excel function essentially the same way, with a few slight differences in the file, print, and command shortcuts. These are due mainly to the absence of a right mouse button for the Macintosh. However, the right button action can be duplicated with a keystroke, and I have described both actions where they differ. Both Macintosh and Windows users will find this book useful.

Nearly all figures shown in this book have been generated using Excel 98 on a Macintosh running Mac OS 8.5. A few were from Excel 97 on a Toshiba notebook running Windows NT. The figures in the Appendix are from Excel 5 on a Macintosh. Students should feel familiar with the look of the Excel interface no matter what the platform.

Which Version of Excel Should I Use?

Naming conventions are slightly confusing because of a plethora of patches, bug fixes, interim releases, and so on. In the Windows environment the main versions used are Excel 5.0, 5.0c, Excel 7.0, 7.0a (for Windows 95), Excel 8.0 (also called

Excel 97), and the recent Excel 2000. For Macintosh, they are Excel 5.0 and 5.0a, and Excel 98.

The major change in the development occurred with Excel 5.0. The statistical tools in Excel 5.0 and Excel 7.0 for Windows and Excel 5.0 for Macintosh function in virtually the same way. These are referred collectively in this book as Excel 5/95. Likewise, Excel 97 for Windows and Excel 98 for Macintosh function similarly and are referred as Excel 97/98.

Excel 97/98 removed some bugs in the Data Analysis ToolPak (but introduced others), improved the interface to the ChartWizard, and replaced the Function Wizard with the Formula Palette. Help was greatly expanded with the introduction of the animated Office Assistant. Minor cosmetic changes were made in the placement of components within some dialog boxes. Otherwise, few substantive changes occurred that might cause differences in execution or statistical capabilities between Excel 5/95 and Excel 97/98.

This book is based on Excel 97/98. However, given the large base of Excel 5/95 users I have tried to make this book equally accessible to them without having two separate editions. Surprisingly, this has not been too difficult. The main differences needing instructional care are the description of the ChartWizard (reduced to four steps in Excel 97/98 from five steps in Excel 5/95) and the implementation of formulas with the Formula Palette (Excel 97/98) instead of the Function Wizard (Excel 5/95). In addition, there are related differences in some pull-down options from the Menu Bar. On the other hand, the Data Analysis ToolPak is virtually the same in each version. Nearly all differences are discussed in Chapters 1 and 2. Since this book uses a step-by-step approach with many figures to assist students visually in following written instructions, I have therefore included a detailed Appendix (with corresponding figures) that parallels for Excel 5/95 users the material in these chapters. In the remaining chapters, whenever there is a technique whose implementation varies between the two versions, I have given separate steps. Usually only a few lines of text (four or fewer) suffice, underscoring the similarity in the versions in using Excel for statistics. It is hoped that this dual approach will make this manual truly equally useful for Excel 5/95 and Excel 97/98.

Do I Need Prior Familiarity with Excel?

The short answer is no. This book is completely self-contained. This introductory chapter contains a summary and description of Excel that should provide enough detail to enable a student to get started quickly in using Excel for statistical calculations. The subsequent chapters give step-by-step details on producing and embellishing graphs, using functions, and invoking the **Analysis Toolbox**. For users without prior exposure to Excel, this book may serve as a gentle exposure to spreadsheets and a starting point for further exploration of their features.

I.2 The Excel Workbook

When you first open Excel, a new file is displayed on your screen. Figure I.1 shows an Excel 98 Macintosh opening workbook.

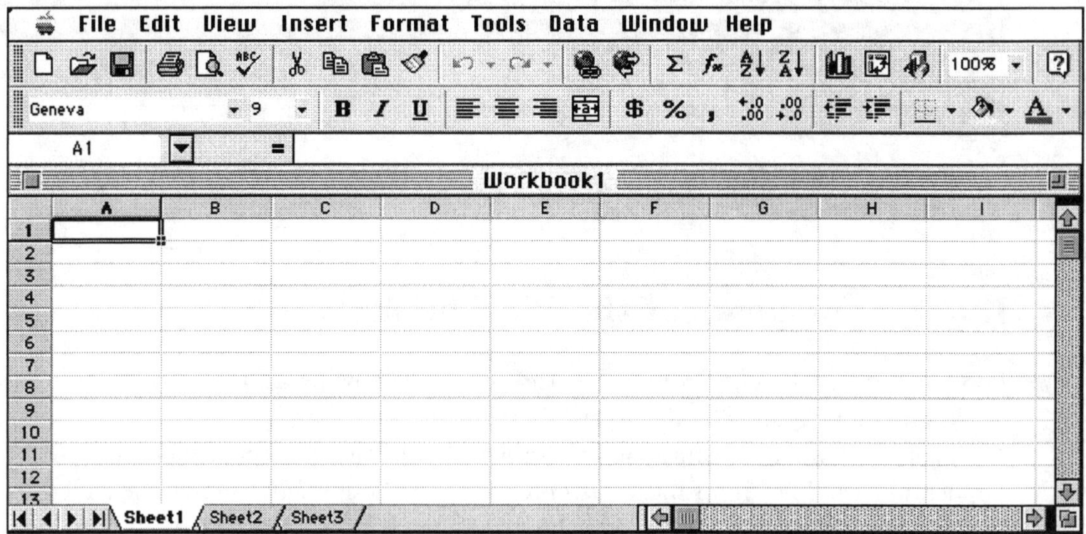

Figure I.1: Excel 98 Macintosh Workbook

A workbook consists of various sheets in which information is displayed, usually related information such as data, charts, or macros. Sheets may be named and their names will appear as tabs at the bottom of the workbook. Sheets may be selected by clicking on their tabs and may be moved within or between workbooks. To keep the presentation simple in this book we have chosen to use one sheet per workbook in each of our examples.

A sheet is an array of cells organized in rows and columns. The rows are numbered from 1 to 65,536 (up from 16,384 in Excel 5/95) while the columns are described alphabetically, as follows,

$$\text{A, B, C,} \ldots \text{, X, Y, Z, AA, AB, AC,} \ldots \text{, IV}$$

for a total of 256 columns.

Each cell is identified by the column and row that intersect at its location. For instance, the selected cell in Figure I.1 has address (or cell reference) A1. When referring to cells in other sheets, we need to also provide the sheet name. Thus Sheet2!D9 refers to cell D9 on Sheet 2.

Information is entered into each cell by selecting an address (use your mouse or arrow keys to navigate among the cells) and entering information, either directly in the cell or else in the **Formula Bar** text entry area. Three types of information can be entered: labels, values, and formulas. We will discuss entering information in detail later in this chapter.

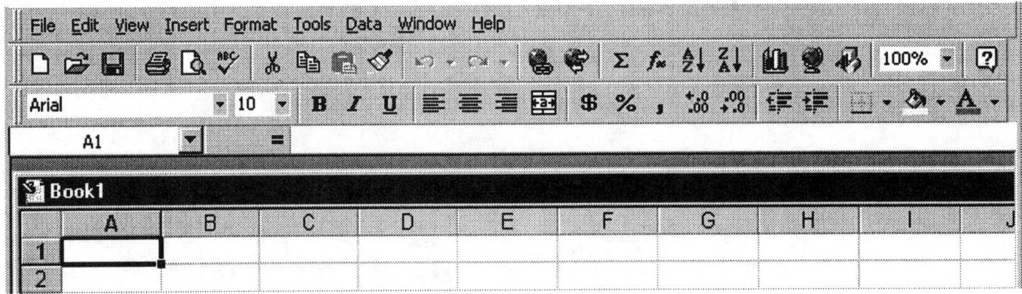

Figure I.2: Excel 97 Windows

I.3 Components of the Workbook

Look at Figure I.1. Then compare with Figure I.2 showing an Excel 97 (Windows) workbook and Figure I.3, an Excel 5 Macintosh workbook.

There are two main components in a workbook: the document window and the application window. Information is entered in the document window identified by the row and column labels. Above the document window are all the applications, functions, tools, formatting features that Excel provides. There are so many commands in a workbook that an efficient system is needed to access them. This is achieved either by pull-down menus invoked from the Menu Bar or from equivalent icons on the toolbars.

The application window is thus the control center from which the user gives instruction to Excel to operate on the data in the document window below it. We examine the main components of the application window in detail.

Menu Bar

The **Menu Bar** (Figure I.4) appears at the top of the screen. It provides access to all Excel commands: **File, Edit, View, Insert, Format, Tools, Data,**

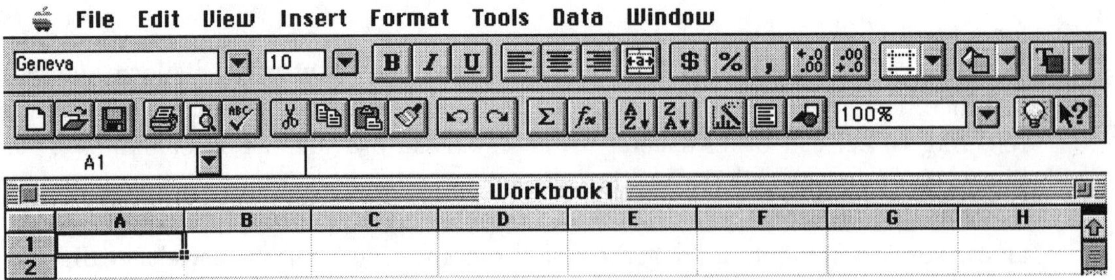

Figure I.3: Excel 5 Macintosh

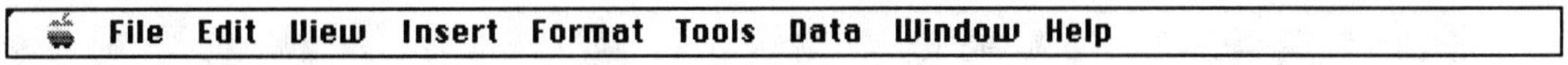

File Edit View Insert Format Tools Data Window Help

Figure I.4: Menu Bar

Table I.1: Menu Bar Pull-Down Options

Menu Bar	Pull-Down Options
File	Open, close, save, print, exit
Edit	Copy, cut, paste, delete, find, etc. (basic editing)
View	Controls which components of workbook are displayed on screen, size, etc.
Insert	Insert rows, columns, sheets, charts, text, etc. into workbook
Format	Format cells, rows, columns
Tools	Access spelling macros, data analysis toolpak (will be used throughout to access the statistical features of Excel)
Data	Database functions such as sorting, filtering
Window	Organize and display open workbooks
Help	Online help (also available on the **Standard Toolbar**)

Window, Help. Each word in the Menu Bar opens a pull-down menu of options familiar to users of any window-based application (there are also keyboard equivalents). As the name implies, this is the main component of the control center and will be elaborated on in various examples throughout this book. Table I.1 summarizes some of the options available.

Toolbars

When Excel is opened, two strips of icons appear below the Menu Bar: the **Standard Toolbar** and the **Formatting Toolbar**. Other toolbars are available by choosing **View – Toolbars** from the Menu Bar and making a selection from the choices available. Existing toolbars can be customized by adding or removing buttons, and new ones can be created. For the purpose of this book you will not need to make such customizations.

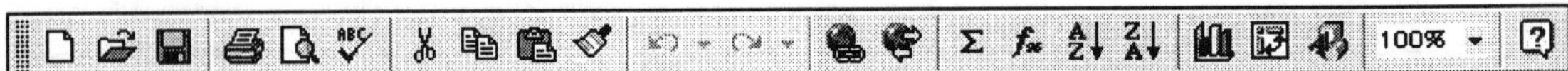

Figure I.5: Excel 98 Standard Toolbar

Standard Toolbar

The (default) **Standard Toolbar** (Figure I.5) provides buttons to ease your access to basic workbook tasks. Included are buttons for the following tasks:

- Start a new workbook
- Open existing workbook
- Save open workbook
- Print, print preview
- Check spelling
- Cut selection and store in clipboard for posting elsewhere.
- Copy selected cells to clipboard, paste data from clipboard
- Copy format
- Undo last action, redo last action
- Insert hyperlink
- World Wide Web interface
- Autosum function (may also be entered directly into cell or **Formula Bar**)
- **Paste Function** (step-by-step dialog boxes to enter a function connected to the **Formula Palette**)
- Sort descending, sort ascending order
- **ChartWizard** (covered in detail in Chapter 1).
- **PivotTable Wizard**
- Drawing toolbar
- Zoom factor for display
- **Office Assistant**

As you pass over a button with your mouse pointer a small text label appears next to the button.

Formatting Toolbar

Figure I.6 shows the **Formatting Toolbar**, by which you can change the appearance of text and data. Features offered:

- Display and select font of selected cell
- Apply bold, italic, underline formatting
- Left, center, or right justify data
- Merge and center

- Apply currency style, percent style, etc.
- Increase or decrease decimal places
- Indent
- Add borders to selected sides of cell
- Change background color of cell, change color of text in cells

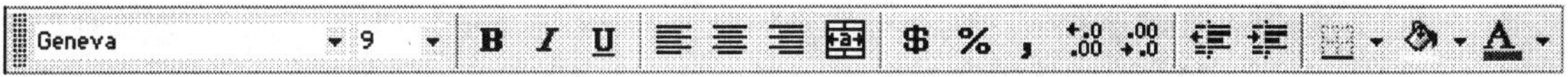

Figure I.6: Formatting Toolbar

Formula Bar

The **Formula Bar** is located just above the document window (Figure I.7). There are six areas in the Formula Bar (from left to right):

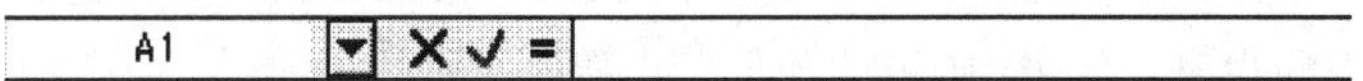

Figure I.7: Formula Bar

- **Name box.** Displays reference to active cell or function.
- **Defined name pull-down.** Lists defined names in workbook.
- **Cancel box.** Click on the red X to delete the contents of the active cell.
- **Enter box.** Click the green check mark to accept the formula bar entry.
- **Formula Palette (Excel 97/98).** Constructs a function using dialog boxes to access Excel's built-in functions, or the function can be entered directly if you know the syntax. (Excel 5/95 has f_x in place of the equal (=) sign to activate the **Function Wizard**).
- **Text/Formula Entry area.** Enter and display the contents of the active cell.

The **Cancel** box and **Enter** box buttons appear only when a cell is being edited. Once the data have been entered, they disappear.

Title Bar

This is the name of your workbook. On a Mac the default is **Workbook1**, which appears just above the document window. With Windows the default name is **Book1**.

Document (Sheet) Window

A sheet in a workbook contains 256 columns by 65,536 rows. Use the mouse pointer or arrow keys to move from cell to cell. The pointer may change appearance depending on what actions are permitted. It might be an arrow, a blinking vertical cursor (I-beam), or an outline plus sign, for instance.

Sheet Tabs

A single workbook can have many sheets, the limit determined by the capacity of your computer; it is sometimes convenient to organize a workbook with multiple sheets, for instance sheets for data, analyses, bar graphs, or Visual Basic macros. Each sheet has a tab located at the bottom of the workbook (Figure I.8), and a sheet is activated when you click on its tab. Tab scrolling buttons allow you to navigate among the sheets. Clicking on a sheet tab on the bottom of the sheet activates it. Each workbook consists initially of three sheets labeled Sheet1, Sheet2, Sheet3 (16 initially in Excel 5). Sheets can be added, deleted, moved, and renamed to achieve a logical organization of data and analyses. To rename, move, delete, or copy a sheet, **right-click (Windows)** on a sheet tab or click and hold down the Control key on a **Macintosh**—we will refer to this as **Control-click**—and a pop-up menu appears from which you can select. A new sheet can also be inserted from the Menu Bar by choosing **Insert–Worksheet**. Note that a new worksheet is added to the left of the current or selected sheet. Sheets can also be deleted, copied, or edited from the Menu Bar using **Edit – Delete Sheet** or **Edit – Move or Copy Sheet...**.

Another way to move a sheet is to grab it by clicking on it and holding the mouse button (left button for Windows). A small icon of a paper sheet will appear under the mouse pointer. As you move the mouse pointer, you will notice a small dark marker moving between the sheet tabs. This marker indicates where the sheet will be moved when you release the mouse button.

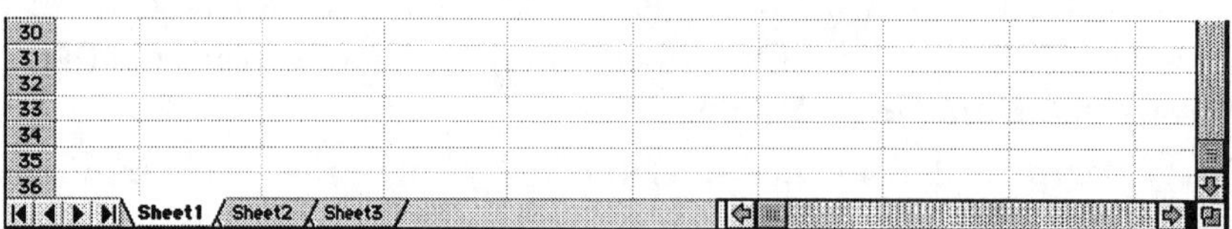

Figure I.8: Sheet Tabs

I.4 Entering and Modifying Information

When a workbook is first opened, cell A1 automatically becomes the active cell. Active cells are surrounded by a dark outline indicating that they are ready to receive data. Use the mouse (or arrow keys on the keyboard) to activate a different cell. Then enter the data and either click on the Enter box or press the enter (return) key.

Labels, Values, and Formulas

Three types of information can be entered into a cell: labels, values, and formulas. **Labels** are character strings such as words or phrases, typically used for headings or descriptions. They are not used in numerical calculations. **Values** are numbers such as 1.3, $1.75, π. **Formulas** are mathematical expressions that use the values or formulas in other cells to create new values or formulas. All formulas begin with an equal (=) sign and are entered directly by hand in the cell or in the text entry area of the **Formula Bar** or by the **Function Wizard**. As an example of how a formula operates, if the formula

$$= A1 + A2 + A3 + A4$$

is entered in cell A5 and if the contents of cells A1, A2, A3, A4 are 11, 12, 19, −6, respectively, then cell A5 will show the value 36 because what is displayed in the cell is the result of the computation, not the formula. The formula in the cell may be viewed in the entry area in the **Formula Bar** if the cursor is placed over the cell.

It is the existence of formulas that makes a spreadsheet such a powerful tool. A formula such as

$$= \texttt{SUM}(A1 : A4)$$

is the Excel equivalent of the mathematical expression

$$\sum_{i=1}^{4} A_i$$

and a complex mathematical expression can be rendered into an Excel workbook in a similar fashion.

Editing Information

There are several ways to edit information. If the data have not yet been entered after typing, then use the backspace or delete key or click on the red X to empty the contents of a cell. After the data have been entered, **activate** the cell by clicking on it. Then move the cursor to the text entry area of the Formula Bar where it turns into a vertical I-beam. Place the I-beam at the point you wish to edit and proceed to make changes.

Cell References, Ranges, and Named Ranges

A **cell reference** such as A10 is a **relative** reference. When a formula containing the reference A10 is copied to another location, the cell address in the new location is changed to reflect the position of the new cell. For instance, if the formula presented earlier

$$= A1 + A2 + A3 + A4$$

that appears in cell A5 is copied to cell D9, then it will become

$$= D5 + D6 + D7 + D8$$

to reflect that the formula sums the values of the four cells just above its location.

This relative addressing feature makes it relatively easy to repeat a formula across a row or column of a sheet, such as adding consecutive rows, by entering the formula once in one cell, then copying its contents to the other cells of interest.

If you need to retain the actual column or row label when copying a formula, then precede the label with a dollar sign ($). This is called an **absolute** cell reference. For instance, $A2 keeps the reference to column A but the reference to row 2 is relative, A$2 leaves A as a relative reference, but fixes the row at 2, $A$2 gives the entire cell (row and column labels) an absolute reference. We will use mixed ($A2 or A$2) references in Chapter 9 with the chi-square distribution.

A group of cells forming a rectangular block is called a **range** and is denoted by something like A2:B4, which includes all the cells {A2 ,A3, A4, B2 B3, B4}. **Named ranges** are names given to individual cells or ranges. The main advantage of named ranges is that they make formulas more meaningful and easy to remember. We will use named ranges repeatedly in this book. To illustrate, suppose we wish to refer to the range A2:A7 by the name "data." Enter the label "data" in cell A1. Then select the range A1:A7 by clicking on A1, holding the mouse button down, dragging to cell A7, and then releasing the mouse button. From the Menu Bar choose **Insert** – **Name** – **Create**, and check the box **Top Row** then click **OK** (Figure I.9).

Perhaps the quickest way to add a named range is to select the cell(s) to be named and then click on the **Name box**. This creates the name and associates it with the selected cells.

You can see which names are in your workbook by clicking on the defined name drop-down list arrow to the right of the **Name box**. Names are displayed alphabetically.

If you type a name from your workbook in the **Name** box and hit Enter then you will be transported to the first entry in the corresponding range, which will appear in the text/formula entry area of the Formula Bar ready for editing. Named ranges are both an aid in remembering and constructing formulas and also a convenient way to move around your workbook. They are used extensively in this book.

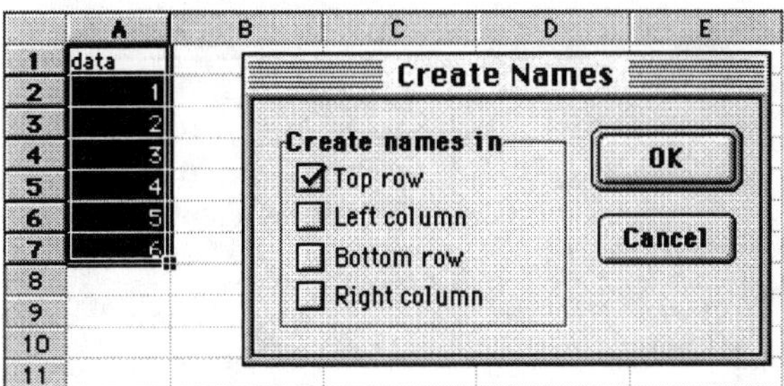

Figure I.9: Named Range

Copying Information

To activate a block of cells, place the cursor in the upper left cell of the block, click and drag to the lower right cell and release. Alternatively activate the upper left cell, then press the **Shift** key and click on the lower right cell. Noncontiguous cells or blocks can be selected by holding down the **Command** key **(Macintosh)** or the **Control** key **(Windows)** while selecting each successive cell or block.

To copy data from a cell or range, activate it, then from the Menu Bar choose **Edit – Copy**. Move the mouse cursor to the new location and from the Menu Bar choose **Edit – Paste.**

An alternative is to activate the range, then place the mouse cursor on the border of the selected range. It will appear as a pointer. Now press the **Control** key and move the cursor to the location for copying. Release the mouse button.

To move data to another location choose **Edit – Cut** from the Menu Bar and then **Edit – Paste** after you move the mouse cursor to the new location. Alternatively, move the mouse cursor to the border of the selected range. It turns into a pointer. "Grab" the border with the mouse pointer and move the cells to the desired location.

Use of the mouse for copying and cutting is a **drag and drop** operation familiar to users of Microsoft Word.

Shortcut Menu

If you activate a range and then **Control-click (Macintosh)** or **right-click (Windows)** on it, a **Shortcut Menu** will pop up next to the range allowing you access to some of the commands in the Menu Bar under **Edit**. This will provide options prior to copying or pasting.

Paste Special

A useful command from the **Shortcut Menu** is the **Paste Special**. This provides a dialog box (Figure I.10) giving a number of options prior to pasting. The

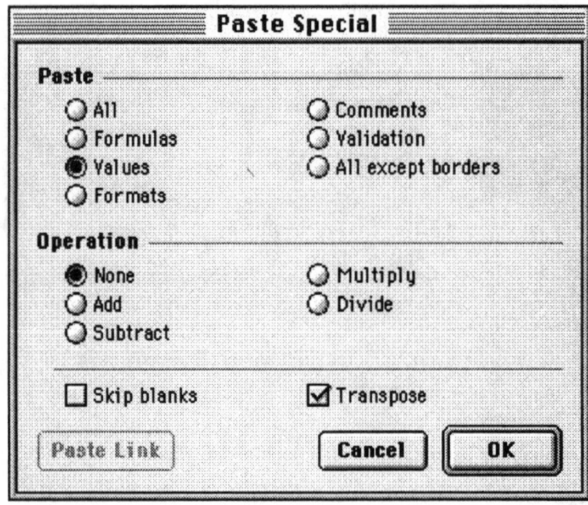

Figure I.10: Paste Special

two options most commonly used in this book are **transpose** check box, which transposes rows and columns, and the **Paste Values** radio button, which is useful if you need to copy a range of values defined by formulas. A straight copy will alter the cell references in the formulas and could produce nonsense. **Paste Special** solves this problem by pasting the values, not the formulas.

Filling

Suppose you need to fill cells A1:A30 with the value 1. Enter the value 1 in cell A1. Activate A1 and move the cursor to the lower right-hand corner (the **fill handle**) of A1. The cursor becomes a cross hair. Drag the fill handle and pull down to cell A30. This copies the value 1 into cells A2:A30.

Alternatively, you can select A1:A30 after entering the value 1 in cell A1. Then choose **Edit – Fill – Down** from the Menu Bar. Other options are available such as **Edit – Fill – Series** if this approach is taken.

I.5 Opening Files

Often you will need to open text or data fields, for instance, the data files on the Student CD-ROM accompanying *IPS*. Other times, the data may be in a binary format.

Binary

Excel can read and open a wide selection of binary files. To see which ones may be imported, choose **File – Open** from the Menu Bar and make the appropriate selection from the drop-down list.

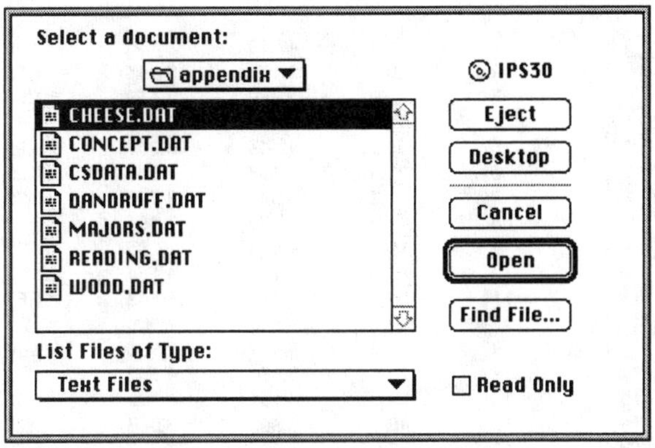

Figure I.11: Importing Files

Text (ASCII)

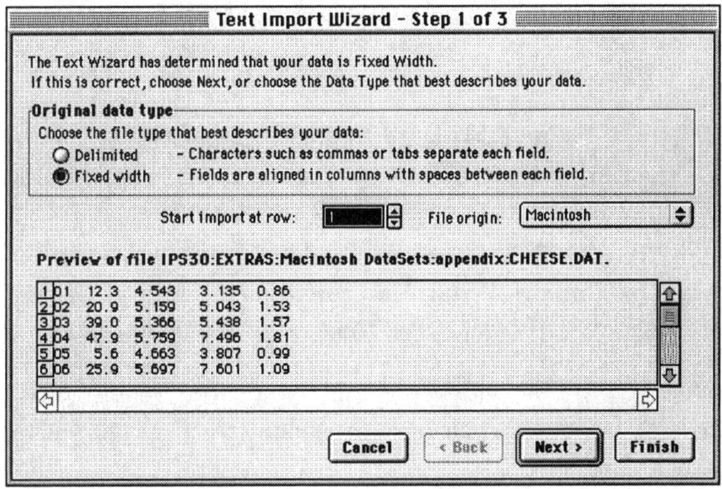

Figure I.12: Text Import Wizard—Step 1

For text (ASCII) files, Excel may start the **Text Import Wizard**, after you make your selection from the drop-down list, once you choose **File – Open** from the Menu Bar. This is a sequence of three dialog boxes allowing you to specify how the text should be imported. The **Text Import Wizard** helps you make intelligent choices.

We illustrate using the "CHEESE" data set on the CD-ROM. Figure I.11 shows that the file "CHEESE.DAT" has been selected following **File – Open – All Files.**

Step 1. The Text Import Wizard (Figure I.12) makes a determination that the

	A	B	C	D	E
1	1	12.3	4.543	3.135	0.86
2	2	20.9	5.159	5.043	1.53
3	3	39	5.366	5.438	1.57
4	4	47.9	5.759	7.496	1.81
5	5	5.6	4.663	3.807	0.99
6	6	25.9	5.697	7.601	1.09
7	7	37.3	5.892	8.726	1.29
8	8	21.9	6.078	7.966	1.78
9	9	18.1	4.898	3.85	1.29
10	10	21	5.242	4.174	1.58
11	11	34.9	5.74	6.142	1.68
12	12	57.2	6.446	7.908	1.9
13	13	0.7	4.477	2.996	1.06
14	14	25.9	5.236	4.942	1.3
15	15	54.9	6.151	6.752	1.52
16	16	40.9	6.365	9.588	1.74
17	17	15.9	4.787	3.912	1.16
18	18	6.4	5.412	4.7	1.49
19	19	18	5.247	6.174	1.63
20	20	38.9	5.438	9.064	1.99
21	21	14	4.564	4.949	1.15
22	22	15.2	5.298	5.22	1.33
23	23	32	5.455	9.242	1.44
24	24	56.7	5.855	10.199	2.01
25	25	16.8	5.366	3.664	1.31
26	26	11.6	6.043	3.219	1.46
27	27	26.5	6.458	6.962	1.72
28	28	0.7	5.328	3.912	1.25
29	29	13.4	5.802	6.685	1.08
30	30	5.5	6.176	4.787	1.25

Figure I.13: Data Imported into Excel

data were **Fixed width**. You may override this with the radio button. Sometimes the Text Import Wizard interprets incorrectly. The default starting position is shown as Row 1. This too can be changed if labels are needed.

Step 2. The next screen depends on whether you chose Delimited or Fixed width in Step 1. If Delimited, you pick the delimiter. If Fixed width, you create line breaks.

Step 3. The final step allows you to select how the imported data will be formatted. Usually the radio button **General**, the default, is appropriate.

The result is shown in Figure I.13. In this case we accepted the defaults of the **Text Import Wizard** and the data were neatly imported. If the data file contains lines of explanatory text at the top of the file, then the **Text Import Wizard** is very handy for formatting correctly upon import.

I.6 Printing

This is generally the last step. In the **Page Setup** dialog box (Figure I.14) accessed from the Menu Bar using **File – Page Setup**, there are four tabs:

- **Page.** Orientation, scaling, paper size, print quality.
- **Margins.** Top, bottom, left, right, preview window.
- **Header/Footer.** Information printed across top or bottom.
- **Sheet.** Print area, column/row title, print order.

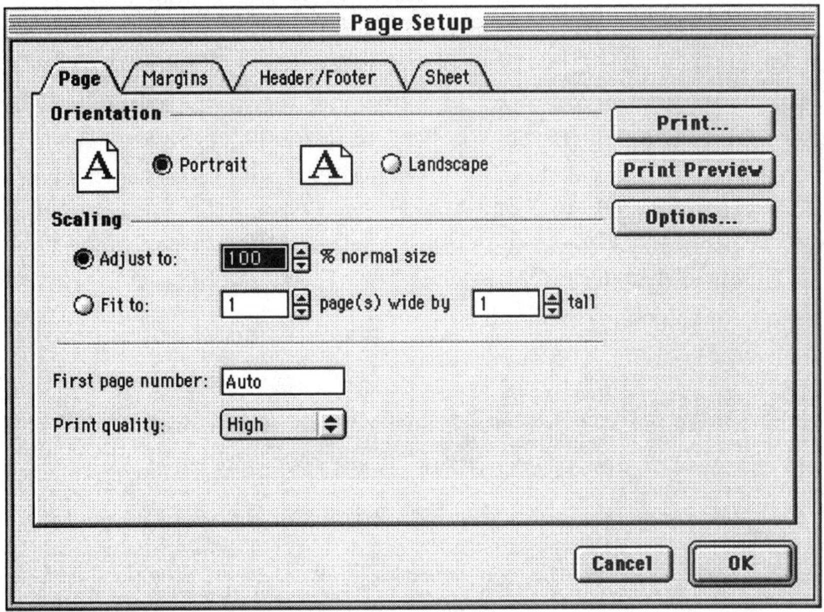

Figure I.14: Page Setup

Figure I.15: Print Preview Options

The **Print Preview** button on the **Standard Toolbar** or **File – Print Preview** from the Menu Bar allows you to see what portion of your document is being printed and where it is positioned. Other options are available (Figure I.15) at the top of the print preview screen to assist you in previewing your output prior to printing.

Finally, when you are satisfied with your output, press the **Print...** button at the top of the preview screen. You can also print directly with the document window open using the **Print** button on the **Standard Toolbar** or **File – Print** from the Menu Bar. This brings up a **Print** dialog box in which you select which pages to print, the number of copies, printer setup, as well as buttons for **Page Setup** and **Print Preview** just discussed.

I.7 Whither?

This brief introductory chapter contains a bare-bones description of Excel, as much as you need to know to access the remainder of this book. In the following chapters, not only will you make use of many of the topics and tips presented here, but you will learn about the **ChartWizard**, enter formulas, copy and paste cells, and so on. By using Excel for statistics you will also obtain a good practical background in spreadsheets.

One way to learn more about Excel is by using the animated **Office Assistant** as in Figure I.16. You can install a gallery of different characters from the Excel installation CD-ROM in the Office:Actors folder. The one shown here is called **Max**. Max can be called up using the **Help** button (the ? at the right of the **Standard Toolbar**). Ask Max a question such as "What's new?" and see how he responds. If you **right-click (Windows)** or **Control-click (Macintosh)** Max, then more options become available.

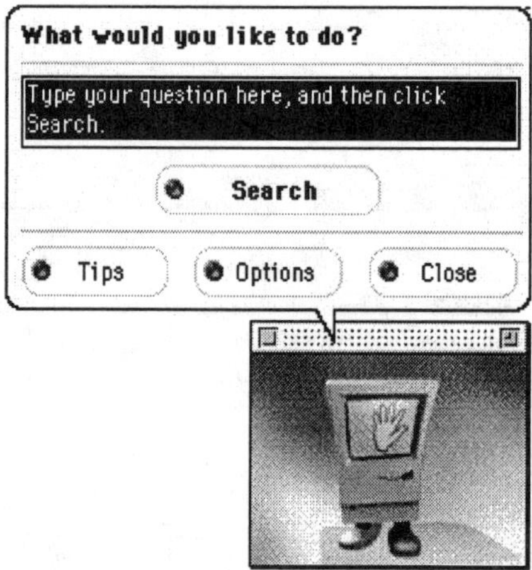

Figure I.16: Office Assistant—Max

Numerous books are available in libraries and in bookstores, but books often contain too much information. Finally, there is a wealth of recent material available on the Internet. Use your favorite search engine or directory and you'll find pointers to macros (Visual Basic programs) and sample workbooks made available by other users for applications of Excel. Excel 97/98 is "Internet ready" and can read html files and carry out "Web queries" to import data directly from the Internet.

If you find interesting Internet resources, please let me know (*hoppe@mcmaster.ca*) and I will place a pointer to them on the Excel statistics Web site, which will be located at http://www.mathematics.net/.

Chapter 1

Looking at Data—Distributions

Excel provides more than 70 functions related to statistics and data analysis as well as tools in the **Analysis ToolPak**. Additionally, the **ChartWizard** gives a step-by-step approach to creating informative graphs.

We first discuss the **ChartWizard**. The figures are from Excel 98 for Macintosh. They differ only cosmetically from Excel 97 for Windows, as illustrated in Figure 1.1.

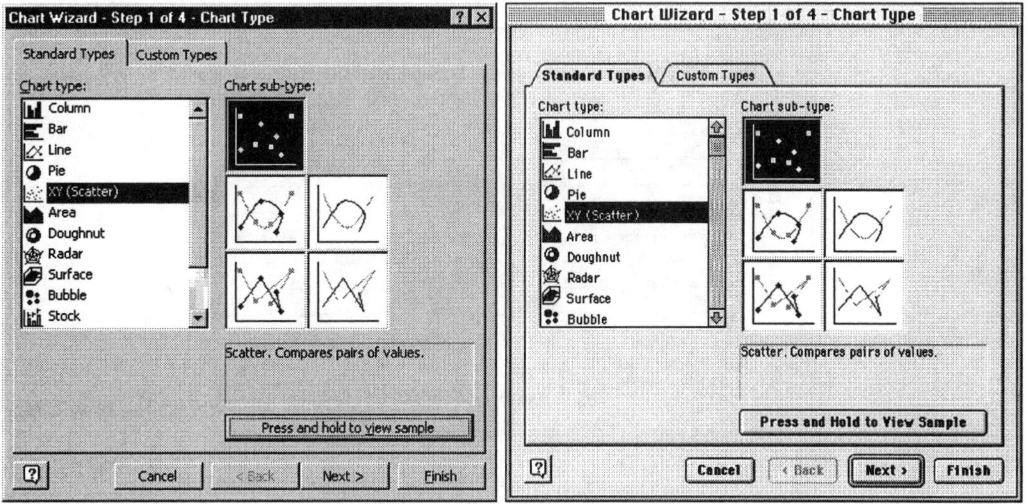

Figure 1.1: Windows (left) and Macintosh (right) ChartWizard Opening Screens

1.1 Displaying Distributions with Graphs

The ChartWizard

The **ChartWizard** is a step-by-step approach to creating informative graphs. Its interface provides a sequence of four steps in **Excel 97 (Windows)** and **Excel 98 (Macintosh)** that guide the user through the creation of a customized graph (called a Chart by Excel). The user supplies details about the chart type, formatting, titles, legends, and so on, in dialog boxes. The ChartWizard can be activated either from the button on the **Standard Toolbar** or by choosing **Insert – Chart** from the Menu Bar. The chart can be inserted in the current sheet or in a new sheet.

The following applies to Excel 97/98. **Users of Excel 5/95** should read Sections A.1 and A.2 in the Appendix instead.

> **Example 1.1.** (Page 6 in the text.) Figure 1.2 shows the marital status for all Americans age 18 and over. Create a bar graph.

	A	B	C
1	*Marital Status*	*Count (millions)*	*Percent*
2	Never married	43.9	22.9
3	Married	116.7	60.9
4	Widowed	13.4	7
5	Divorced	17.6	9.2

Figure 1.2: Marital Status

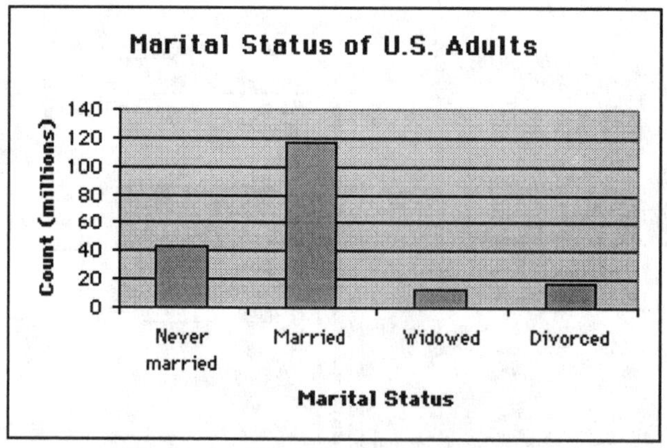

Figure 1.3: Excel Bar Chart

Solution. Figure 1.3 is a bar graph produced by Excel that displays the same information as in Figure 1.2. For other types of graphical displays, make appropriate choices from the same sequence of dialog boxes. The following steps describe how it is obtained. First enter the data and labels in cells A1:B5 and format the display as in Figure 1.2.

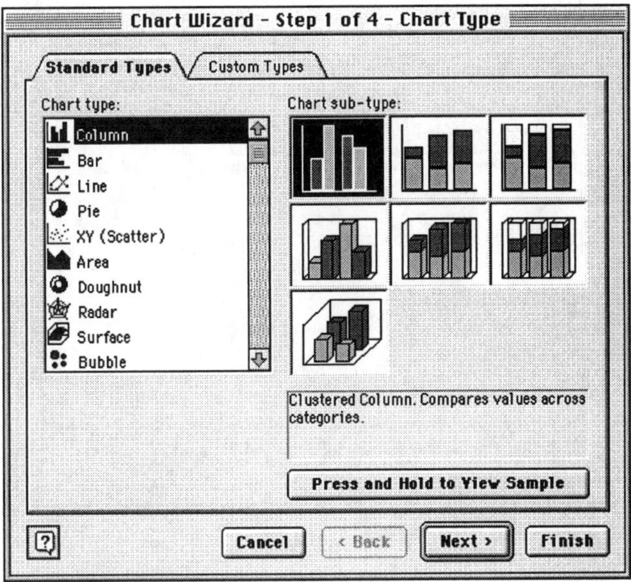

Figure 1.4: ChartWizard—Step 1

Step 1. Select cells A1:B5 and click on the **ChartWizard**. The ChartWizard (Figure 1.4) displays the types of graphs that are available. In the left field select **Column** for Chart type.

In the right field select **Clustered Column** for Chart sub-type, which is the first choice in the top row. When you select a sub-type, an explanation of the chart appears in the box below all the choices, and you can preview your chart's appearance using the "Press and Hold to View Sample" button. Click Next.

Step 2. The next dialog box (Figure 1.5) with title **Chart Source Data** previews your chart and allows you to select the data range for your chart. Since you had already selected cells A1:B5 prior to invoking the ChartWizard, this block appears in the text area **Data range**. You can make any corrections. Had you not selected the data range, then you would input the range now. Click Next.

Step 3. A dialog box **Chart Options** (Figure 1.6) appears with the default chart. Rarely is the default satisfactory; you will generally need to make cosmetic changes to its appearance.

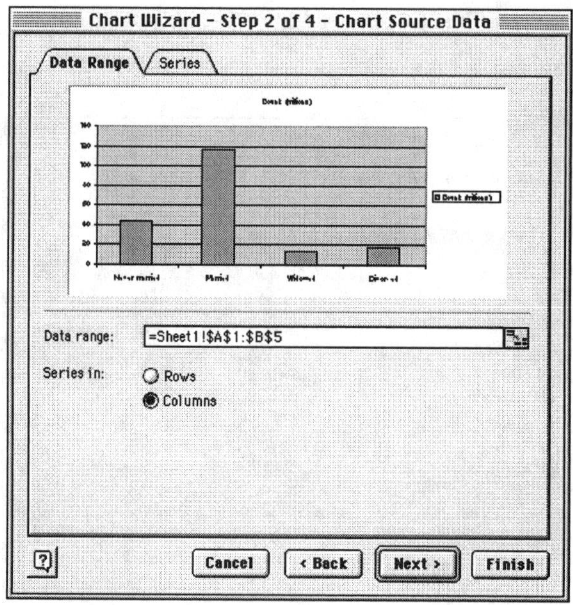

Figure 1.5: ChartWizard—Step 2

- Click the **Titles** tab. Enter "Marital Status of U.S. Adults" for Chart title, "Marital Status" for Category (X) axis and "Count (millions)" for Value (Y) axis.
- Click the **Legend** tab. We don't require a legend since only one variable is plotted, so make sure the check box **Show Legend** is cleared.
- Additional tabs are available to customize other types of charts. They are not required here. Click Next.

Step 4. The final step lets you decide if you want the chart placed on the same worksheet as the data or in another worksheet. With each choice there is a field for entering the worksheet name. We will embed the chart on the same worksheet, so we select the radio button **As object in:** (Figure 1.7). As our current workbook only has one sheet, Excel has used the default name Sheet1. We could also embed the chart on another sheet in the same workbook. Click the Finish button.

The chart appears with eight handles indicating that it is selected. The chart can be resized by selecting a handle and then dragging the handle to the desired size. The chart can also be moved. Click the interior of the chart and drag it to another location (holding the mouse button down). Then click outside the chart to deselect.

You will also find the **Chart Toolbar** (Figure 1.8) embedded on the worksheet. This is used for embellishments of the chart. Use of the this toolbar is described

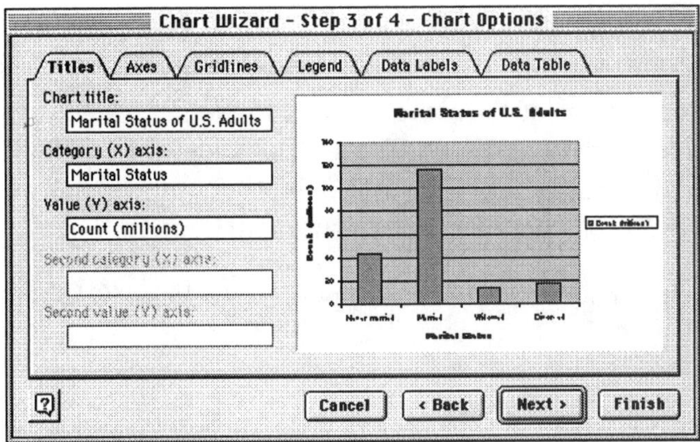

Figure 1.6: ChartWizard—Step 3

Figure 1.7: ChartWizard—Step 4

in the next section on creating histograms. Note that the Chart Toolbar may also be called from the Menu Bar by **View – Toolbars – Chart**. Also, if you select the Chart by clicking once within its area, the Menu Bar will change. In place of the word **Data** there will now appear **Chart** from which a pull-down menu will provide the same tools as are displayed with icons on the Chart Toolbar.

Note: You might get an error message "Cannot add chart to shared workbook" even if you are not sharing your workbook. This is a bug in Excel 97/98 introduced when Shared Workbooks were implemented. It occurs under certain conditons if you try to create a chart using the data analysis tools.

You can still output your chart to a new workbook, and then, if desired, copy

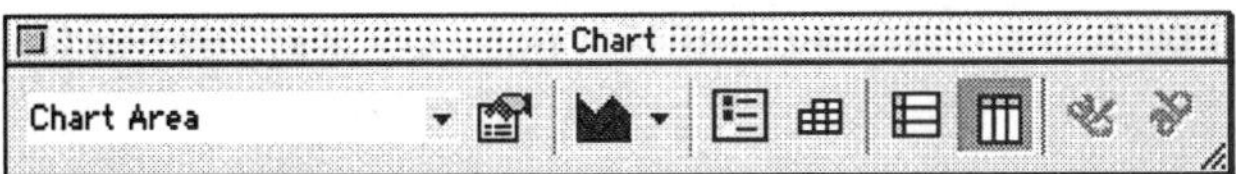

Figure 1.8: Chart Toolbar

the chart to the existing workbook. This is one workaround.

Fortunately, this problem can be fixed by installing an updated file ProcDBRes to replace the existing one of the same name in the folder/directory "Microsoft Office 98:Office:Excel Add-Ins:Analysis Tools" (the folder for a default installation—your location may differ). This file is available for download at the Microsoft Software Library. Details may be found at the URL

http://support.microsoft.com/support/kb/articles/Q183/1/88.asp

Pie Charts

It is easy as pie to produce a chart as in Figure 1.9 using Excel. Select Pie in place of Column in Step 1 of the ChartWizard and follow the remaining steps.

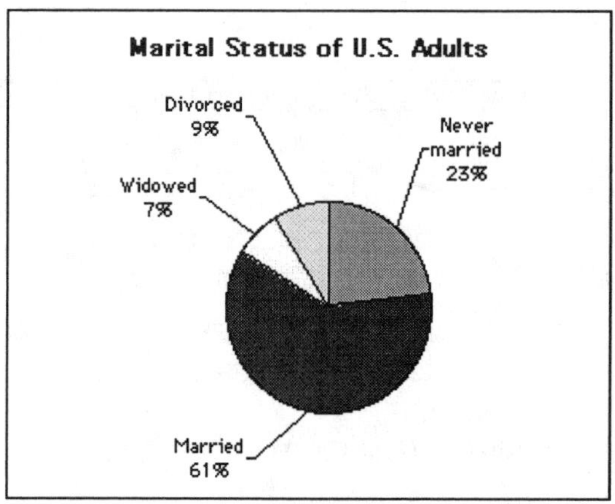

Figure 1.9: Pie Chart

Alternatively, since we have already created a bar graph, it is instructive to use the **Chart Toolbar** as an illustration of how easily modifications may be made. This interface is a vast improvement over the previous version of Excel.

1. Select the completed bar graph by clicking once within its border.
2. From the Menu Bar select **Chart – Chart Type. . . .** You will be presented with a box that is identical to Figure 1.4 but for the title, which contains only the words **Chart Type** without mention of Step 1 of the ChartWizard. In the left field, referring to Figure 1.4, select **Pie** for Chart type and in the right field select **Pie** for Chart sub-type, which is the first choice in the top row.
3. From the Menu Bar select **Chart – Chart Options. . . .** Figure 1.10 appears with three tabs: Titles, Legend, and Data Labels. Click the tab **Data Labels** and then select the radio button **Show label and percent** and

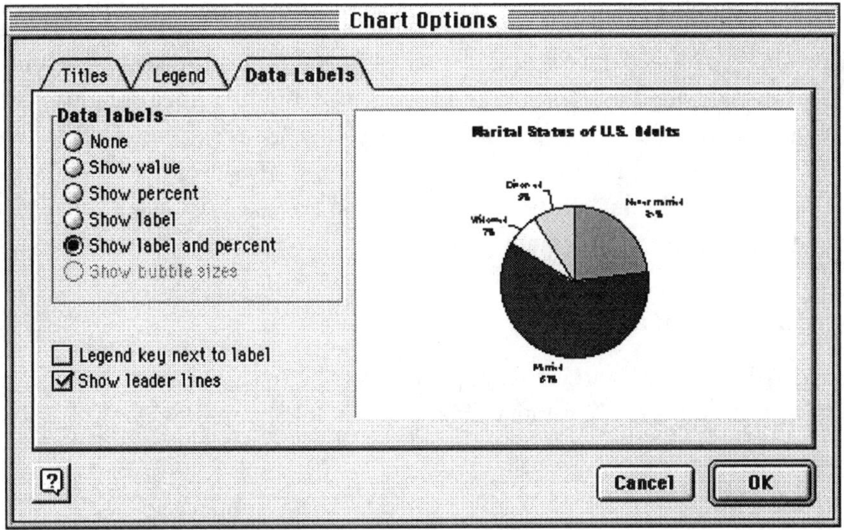

Figure 1.10: Chart Options

check the box **Show leader lines**. Click OK. A pie chart now replaces the bar graph.

Histograms

The ChartWizard is designed for use with data that are already grouped, for instance, categorical variables, and can therefore be used to construct a histogram of quantitative variables that have been grouped into categories or intervals. However, for raw numerical data, Excel provides additional commands within the **Analysis ToolPak** for constructing histograms.

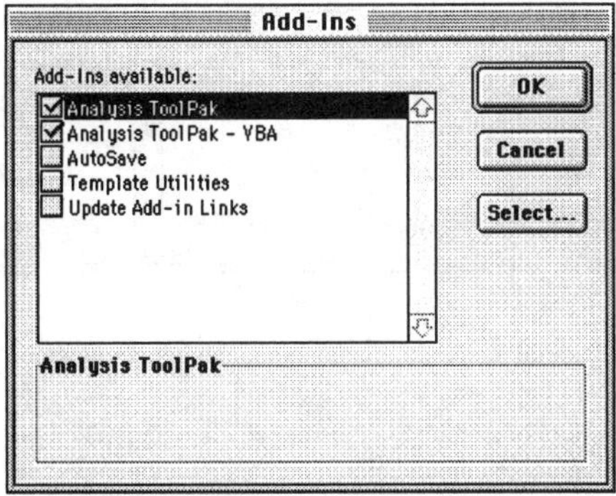

Figure 1.11: Data Analysis Add-In

To determine whether this toolpak is installed, choose **Tools – Add-Ins** from the Menu Bar. The **Add-Ins** dialog box (Figure 1.11) appears. Depending on whether other Add-Ins have been loaded, your box might appear slightly different. If the Analysis ToolPak box is not checked, then select it and click OK. The **Analysis ToolPak** will now be an option in the pull-down menu when you choose **Tools – Data Analysis**. Note that you can also use the Select button to add customized add-ins to complement Excel.

Histogram from Raw Data

Example 1.2. (Exercise 1.27, page 32 in the text.) Make a histogram of survival times of 72 guinea pigs (Figure 1.12).

	A	B	C	D	E	F	G	H	I	J
1	*Survival Times of Guinea Pigs*									
2	43	45	53	56	56	57	58	66	67	73
3	74	79	80	80	81	81	81	82	83	83
4	84	88	89	91	91	92	92	97	99	99
5	100	100	101	102	102	102	103	104	107	108
6	109	113	114	118	121	123	126	128	137	138
7	139	144	145	147	156	162	174	178	179	184
8	191	198	211	214	243	249	329	380	403	511
9	522	688								

Figure 1.12: Survival Times of Guinea Pigs

Solution. Excel requires a contiguous block of data for the histogram tool.

1. Enter the data in a block (cells A2:A73) with the label "Times" in cell A1.
2. From the Menu Bar choose **Tools – Data Analysis** and scroll to the choice Histogram (Figure 1.13). Click OK.

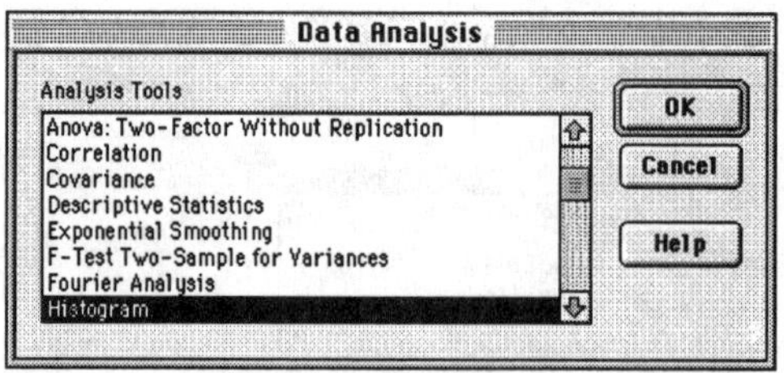

Figure 1.13: Data Analysis Tools

3. In the dialog box (Figure 1.14) type the reference for the range A1:A73 in the **Input range** area, which is the location on the workbook for the data. As with the Bar Chart, you may instead click and drag from cell

A1 to A73. The choice depends on whether your preference is for strokes (keyboard) or clicks (mouse). Leave the **Bin range** blank to allow Excel to select the bins, check the **Labels** box because A1 has been included in the **Input range**, type C1 for **Output range** to denote the upper left cell of the output range, and check the box **Chart output** to obtain a histogram on the same sheet of the workbook as the data. The option Pareto (sorted histogram) constructs a histogram with the vertical bars sorted from left to right in decreasing height. If Cumulative Percentage is checked, the output will include a column of cumulative percentages.

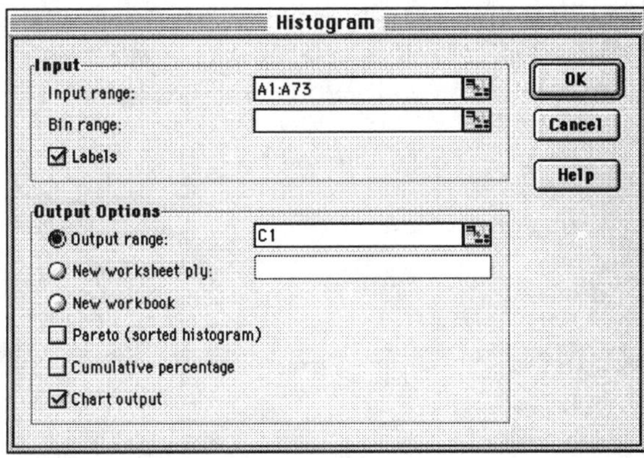

Figure 1.14: Histogram Tool

4. The output appears in Figure 1.15. The entries under Bin in C2:C10 are not the midpoints of the bin intervals, as you might expect. Rather they are the **upper limits** of the boundaries for each interval. The corresponding frequencies appear in cells D2:D10 with the histogram to the right. We shall shortly modify the histogram by changing the labels and allowing adjacent bars to touch. But first, we explain how to customize the selection of bins.

	C	D
1	*Bin*	*Frequency*
2	43	1
3	123.625	45
4	204.25	16
5	284.875	4
6	365.5	1
7	446.125	2
8	526.75	2
9	607.375	0
10	More	1

Histogram

Frequency

50

0

Frequency

Bin

Figure 1.15: Output Table and Default Histogram

Changing the Bin Intervals

If the bin intervals are not specified, then Excel creates them automatically, choosing the number of bins roughly equal to the square root of the number of observations beginning and ending at the minimum and maximum, respectively, of the data set. In creating a histogram from raw data, we let Excel choose the default bins. Here we select our own bin intervals.

1. Type "New Bin" (or another appropriate label) in cell B1. Then enter the values 100, 200, 300, 400, 500, 600, and 700 in cells B2:B8. An easy way to accomplish this is to type 100 and 200 in cells B2 and B3 respectively, then select B2 and B3, click the fill handle in the lower right corner of B3, then drag the fill handle down to cell B8 and release the mouse button.

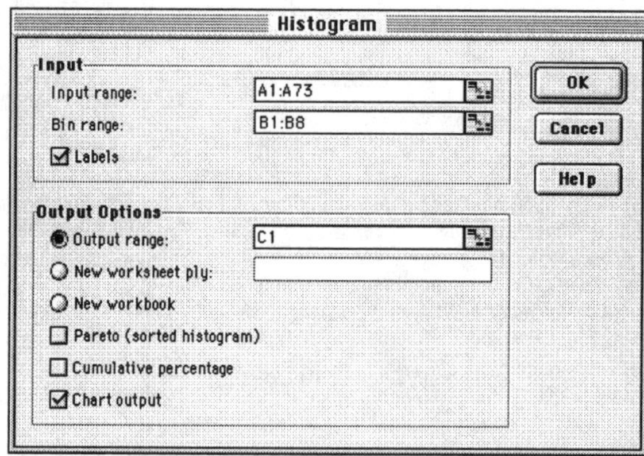

Figure 1.16: Histogram Tool—Specified Bin Intervals

2. Repeat the earlier procedure for creating a histogram, but this time type B1:B8 in the text area for **Bin range** (Figure 1.16). Before the output appears you will be prompted with a warning that you are overwriting existing data. Continue and the new output (Figure 1.17) will replace Figure 1.15 with your selected bin intervals.

Enhancing the Histogram

While the default histogram captures the overall features of the data set, it is inadequate for presentation. Excel provides a set of tools for enhancing the histogram. These are too numerous for all to be mentioned here, but a few will be discussed with reference to the example. The other options may be invoked analogously.

Legend. Remove the legend (which is not needed here) by clicking on it (the word "Frequency" on the right in the chart in Figure 1.17), select **Chart – Chart**

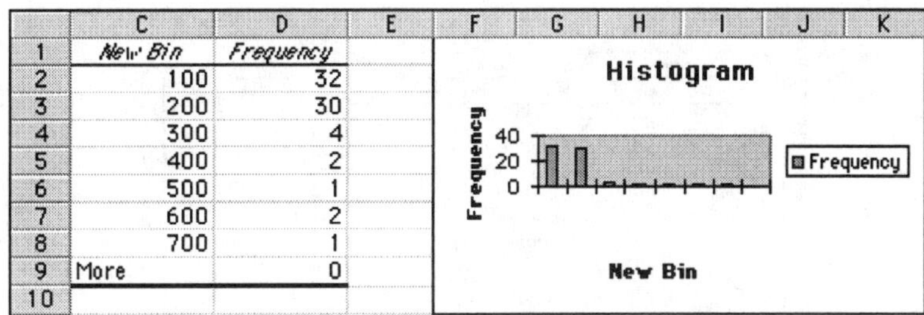

	C	D
1	New Bin	Frequency
2	100	32
3	200	30
4	300	4
5	400	2
6	500	1
7	600	2
8	700	1
9	More	0
10		

Figure 1.17: Output Table and New Bin Histogram

Options... from the Menu Bar, click the **Legend** tab, and clear the box **Show legend**. Alternatively you can choose **Edit – Clear – All** from the Menu Bar after selecting "Frequency" on the chart.

Resize. Both the histogram (called the **Plot Area**) and the box (called the **Chart Area**) that contains it can be resized and moved. Select the Chart Area by clicking once within its boundary, resize using any of the eight handles that appear, or move it by dragging or cutting and pasting from **Edit** on the Menu Bar to a new location. Likewise, select the Plot Area by clicking once within its boundary and then resize or move as with the Chart Area. The X axis labels may appear placed by default horizontally, vertically or diagonally to accommodate the selected size. This can also be changed. After removing the Legend and resizing you will obtain something similar to Figure 1.18, which also shows the resize handles. Click **outside** the Chart Area to deselect.

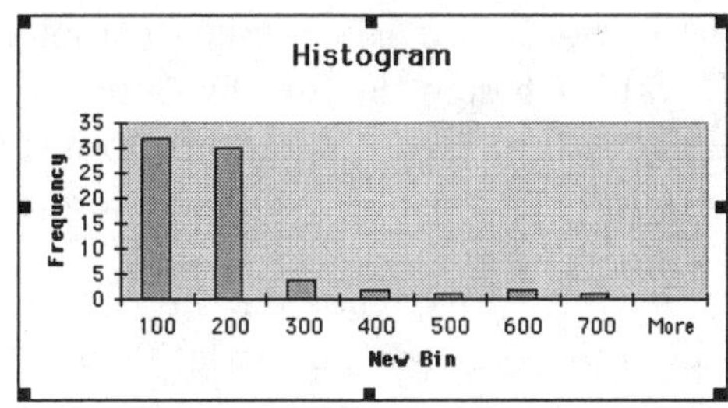

Figure 1.18: Resizing the Histogram

Bar Width. Adjacent bars do not touch in the default, which looks more like a bar chart for categorical data. To adjust the bar width, double-click one of the bars to bring up the **Format Data Series** dialog box and select the Options tab (Figure 1.19). Change the Gap width from 150% to 0%.

Chart Title. Click on the title word Histogram. A rectangular grey border with handles will surround the word, indicating that it is selected for editing.

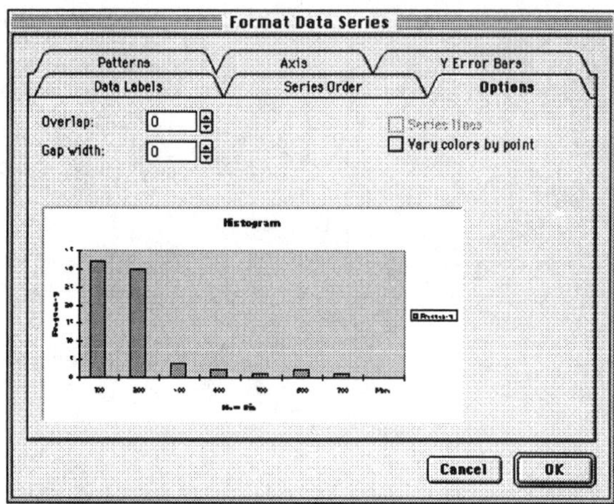

Figure 1.19: Format Data Series

Begin typing "Survival Times (Days) of Guinea Pigs," hold down the **Alt** key (**Windows**) or the **Command** key (**Macintosh**), and press Enter. You may now type a second line of text in the **Formula Bar** entry area. Continue typing "in a Medical Experiment," then press Enter. If you want to move the title within the Chart Area, use the handles. To change the font of your title, select the title, then from the Menu Bar choose **Format – Selected Chart Title. . . .** The dialog box has three tabs: Patterns, Font, and Alignment. Select the **Font** tab and pick a font face, style, and size.

X axis Title. Click on the word New Bin at the bottom of the chart and type "Survival Time (Days)." Change the font by selecting the X axis title, then from the Menu Bar choose **Format – Selected Axis Title. . .** and complete the dialog box as desired in the same fashion as for the Chart Title.

Y axis Title. Click on the word Frequency on the left side, and then from the Menu Bar choose **Format – Selected Axis Title. . .** for any desired formatting.

X axis Format. Double-click the X axis, and in the **Format Axis** dialog box (Figure 1.20) you can click on various tabs to change the appearance of the X axis. If you click on the **Alignment** tab you can change the orientation of the X axis labels.

Y axis Format. Double-click the Y axis, and in the **Format Axis** dialog box you can click on various tabs to change the appearance of the Y axis. Click on the **Scale** tab and change the **Maximum** to 40. Sometimes if you resize the chart you will need to experiment with the **Major** or **Minor** units to achieve a pleasing result. Click OK.

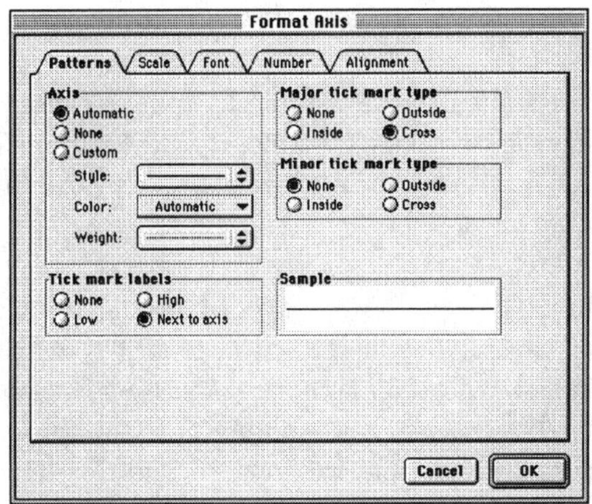

Figure 1.20: Format X axis

More Interval. The "More" interval with 0 counts is unattractive, especially if it appears on your graph (which it may, depending on your default settings, prior to enhancement). In the workbook in cell C9 (refer to Figure 1.17), change the label More to 800. The histogram is dynamically linked to the data in columns C and D and the label More on the X axis becomes 800. Of course, knowing that the count is zero in the bin labeled 800, we could redo the histogram ignoring this bin, if we wished to exclude it.

Plot Area Pattern. The default histogram has a border around the Plot Area and the Plot Area is shaded grey. Both defaults can be changed by double-clicking the **Plot Area** to bring up the **Format Plot Area** dialog box. If desired, select the radio button **None** for Border and also select the radio button **None** for Area.

At the conclusion of the formatting, the histogram will look like Figure 1.21.

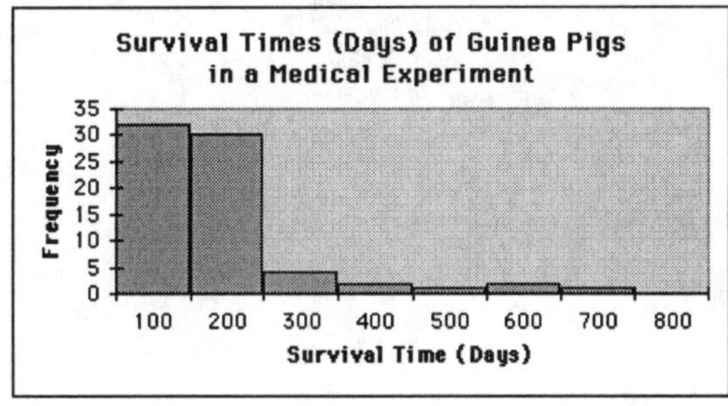

Figure 1.21: Final Histogram after Editing

Histogram from Grouped Data

The **Histogram** tool requires the raw data as input. When numerical data have already been grouped into a frequency table, it is the **ChartWizard** that is the appropriate tool. First use it to obtain a bar chart, and then modify it exactly as you would enhance a histogram.

> **Example 1.3.** (Example 1.8, page 14 in the text.) Figure 1.22 gives the frequencies of vocabulary scores of all 947 seventh graders in Gary, Indiana, on the vocabulary part of the Iowa Test of Basic Skills. Column A is the bin interval and column B is the score. Construct a histogram. The final histogram is shown in Figure 1.23.

	A	B	C
1	Class	Bin	Number of Students
2	2.0 - 2.9	3	9
3	3.0 - 3.9	4	28
4	4.0 - 4.9	5	59
5	5.0 - 5.9	6	165
6	6.0 - 6.9	7	244
7	7.0 - 7.9	8	206
8	8.0 - 8.9	9	146
9	9.0 - 9.9	10	60
10	10.0 - 10.9	11	24
11	11.0 - 11.9	12	5
12	12.0 - 12.9	13	1
13	Total		947

Figure 1.22: Vocabulary Scores—Grouped Data

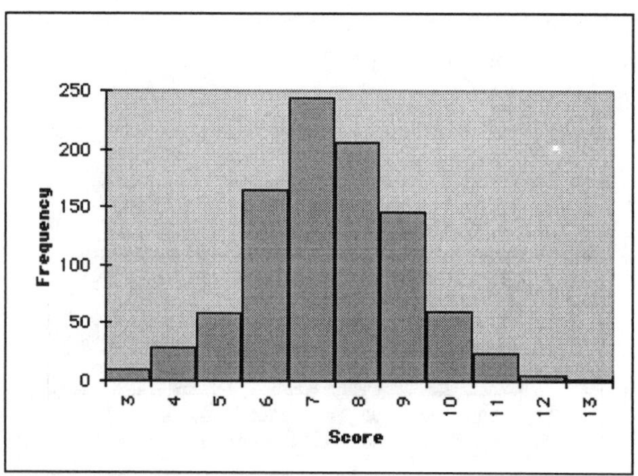

Figure 1.23: Histogram from Grouped Data

1.2 Describing Distributions with Numbers

The most direct way to obtain the common summary statistics is through the **Descriptive Statistics Tool**, which provides preformatted output very quickly. It is explained in this section. An alternative is the **Formula Palette** (replacing the **Function Wizard** of **Excel 5/95**) which provides greater flexibility of output and many more functions and formulas over its predecessor. We first describe the Descriptive Statistics Tool and then the Formula Palette.

The Descriptive Statistics Tool

Example 1.4. (Table 1.8, page 40 in the text.) Figure 1.24 shows the calories and sodium levels measured in three types of hot dogs: beef, meat (mainly pork and beef), and poultry. Data is from *Consumer Reports*, June 1986, pp. 366-367. Describe the data using the Descriptive Statistics Tool.

	A	B	C	D	E	F	G	H
1	Beef hot dogs			Meat hot dogs			Poultry hot dogs	
2								
3	Calories	Sodium		Calories	Sodium		Calories	Sodium
4	186	495		173	458		129	430
5	181	477		191	506		132	375
6	176	425		182	473		102	396
7	149	322		190	545		106	383
8	184	482		172	496		94	387
9	190	587		147	360		102	542
10	158	370		146	387		87	359
11	139	322		139	386		99	357
12	175	479		175	507		170	528
13	148	375		136	393		113	513
14	152	330		179	405		135	426
15	111	300		153	372		142	513
16	141	386		107	144		86	358
17	153	401		195	511		143	581
18	190	645		135	405		152	588
19	157	440		140	428		146	522
20	131	317		138	339		144	545
21	149	319						
22	135	298						
23	132	253						

Figure 1.24: Hot Dog Data

Solution. For illustration purposes we consider only the beef calories data.

1. From the Menu Bar choose **Tools – Data Analysis** and double-click **Descriptive Statistics** (or, equivalently, select **Descriptive Statistics** and click OK) in the **Data Analysis Dialog** box. A dialog box **Descriptive Statistics** appears (Figure 1.25) which prompts for user input.

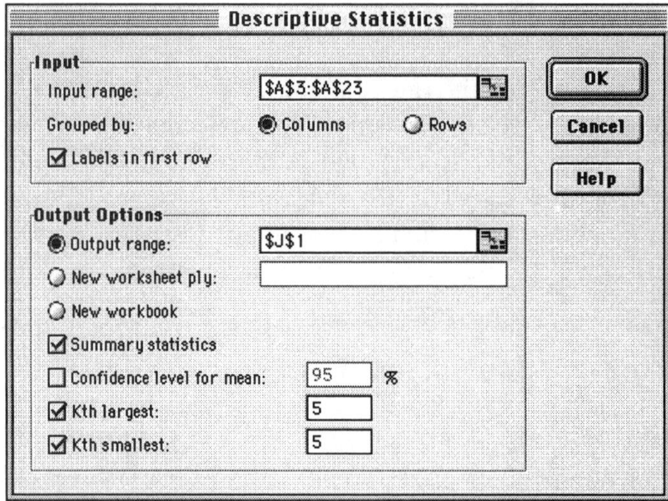

Figure 1.25: Descriptive Statistics Dialog Box

2. Complete the input as indicated in Figure 1.25. The **Input range:** is A3:A23, corresponding to the beef calories, including labels. (If you selected this range prior to invoking Descriptive Statistics, it will already be inserted by Excel.) Check the box **Labels in first row**. The **Confidence level for mean:** is not needed at this time (it gives the half-width). Check the **Kth largest:** or **Kth smallest:** boxes if needed. We have selected K = 5 for illustration.

3. The **Output Options** tell Excel where to place the output. Select cell J1. Finally check the box Summary Statistics and click OK. The output appears in Figure 1.26. We have formatted the output by reducing the number of decimal points using the **Decimal** button in the **Formatting Toolbar**. We can read off the summary statistics:

   mean = 156.85
   standard deviation = 5.06
   median = 152.50
   minimum = 111   maximum = 190
   5th smallest = 139   5th largest = 181

Formula Palette

In **Excel 97/98** the **Formula Palette** replaced the **Function Wizard**. It assists in entering formulas and functions included in Excel, particularly complex ones.

	A	I	J	K
1	Beef hot dogs		*Calories*	
2				
3	Calories		Mean	156.85
4	186		Standard Error	5.063
5	181		Median	152.5
6	176		Mode	149
7	149		Standard Devia	22.642
8	184		Sample Varianc	512.661
9	190		Kurtosis	-0.8131
10	158		Skewness	-0.0313
11	139		Range	79
12	175		Minimum	111
13	148		Maximum	190
14	152		Sum	3137
15	111		Count	20
16	141		Largest(5)	181
17	153		Smallest(5)	139
18	190			
19	157			
20	131			
21	149			
22	135			
23	132			

Figure 1.26: Descriptive Statistics Output

The functions can perform decision-making, action-taking, or value-returning operations. The Formula Palette simplifies this process by guiding you step by step.

It can be fired up in one of two ways. When you select a cell and press the **Paste Function** button f_x next to the autosum button Σ on the **Standard Toolbar** (or, equivalently, choose **Insert – Function...** from the Menu Bar), an equal sign (=) appears both in the cell and in the **Formula Bar**. The **Paste Function** dialog box (Figure 1.27) appears showing all available functions grouped by category on the left and the function name on the right. Both lists have scroll bars for choices not directly visible on the screen. At the bottom of the box, the selected function is shown with the arguments it takes and a brief description. (In previous versions of Excel, a similar dialog box called **Function Wizard – Step 1 of 2** appeared.) When you click OK in the **Paste Function** box, the **Formula Palette** box appears below the **Formula Bar**, requesting parameters and an input range for the function you selected. In addition, the Formula Bar is now activated showing the Formula Palette's drop-down list control with the 10 most recently used functions, and an equal (=) sign appears in the Formula Bar showing the selected function partially constructed and awaiting completion of its arguments. You may enter these either directly into the Formula Bar or in the Formula Palette box.

The **Formula Palette** is usually invoked in a second, more direct way. Select a cell and press (=) on the **Formula Bar** to open the **Formula Palette** dialog box (Figure 1.28). On the far left side of the **Formula Toolbar** is a button with the most recently used function, in this case `AVERAGE`. If this is the function you need then click on the word `AVERAGE` and the **Formula Palette** dialog box will

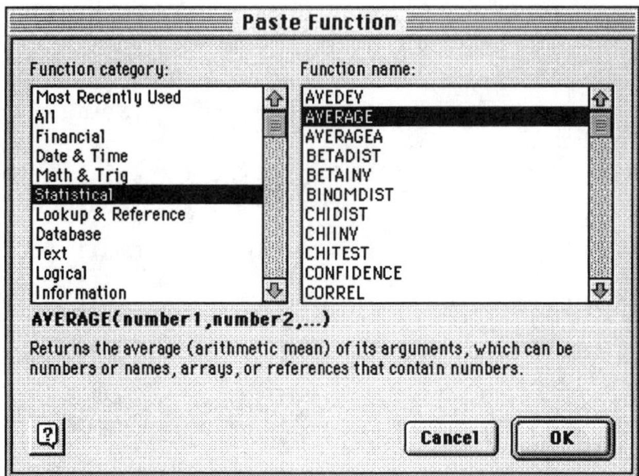

Figure 1.27: Paste Function

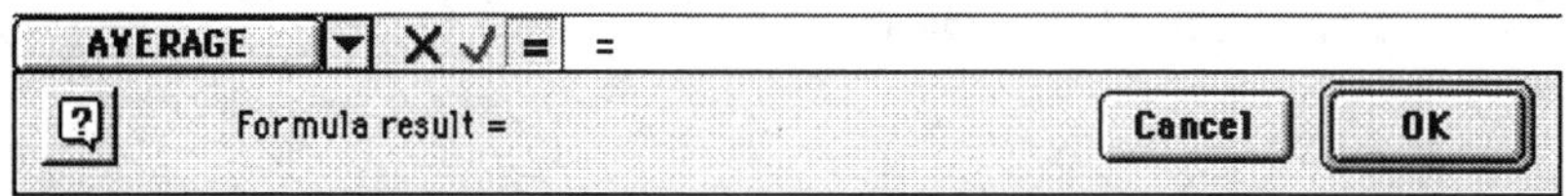

Figure 1.28: Formula Palette—Default

expand (Figure 1.29) requesting the required parameter or the data range for the function (which can be typed directly or *referenced* by using the mouse to point to the data by clicking and dragging over cells in the data range). As you input this information, Excel will correspondingly build the function both in the **Formula Bar** and in the cell you had selected in the workbook. When you have completed entering the requested input, click OK to complete the function. If you want some other function than the default, click the small arrow to the right of the function name. Select from the drop-down list of your 10 most recently used functions or select **More functions. . . .** If you select the latter, then the **Paste Function** dialog box appears. An OK (checkmark symbol) and a cancel (X) button appear to the right of this arrow. Click the checkmark and the formula is entered into the active cell. Click the cancel to discard the formula without making changes.

Recommendation. The **Paste Function** button on the Standard Toolbar duplicates the actions of the **Formula Palette.** Since Excel formulas start with an equal (=) sign, we recommend that you begin your formulas by pressing the (=) symbol on the Formula Toolbar instead of using the Paste Function. This activates the **Formula Palette**, and you can either type the formula by hand into the Formula Bar or order up a function from the **Paste Function** box, if required. Experienced users of Excel often **customize** the **Standard Toolbar** and replace the Paste Function button with some other one.

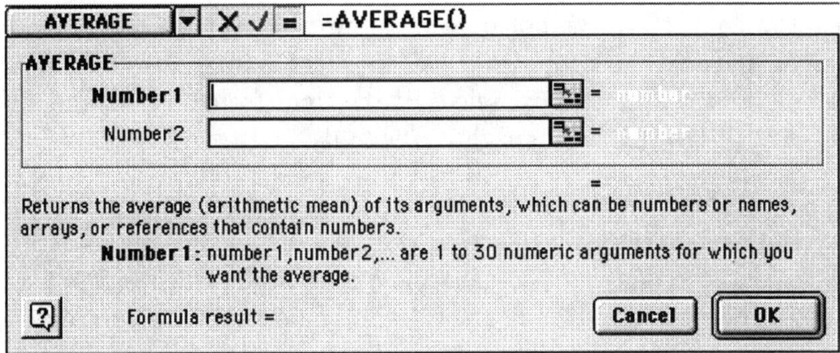

Figure 1.29: Formula Palette—Expanded

The Five-Number Summary

Example 1.4 continued. (Table 1.8, page 40 in the text.) Find the five-number summary {minimum, first quartile, median, third quartile, maximum} for the calorie distribution of the hot dogs shown in Figure 1.30.

	A	B	C	D
1	Beef Hot Dogs	*Five Number Summary*		
2				
3	Calories			
4	186	Min	111	=MIN(A4:A23)
5	181	Q1	140.5	=QUARTILE(A4:A23,1)
6	176	Med	152.5	=MEDIAN(A4:A23)
7	149	Q3	177.25	=QUARTILE(A4:A23,3)
8	184	Max	190	=MAX(A4:A23)
9	190			
10	158			
11	139			
12	175			
13	148			
14	152			
15	111			
16	141			
17	153			
18	190			
19	157			
20	131			
21	149			
22	135			
23	132			

Figure 1.30: Five-Number Summary

Solution

1. Enter the labels "Min," "Q1," "Med," "Q3," and "Max" as shown in cells B4:B8.

2. Click the equal (=) symbol on the **Formula Bar** to start the **Formula Palette** and use the drop-down list to select **More functions....** In the **Paste Function** dialog box, select **Statistical** from the left and scroll down and select `QUARTILE` on the right. Click OK.

3. The **Formula Palette** dialog box appears. Move it out of the way and enter the data **Array** by selecting cells A4:A23 with your mouse (or more mundanely by typing A4:A23 into the dialog box). Click in the text area for **Quart** and type "1" to indicate the first quartile. The completed formula appears in the **Formula Toolbar** and value of the formula 140.5 shows in the dialog box (Figure 1.31). Click OK and the value 140.5 is printed in C5.

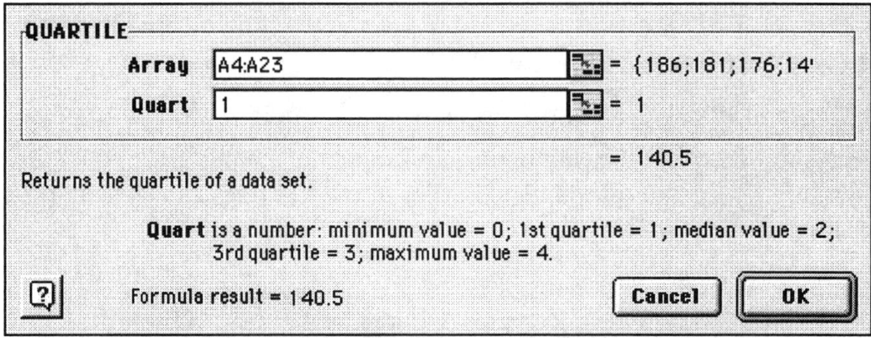

Figure 1.31: Quartile Formula

4. Continue in this fashion using the Formula Palette to complete the five number summary. Of course you can still enter the formulas by hand in the Formula Bar once you are familiar with them. Cells C4:C8 present the syntax while the values are in B4:B8.

The five-number summary is {111, 140.5, 152.5, 177.25, 190}. Note that Excel uses a slightly different definition of quartiles for a finite data set than the text.

1.3 The Normal Distribution

Areas under a normal curve can be found using the `NORMDIST` function. The syntax is = `NORMDIST`$(x, \mu, \sigma,$ cumulative), where μ is the mean and σ is the standard deviation. The parameter cumulative indicates whether the density (set cumulative = "false" or "0") or whether the cumulative distribution (set cumulative = "true" or "1") is wanted. The formula = `NORMDIST`$(x, \mu, \sigma, 1)$ returns $F(x)$, which is the area to the left of x under an $N(\mu, \sigma)$ density and can be used to produce a table of normal areas as found in many statistics texts. Another formula, =

NORMINV(p, μ, σ), returns the inverse $F^{-1}(x)$ of the cumulative, that is a value x such that the area to the left of x is the specified p. For $N(0,1)$, use **NORMSDIST** and **NORMSINV** instead.

Normal Distribution Calculations

Example 1.5. (Examples 1.25 and 1.36, pages 76–78 in the text.) The level of cholesterol in the blood is important because high cholesterol levels increase the risk of heart disease. The distribution of blood cholesterol levels in a large population of people of the same age and sex is roughly normal. For 14-year-old boys the mean $\mu = 170$ milligrams of cholesterol per deciliter of blood (mg/dl) and the standard deviation is $\sigma = 30$ mg/dl. Levels above 240 mg/dl may require medical attention.
(a) What percent of 14-year-old boys have more than 240 mg/dl of cholesterol?
(b) What percent of 14-year-old boys have blood cholesterol between 170 and 240 mg/dl?

Solution

1. Click on a cell (activate it) where you want to locate the answer, say A1.
2. The syntax is
$$\Phi(x) = \texttt{NORMDIST}(\mu, \sigma, x, 1)$$
so enter the formula $= 1 - \texttt{NORMDIST}(240, 170, 30, 1)$ since the upper tail area is wanted. The answer 0.00982 appears in cell A1. (Users of **Excel 5/95** may also use the **Function Wizard** while users of **Excel 97/98** may use the **Formula Palette** instead to enter the formula.)

Example 1.6. (Example 1.27, page 78 in the text.) Scores on the SAT verbal test in recent years follow approximately the $N(505, 110)$ distribution. How high must a student score in order to place in the top 10% of all students taking the SAT?

Solution

1. Activate cell B1 for entry of the function.
2. The syntax is
$$\Phi^{-1}(x) = \texttt{NORMINV}(x, \mu, \sigma)$$
for the inverse of the cumulative normal distribution. Enter the formula $= \texttt{NORMINV}(0.90, 505, 110)$ and read off 645.971, the 90th percentile of the SAT scores, in cell B1.

Problems involving $\Phi(x)$ and $\Phi^{-1}(x)$ occur repeatedly, and it may be convenient to use a template. Figure 1.32 shows a workbook with the required formulas.

	A	B	C	D	E
1		*Calculating Normal Areas and Inverse Probabilities*			
2	*Parameters*				
3	mean				
4	sigma				
5					
6	*Cumulative*				
7	x				
8	Area to left of x	=NORMDIST(B7, B3, B4, 1)			
9	Area of right of x				
10	Standard Score Z				
11					
12	*Interval*				
13	a				
14	b				
15	Area between a and b	=NORMDIST(B14, B3, B4, 1)-NORMDIST(B13, B3, B4, 1)			
16					
17	*Inverse*				
18	area A to left				
19	percentile	=NORMINV(B18,B3,B4)			

Figure 1.32: Calculating Normal Areas and Percentiles

Graphing the Normal Curve

By combining the **ChartWizard** and the `NORMDIST` function, we can create a graph of any normal curve. *In fact, the procedure described here can be used to plot the graph of any function Excel can evaluate.*

Constructing a Graph of the Standard Normal Curve

1. Enter the labels z and $f(z)$ in cells A2, B2. (See Figure 1.33.)

2. Enter -3.5 and -3.4 in cells A3 and A4, respectively. We are going to create a column of z values at which the standard normal density will be calculated. Select A2:A3, check the fill handle in the lower right corner of A3, and drag to cell A73 to fill the column with decreasing values of z decremented by 0.1. Format the values with two decimal places.

3. Select cell B3 and enter = `NORMDIST`(A3, 0, 1, 0) in the **Formula Bar.** Cell B3 now contains the value 0.00087268, the standard normal density evaluated at $z = -3.50$.

4. Select cell B3, click the fill handle, and drag down to B73. This copies the formula you just entered in B3 into cells B4:B73 relative to the corresponding cell references in column A. Column B is filled with values $f(z)$ of the standard normal density corresponding to each value of z in column A.

5. **Users of Excel 5/95.** Select cells A2:B73, click the **ChartWizard** button, and then click in cell C2 and drag to I26 to locate the graph. A dialog box **ChartWizard – Step 1 of 5** appears.

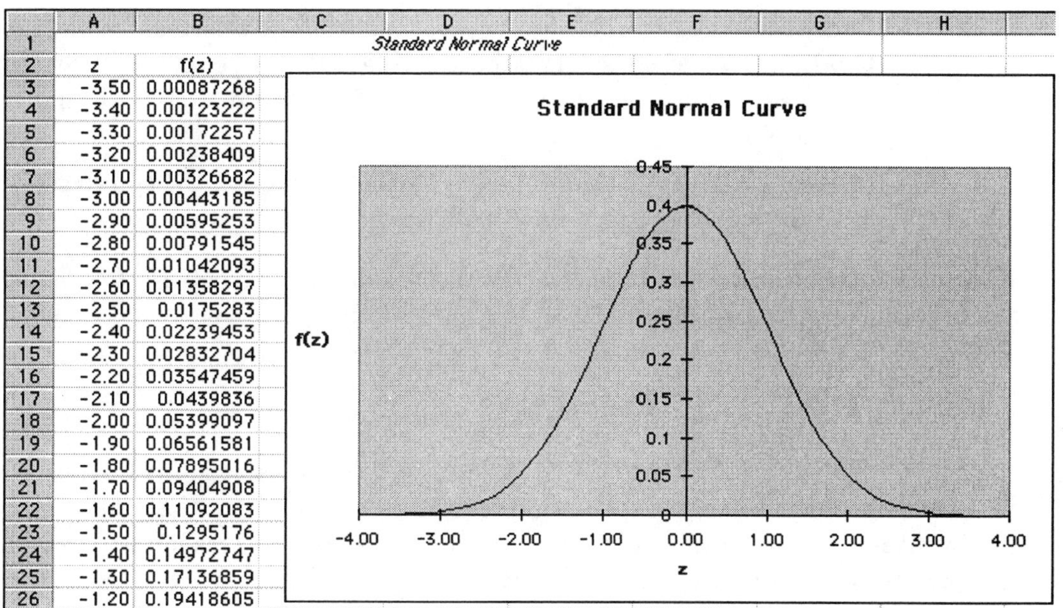

	A	B
1		
2	z	f(z)
3	-3.50	0.00087268
4	-3.40	0.00123222
5	-3.30	0.00172257
6	-3.20	0.00238409
7	-3.10	0.00326682
8	-3.00	0.00443185
9	-2.90	0.00595253
10	-2.80	0.00791545
11	-2.70	0.01042093
12	-2.60	0.01358297
13	-2.50	0.0175283
14	-2.40	0.02239453
15	-2.30	0.02832704
16	-2.20	0.03547459
17	-2.10	0.0439836
18	-2.00	0.05399097
19	-1.90	0.06561581
20	-1.80	0.07895016
21	-1.70	0.09404908
22	-1.60	0.11092083
23	-1.50	0.1295176
24	-1.40	0.14972747
25	-1.30	0.17136859
26	-1.20	0.19418605

Figure 1.33: Graphing a Normal Density

- In Step 1 you are given an opportunity to correct or confirm your range.
- In Step 2 select **XY (Scatter)** chart.
- In Step 3 select format **6**.
- In Step 4 select the radio button **Columns** for Data Series, enter "1" for Column for Category (X) Axis Labels, and enter "1" for Row for Legend Text.
- In Step 5 select the radio button **No** for Add a Legend?, type "Standard Normal Curve" as the Chart Title, and type z and $f(z)$ for Category (X) and Value (Y) titles, respectively. Finally, click Finish.

Users of Excel 97/98. Select cells A2:B73 and click the **ChartWizard.** A dialog box **ChartWizard – Step 1 of 4 – Chart Type** appears.

- In Step 1 select **XY (Scatter)** for Chart Type and the lower right Chart sub-type **Scatter without markers.**
- In Step 2 under the **Data Range** tab, the range will already be indicated and the Series radio button for **Columns** will be selected. You may edit the range if it is incorrect. Under the **Series** tab no changes are necessary.
- In Step 3 under the **Titles** tab, type "Standard Normal Curve" as the Chart Title, z for Value (X) Axis, and $f(z)$ for Value (Y) Axis.

Under the **Axes tab**, both check boxes should be selected. Under the **Gridlines** tab, clear all check boxes. Under the **Legend** tab, clear the Show legend. Finally, under the **Data Labels** tab, select the radio button **None**.

- In Step 4 embed the graph in the current workbook by selecting the radio button **As object in**. Finally, click Finish.

6. Activate the graph for editing and format the display as you wish to present it using the editing features discussed previously.

Constructing a Normal Table

It is very easy in Excel to produce a table of normal areas. Figure 1.34 shows the formulas behind the workbook in Figure 1.35. Remember that the $ sign prefix makes the corresponding row or column label absolute. This method can be adapted to produce tables of other continuous distributions.

	A	B	C	D
1	z	0.00	0.01	0.02
2	0.00	=NORMSDIST($A2+B$1)	=NORMSDIST($A2+C$1)	=NORMSDIST($A2+D$1)
3	0.10	=NORMSDIST($A3+B$1)	=NORMSDIST($A3+C$1)	=NORMSDIST($A3+D$1)
4	0.20	=NORMSDIST($A4+B$1)	=NORMSDIST($A4+C$1)	=NORMSDIST($A4+D$1)

Figure 1.34: Formulas for a Normal Table

Figure 1.35 gives areas under a standard normal curve for values of $z \geq 0$. It is obtained as follows.

1. Enter the label and values in column A and row 1. Column A gives the first decimal of z while row 1 gives the second decimal.

2. Enter the formula

$$= \texttt{NORMDIST}(\$A2 + B\$1)$$

in cell B2. Select cell B2, click the fill handle in the lower right corner of B2, and drag to K2.

3. Select cells B2:K2, click the fill handle in the lower right corner of K2, and drag to K41 to fill the block B2:K41.

Normal Quantile Plots

Excel does not provide a normal quantile (probability) plot, but it is very easy to construct such a graph. We defer this to the next chapter, where it will be discussed as an application of the **ChartWizard** as a tool for constructing scatterplots.

	A	B	C	D	E	F	G	H	I	J	K
1	z	**0.00**	**0.01**	**0.02**	**0.03**	**0.04**	**0.05**	**0.06**	**0.07**	**0.08**	**0.09**
2	**0.00**	0.5000	0.5040	0.5080	0.5120	0.5160	0.5199	0.5239	0.5279	0.5319	0.5359
3	**0.10**	0.5398	0.5438	0.5478	0.5517	0.5557	0.5596	0.5636	0.5675	0.5714	0.5753
4	**0.20**	0.5793	0.5832	0.5871	0.5910	0.5948	0.5987	0.6026	0.6064	0.6103	0.6141
5	**0.30**	0.6179	0.6217	0.6255	0.6293	0.6331	0.6368	0.6406	0.6443	0.6480	0.6517
6	**0.40**	0.6554	0.6591	0.6628	0.6664	0.6700	0.6736	0.6772	0.6808	0.6844	0.6879
7	**0.50**	0.6915	0.6950	0.6985	0.7019	0.7054	0.7088	0.7123	0.7157	0.7190	0.7224
8	**0.60**	0.7257	0.7291	0.7324	0.7357	0.7389	0.7422	0.7454	0.7486	0.7517	0.7549
9	**0.70**	0.7580	0.7611	0.7642	0.7673	0.7704	0.7734	0.7764	0.7794	0.7823	0.7852
10	**0.80**	0.7881	0.7910	0.7939	0.7967	0.7995	0.8023	0.8051	0.8078	0.8106	0.8133
11	**0.90**	0.8159	0.8186	0.8212	0.8238	0.8264	0.8289	0.8315	0.8340	0.8365	0.8389
12	**1.00**	0.8413	0.8438	0.8461	0.8485	0.8508	0.8531	0.8554	0.8577	0.8599	0.8621
13	**1.10**	0.8643	0.8665	0.8686	0.8708	0.8729	0.8749	0.8770	0.8790	0.8810	0.8830
14	**1.20**	0.8849	0.8869	0.8888	0.8907	0.8925	0.8944	0.8962	0.8980	0.8997	0.9015
15	**1.30**	0.9032	0.9049	0.9066	0.9082	0.9099	0.9115	0.9131	0.9147	0.9162	0.9177
16	**1.40**	0.9192	0.9207	0.9222	0.9236	0.9251	0.9265	0.9279	0.9292	0.9306	0.9319
17	**1.50**	0.9332	0.9345	0.9357	0.9370	0.9382	0.9394	0.9406	0.9418	0.9429	0.9441
18	**1.60**	0.9452	0.9463	0.9474	0.9484	0.9495	0.9505	0.9515	0.9525	0.9535	0.9545
19	**1.70**	0.9554	0.9564	0.9573	0.9582	0.9591	0.9599	0.9608	0.9616	0.9625	0.9633
20	**1.80**	0.9641	0.9649	0.9656	0.9664	0.9671	0.9678	0.9686	0.9693	0.9699	0.9706
21	**1.90**	0.9713	0.9719	0.9726	0.9732	0.9738	0.9744	0.9750	0.9756	0.9761	0.9767
22	**2.00**	0.9772	0.9778	0.9783	0.9788	0.9793	0.9798	0.9803	0.9808	0.9812	0.9817
23	**2.10**	0.9821	0.9826	0.9830	0.9834	0.9838	0.9842	0.9846	0.9850	0.9854	0.9857
24	**2.20**	0.9861	0.9864	0.9868	0.9871	0.9875	0.9878	0.9881	0.9884	0.9887	0.9890
25	**2.30**	0.9893	0.9896	0.9898	0.9901	0.9904	0.9906	0.9909	0.9911	0.9913	0.9916
26	**2.40**	0.9918	0.9920	0.9922	0.9925	0.9927	0.9929	0.9931	0.9932	0.9934	0.9936
27	**2.50**	0.9938	0.9940	0.9941	0.9943	0.9945	0.9946	0.9948	0.9949	0.9951	0.9952
28	**2.60**	0.9953	0.9955	0.9956	0.9957	0.9959	0.9960	0.9961	0.9962	0.9963	0.9964
29	**2.70**	0.9965	0.9966	0.9967	0.9968	0.9969	0.9970	0.9971	0.9972	0.9973	0.9974
30	**2.80**	0.9974	0.9975	0.9976	0.9977	0.9977	0.9978	0.9979	0.9979	0.9980	0.9981
31	**2.90**	0.9981	0.9982	0.9982	0.9983	0.9984	0.9984	0.9985	0.9985	0.9986	0.9986
32	**3.00**	0.9987	0.9987	0.9987	0.9988	0.9988	0.9989	0.9989	0.9989	0.9990	0.9990
33	**3.10**	0.9990	0.9991	0.9991	0.9991	0.9992	0.9992	0.9992	0.9992	0.9993	0.9993
34	**3.20**	0.9993	0.9993	0.9994	0.9994	0.9994	0.9994	0.9994	0.9995	0.9995	0.9995
35	**3.30**	0.9995	0.9995	0.9995	0.9996	0.9996	0.9996	0.9996	0.9996	0.9996	0.9997
36	**3.40**	0.9997	0.9997	0.9997	0.9997	0.9997	0.9997	0.9997	0.9997	0.9997	0.9998
37	**3.50**	0.9998	0.9998	0.9998	0.9998	0.9998	0.9998	0.9998	0.9998	0.9998	0.9998
38	**3.60**	0.9998	0.9998	0.9999	0.9999	0.9999	0.9999	0.9999	0.9999	0.9999	0.9999
39	**3.70**	0.9999	0.9999	0.9999	0.9999	0.9999	0.9999	0.9999	0.9999	0.9999	0.9999
40	**3.80**	0.9999	0.9999	0.9999	0.9999	0.9999	0.9999	0.9999	0.9999	0.9999	0.9999
41	**3.90**	1.0000	1.0000	1.0000	1.0000	1.0000	1.0000	1.0000	1.0000	1.0000	1.0000

Figure 1.35: Constructing a Normal Table

1.4 Boxplots

Excel does not provide a boxplot. However the Microsoft Personal Support Center has a Web page "How to Create a BoxPlot – Box and Whisker Chart" located at

http : //support.microsoft.com/support/kb/articles/q155/1/30.asp

with instructions for creating a reasonable boxplot in Excel 5/95. We have modified these for use with Excel 97/98 and we illustrate by constructing side-by-side boxplots of the calorie data for beef, meat, and poultry hot dogs in Example 1.4.

Step 1. Enter the calorie data into three columns of a workbook (Figure 1.36), then find and enter the five-number summary into another three columns **in the order** median, first quartile, minimum, maximum, third quartile. We have entered this information, including labels in block D3:G7 in Figure 1.36.

	A	B	C	D	E	F	G
1			*Boxplots of Hot Dog Calories*				
2	Beef	Meat	Poultry		Beef	Meat	Poultry
3	186	173	129	median	152.5	153	129
4	181	191	132	Q1	140.5	139	102
5	176	182	102	min	111	107	86
6	149	190	106	max	190	195	170
7	184	172	94	Q3	177.25	179	143
8	190	147	102				
9	158	146	87				
10	139	139	99	*Formulas for Beef Column*			
11	175	175	170	median	=MEDIAN(A3:A22)		
12	148	136	113	Q1	=QUARTILE(A3:A22,1)		
13	152	179	135	min	=MIN(A3:A22)		
14	111	153	142	max	=MAX(A3:A22)		
15	141	107	86	Q3	=QUARTILE(A3:A22,3)		
16	153	195	143				
17	190	135	152				
18	157	140	146				
19	131	138	144				
20	149						
21	135						
22	132						

Figure 1.36: Boxplot—Data and Preparation

Step 2. Select cells D2:G7 and click on the **ChartWizard** button. Choose the **Stock** Chart type from the selections on the left and then from the Chart sub-types on the right select **Volume-Open-High-Low-Close Chart** (at the bottom right in Figure 1.37). Click Next.

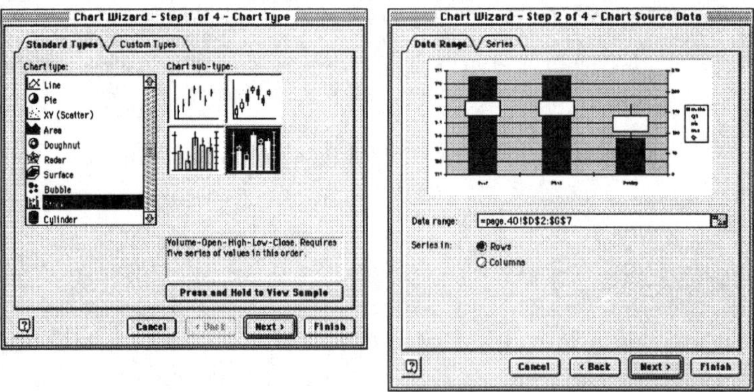

Figure 1.37: Boxplot—ChartWizard Steps 1 and 2

Step 3. Check the radio button **Rows** for Series in: and click Next.

Step 4. In the next dialog box (Figure 1.38), under the **Titles** tab enter "Boxplots of Calories Data" for the Chart title, under the **Axes** tab clear the check box next to Value (Y) Axis under **Secondary Axis**, under the

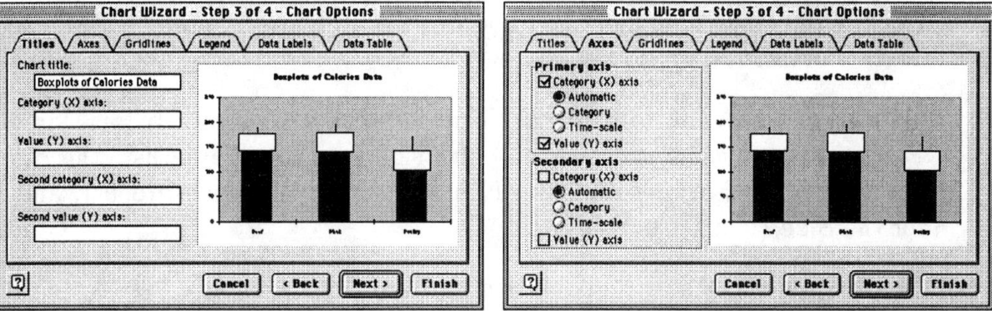

Figure 1.38: Boxplot—ChartWizard Step 3

Gridlines tab clear all boxes, and finally under the **Legend** tab clear the Show Legend box. Click Next.

Step 5. In the final dialog box, locate the boxplot on your sheet.

Next we edit this chart.

1. Click the chart to activate it. Click once on any one of the colored columns to select the series. Do not click on the white columns. From the Menu Bar choose **Chart – Chart Type...**, select Chart type **Line** on the left and select Chart sub-type **Line** (upper left selection on the right side of the dialog box). A line that connects the three white columns appears in the chart.

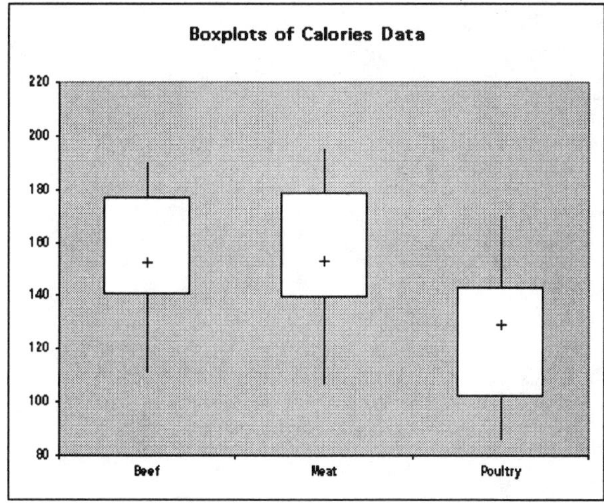

Figure 1.39: Boxplot of Calories Data

2. Click once on the line, and from the Menu Bar choose **Format – Selected Data Series. . . .**

3. Under the **Patterns** tab, select **None** for Line and **Custom** for Marker. For the custom marker choose the plus sign from the **Style** list, the color black from the **Foreground** list, **None** from the **Background** list, and 5 for the **Size** (font). Click OK.

4. Double-click the Y axis (or equivalently, click the Y axis once to select it and then from the Menu Bar choose **Format – Selected Axis. . .**) and in the **Format Axis** dialog box under the **Scale** tab set the Minimum to 80. Click OK. The final boxplot appears on your sheet (Figure 1.39).

These side-by-side boxplots give a vivid graphical summary showing that beef and meat hot dogs are similar in calories and that poultry hot dogs are lower in general.

Users of Excel 5/95 are referred to Appendix Section A.5.

Chapter 2

Looking at Data–Relationships

In many studies, both independent and dependent variables typically arise either as part of a controlled experiment or as an observational study. Before any specific model is imposed that can be tested statistically, it is important to judge graphically whether any relationship is justified.

If we let x denote the explanatory (independent) variable and y the response (dependent) variable, then we might plot in Cartesian coordinates all pairs (x_i, y_i) of observed values. This is called a **Scatterplot**.

2.1 Scatterplot

The steps involved in creating a scatterplot are similar to those for producing a **Histogram** using the **ChartWizard**. The following instructions are based on **Excel 97/98**. **Excel 5/95** users should refer to the Appendix.

Table 2.1: Assets and Income for Banks

Bank	1	2	3	4	5	6	7	8	9	10
Assets	49.0	42.3	36.3	16.4	14.9	14.2	13.5	13.4	13.2	11.8
Income	218.8	265.6	170.9	85.9	88.1	63.6	96.9	60.9	144.2	53.6

Bank	11	12	13	14	15	16	17	18	19	20
Assets	11.6	9.5	9.4	7.5	7.2	6.7	6.0	4.6	3.8	3.4
Income	42.9	32.4	68.3	48.6	32.3	42.7	28.9	40.7	13.8	22.2

Example 2.1. (Exercise 2.9, page 121 in the text.) In 1974 the Franklin National Bank failed. Franklin was one of the 20 largest banks in the nation and the largest to fail. Could Franklin's weakened condition have been detected in advance by simple data analysis? Table 2.1

gives the total assets (in billions of dollars) and net income (in millions of dollars) for the 20 largest banks in 1973, the year before Franklin failed. Franklin is bank number 19. Make a scatterplot of these data that displays the relation between assets and income. Mark Franklin with a separate symbol.

Creating a Scatterplot

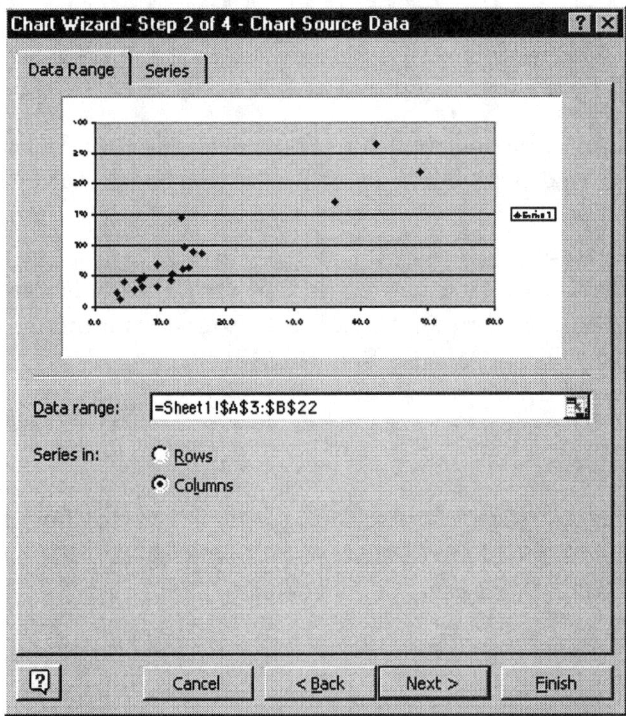

Figure 2.1: Windows ChartWizard—Step 2

Figures 2.1–2.3 show the **ChartWizard** interface using Excel 97 on a Toshiba notebook running Windows NT 4.0, rather than on a Macintosh, only for illustration purposes, because a user of Excel should be familiar with the software interface no matter the operating system.

Enter the data from Table 2.1 into cells A3:A22 and B3:B22 of a workbook with the labels "Assets" and "Income" in A2:B2.

Step 1. Select cells A3:B22 and click on the **ChartWizard**. From the choice of charts select **XY (Scatter)** for Chart type on the left and select the top Chart sub-type **Scatter** on the right. Click Next.

Step 2. The next dialog box previews the chart and allows any changes to be made to the data range (Figure 2.1). Click Next.

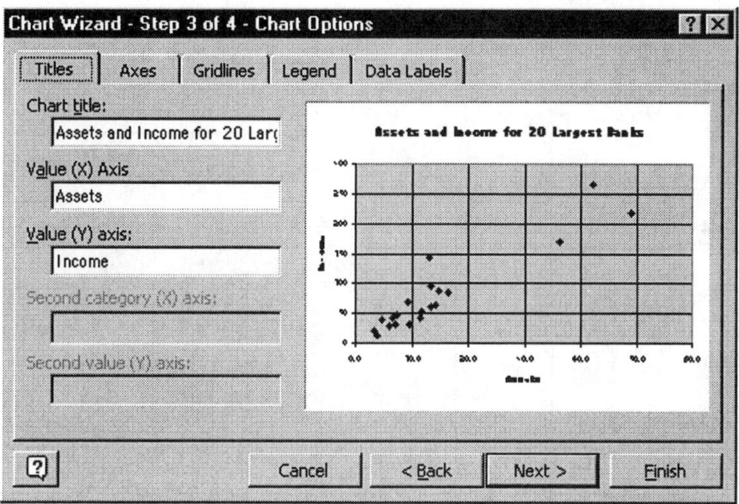

Figure 2.2: Windows Preview of Scatterplot—Step 3

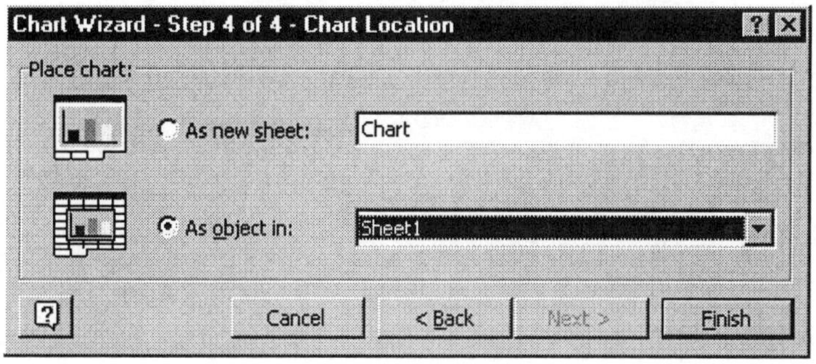

Figure 2.3: Windows Preview of Scatterplot—Step 4

Step 3. The **Chart Option** dialog box appears (Figure 2.2).

- Click the **Titles** tab and enter "Assets and Income for 20 Largest Banks" for Chart title, "Assets" for Category (X) axis, and "Income" for Value (Y) axis.
- Click the **Legend** tab. Clear the Show Legend check box. Click Next.
- Click the **Gridlines** tab and make sure that the Major gridlines box is checked for both axes. Click Next.

Step 4. In the last step, Figure 2.3, select the radio button to enter the chart on the current sheet. Click Finish. The scatterplot appears embedded on your workbook (Figure 2.4).

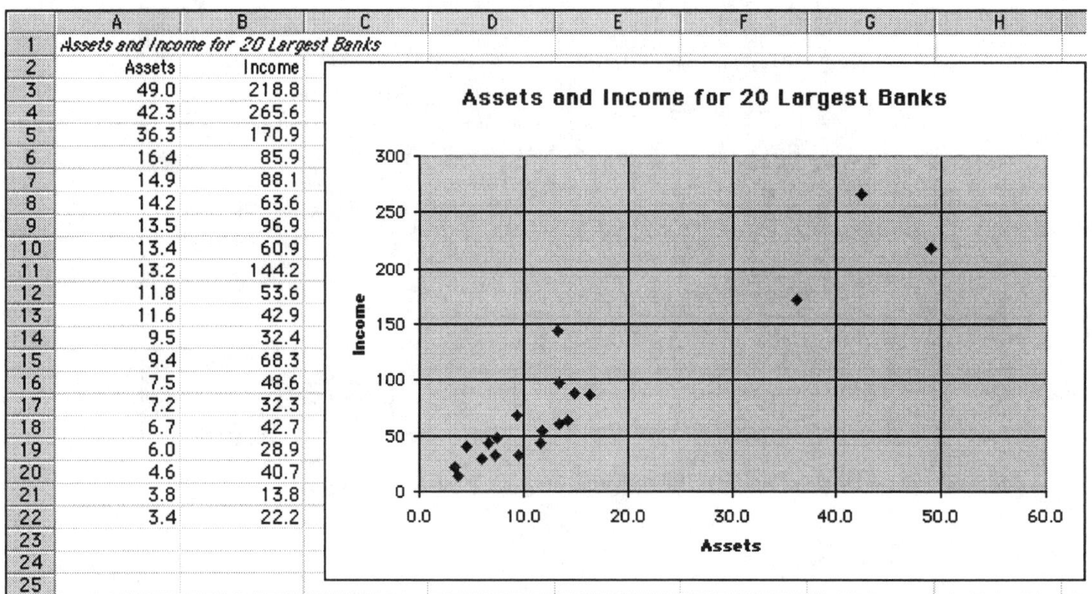

	A	B
1	*Assets and Income for 20 Largest Banks*	
2	Assets	Income
3	49.0	218.8
4	42.3	265.6
5	36.3	170.9
6	16.4	85.9
7	14.9	88.1
8	14.2	63.6
9	13.5	96.9
10	13.4	60.9
11	13.2	144.2
12	11.8	53.6
13	11.6	42.9
14	9.5	32.4
15	9.4	68.3
16	7.5	48.6
17	7.2	32.3
18	6.7	42.7
19	6.0	28.9
20	4.6	40.7
21	3.8	13.8
22	3.4	22.2
23		
24		
25		

Figure 2.4: Assets and Income Data and Scatterplot Embedded on Sheet

Enhancing a Scatterplot

The scatterplot may be enhanced using editing tools, some of which were described in Chapter 1. Activate the scatterplot by clicking once within its border to access new commands that become available under the Menu Bar.

Changing Scale

Excel uses a range from 0 to 100% as the default, and sometimes the scatterplot will show unwanted blank space. For instance, there are no assets larger than $50 billion and we can edit the scatterplot so that the maximum X axis value is 50.0. Double click the X axis, and in the **Format Axis** dialog box click on the **Scale** to change the maximum to 50 on the X axis. If desired, the Y axis may similarly be selected for editing. Refer to Figure 1.20 and the discussion for enhancing a histogram in Section 1.1, which applies to any chart, histogram, scatterplot, or other type.

Changing Titles

You can change titles on the scatterplot after you have completed it. Click on the X axis title "Assets" to select it, and enter "Assets (Billions of Dollars)." Click on the Y axis title "Income" and enter "Income (Millions of Dollars)."

Labeling a Data Point

By default Excel uses diamonds to plot the points. Suppose, for presentation purposes, you wish to use a different shape (and color) to represent Franklin and also to label Franklin on the scatterplot. The following steps describe how to achieve this.

1. Activate the chart and click on the observation for Franklin.
2. Hold down the **Control** key **(Windows)** or **Command** key **(Macintosh)** and with your mouse pointer **select** the point representing Franklin. The pointer becomes a four-pointed plus sign (Figure 2.5).

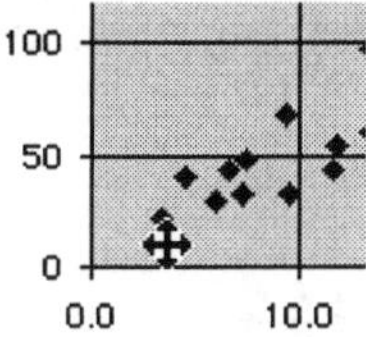

Figure 2.5: Editing a Point

3. For **Excel 5/95** choose **Insert – Data Labels** from the Menu Bar to open the **Format Data Point** dialog box. For **Excel 97/98** choose **Format – Selected Data Point...** from the Menu Bar to open a corresponding **Format Data Point** dialog box. Under the **Data Labels** tab select the radio button for Show Value. Click OK. Excel attaches the y value 13.8 to the point for Franklin on the scatterplot and encloses it within a grey bordered selection box ready for editing. Type "Franklin" (which appears in the **Formula Bar**) and press enter. The selection box now contains the word "Franklin". Move it to a convenient place and **deselect** by clicking elsewhere.

Changing the Marker and Color of a Data Point

To make the point for Franklin stand out more, we will now change the symbol and color of its point on the scatterplot.

1. Activate the chart and click on the observation for Franklin.

2. Hold down the **Control** key **(Windows)** or **Command** key **(Macintosh)** and select the point representing Franklin. The pointer again becomes a four-pointed plus sign.

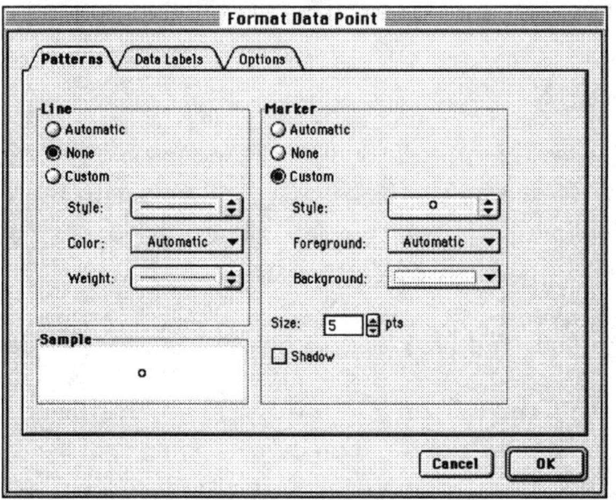

Figure 2.6: Changing the Default Marker

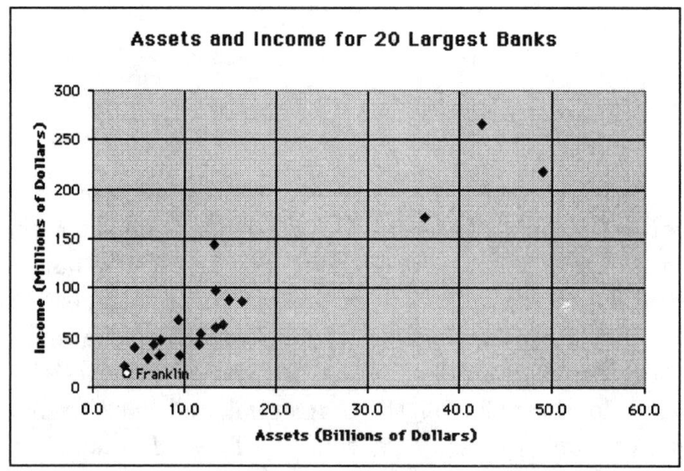

Figure 2.7: Scatterplot with Labeled Point

3. For **Excel 5/95**, choose **Insert – Data Labels** from the Menu Bar to open the **Format Data Point** dialog box. For **Excel 97/98**, choose **Format – Selected Data Point...** from the Menu Bar to open a corresponding **Format Data Point** dialog box. Under the **Patterns** tab, leave the **Line** selection as **None**. Under **Marker**, select a marker type from the pull-down

list for **Style**, and also select a Foreground and Background color (Figure 2.6). In **Excel 5/95** the size of the marker cannot be changed and there is no **Options** tab. Your selection is previewed in the small **Sample** box in the lower portion of the dialog box. Click OK. The final scatterplot appears in Figure 2.7.

Normal Quantile Plots

There are several ways to assess whether a data set is normal. An analytic approach beyond the level of this book was developed by S. Shapiro and M. B. Wilk (An analysis of variance test for normality, *Biometrika* **52**, pp. 591–611, 1965).

	A	B	C	D	E	F
1	*Newcomb Measurements*					
2	*Speed of Light*					
3						
4	28	26	33	24	34	-44
5	27	16	40	-2	29	22
6	24	21	25	30	23	29
7	31	19	24	20	36	32
8	36	28	25	21	28	29
9	37	25	28	26	30	32
10	36	26	30	22	36	23
11	27	27	28	27	31	27
12	26	33	26	32	32	24
13	39	28	24	25	32	25
14	29	27	28	29	16	23

Figure 2.8: Newcomb Data

A simple graphical approach is to construct a histogram and compare the observed counts with the 68-95-99.7% rule. A more sensitive version of this idea is to order the observations and examine their distribution visually, using a scatterplot involving the corresponding expected quantiles of a normal curve. Normal data will tend to fall on a straight line. (This is the basis for the Shapiro-Wilk test.)

Although Excel does not provide a normal quantile plot, one can easily be constructed. The expected value of the ith order statistic (the ith largest in increasing magnitude) of a sample of size n from a $N(0,1)$ distribution can be approximated by the percentile

$$z_{(i)} = \texttt{NORMSINV}\left(\frac{i-\frac{3}{8}}{n+\frac{1}{4}}\right)$$

which is the value of a standard normal such that the area to the left is $\frac{i-\frac{3}{8}}{n+\frac{1}{4}}$.

Then plot $z_{(i)}$ on the vertical axis against $x_{(i)}$ on the horizontal axis where $x_{(i)}$ is the ith largest from the data set $\{x_1, x_2, \ldots, x_n\}$. This is accomplished using the **ChartWizard**.

Example 2.2. (Example 1.28, page 81 in the text.) Figure 2.8 contains 66 measurements of the speed of light made by Simon Newcomb

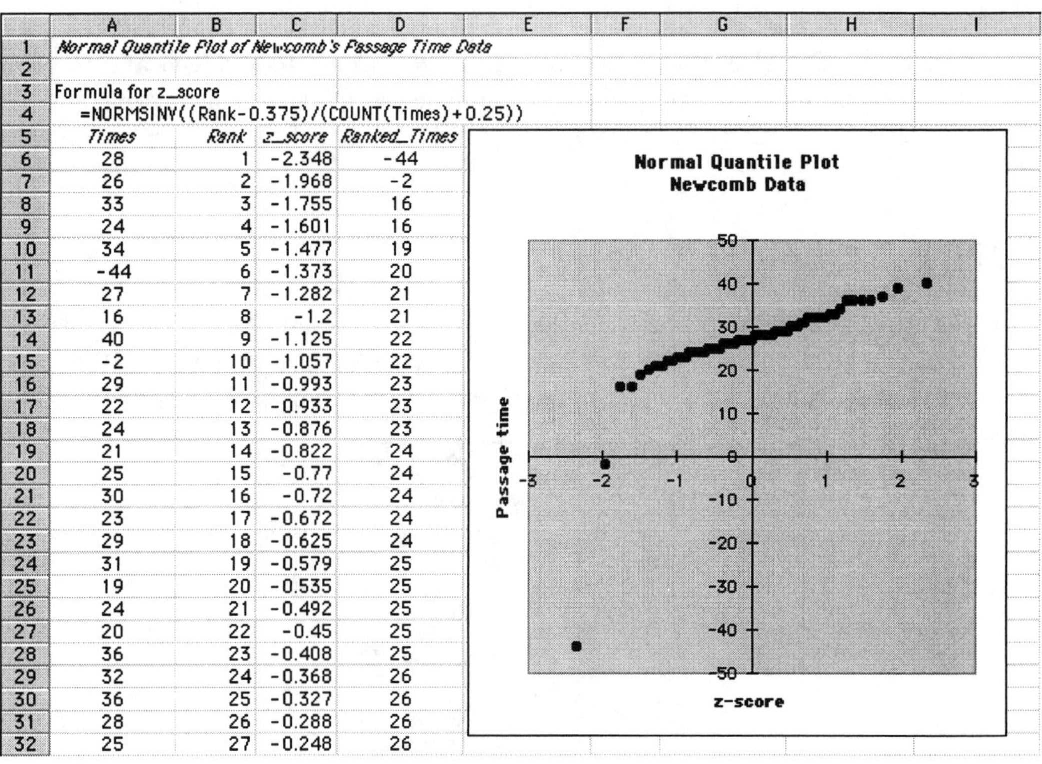

	A	B	C	D
1	*Normal Quantile Plot of Newcomb's Passage Time Data*			
2				
3	Formula for z_score			
4	=NORMSINV((Rank-0.375)/(COUNT(Times)+0.25))			
5	*Times*	*Rank*	*z_score*	*Ranked_Times*
6	28	1	-2.348	-44
7	26	2	-1.968	-2
8	33	3	-1.755	16
9	24	4	-1.601	16
10	34	5	-1.477	19
11	-44	6	-1.373	20
12	27	7	-1.282	21
13	16	8	-1.2	21
14	40	9	-1.125	22
15	-2	10	-1.057	22
16	29	11	-0.993	23
17	22	12	-0.933	23
18	24	13	-0.876	23
19	21	14	-0.822	24
20	25	15	-0.77	24
21	30	16	-0.72	24
22	23	17	-0.672	24
23	29	18	-0.625	24
24	31	19	-0.579	25
25	19	20	-0.535	25
26	24	21	-0.492	25
27	20	22	-0.45	25
28	36	23	-0.408	25
29	32	24	-0.368	26
30	36	25	-0.327	26
31	28	26	-0.288	26
32	25	27	-0.248	26

Figure 2.9: Normal Quantile Plot

between July and September 1882. The measurements give the deviation from 24,800 nanoseconds baseline. Construct a normal quantile plot of the data.

Solution

1. Referring to columns A:D in Figure 2.9, reenter the data in cells A6:A71 of a workbook and the label "Times" in A5. From the Menu Bar, choose **Data – Sort** to sort the data in increasing order and enter the sorted data in D6:D71.

2. Enter the label "Rank" in B5 followed by the integers $\{1, 2, \ldots, 66\}$ in B6:B71.

3. Enter the label "z_score" in C5. Name the ranges "Times" and "Rank." Then select cell C6 and enter

$$= \texttt{NORMSINV}((\text{Rank} \;- 0.375)/(\texttt{COUNT}(\text{Times}) + 0.25))$$

Click the fill handle at the lower right corner of C6 and fill to C71.

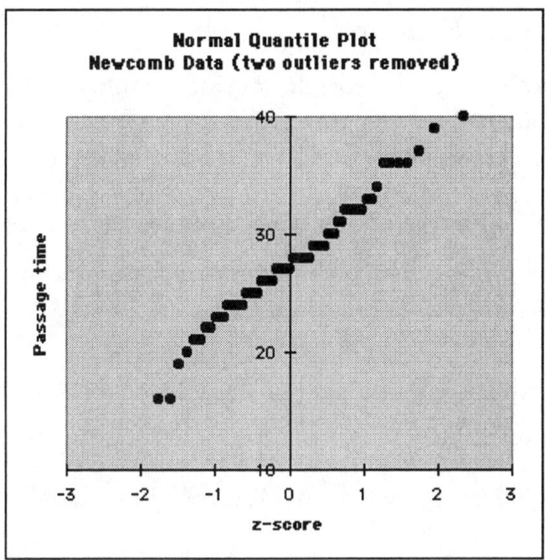

Figure 2.10: Normal Quantile Plot—Outliers Removed

4. Select a cell to locate the quantile plot, then click the **ChartWizard** button. Choose scatterplot as the chart, enter the data range C5:D71, and complete the remaining steps. Enhance the chart to reproduce the scatterplot shown in Figure 2.9.

Two outliers are apparent $\{-44, -2\}$. Remove them from the plot by selecting rows 6 and 7 and choosing from the Menu Bar **Format – Row – Hide**. Since the chart is *linked* to the data, the scatterplot instantly changes to that shown in Figure 2.10, whose straight line appearance is an indication that a normal distribution fits the censored data quite well.

2.2 Correlation

The correlation between two variables x and y measures the strength of the linear association between them. For n pairs (x_i, y_i), $1 \leq i \leq n$, of data points the sample correlation coefficient is defined to be

$$r = \frac{1}{n-1}\sum_{i=1}^{n}\left(\frac{x_i - \bar{x}}{s_x}\right)\left(\frac{y_i - \bar{y}}{s_y}\right)$$

where $\bar{x}$ and $\bar{y}$ are the sample means of the $\{x_i\}$ and $\{y_i\}$, respectively, and s_x and s_y are the corresponding sample standard deviations.

Using the CORREL Function

The most direct way to find the correlation for Example 2.1 is by the **Formula Palette** for **Excel 97/98** or the **Function Wizard** for **Excel 5/95** with the function `CORREL`, which computes the correlation coefficient.

> **Example 2.3.** Find the correlation between Assets and Income for the 20 largest banks in Example 2.1.

Solution

1. Open the workbook containing the data and labels you used in Example 2.1. (Refer to Figure 2.4.) Select an empty cell where you want the correlation to appear and invoke either the **Function Wizard** or the **Formula Palette**. In each case, select **Statistical** for Function Category and `CORREL` for Function Name.

2. Enter A3:A22 for **Array1** and B3:B22 for **Array2**. You may enter by hand or click and drag on the workbook over the range A3:A22, press Tab on the keyboard, then click and drag over the range B3:B22, and finally click OK (or Finish, for **Excel 5/95**). The answer 0.9328 appears in the cell you selected.

Using the ToolPak

Correlation between two variables can also be calculated using the **Correlation** tool in the **Analysis ToolPak**. This tool is most effective, however, for determining pairwise correlations for multivariate data sets for which repeated use of the `CORREL` function would be inefficient.

This tool prints out a matrix of correlations. Such a matrix is helpful in multiple regression in deciding which variables to include in a model.

> **Example 2.4.** (CONCEPT data set from the Student CD-ROM.) Darlene Gordon of the Purdue University School of Education provided the data in the CONCEPT data set. The data were collected on 78 seventh-grade students in a rural Midwestern school. The research concerned the relationship between the students' "self-concept" and their academic performance. The variables are OBS, a subject identification number; GPA, grade point index; IQ, score on an IQ test; AGE, age in years; SEX, female (F) or male (M); SC, overall score on the Piers-Harris self-concept scale; and C1–C6, "cluster scores" for specific aspects of self-concept: C1 = behavior, C2 = school status, C3 = physical, C4 = anxiety, C5 = popularity, and C6 = happiness.
>
> Find the correlations between the response variable GPA and each of the explanatory variables IQ, AGE, SEX, SC, and C1–C6. Of all the

	A	B	C	D	E	F	G	H	I	J	K	L
1	CONCEPT DATA SET											
2	*OBS*	*GPA*	*IQ*	*AGE*	*SEX*	*SC*	*C1*	*C2*	*C3*	*C4*	*C5*	*C6*
3	1	7.9	111	13	2	67	15	17	13	13	11	9
4	2	8.3	107	12	2	43	12	12	7	7	6	6
5	3	4.6	100	13	2	52	11	10	5	8	9	7
6	4	7.5	107	12	2	66	14	15	11	11	9	9
7	5	8.9	114	12	1	58	14	15	10	12	11	6
8	6	7.6	115	12	2	51	14	11	7	8	6	9
9	7	7.7	111	13	2	71	15	17	12	14	11	10
10	8	2.4	97	13	2	51	10	12	5	11	5	6
11	9	6	100	13	1	49	12	9	6	9	6	7
12	10	8.8	112	13	2	51	15	16	4	9	5	8
13	11	7.5	104	12	1	35	12	5	3	2	4	7
14	12	5.5	89	13	1	54	16	8	3	11	7	7
15	13	7.2	104	13	2	54	16	14	6	7	2	7
16	14	7.6	102	13	1	64	14	12	8	10	12	9
17	15	4.7	91	14	1	56	14	13	8	10	7	8
18	16	8.2	114	13	1	69	15	15	9	12	11	9
19	17	7.8	114	13	1	55	14	11	6	11	11	9
20	18	7.6	103	12	1	65	16	16	5	12	11	9
21	19	4	106	13	2	40	8	5	6	5	4	8
22	20	6.2	105	12	1	66	15	15	11	10	10	10
23	21	7.6	113	12	2	55	13	12	6	9	7	8
24	22	1.8	109	12	2	20	5	2	2	4	2	1
25	24	6.4	108	14	1	56	10	13	11	8	10	10
26	26	9.6	113	13	2	68	13	14	13	13	11	9
27	27	11	130	12	1	69	15	17	13	11	11	10
28	28	11	128	13	2	70	15	16	11	11	11	10
29	29	9.4	128	12	2	80	16	17	13	14	11	10
30	30	8	118	13	2	53	5	13	9	12	7	9
31	31	9.6	113	13	2	65	15	14	9	12	10	10
32	32	9.6	120	13	1	67	14	15	11	12	11	9

Figure 2.11: Concept Data Set

explanatory variable, IQ does the best job of explaining GPA in a simple linear regression. How do you know this without doing all the regressions?

Solution. Figure 2.11 shows the first 30 sets of observations in this data set, to which the following instructions apply. For the full set merely select the appropriate Input range.

1. From the Menu Bar choose **Tools – Data Analysis** and in the dialog box highlight **Correlation** and click OK.

2. In the next dialog box, **Correlation**, enter A2:L32 for **Input range** (most conveniently done by clicking and dragging over this range on the workbook and pressing the Tab key). Check the box **Labels in first row** and point to cell N1 for **Output range**. Click OK.

Excel Output

The output appears in N1:Z13, as shown in Figure 2.12. In view of symmetry, only half the correlation matrix is required. From Column P we read off the correlations between GPA and the explanatory variables. The largest correlation involving GPA is with IQ and is 0.709.

	N	O	P	Q	R	S	T	U	V	W	X	Y	Z
1		OBS	GPA	IQ	AGE	SEX	SC	C1	C2	C3	C4	C5	C6
2	OBS	1											
3	GPA	0.318	1										
4	IQ	0.520	0.709	1									
5	AGE	0.110	-0.164	-0.299	1								
6	SEX	-0.087	-0.051	0.211	-0.103	1							
7	SC	0.255	0.654	0.415	0.094	-0.098	1						
8	C1	-0.116	0.562	0.099	-0.080	-0.248	0.685	1					
9	C2	0.109	0.679	0.450	0.050	0.020	0.874	0.633	1				
10	C3	0.292	0.601	0.578	0.068	0.027	0.790	0.316	0.733	1			
11	C4	0.192	0.484	0.318	0.150	-0.021	0.857	0.474	0.787	0.662	1		
12	C5	0.230	0.590	0.399	0.071	-0.320	0.828	0.480	0.667	0.717	0.760	1	
13	C6	0.325	0.644	0.396	0.209	-0.124	0.812	0.517	0.685	0.676	0.572	0.681	1

Figure 2.12: Correlation ToolPak Output

2.3 Least-Squares Regression

We have seen how to plot two variables against each other in a scatterplot and have calculated the correlation coefficient to measure the strength of the linear association between them. It is useful to have an analytic relationship between the explanatory variable x and the response variable y of the form

$$y = f(x)$$

for predicting y from x. Such a relationship is called a simple (meaning one explanatory variable) **regression curve**. The simplest curve is a straight line

$$y = a + bx$$

called the regression line of y on x. The regression line represents, under certain assumptions, the mean response at each specified value x.

The method used to determine the coefficients a and b goes back at least to the great mathematician Gauss and is called the **Principle of Least-Squares.** Gauss himself recognized that the criterion was arbitrary and he used it because the coefficients a and b were then solvable in closed form. (Additional reasons connected with the errors being normal are presented in more advanced treatments.)

For a given x_i, we call

$$\hat{y}_i = a + bx_i$$

the **predicted** value and

$$e_i = y_i - \hat{y}_i$$

the **residual.** The **error sum of squares** is defined to be

$$\sum_{i=1}^{n} e_i^2 = \sum_{i=1}^{n} (y_i - a - bx_i)^2$$

By differentiating with respect to a and b, we can solve for the values that minimize $\sum_{i=1}^{n} e_i^2$. These are the values used in the regression line. They are given by the formulas

$$\begin{aligned} \text{slope} \quad b &= r\,\frac{s_y}{s_x} \\ \text{intercept} \quad a &= \bar{y} - b\bar{x} \end{aligned}$$

where $\bar{x} = \frac{1}{n}\sum_{i=1}^{n} x_i$, $\bar{y} = \frac{1}{n}\sum_{i=1}^{n} y_i$, r is the correlation coefficient, and

$$\begin{aligned} (n-1)s_x^2 &= \sum_{i=1}^{n}(x_i - \bar{x})^2 = \sum_{i=1}^{n} x_i^2 - \frac{1}{n}\left(\sum_{i=1}^{n} x_i\right)^2 \\ (n-1)s_y^2 &= \sum_{i=1}^{n}(y_i - \bar{y})^2 = \sum_{i=1}^{n} y_i^2 - \frac{1}{n}\left(\sum_{i=1}^{n} y_i\right)^2 \end{aligned}$$

Fitting a Line to Data

Excel provides three built-in methods for regression analysis: **Trendline**, the **Regression** tool in the **Analysis ToolPak**, and regression functions such as `FORECAST` and `TREND`. For merely graphing a regression line and providing its equation and the coefficient of determination r^2, the **Trendline** command suffices. We will consider the **Regression** tool in Chapters 10 and 11, as well as regression functions.

Linear Trendline

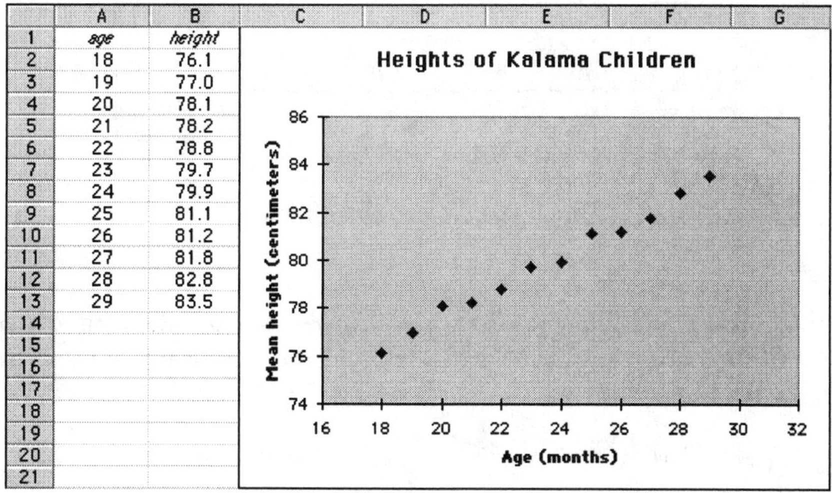

	A	B
1	age	height
2	18	76.1
3	19	77.0
4	20	78.1
5	21	78.2
6	22	78.8
7	23	79.7
8	24	79.9
9	25	81.1
10	26	81.2
11	27	81.8
12	28	82.8
13	29	83.5

Figure 2.13: Kalama Data and Scatterplot

We use the **Linear Trendline** to insert a curve on a scatterplot. The trendline can be added to any scatterplot even after the **Regression** tool is used.

> **Example 2.5.** (Example 2.11, page 135 in the text.) Columns A, B of Figure 2.13 give the data for a group of children in Kalama, an Egyptian village that was the site of a study of nutrition in developing countries. The explanatory variable x is age, the response variable y is height. Fit the least-squares regression line to the data.

Solution. We first construct a scatterplot of the data (also shown in Figure 2.13) to verify that a linear model is appropriate.

1. Enter the data in cells A2:B13 of a workbook. Use the **ChartWizard** to create a scatterplot and edit it (primarily to change the horizontal scale), as discussed previously, so that it appears as shown in Figure 2.13. The scatterplot shows an approximate linear relationship, so it is appropriate to fit the data pairs with a straight line.

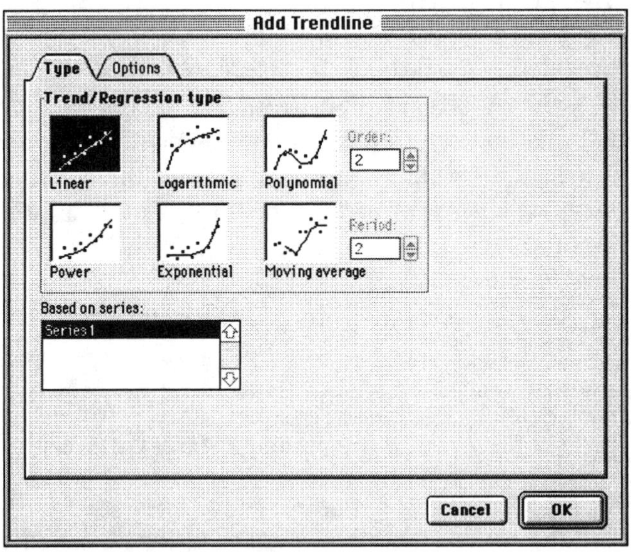

Figure 2.14: Trendline Type

2. Activate the chart for editing and select the data by **clicking on one of the points**. The points appear highlighted, the **Name** box in the **Formula Bar** shows S1, and in the text entry area we can read

$$= \texttt{SERIES}(\text{,Sheet1!\$}A\text{\$2} : \text{\$}A\text{\$13},\ \text{Sheet1!\$}B\text{\$2} : \text{\$}B\text{\$13}, 1)$$

   meaning that the series has been selected. (Refer to the online help for more information on this function and the Introduction for the meaning of Sheet1! notation.)

3. For **Excel 5/95**, choose **Insert – Trendline** from the Menu Bar; for **Excel 97/98**, choose **Chart – Add Trendline...** from the Menu Bar. Then proceed as follows. Click the **Type** tab and select **Linear** (Figure 2.14). **Excel 97/98** has an additional text area (**Based on series**) in the blank space at the bottom of this figure. Click the **Options** tab and select the radio button **Automatic:Linear (Series1)**. Check the boxes **Display Equation on Chart** and **Display r-squared Value on Chart**. Make sure that the **Set Intercept** box is clear. Click OK. The regression line is superimposed on the scatterplot, its equation $y = 0.635x + 64.928$ is displayed, and the coefficient of determination $R^2 = 0.9888$ (in Excel's notation) is inserted on the scatterplot.

4. The output may be edited as previously with other charts for presentation purposes. For instance, activate the chart, and click on the rectangular box surrounding the equation; the border turns a darker grey. Use the **Decimal** tool to increase or decrease the number of decimal points. Edit the text by replacing x with "age" and y with "height." Move $R^2 = 0.9888$ from its location on the graph to a more convenient place. The final result with the regression line appears as Figure 2.15.

Residuals

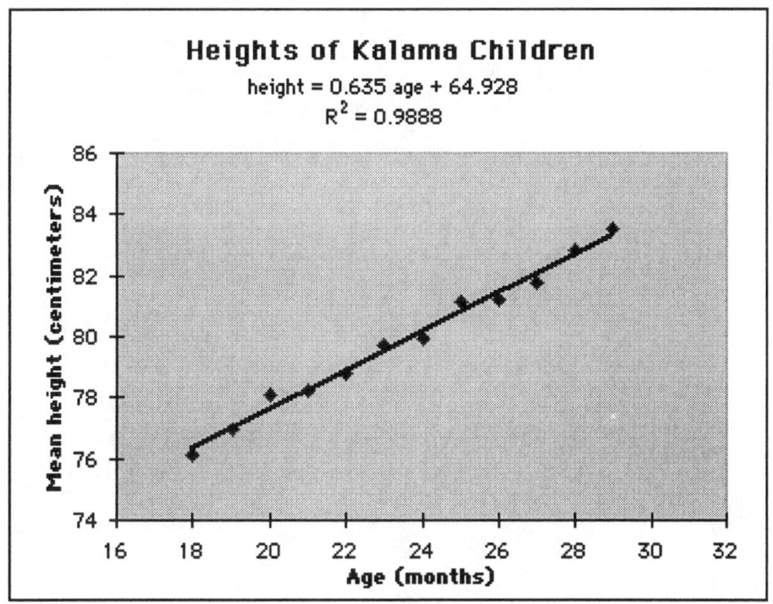

Figure 2.15: Regression Line and Scatterplot

No discussion of regression is complete without an analysis of residuals, which provide evidence of how well the regression model fits. We defer discussion of this topic to Chapter 10 where the **Regression** tool will be introduced. However, in Figure 2.16 we show a scatterplot of residuals against age for the Kalama example. There does not appear to be any discernable pattern in the plot, indicating that a straight-line fit is appropriate.

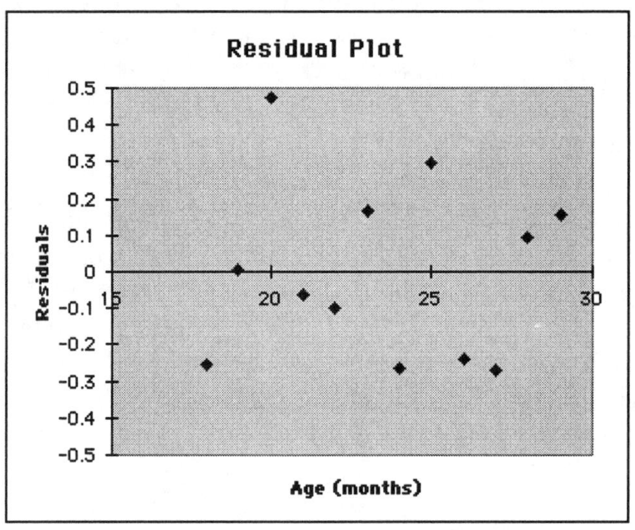

Figure 2.16: Residual Plot

Chapter 3

Producing Data

Excel provides tools for sampling from a specified population, for calculating probabilities associated with the standard models and their inverse cumulative distributions using the **Formula Palette**, and for simulating values from probability distributions using both the `RAND()` function and the **Random Number Generation** tool. In this chaper we consider both sampling without replacement and sampling with replacement (SRS) from a specified population.

3.1 Samples with Replacement

Using the Sampling Tool

> **Example 3.1.** Find a sample with replacement from a finite population.

Solution. Figure 3.1 shows a sample of size 10 from $\{1, 2, 3, 4\}$. Following are the steps to obtain such a sample.

1. Enter the values $\{1, 2, 3, 4\}$ in A2:A5.

	A	B
1	*Population*	*Sample*
2	1	1
3	2	2
4	3	2
5	4	1
6		1
7		4
8		4
9		2
10		4
11		2

Figure 3.1: Sample with Replacement

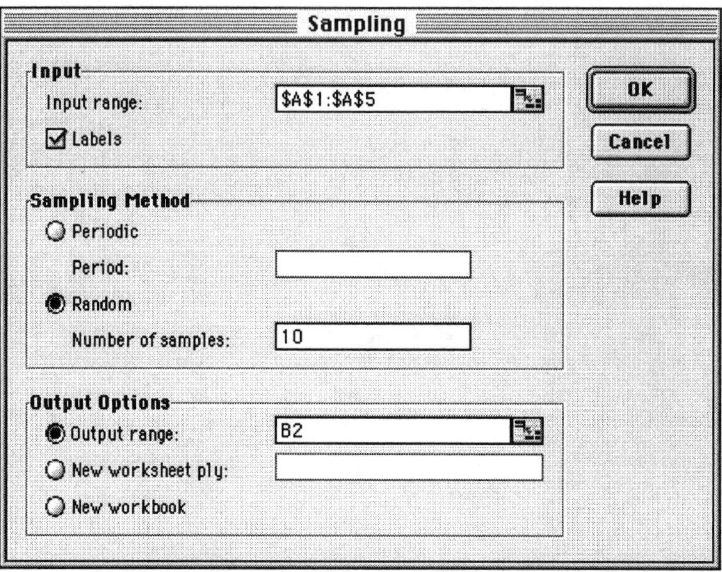

Figure 3.2: Sampling Dialog Box

2. From the Menu Bar choose **Tools – Data Analysis** and select **Sampling** from the dialog box. Click OK.

3. Complete the **Sampling** dialog box as shown in Figure 3.2 and click **OK**. A random sample of size 10 appears in cells B2:B11.

3.2 Simple Random Samples (SRS)

The Excel function RAND() picks a number uniformly on the interval (0, 1). We can repeatedly select random uniform (0, 1) numbers and assign them to members of a population. Then by sorting the random numbers, we obtain a random permutation of the population that provides an SRS of any desired size.

> **Example 3.2.** (Examples 3.6 and 3.7, pages 241–244 in the text.) A food company assesses the nutritional quality of a new "instant breakfast" product by feeding it to newly weaned male white rats and measuring their weight gain over a 28-day period. A control group of rats receives a standard diet for comparison. This nutrition experiment has a single factor (the diet) with two levels. The researchers use 30 rats for the experiment and so must divide them into two groups of 15. To do this in a completely unbiased fashion, they put the cage numbers of the 30 rats in a hat, mix them up, and draw 15. These rats form the experimental group and the remaining 15 make up the control group. Show how to carry out the randomization.

	A	B
1	*Rats*	*Sample*
2	1	0.087079417
3	2	0.170785871
4	3	0.499370679
5	4	0.530765263
6	5	0.856979031
7	6	0.602390579
8	7	0.289199918
9	8	0.443723483
10	9	0.019658386
11	10	0.276338967
12	11	0.136321585
13	12	0.02561087
14	13	0.735682793
15	14	0.352039927
16	15	0.595454284

Figure 3.3: Original Labels

Solution.

1. Give each rat a unique numerical label from the set $\{1, 2, 3, \ldots, 30\}$ and enter the values in cells A2:A31 of a workbook. Enter the label "Rats" in cell A1 and "Sample" in cell B1.

2. Enter = `RAND()` in cell B2 and fill down to B31. The function `RAND()` selects a number uniformly in (0,1). Figure 3.3 shows a portion of the workbook.

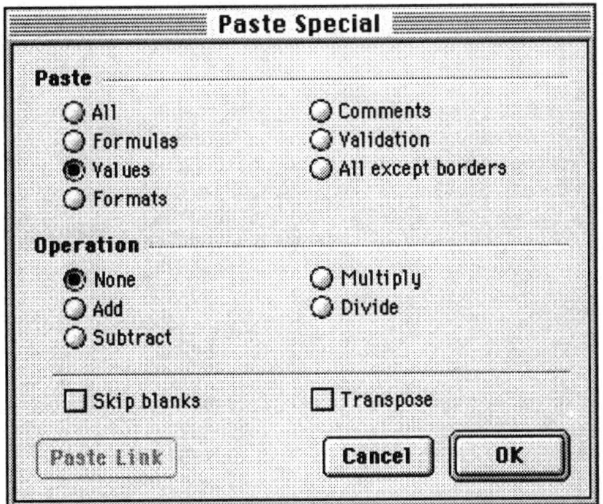

Figure 3.4: Paste Special Dialog Box

3. Select cells B2:B31 and from the Menu Bar choose **Edit** – **Copy**. Then, with B2:B31 **still selected**, choose **Edit** – **Paste Special** from the Menu Bar. (**Windows** users can click the **right mouse button** while **Macintosh** users should hold down the **Option** – **Command** keys and click to get the

Shortcut Menu box.) Select **Paste – Special**, and in the dialog box select the radio buttons for **Values** and **None** (Figure 3.4), which replaces the formulas in the cells of column B by the actual values they take.

4. Select cells A1:B31, from the Menu Bar choose **Data – Sort**, and in the Sort By drop-down list, click the arrow and select **Sample.** Also select the radio button for **Header Row** (Figure 3.5). Excel sorts the data in ascending order in column B and carries the order to column A, which gives a random permutation of column A.

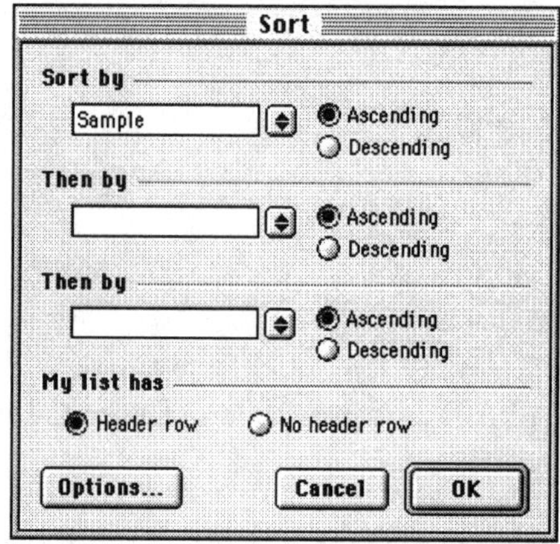

Figure 3.5: Sort Dialog Box

	A	B
1	*Rats*	*Sample*
2	9	0.019658386
3	27	0.024283844
4	12	0.02561087
5	1	0.087079417
6	21	0.116760743
7	18	0.129875827
8	11	0.136321585
9	16	0.167847609
10	2	0.170785871
11	10	0.276338967
12	20	0.285602834
13	7	0.289199918
14	30	0.331929106
15	14	0.352039927
16	8	0.443723483

Figure 3.6: Sorted Sample

5. Designate cells A2:A16 to label the rats in the control group, cells A17:A31 for the rats in the experimental group. A portion of the data appears in Figure 3.6, which may be compared with the original data in Figure 3.3.

3.3 Random Digits

Figure 3.7 is a table of random digits, a list of the digits $\{0, 1, \ldots, 9\}$ that has the following properties:

1. The digits in all positions in the list have the same chance of being any one of $\{0, 1, \ldots, 9\}$.

2. The digits in different positions are independent in the sense that the value of one has no influence on the value of any other.

You can imagine asking an assistant (or computer) to mix the digits $\{0, 1, \ldots, 9\}$ in a hat, draw one, then replace the digit drawn, mix again, draw a second digit, and so on. We did something like this in the previous section. Now we produce a table of random digits that dynamically changes when the F9 key on the keyboard is pressed.

	A	B	C	D	E	F	G	H	I	J	K	L	M	N	O	P	Q	R	S	T
1	*Table of Random Digits*																			
2	=INT(10*(RAND()))																			
3																				
4	0	4	5	4	2	3	4	4	7	6	3	5	9	7	7	4	5	2	0	7
5	6	2	9	7	0	9	3	5	2	6	9	8	6	1	4	2	7	9	8	2
6	7	8	9	1	5	4	2	0	7	5	2	4	3	3	2	1	7	1	3	8
7	9	5	7	8	9	2	6	3	4	3	6	6	5	0	5	3	0	9	1	8
8	3	4	9	2	5	1	2	4	2	1	2	0	4	1	4	5	3	4	9	8
9	4	6	8	7	6	9	2	8	8	2	0	6	5	5	1	4	1	5	1	6
10	0	5	0	7	4	3	5	8	9	7	9	1	9	3	3	5	3	7	5	2
11	1	3	0	5	1	6	7	2	0	9	3	8	3	5	8	5	3	9	8	6
12	6	3	4	7	2	4	4	6	1	7	0	3	4	4	6	7	2	2	0	1
13	8	3	5	1	9	1	1	0	3	2	9	9	4	9	2	4	7	5	4	5
14	7	4	4	3	4	4	3	1	4	0	6	4	5	0	5	9	0	3	2	7
15	1	1	4	1	1	2	8	2	4	6	6	5	4	8	8	4	2	4	4	8
16	8	7	3	3	2	0	0	8	8	8	3	8	4	5	6	7	3	3	2	9
17	0	4	5	5	5	8	6	6	6	0	6	4	8	0	5	5	3	9	7	3
18	1	6	3	3	7	9	8	9	1	1	6	5	7	9	2	1	1	0	4	1
19	2	8	1	4	4	5	6	5	7	3	6	4	2	4	9	3	6	8	0	7
20	0	5	0	8	7	1	1	6	8	4	0	1	4	2	5	2	0	2	8	4
21	7	0	2	5	1	6	7	0	8	3	4	9	5	9	9	2	8	5	0	8
22	1	7	8	9	8	1	8	3	5	3	4	5	0	3	3	7	1	7	1	6
23	4	0	5	7	6	7	0	3	2	7	6	1	1	8	7	5	5	6	4	3
24	4	2	5	9	3	4	0	4	5	4	8	9	6	2	2	4	8	6	1	6
25	6	2	6	1	9	2	8	8	2	5	4	3	4	1	8	9	4	7	8	6
26	1	0	4	1	6	5	7	5	1	8	5	7	0	9	0	2	7	3	9	0
27	1	2	0	4	1	6	8	7	6	3	7	4	9	1	4	2	4	4	2	9
28	8	2	8	0	7	6	0	1	5	5	9	8	4	8	9	9	4	1	9	4

Figure 3.7: Table of Random Digits

Using the RAND() Function

The Excel function `INT` truncates a real number to its integer value. For instance, the formula = `INT(3.82)` produces the value 3. By combining `INT` and `RAND()` as = `INT(10*RAND())` we can produce random digits from $\{0, 1, \ldots, 9\}$.

Example 3.3. Produce a table of 500 random digits.

Solution

1. In cell A4 of a workbook, enter the formula = `INT(10*RAND())`.
2. Select cell A4, click the fill handle in the lower right corner, and drag across to cell T4.
3. Select cells A4:T4, click the fill handle in the lower right corner, and drag down to cell T28.

The result is shown in Figure 3.7. Your digits will necessarily be different as they are random. Press the **F9** key and see what happens. The entire table changes as new random digits are now **dynamically selected**.

Chapter 4

Probability: The Study of Randomness

Probability models are used to describe and analyze real-world phenomena involving randomness. One way to develop an intuition for randomness is to observe random behavior. Computer simulations allow visual penetration into the concept of random variation.

4.1 Randomness

Simulating Bernoulli Random Variables

A real-world probability can only be estimated through the observation of data. Computer simulations are useful because they help develop insight into the meaning of random variation. Excel is well suited for simulation and provides both a `RAND()` function and a **Random Number Generation** tool for such purpose.

> **Example 4.1.** (Exercise 4.9, page 294 in the text.) The basketball player Shaquille O'Neal makes about half of his free throws over an entire season. Use Excel to simulate 100 free throws shot independently by a player who has probability 0.5 of making each shot. The technical term for independent trials with yes/no outcomes is Bernoulli trials. Our outcomes here are hit or miss.
> (a) What percent of the 100 shots did he hit in the simulation?
> (b) Examine the sequence of hits and misses. How long was the longest run of shots made? Of shots missed? (Sequences of random outcomes often show runs longer than our intuition thinks likely.)

Solution. Again we will use the `RAND()` function to generate a sequence of 100 free throws and then invoke the **ChartWizard** to dramatically display the results.

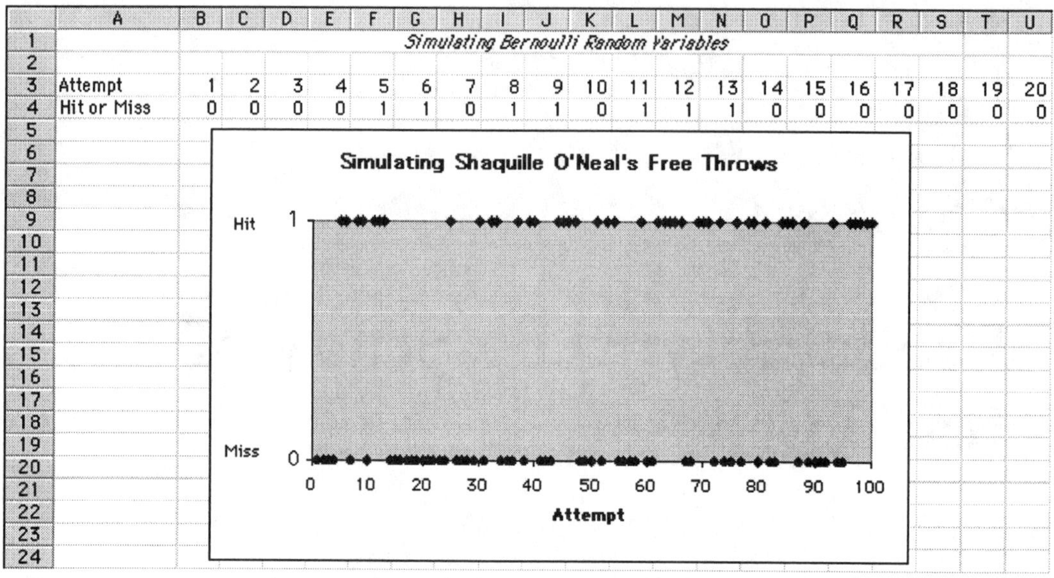

Figure 4.1: Simulating Shaq's Free Throws

1. Enter the labels "Attempt" in Cell A3 of a new workbook and "Hit" or "Miss" in cell A4. Enter the number "1" in cell B3 and fill to cell CW3 with successive integers $\{1, 2, \ldots, 100\}$. This can be achieved efficiently by selecting cell B3, then choosing **Edit – Fill – Series...** from the Menu Bar. Complete the **Series** dialog box with **Series in** Rows, **Type** Linear, **Step value** 1, and **Stop value** 100. Click OK.

2. Enter the = `INT(2*RAND())` in cell B4. Select B4, click the fill handle, and drag to cell CW4 to generate 100 independent Bernoulli random variables.

In Figure 4.1 we show our workbook with the first 20 free throw simulations together with a graph of Shaq's hits or misses in 100 attempts. Notice how much more illuminating the graph is than the numerical sequence of hits and misses in showing random variation, including the presence of hot and cold streaks based on chance alone. The following steps explain how this chart is constructed.

Click the **ChartWizard** button.

Users of Excel 5/95

- In Step 1 enter the data range A3:CW4.
- In Step 2 click the **Scatter** chart type.
- In Step 3 select Format **1**.

- In Step 4 click the button for Data Series in **Rows.** Enter "1" for Use First Row for Category(X) Axis Labels and enter "1" for Use First 1 Column for Legend Text.
- In Step 5 select the radio button **No** for Add a legend?, and label the chart and X axis as shown in Figure 4.1.
- Click Finish.

Users of Excel 97/98

- In Step 1 click the **XY (Scatter)** Chart type and the first Chart subtype (upper left on right side).
- In Step 2 on the **Data Range** tab, enter A3:CW4 for the range and check the radio button **Rows** for Series in:.
- In Step 3 on the **Titles** tab, enter the title and labels of the axes, on the **Axes** tab check both Category (X) axis and Category (Y) axis, on the **Gridlines** tab turn off all gridlines, on the **Legend** tab clear the legend, and finally on the **Data Labels** tab select the radio button **None.**
- In Step 4 embed the graph in the current workbook.
- Click Finish.

After the chart appears embedded on your workbook select it for editing and add the text "Hit" and "Miss" on the vertical axis.

Answers to Example 4.1 questions:
(a) By entering = `SUM`(B4:CW4)/100 in an empty cell, we find that the player hit 45% of his shots in the simulation.
(b) From Figure 4.1, we read that the longest run of hits is 5 and the longest run of misses is 11.
To simulate an additional 100 free throws, press the F9 key.

4.2 Probability Models

A probability model consists of a list of possible outcomes and a probability for each outcome (or interval of outcomes, in the case of continuous models). The probabilities are determined by the experiment that leads to the occurrence of one or more of the outcomes in the specified list.

Excel provides many distributions that may be constructed in a common fashion with the **Formula Palette** or the **Function Wizard**. The meaning of the required parameters is available online through Excel's help feature. Because of its prominence, the normal distribution was already discussed in Chapter 1. The binomial model will be considered in Chapter 5. Here we discuss some other distributions of particular interest in statistics.

Hypergeometric

HYPERGEOMDIST$(x, n, M, N,)$ provides probabilities for an experiment in which a simple random sample of size n is taken from a finite population of N individuals of which M are in a so-called "preferred category" called "success" or "1," while the remaining $N - M$ are deemed "failure" or "0." The function returns the probability of x successes in the sample of size n.

> **Example 4.2.** (Lotto 6/49) Suppose a box contains 49 balls in which one and only one ball is marked with an integer taken from $\{1, 2, \ldots, 49\}$. The balls are identical otherwise. Suppose that the balls numbered $\{1, 2, 3, 4, 5, 6\}$ are considered "successes." If an SRS of six balls is taken at random (without replacement), what is the probability that the sample contains k successes (for $k = 0, 1, 2, 3, 4, 5, 6$)?

	A	B	C
1		*Lotto 6/49 Probabilities*	
2	k	P(X=k) value	P(K=k) formula
3	0	0.435964976	=HYPGEOMDIST(A3,6,6,49)
4	1	0.41301945	=HYPGEOMDIST(A4,6,6,49)
5	2	0.132378029	=HYPGEOMDIST(A5,6,6,49)
6	3	0.017650404	=HYPGEOMDIST(A6,6,6,49)
7	4	0.00096862	=HYPGEOMDIST(A7,6,6,49)
8	5	1.84499E-05	=HYPGEOMDIST(A8,6,6,49)
9	6	7.15112E-08	=HYPGEOMDIST(A9,6,6,49)

Figure 4.2: Calculating Lotto Probabilities

Solution. The answer is provided in Figure 4.2 where column B gives the probabilities and column C the corresponding Excel formulas.

Student t-Distribution

The Student t-distribution arises as the distribution of the *Studentized* score (similar to a standardized score)

$$t = \frac{\bar{x} - \mu}{s/\sqrt{n}}$$

where $\bar{x}, s$ are the sample mean and sample standard deviation from a sample of size n taken from a $N(\mu, \sigma)$ population. It is determined by a single parameter called the degrees of freedom ν. In the above ratio, ν takes the value $n - 1$.

Excel has an unusual definition of the c.d.f. and inverse c.d.f. for the t-distribution. The Excel function TDIST returns the tail of the distribution, that is, if $t(\nu)$ is a random variable with a t-distribution on ν d.f., then

$$\texttt{TDIST}(x, \nu, 1) = P[t(\nu) > x]$$

and

$$\texttt{TDIST}(x, \nu, 2) = 2P[t(\nu) > x]$$

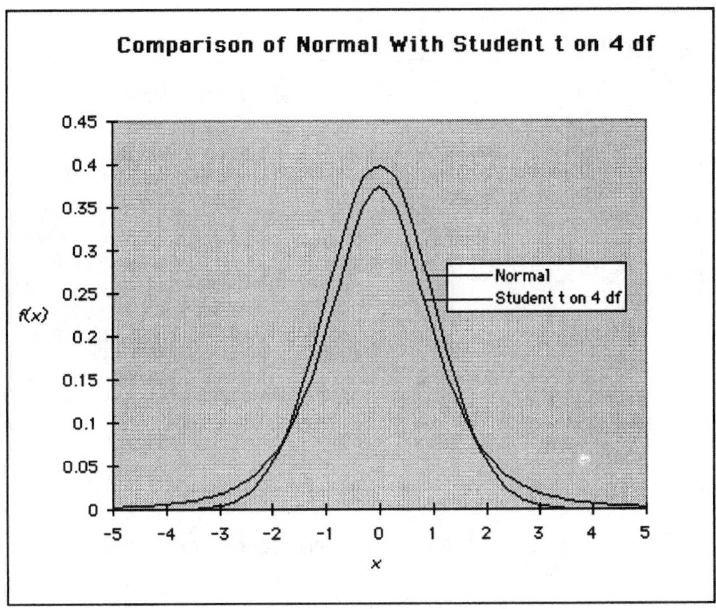

Figure 4.3: Comparing the Normal and the t Curves

The argument x in TDIST must be positive. Thus the c.d.f. takes a rather complicated expression using the logical IF function:

$$P(t(\nu) \leq x) = \texttt{IF}(x < 0, \texttt{TDIST}(\texttt{ABS}(x), \nu, 1), 1 - \texttt{TDIST}(x, \nu, 1))$$

where $\texttt{ABS}(x) = |x|$, and similarly

$$P(|t(\nu)| \leq x) = 1 - \texttt{TDIST}(x, \nu, 2)$$

The inverse function TINV is defined by

$$P[t(\nu) > \texttt{TINV}(\alpha, \nu)] = \frac{\alpha}{2}$$

so $\texttt{TINV}(\alpha, \nu)$ is the critical value for a two-sided significance test at level α of a normal mean (to be discussed in Chapter 7).

> **Exercise.** Using NORMSDIST and TDIST, graph on the same figure and to the same scale the densities of a $N(0,1)$ and the Student t-distribution on 4 d.f. (Figure 4.3).

Uniform

A uniform random number is one whose values are spread out uniformly across the interval from 0 to 1. Its density curve has height 1 over the interval 0 to 1.

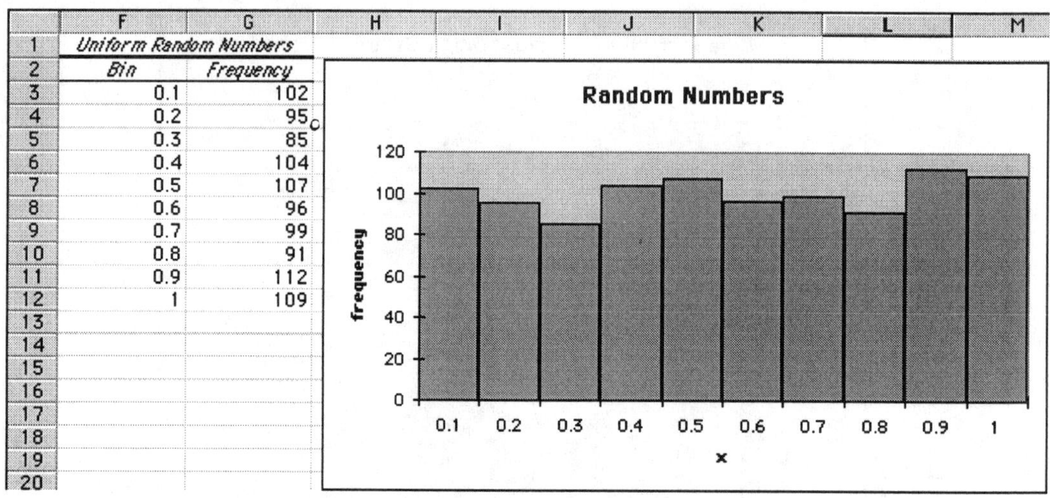

Bin	Frequency
0.1	102
0.2	95
0.3	85
0.4	104
0.5	107
0.6	96
0.7	99
0.8	91
0.9	112
1	109

Figure 4.4: Simulating Uniform Random Variables

Example 4.3. (Based on Exercise 4.44, page 324 in the text.) Let X be a uniform random number between 0 and 1. Use Excel to generate 1000 random uniform numbers, and from your simulations estimate the following probabilities and compare them with the theoretical values.

(a) $P(0 \leq X \leq 0.4)$

(b) $P(0.4 \leq X \leq 1)$

(c) $P(0.3 \leq X \leq 0.5)$

(d) $P(0.3 < X < 0.5)$

Solution. Use `RAND()` to generate 1000 uniform random variables in a column and construct a histogram with bin intervals of width 0.10 beginning at 0 and ending at 1. Figure 4.4 shows the sample output from a workbook where this has been done. The frequencies shown are the number of times the random number generator produced a number X in the specified interval. The values listed under the heading *Bin* are the right endpoints of the intervals. We count the number of observations in the relevant intervals and divide by 1000 to convert to a probability.

(a) $P(0 \leq X \leq 0.4) = 0.386$

(b) $P(0.4 \leq X \leq 1) = 0.614$

(c) $P(0.3 \leq X \leq 0.5) = 0.211$

(d) $P(0.3 < X < 0.5) = 0.211$

The theoretical values are 0.4, 0.6, 0.2, and 0.2, respectively.

Triangular—Adding Random Numbers

Example 4.4. (Based on Exercise 4.46, page 325 in the text.) Generate two random numbers between 0 and 1 and take Y to be their sum. Clearly the sum Y can take any number between 0 and 2. It is known that the idealized density curve of Y is a triangle. Use Excel to generate 1000 pairs of uniform random numbers, add them, and from your simulations estimate the following probabilities and compare them with the theoretical values.

(a) $P(0 \leq X \leq 0.5)$

(b) $P(0.5 \leq X \leq 1.5)$

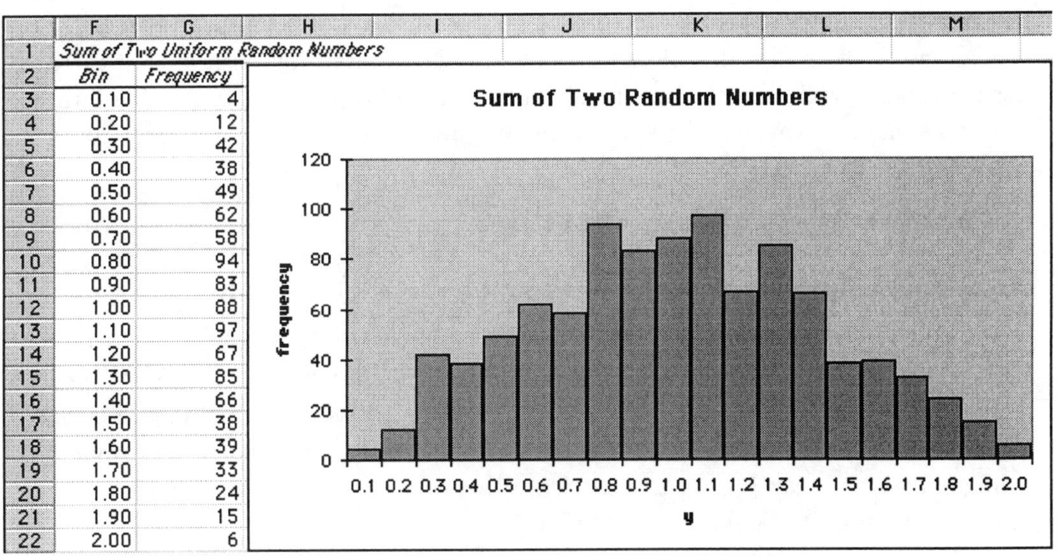

Sum of Two Uniform Random Numbers

Bin	Frequency
0.10	4
0.20	12
0.30	42
0.40	38
0.50	49
0.60	62
0.70	58
0.80	94
0.90	83
1.00	88
1.10	97
1.20	67
1.30	85
1.40	66
1.50	38
1.60	39
1.70	33
1.80	24
1.90	15
2.00	6

Figure 4.5: Simulating Triangular Random Variables

Solution. Again use `RAND()` to generate 1000 pairs of uniform random variables in two columns, add the columns and construct a histogram with bin intervals of width 0.10 beginning at 0 and ending at 2. Figure 4.5 shows the sample output from a workbook where this has been done. The frequencies shown are the number of times the random number generator produced a number in the specified interval.

(a) $P(0 \leq X \leq 0.5) = 0.145$

(b) $P(0.5 \leq X \leq 1.5) = 0.738$

The corresponding theoretical values are 0.125 and 0.750, respectively.

4.3 Random Variables

A random variable is completely prescribed by its probability distribution. This distribution has a long-run frequency interpretation associated with the **Law of Large Numbers.**

> Draw independent observations at random from any population with finite mean μ. Decide how accurately you would like to estimate μ. As the number of observations drawn increases, the mean $\bar{x}$ of the observed values eventually approaches the mean μ of the population as closely as you specified and then stays that close.

Applying this phenomenon to a discrete random variable X, suppose that

$$P(X = 3) = 0.5$$

In repeated trials, consider the proportion $\hat{p}$ of times that X takes the value 3. The random variable $\hat{p}$ is the sample mean of a sequence of random variables taken from a Bernoulli population with probability of success = 0.5 and whose population mean is therefore also 0.5. The Law of Large Numbers then asserts that in some sense (made precise by the theory of probability) $\hat{p}$ approaches 0.5 as the number of trials increases, which gives a relative frequency interpretation of probability.

Simulating Random Variables

The Law of Large Numbers Using the RAND() Function

The Excel function **RAND**() picks a number uniformly on the interval (0,1). To generate a uniform random variable on (a, b) use $a +$ **RAND**() $* (b - a)$. Using the inverse probability function $h(a) = \inf\{x : F(x) \geq a\}$, we can then generate other distributions. Thus **NORMINV(RAND**(), Mean, StDev) returns a random normal with mean given by Mean (either a numerical value or a named reference to a numerical value) and standard deviation by StDev.

By examining the list of functions available (clicking the **Paste Function** button f_x on the Standard Toolbar), you can determine which distributions Excel can simulate this way and how to describe the required parameters.

> **Example 4.5.** Using the **RAND**() function, simulate 1000 independent Bernoulli trials based on tossing a fair coin, calculate the cumulative proportion of heads $\hat{p}$ after each trial, and construct a graph that demonstrates the law of large numbers in action. Also show on the same graph a horizontal line at the height 0.5

Solution. The **RAND**() function produces a number uniformly distributed on the interval (0,1). This can be converted into integers taking the values 0 or 1 with equal probability if this uniform random number is multiplied by 2 and then the integer part is taken. The Excel formula for these operations is = **INT(2*RAND())**.

1. Enter the formula = `INT(2*RAND())` in cell A5 of a new workbook and copy this formula down to cell A1003 by selecting cell A4, then clicking the fill handle and dragging to cell A1003 to generate 1000 tosses of a fair coin (0 representing tails and 1 representing heads).

2. Enter the value "0" in cell B3 followed by the formula = A4+B3 in cell B4. Copy the formula in cell A4 down to cell A1003. Column B tracks the cumulative number of heads.

3. Enter the number 1 in cell C4 and fill to cell C1003 with successive integers $\{1, 2, \ldots, 1000\}$. This can be achieved efficiently by selecting cell C4 and then choosing **Edit – Fill – Series** from the Menu Bar. Complete the **Series** dialog box with Series in **Columns**, Type **Linear**, and **Step Value** 1, **Stop Value** 1000. Click OK. Column C will label the 1000 tosses.

4. Fill cells D4 to D1003 with the value 0.5. This will represent the horizontal line at height 0.5 on the graph.

5. Enter the formula = B4/C4 in cell E4 and copy to cell E1003.

Figure 4.6 shows part of the workbook with the required formulas.

	A	B	C	D	E
1	*Simulation of 1000 Tosses of a Fair Coin*				
2					
3		0			
4	=INT(2*RAND())	=A4+B3	1	0.5	=B4/C4
5	=INT(2*RAND())	=A5+B4	2	0.5	=B5/C5
6	=INT(2*RAND())	=A6+B5	3	0.5	=B6/C6

Figure 4.6: Simulating 1000 Tosses of a Fair Coin

We next construct a graph displaying the same results. Click the **ChartWizard** button.

Users of Excel 5/95

- In Step 1 enter the data range C4:E1003.
- In Step 2 click the **Line** chart type.
- In Step 3 select Format **2**.
- In Step 4 click the button for Data Series in **Columns**. Enter "1" for Use First 1 Column for Category(X) axis labels and enter "0" for Use First 0 Column for Legend Text.
- In Step 5 click the radio button **No** for Add a legend? and label the chart and X axis as shown in Figure 4.7. Click Finish.

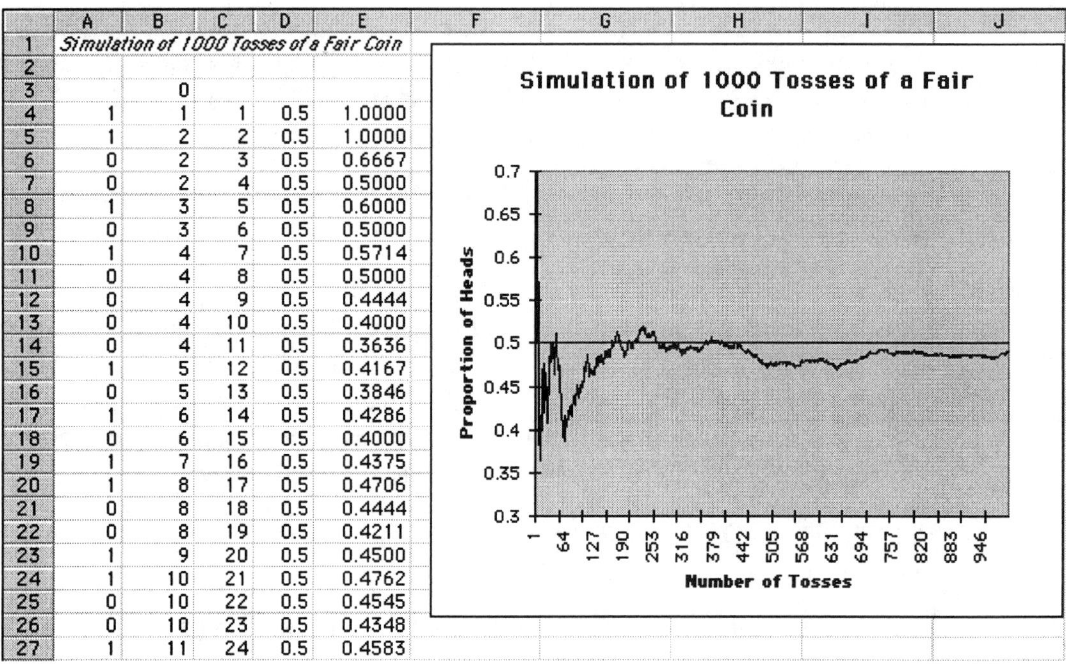

	A	B	C	D	E
1	Simulation of 1000 Tosses of a Fair Coin				
2					
3		0			
4	1	1	1	0.5	1.0000
5	1	2	2	0.5	1.0000
6	0	2	3	0.5	0.6667
7	0	2	4	0.5	0.5000
8	1	3	5	0.5	0.6000
9	0	3	6	0.5	0.5000
10	1	4	7	0.5	0.5714
11	0	4	8	0.5	0.5000
12	0	4	9	0.5	0.4444
13	0	4	10	0.5	0.4000
14	0	4	11	0.5	0.3636
15	1	5	12	0.5	0.4167
16	0	5	13	0.5	0.3846
17	1	6	14	0.5	0.4286
18	0	6	15	0.5	0.4000
19	1	7	16	0.5	0.4375
20	1	8	17	0.5	0.4706
21	0	8	18	0.5	0.4444
22	0	8	19	0.5	0.4211
23	1	9	20	0.5	0.4500
24	1	10	21	0.5	0.4762
25	0	10	22	0.5	0.4545
26	0	10	23	0.5	0.4348
27	1	11	24	0.5	0.4583

Figure 4.7: Law of Large Numbers

Users of Excel 97/98

- In Step 1 click the **Line** chart type and the second Chart sub-type **Stacked Line.**
- In Step 2 on the **Data Range** tab enter D4:E1003 for the Data range check the radio button **Series in: Columns.**
- In Step 3 on the **Titles** tab, enter the title and labels of the axes, on the **Axes** tab check radio button **Automatic** for Category (X) axis and check the Value (Y) axis box, on the **Gridlines** tab turn off all gridlines, on the **Legend** tab clear the legend, and finally on the **Data Labels** tab select the radio button **None.**
- In Step 4 embed the graph in the current workbook. Click Finish.

Format the X and Y axes as shown in Figure 4.7, for instance by changing the number of categories between tick marks and reorienting the X axis labels.

Figure 4.7 shows a segment of the completed workbook with the embedded graph. Press the **F9** key to reevaluate all functions and the graph will dynamically change. From column A you can see the random sequence of heads and tails generated, while column E exhibits the proportions of heads. These are quite variable at first but then settle down, appearing to approach the value 0.5 (shown by the horizontal line). This behavior is known in statistics as a law of large numbers, commonly referred to as the "law of averages."

Simulating Random Variables Using Random Number Generation

In addition to the function `RAND()`, Excel has a **Random Number Generation** tool built into the **Analysis ToolPak** that provides an alternative and more systematic approach to simulation.

The **Random Number Generation** tool creates columns of random numbers, as specified by the user, from any of six probability models (uniform, normal, Bernoulli, binomial, Poisson, discrete) as well as having an option for patterned that creates not random data but rather data according to a specified pattern.

All options are invoked from a common dialog box (as in Figure 4.8 for discrete) following the choice **Tools – Data Analysis – Random Number Generation** from the Menu Bar. Select the distribution of interest using the drop-down arrow and the Parameters sub-box will automatically change, prompting input of parameters.

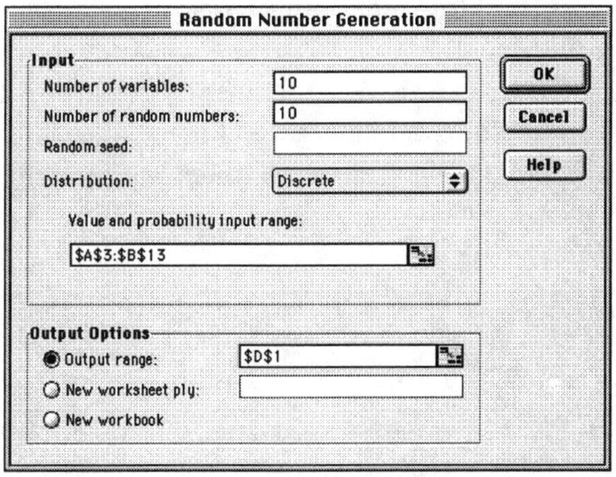

Figure 4.8: Random Number Generation Tool—Discrete

Number of Variables. Enter the number of columns of random variables. The default is all columns.

Number of Random Numbers. Enter the number of rows (cases) of random variables.

Distributions. Use the drop-down arrow to open a list of choices with requested parameters.

> **Uniform.** Upper and lower limits
>
> **Normal.** μ, σ

Bernoulli. $p =$ probability of success; Excel unfortunately refers to this as a p Value.

Binomial. p, n

Poisson. λ

Discrete. Specify the possible values and their corresponding probabilities. Before using this option enter the values and probabilities in adjacent columns in the workbook.

Patterned. This option creates data according to a prescribed pattern of values repeated in specified steps. This is useful if a linear array of data needs to be coded using another variable.

Example 4.6. To generate 100 tosses of a pair of fair dice enter $\{2, 3, \ldots, 11\}$ into cells A3:A13 and enter

$$\{1/36, 2/36, \ldots, 6/36, 5/36, \ldots, 2/36, 1/36\}$$

into cells B3:B12 (Figure 4.9). Excel may interpret the value 1/36 as a date Jan 1936. If this happens then format the cells by choosing **Format – Cells** from the Menu Bar and selecting **Number**. Then choose **Tools – Data Analysis – Random Number Generation** from the Menu Bar and complete as in Figure 4.8. The output will appear in cells D1:M10. Since these numbers are random your output will of course be different.

	A	B	C	D	E	F	G	H	I	J	K	L	M
1	*Simulation of a Pair of Fair Dice*			4	9	3	9	4	11	8	12	6	2
2	k	P(X=k)		10	8	8	4	5	7	3	7	4	6
3	2	0.0277778		3	8	8	11	4	5	5	12	7	5
4	3	0.0555556		6	5	7	6	5	8	9	5	11	6
5	4	0.0833333		4	7	10	8	10	12	5	5	5	7
6	5	0.1111111		7	2	3	6	11	9	5	8	8	9
7	6	0.1388889		5	11	6	7	6	6	8	3	7	9
8	7	0.1666667		5	10	8	4	8	9	5	7	2	4
9	8	0.1388889		9	5	10	3	11	6	7	7	5	3
10	9	0.1111111		7	3	5	4	7	4	11	7	2	10
11	10	0.0833333											
12	11	0.0555556											
13	12	0.0277778											

Figure 4.9: Simulating a Pair of Fair Dice

Chapter 5

From Probability to Inference

The probability distribution of a statistic obtained from an experiment is called its sampling distribution. An important class of statistics arises when the observations are counts of some variable. This leads to the binomial model for sample counts and sample proportions. These sampling distributions can be approximated by normal curves, and they directly demonstrate several important results about sample means $\bar{x}$ in general:

1. $\bar{x}$ is an unbiased estimate of the population mean μ.
2. The standard deviation of $\bar{x}$ is equal to $\frac{\sigma}{\sqrt{n}}$ where n is the sample size and σ the population standard deviation.
3. The sampling distribution of $\bar{x}$ is approximately $N(\mu, \frac{\sigma}{\sqrt{n}})$.

5.1 Sampling Distributions for Counts and Proportions

The Binomial Distribution

A binomial distribution is associated with an experiment comprising n independent trials each of which has the same success probability p. The random variable X counts the number of successes.

It is known that

$$P(X = k) = \binom{n}{k} p^k (1-p)^{n-k} \qquad k = 0, 1, 2, \ldots, n$$

$$\text{mean} = \mu_X = np$$

and

$$\text{standard deviation} = \sigma_X = \sqrt{np(1-p)}$$

The corresponding Excel function is BINOMDIST(k, n, p, cumulative). If the parameter cumulative is set to "false," Excel returns the probabilities $P(X = k)$, while if it is set to "true," Excel returns the cumulative probabilities $P(X \leq k)$.

Example 5.1. Construct a binomial table for $n = 15$ and $p = 0.03$, including both individual and cumulative probabilities.

Solution

1. Enter the label k in cell A1 and the label $P(X = k)$ in cell B1 of a new workbook. In A2:A17 enter the values $\{0, 1, 2, \ldots, 15\}$.

2. Activate cell B2. Using either the **Formula Palette** or the **Function Wizard**, construct the binomial function by selecting **Statistical** for Function Category and BINOMDIST for Function Name.

3. Input the following into the dialog box.

   number_s. Enter the cell address A2.
   trials. Enter the value 15.
   probability_s. Enter the value 0.3.
   cumulative. Enter the value 0.

   Click Finish or OK.

4. Activate cell B2, click the fill handle in the lower right corner, and drag to cell B17 to fill the column with individual binomial probabilities (Figure 5.1).

	A	B	C
1	x	P(X=k)	P(X<=k)
2	0	0.00475	0.00475
3	1	0.03052	0.03527
4	2	0.09156	0.12683
5	3	0.17004	0.29687
6	4	0.21862	0.51549
7	5	0.20613	0.72162
8	6	0.14724	0.86886
9	7	0.08113	0.94999
10	8	0.03477	0.98476
11	9	0.01159	0.99635
12	10	0.00298	0.99933
13	11	0.00058	0.99991
14	12	0.00008	0.99999
15	13	0.00001	1.00000
16	14	0.00000	1.00000
17	15	0.00000	1.00000

Figure 5.1: Binomial Probabilities

5. Next label cell C1 as $P(X <= k)$ and repeat Steps 2, 3, and 4. Activate C2 instead of B2 in Steps 2 and 4 and enter the value 1 for the cumulative distribution in Step 3.

The resulting table of individual and cumulative binomial probabilities appears in Figure 5.1.

Binomial Distribution Chart

We can quickly construct a histogram using the **ChartWizard** displaying the binomial probabilities just calculated. As the procedure is identical to earlier constructions of charts, we omit the details. This histogram appears in Figure 5.2.

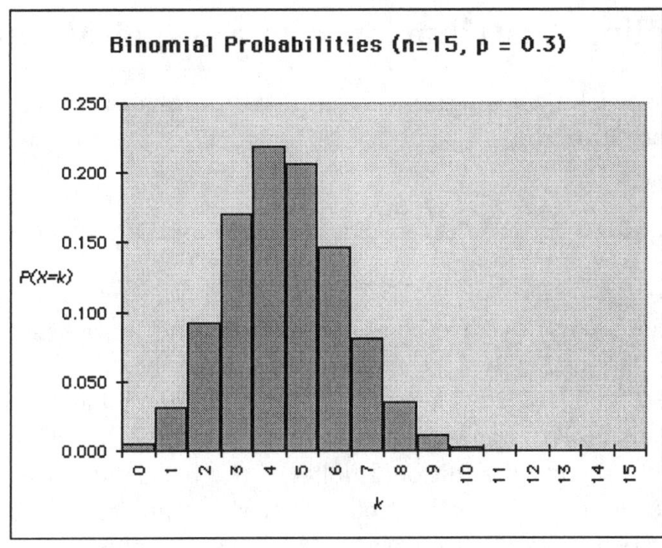

Figure 5.2: Binomial Histogram

Inverse Cumulative Binomial

`CRITBINOM`(trials, probability_s, alpha) returns the smallest x for which the binomial cumulative distribution function (c.d.f.) is greater than or equal to alpha; that is, if $B(x)$ represents the binomial c.d.f., then = `CRITBINOM`(n, p, α) returns

$$B^{-1}(\alpha) = \inf\{x : B(x) \geq \alpha\}, \quad 0 < \alpha \leq 1$$

For $\alpha = 0$, this definition gives $-\infty$ and Excel gives the error message **`#NUM!`**.

Inverse probabilities are useful for finding P-values and in **simulation** because from the definition, if U is uniform $(0, 1)$ and $F(x)$ is an arbitrary c.d.f. with inverse defined by

$$F^{-1}(\alpha) = \inf\{x : F(x) \geq \alpha\}$$

then

$$X = F^{-1}(U)$$

has the specified distribution $F(x)$. Thus, for instance

$$= \texttt{CRITBINOM}(u, p, U)$$

is a binomial random variable on n trials and success probability p, and

$$= \texttt{NORMINV}(U, \mu, \sigma)$$

is a $N(\mu, \sigma)$ random variable.

5.2 Sampling Distribution of a Sample Mean

Simulation followed by a histogram of the results provides an insightful view of the *central limit theorem.*

Simulating the Central Limit Theorem

Example 5.2. (Exercise 5.31, page 409 in the text.) A roulette wheel has 38 slots – 18 are black, 18 are red, and 2 are green. When the wheel is spun, a ball is equally likely to come to rest in any of the slots. Gamblers can place a number of different bets in roulette. One of the simplest wagers chooses red or black. A bet of one dollar on red will pay off an additional dollar if the ball lands in a red slot. Otherwise the player loses his dollar. When a gambler bets on red or black, the two green slots belong to the house. A gambler's winnings on a \$1 bet are either \$1 or −\$1.

(a) Simulate a gambler's winnings after 50 bets and compare the gambler's mean winnings per bet with the theoretical results.

(b) Compare the results with the normal approximation.

Solution. The number of wins after 50 bets X is a binomial $B(50, 10/38)$ random variable with

$$\text{mean} \qquad \mu_X = 50\left(\frac{18}{38}\right) = 23.684$$

$$\text{standard deviation} \qquad \sigma_X = \sqrt{50\left(\frac{18}{38}\right)\left(\frac{20}{38}\right)} = 3.5306$$

The proportion of wins after 50 bets is $\hat{p} = X/50$ with

$$\text{mean} \qquad \mu_{\hat{p}} = \frac{18}{38} = 0.4737$$

$$\text{standard deviation} \qquad \sigma_{\hat{p}} = \sqrt{\left(\frac{18}{38}\right)\left(\frac{20}{38}\right)\Big/50} = 0.0706$$

Since the gambler wins \$1 or loses \$1, his average winnings per game, denoted by $\bar{w}$, are $\bar{w} = \hat{p}(1) + (1 - \hat{p})(-1) = 2\hat{p} - 1$ with

$$\begin{array}{rl} \text{mean} & \mu_{\bar{w}} = 2\mu_{\hat{p}} - 1 = -0.0527 \\ \text{standard deviation} & \sigma_{\bar{w}} = 2\sigma_{\hat{p}} = 0.14123 \end{array}$$

By the **Central Limit Theorem** $\hat{p}$ is approximately normal, and therefore so is $\bar{w}$ and

$$\bar{w} \text{ is approximately } N(-0.0527, 0.14123)$$

We can simulate a binomial random variable X, convert it first to $\hat{p} = \frac{X}{n}$ and then to $\bar{w} = 2\hat{p} - 1$, after which we construct a histogram of the simulation results.

The following steps, referring to Figure 5.3, show how to develop a workbook to simulate 500 replications of 50 games. We will use the **RAND** function, which *links* the output to a histogram. By using the F9 key you can repeat the simulation and watch how the results vary.

	A	B	C	D	E	F	G
1		*Demonstrating the Central Limit Theorem by Simulation*					
2			mean	-0.0526	true mean =	-0.0527	
3	*Simulation*	*Average*	st_dev	0.14123	true st_dev =	0.14123	
4	*Number*	*Per Game*	*Formula entered in column B*	=2*CRITBINOM(50,18/38,RAND())/50 - 1			
5	1	0.160	*Bin Formulas*	*Bin*	*Freq.*		
6	2	0.040	=mean-3.5*st_dev	-0.55	0		
7	3	-0.080	=mean-3*st_dev	-0.48	2		
8	4	-0.080	=mean-2.5*st_dev	-0.41	1		
9	5	0.040	=mean-2*st_dev	-0.34	10		
10	6	0.040	=mean-1.5*st_dev	-0.26	19		
11	7	0.080	=mean-st_dev	-0.19	55		
12	8	-0.080	=mean-0.5*st_dev	-0.12	46		
13	9	-0.120	=mean	-0.05	113		
14	10	-0.160	=mean+0.5*st_dev	0.02	104		
15	11	-0.120	=mean+st_dev	0.09	81		
16	12	0.000	=mean+1.5*st_dev	0.16	23		
17	13	-0.240	=mean+2*st_dev	0.23	29		
18	14	0.240	=mean+2.5*st_dev	0.30	15		
19	15	-0.160	=mean+3*st_dev	0.37	2		
20	16	0.200	=mean+3.5*st_dev	0.44	0		

Figure 5.3: Simulating the Central Limit Theorem

We will use bin intervals determined by the simulation and boundaries located at multiples of the standard deviation from the mean in order to allow comparison with the 68-95-99.7% rule mentioned in Chapter 1. We use the sample mean $\bar{w}$ and sample standard deviation $s_{\bar{w}}$ rather than the theoretical values $\mu_{\bar{w}}$ and $\sigma_{\bar{w}}$, respectively, to show how $\bar{w}$ and $s_{\bar{w}}$ vary each time the F9 key is pressed.

1. Prepare a new workbook by entering "Simulating the Central Limit Theorem" in cell A1 and centering the heading across A1:E1. Enter "Simulation Number" in A3:A4 and "Average per Game" in B3:B4. Enter "mean" in

C2, "st_dev" in C3, select C2:D3, and from the Menu Bar choose **Insert – Name – Create** and check the box **Left Column** in the dialog box. The formulas required are

$$\text{mean:} = \texttt{AVERAGE}(\text{B5:B1005})$$

and

$$\text{st\_dev:} = \texttt{SQRT}((1 - \text{mean*mean})/50)$$

2. Enter the labels "Bin" in D5 and "Freq." in E5.

3. Next describe the bin endpoints in cells D6:D19 based on the sample mean and sample standard deviation. Enter "= mean −3.5∗ st_dev" in D6, " = mean −3.0∗ st_dev" in D7 and so on. Refer to Figure 5.3 where we have shown in cells C6:C19 the formulas to be entered in D6:D19. Note that you do not require a column C.

4. Enter the values $1, 2, \ldots, 500$ in cells A5:A504 as follows: Enter "1" in A5. Select A5 and choose **Edit – Fill – Series...** from the Menu Bar. In the **Series** dialog box, check Series in **Columns** and Type **Linear**. Clear the **Trend** box and type "1" and "600" for the **Step** and **Stop** values, respectively.

5. In cell B5 enter = 2*CRITBINOM(50,18/38, RAND())/(50 − 1) to generate the random variable $\bar{w}$. We have shown the formula on line 4 beginning in column D. Select B5, click the fill handle at the lower right corner of B5 and drag down to cell B504. Cells B5:B504 are now filled with 500 replications of the gambler's average net gain per game after 50 games.

6. Select E6:E19. Then type = FREQUENCY(B5:B504,D6:D19) in the entry area of the **Formula Bar**. Hold down the **Shift and Control** keys (either **Macintosh** or **Windows**) and press enter/return to **array-enter** the formula. The formula will appear **surrounded by braces { }** in the **Formula Bar**, and the bin frequencies will appear in cells E6:E19.

7. Select cells D6:D19 and then complete the sequence of steps in the **ChartWizard** as discussed previously. The resulting histogram appears in Figure 5.4.

Excel Output

The sample mean and sample standard deviation appear in D2:D3, and the population mean and standard deviation appear in F2:F3 for comparison purposes. For the simulation shown

$$\bar{w} = -0.0536 \qquad \mu_{\bar{w}} = -0.0527$$
$$s_{\bar{w}} = 0.14122 \qquad \sigma_{\bar{w}} = 0.14123$$

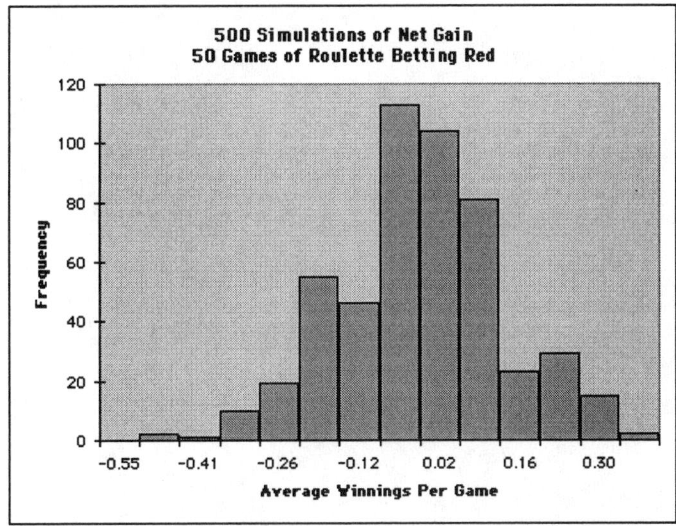

Figure 5.4: Graphing Outcomes of 50 Roulette Games

The table of frequency counts appears in E5:E19 with the corresponding histogram in Figure 5.4. The histogram appears normal shaped with no unusual features.

Recalling that the bin entries are the right endpoints of the bin interval, we can determine the proportion of counts within 1, 2, and 3 standard deviation units of the mean. The simulation results are alarmingly good.

	Actual	Theoretical
Within 1 s	.642	.68
Within 2 s	.944	.95
Within 3 s	.996	.997

Chapter 6

Introduction to Inference

This chapter discusses procedures for finding confidence intervals and carrying out significance tests for the mean of a population. The methods require use of the normal distribution and hence are applicable only when the underlying population may be assumed to be approximately normal or when the sample size is so large that the normal approximation given by the central limit theorem may be invoked.

6.1 Estimating with Confidence

The data $\{x_1, \ldots, x_n\}$ are assumed to come from a $N(\mu, \sigma)$ population with mean μ and a known standard deviation σ. A level C confidence interval for the mean μ is given by

$$\bar{x} \pm z^* \frac{\sigma}{\sqrt{n}}$$

where $\bar{x} = \frac{1}{n}\sum_{i=1}^{n} x_i$ is the sample mean, n is the sample size, and z^* is the value such that the area between $-z^*$ and z^* under a standard normal curve equals C.

The Excel function required is `NORMSINV`(a), which returns the inverse of the standard normal cumulative distribution function $\Phi^{-1}(a)$. Thus for level C confidence we use

$$z^* = \texttt{NORMSINV}\left(0.50 + \frac{C}{2}\right)$$

Example 6.1. (Example 6.2, page 440 in the text.) Tim Kelley has been weighing himself once a week for several years. Last month his four measurements (in pounds) were

190.5 189.0 195.5

Give a 95% confidence interval for his mean weight last month.

Solution. In Figure 6.1 the Excel formulas required are given in column C. These are entered into the adjacent cells in column B to create the workbook template

	A	B	C	D
1		*Z Interval for a Normal Mean - Values and Formulas*		
2				
3		values	formulas	
4	User Input			
5	sigma	3		Data
6	conf	0.90		190.5
7	Summary Statistics			189.0
8	n	4	=COUNT(Data)	195.5
9	xbar	190.5	=AVERAGE(Data)	187.0
10	Calculations			
11	SE	1.500	=sigma/SQRT(n)	
12	z	1.64	=NORMSINV(0.5+conf/2)	
13	ME	2.47	=z*SE	
14	Excel ME	2.47	=CONFIDENCE(1-conf,sigma,n)	
15	Confidence Limits			
16	lower	188.03	=xbar-ME	
17	upper	192.97	=xbar+ME	

Figure 6.1: Confidence Interval for a Normal Mean

to solve this problem. The Excel output is in column B. The user inputs required are the standard deviation and the confidence level. If the data have already been summarized, then you can enter the values for the sample size n and the average in cells B8:B9, respectively. Otherwise Excel will read the data and calculate n and $\bar{x}$. The data can be located in a convenient place on the same sheet or it can be located on another sheet. The latter is particularly useful for large data sets. In this example, with only three data points, we have recorded them on the same sheet as the calculations.

The following steps describe how to construct the workbook.

1. Enter the labels as shown in column A.

2. **Name** the cell ranges to be used. Select cells A5:B6, A8:B9, A11:B13 and A16:B17. To select **noncontiguous blocks** of cells, make the first selection A5:B6, then hold down the **Control (Windows)** or **Command (Macintosh)** key while selecting the other ranges. From the Menu Bar choose **Name – Create**, select **Left Column**, and then click OK. Next, select D5:D9, and from the Menu Bar choose **Name – Create**, select **Top Row**, and then click OK to name the data range.

3. Enter the formulas shown Figure 6.1 into columns B8:B9, B11:B14, and B16:B17, and enter the data in D5:D9. Since you have named the data range you can refer to the cells D5:D9 as "Data," for instance as in the formula = `COUNT`(Data). Otherwise, you would type = `COUNT`(D5:D9) giving the actual locations. These formulas are sufficiently simple that you can enter them by hand rather than use the Formula Palette or the Function Wizard.

The only input needed once the workbook has been constructed are the population standard deviation (sigma) and the confidence level. Type "3" and "0.90" into cells B5 and B6, respectively. The results are immediately recorded in cells B16:B17, showing a lower confidence limit 188.03 and an upper confidence limit 192.97 for the population mean μ.

Explanation

The formula = `COUNT`(Data) gives the sample size by counting the number of cells named by the variable Data. You could also type the integer "4" instead. Likewise, = `AVERAGE`(Data) is the Excel formula for the sample mean $\bar{x}$. We have also included, for comparison purposes, the Excel formula = `CONFIDENCE`(α, σ, n), which calculates the margin of error (half width of interval) associated with a $C = 1 - \alpha$ level confidence interval for the mean of a normal distribution with standard deviation σ based on a sample of size n. This function can be used in a less formal setting.

How Confidence Intervals Behave

A confidence interval is a random interval that has a specified probability of containing an unknown parameter. Thus, a 90% confidence interval for a population mean has probability 0.90 of containing the mean. So, in repeated confidence intervals, in the long run approximately 90% of these confidence intervals would contain the population mean.

> **Example 6.2.** Take 100 SRS of size 3 from an $N(3.0, 0.2)$ population and construct a 90% confidence interval for the mean. Count how many times the confidence interval contains the mean 3.0.

Solution

1. Following the instructions given in Example 4.6 for simulating samples from a specified distribution, choose **Tools – Data Analysis – Random Number Generation** from the Menu Bar, complete a box like the one shown in Figure 4.8, but for normal not discrete random numbers, with "3" for the **Number of Variables**, "100" for the **Number of Random Numbers**, "3.0" for the **Mean**, "0.2" for the **Standard Deviation**, and choose a convenient range for the output. We have selected the range A8:C107.

2. In cell E8 enter
   = `AVERAGE`(A8:C8) – `NORMSINV`(0.5+0.9/2)*0.2/`SQRT`(3)
   In cell F8 enter
   = `AVERAGE`(A8:C8) + `NORMSINV`(0.5+0.9/2)*0.2/`SQRT`(3)

3. Select cells E8:F8, click the fill handle and drag the contents to F107. The cells in column F will contain the value 1 if the confidence interval for the data in the corresponding row contains the true value 3.0, while the cells will contain 0 otherwise.

4. Count the number of times 1 appears by entering $=$ SUM(G8:G107) in an empty cell (H8, for example).

Figure 6.2 shows a portion of a workbook with the simulation for which 92 times out of 100 the true mean was within the 90% confidence limits.

	A	B	C	D	E	F	G	H
1		*Behavior of Repeated Confidence Intervals*						
2								
3	lower limit=AVERAGE(A8:C8) -NORMSINV(0.5+0.90/2)*0.2/SQRT(3)							
4	upper limit=AVERAGE(A8:C8) +NORMSINV(0.5+0.90/2)*0.2/SQRT(3)							
5	G8 contains =IF(AND(E8<3, 3<F8), 1,0)							
6					lower	upper		
7								
8	3.1772	2.7218	3.3097		2.880	3.259	1	92
9	3.0863	3.0417	2.8220		2.793	3.173	1	
10	2.8207	2.8480	2.9353		2.678	3.058	1	
11	2.9131	3.2380	3.0292		2.870	3.250	1	
12	3.0904	3.1497	2.9295		2.867	3.246	1	
13	2.8767	3.0868	3.2555		2.883	3.263	1	
14	2.8937	2.7254	2.9995		2.683	3.063	1	
15	3.1976	3.0303	2.8750		2.844	3.224	1	
16	2.8378	2.9206	2.7565		2.648	3.028	1	
17	2.7972	2.9133	3.1956		2.779	3.159	1	

Figure 6.2: Repeated Confidence Intervals

6.2 Tests of Significance

Significance tests are used to judge whether a specified (null) hypothesis is consistent with a data set.

We create a workbook for testing the null hypothesis $H_0 : \mu = \mu_0$ for a specified null value μ_0 against one-sided or two-sided alternatives. The data $\{x_1, x_2, \ldots, x_n\}$ are assumed to come from an $N(\mu, \sigma)$ population where σ is known. The same procedure can also be used to carry out a large sample test. In the workbook in Figure 6.3, the user can either test at a specified level of significance or determine a P-value.

The user inputs are the sample size, sample mean, standard deviation (which may be input as values, as formulas, or as named references depending on the context), null hypothesis, and level of significance.

> **Example 6.3.** (Example 6.16, page 467 in the text.) Bottles of a popular cola drink are supposed to contain 300 ml of cola. There is some variation from bottle to bottle because the filling machinery is not precise. The distribution of the contents is normal with standard deviation $\sigma = 3$ ml. A student who suspects that the bottle is underfilling measures the contents of six bottles. The results are 299.4, 297.7, 310.0, 298.9, 300.2, 297.0. Is this convincing evidence that the mean content of cola bottles is less than the advertised 300 ml?

	A	B	C	D
1		Z Test for a Normal Mean - Values and Formulas		
2				
3		values	formulas	
4	User Input			
5	sigma	3.0		Data
6	null	300.0		299.4
7	alpha	0.05		297.7
8				301.0
9	Summary Statistics			298.9
10	n	6	=COUNT(Data)	300.2
11	xbar	299.03	=AVERAGE(Data)	297.0
12	Calculations			
13	SE	1.225	=sigma/SQRT(n)	
14	z	-0.789	=(xbar-Null)/SE	
15	Lower Test			
16	lower_z	-1.645	=NORMSINV(alpha)	
17	Decision	Do Not Reject H0	=IF(z<lower_z,"Reject H0","Do Not Reject H0")	
18	Pvalue	0.215	=NORMSDIST(z)	
19	Upper Test			
20	upper_z		=-NORMSINV(alpha)	
21	Decision		=IF(z>upper_z,"Reject H0","Do Not Reject H0")	
22	Pvalue		=1-NORMSDIST(z)	
23	Two-Sided Test			
24	two_z		=ABS(NORMSINV(alpha/2))	
25	Decision		=IF(ABS(z)>two_z,"Reject H0","Do Not Reject H0")	
26	Pvalue		=2*(1-NORMSDIST(ABS(z)))	

Figure 6.3: Significance Test for a Normal Mean

Solution. The workbook template in Figure 6.3 shows all formulas required in column C for for lower, upper, and two-sided tests, respectively. These go into the adjacent cells of column B. Then enter the values sigma = 3, alpha = 0.05, and null hypothesis = 300.0. The Excel output is shown in column B for this lower test $H_a : \mu < \mu_0$ giving the critical value of $-z^* = -1.645$ at the 5% level of significance and also the P-value 0.215. The conclusion is not to reject H_0.

Explanation

We encountered the function **NORMSINV** previously. The formula = **NORMSDIST** returns the cumulative normal distribution function $\Phi(z)$. For a one-sided lower test, the P-value is the area to the left of the computed z score $\frac{\bar{x}-\mu_0}{\sigma/\sqrt{n}}$ and is thus given by = **NORMSDIST**(z). For an upper test, the P-value is the area to the right of $\frac{\bar{x}-\mu_0}{\sigma/\sqrt{n}}$ and is given by = 1 − **NORMSDIST**(z). For a two-sided test, the formula for the P-value is = 2*(1 − **NORMSDIST**($|z|$)). The formula = **ABS**(z) returns the absolute value of z. The decision rule uses the logical = **IF**(statement, true, false), which returns the string designated as true if the statement is true or else it returns false.

Chapter 7

Inference for Distributions

7.1 Inference for the Mean of a Population

The workbook in Section 6.1 for the mean of a normal can easily be modified to produce a level C confidence interval based on the Student t distribution,

$$\bar{x} \pm t^* \frac{s}{\sqrt{n}} .$$

As before n, $\bar{x}$ are the sample size and sample mean respectively, s is the sample standard deviation, and t^* is the critical t value such that the area between $-t^*$ and t^* under the curve of a t density with $n - 1$ degrees of freedom equals C.

The Excel formula required is

$$= \texttt{TINV}(\alpha, \nu)$$

which returns the critical value for a level $C = 1 - \alpha$ confidence interval based on a t distribution with ν degrees of freedom.

The One-Sample t Confidence Interval

Example 7.1. (Example 7.1, page 507 in the text.) In fiscal 1996, the United States Agency for International Development provided 238,300 metric tons of corn soy blend (CSB) for development programs and emergency relief in countries throughout the world. CSB is a highly nutritious, low-cost fortified food that is partially precooked and can be incorporated into different food preparations by the recipients. As part of a study to evaluate appropriate vitamin C levels in this commodity, measurements were taken on samples of CSB produced in a factory. The following data are the amounts of vitamin C, measured in milligrams per 100 grams (mg/100 g) of blend, for a random sample of size 8 from a production run:

26 31 23 22 11 22 14 31

Find a 95% confidence interval for μ, the mean vitamin C content of the CSB produced during the run.

	A	B	C	D
1	*T Interval for a Normal Mean*			
2				
3		values	formulas	
4	User Input			
5	conf	0.95		Data
6	Summary Statistics			26
7	n	8	=COUNT(Data)	31
8	xbar	22.5	=AVERAGE(Data)	23
9	Calculations			22
10	s	7.19	=STDEV(Data)	11
11	SE	2.54	=s/SQRT(n)	22
12	df	7	=n-1	14
13	t	2.365	=TINV(1-conf, df)	31
14	ME	6.01	=t*SE	
15	Confidence Limits			
16	lower	16.49	=xbar-ME	
17	upper	28.51	=xbar+ME	

Figure 7.1: Confidence Interval for a Normal Mean

Solution. Create the workbook shown in Figure 7.1 using steps analogous to the production of Figure 6.1. In place of an assumed standard deviation σ, we calculate the sample standard deviation s using the Excel formula = STDEV(Data). The critical value t^* (denoted by t on the workbook) is obtained from the formula = TINV$(1 - conf, df)$, where $conf$ is the confidence level and $df = n - 1$ are the degrees of freedom. With this template in hand, it is only necessary to input and name the data. We have typed the eight data points in cells D6–D13 and **Named** them "Data" in cell D5. Column C has the formulas that you must enter in the respective cells in column B. The output provided by Excel gives a 95% confidence interval for μ as (16.49, 28.51) from cells B16:B17.

The One-Sample t Test

Again, the workbook based on the normal distribution in Section 6.2 is easily modified for use with Student's t distribution. For the significance test

$$H_0 : \mu = \mu_0$$

the test statistic is

$$t = \frac{\bar{x} - \mu_0}{s/\sqrt{n}}$$

which has Student's t distribution on $n - 1$ degrees of freedom. We remind the reader of the *unusual* Excel definition for the function that calculates the cumulative t, namely,

$$\texttt{TDIST}(x, \nu, 1) = P[t(\nu) > x]$$

where $t(\nu)$ is a t distribution on ν degrees of freedom. The argument x must be positive, and this accounts for the more complicated syntax in the decision rule in the corresponding template (Figure 7.2).

	A	B	C	D
1			*T Test for a Normal Mean - Formulas and Values*	
2				
3		values	formulas	
4	User Input			
5	null	40.0		*Data*
6	alpha	0.1		26
7				31
8	Summary Stats			23
9	n	8	=COUNT(Data)	22
10	xbar	22.5	=AVERAGE(Data)	11
11	Calculations			22
12	s	7.2	=STDEV(Data)	14
13	SE	2.542	=s/SQRT(n)	31
14	t	-6.883	=(xbar-null)/SE	
15	df	7	=n-1	
16	Lower Alternative			
17	lower_t		=-TINV(2*alpha, df)	
18	Decision		=IF(t<lower_t,"Reject H0", "Do Not Reject H0")	
19	Pvalue		=IF(t<0, TDIST(ABS(t),df,1), 1-TDIST(t,df,1))	
20	Upper Alternative			
21	upper_t		=TINV(2*alpha, df)	
22	Decision		=IF(t>upper_t,"Reject H0","Do Not Reject H0")	
23	Pvalue		=IF(t>0, TDIST(t,df,1), 1-TDIST(ABS(t),df,1))	
24	Two-Sided Alternative			
25	two_t	1.895	=TINV(alpha, df)	
26	Decision	Reject H0	=IF(ABS(t)>two_t,"Reject H0","Do Not Reject H0")	
27	Pvalue	0.00023	=TDIST(ABS(t), df, 2)	

Figure 7.2: One-Sample Student t Test

Example 7.2. (Example 7.2, page 508 in the text.) The specifications for the CSB described in Example 7.1 state that the mixture should contain 2 pounds of vitamin premix for every 2000 pounds of product. These specifications are designed to produce a mean (μ) vitamin C content in the final product of 40 mg/100 g. We test the null hypothesis that the mean vitamin C content is 40 mg/100 g.

$$\begin{aligned} H_0 : \mu &= 40 \\ H_a : \mu &\neq 40 \end{aligned}$$

Solution. Figure 7.2 gives the formulas for carrying out a t test on a population mean. These are shown in column C but are to be entered in column B where Excel evaluates them. This problem involves a two-sided alternative, and column B shows only this output. The user inputs are the null value 40, $\alpha = 0.01$, and the data, entered in cells D6:D13. The calculated t value is -6.883, which is larger in absolute value than the 0.01 critical value for a two-sided test, namely 3.499. Therefore the conclusion is to reject H_0. The P-value is provided in cell B27. Its value of 0.00023 shows how extreme the evidence is against H_0.

Matched Pairs t Procedure

In order to reduce variability in a data set scientists sometimes use paired data matched on characteristics believed to affect the response. This is equivalent to a randomized block design. Such data are best analyzed if one-sample procedures are applied to differences between the pairs. There is a loss in degrees of freedom for error, but if the matching is effective, then this will be more than offset by the gain in reduced variance of the differences. The same method applies to before-after measurements on the same subjects.

Thus, we may apply the previous one-sample workbooks to the differences. Excel also provides a direct method for the matched pairs t test using the **Analysis ToolPak.** We will describe both approaches applied to the same data set.

Table 7.1: Pretest and Posttest Scores for French Immersion

Teacher	Pretest	Posttest	Teacher	Pretest	Posttest
1	32	34	11	30	36
2	31	31	12	20	26
3	29	35	13	24	27
4	10	16	14	24	24
5	30	33	15	31	32
6	33	36	16	30	31
7	22	24	17	15	15
8	25	28	18	32	34
9	32	26	19	23	26
10	20	26	20	23	26

Example 7.3. (Example 7.7, page 513 in the text.) The National Endowment for the Humanities sponsors a summer institute to improve the skills of high school teachers of foreign languages. One such institute hosted 20 French teachers for 4 weeks. At the beginning of the period, the teachers were given the Modern Languages Association's listening test of understanding of spoken French. After 4 weeks of immersion in French in and out of class, the listening test was given again. The pretest and posttest scores are provided in Table 7.1. Let μ denote the mean improvement that would be achieved if the entire population of French teachers attended a summer institute. We wish to test at the 0.01 level of significance

$$\begin{aligned} H_0 : \mu &= 0 \\ H_a : \mu &> 0 \end{aligned}$$

	A	B	C	D	E	F
1			*Paired T Test for a Normal Mean (Direct Approach)*			
2						
3				Pretest	Posttest	Difference
4	User Input			32	34	2
5	null	0.0		31	31	0
6	alpha	0.01		29	35	6
7				10	16	6
8	Summary Stats			30	33	3
9	n	20	=COUNT(Difference)	33	36	3
10	xbar	2.5	=AVERAGE(Difference)	22	24	2
11	Calculations			25	28	3
12	s	2.9	=STDEV(Difference)	32	26	-6
13	SE	0.647	=s/SQRT(n)	20	26	6
14	t	3.865	=(xbar-null)/SE	30	36	6
15	df	19	=n-1	20	26	6
16	Upper Alternative			24	27	3
17	upper_t	2.539	=TINV(2*alpha, df)	24	24	0
18	Decision	Reject H0	=IF(t>upper_t,"Reject H0","Do Not Reject H0")	31	32	1
19	Pvalue	0.00052	=IF(t>0, TDIST(t,df,1), 1-TDIST(ABS(t),df,1))	30	31	1
20				15	15	0
21				32	34	2
22				23	26	3
23				23	26	3

Figure 7.3: Paired t Test—Direct Approach

To analyze these data, subtract the pretest scores from the posttest scores for each teacher. The 20 differences then form a single sample to which a one-sample test applies.

Matched Pairs t Test—Direct Approach

1. Adapt the template shown in Figure 7.2 by customizing it for this upper test (Figure 7.3).

2. Enter 0 for null and 0.01 for alpha, and record the data in a convenient place, say columns D, E, and F. Enter the label "Pretest" in cell D3 followed by the pretest scores in cells D4:D23. Enter the label "Posttest " in cell E3 followed by the posttest scores in cells E4:E23. Enter the label "Difference " in cell F3. Then in cell F4 enter the difference D4 – E4 in the **Formula Bar**. Select cell F4 and then move the mouse pointer over the fill handle in the lower right corner of cell F4. The pointer changes from an outline plus sign to a cross hair + when you are in position on the fill handle. Click the fill handle and drag it down so that the range F4:F23 is selected. Release the mouse button and Excel fills the contents of cells F4:F23 with the corresponding differences between posttest and pretest scores.

3. **Name** the difference range by selecting F3:F23, and from the Menu Bar choose **Insert – Name – Create** and then check the Top Row box in the **Create Names** dialog box. We can now refer to the data in F4:F23 by the

	A	B	C	D	E	F
1			*T Test for a Normal Mean (Using the Analysis ToolPak)*			
2						
3	Pretest	Posttest	Difference	t-Test: Paired Two Sample for Means		
4	32	34	2			
5	31	31	0		*Posttest*	*Pretest*
6	29	35	6	Mean	28.3	25.8
7	10	16	6	Variance	35.379	39.747
8	30	33	3	Observations	20	20
9	33	36	3	Pearson Correlation	0.890	
10	22	24	2	Hypothesized Mean Difference	0	
11	25	28	3	df	19	
12	32	26	-6	t Stat	3.865	
13	20	26	6	P(T<=t) one-tail	0.00052	
14	30	36	6	t Critical one-tail	2.5395	
15	20	26	6	P(T<=t) two-tail	0.0010	
16	24	27	3	t Critical two-tail	2.8609	
17	24	24	0			
18	31	32	1			
19	30	31	1			
20	15	15	0			
21	32	34	2			
22	23	26	3			
23	23	26	3			

Figure 7.4: Paired t Test—Analysis ToolPak

name "Difference." (This is convenient but not necessary; we could equally use the cell reference F4:F23.)

4. Finally, where the name "Data" is used in formulas in the template of Figure 7.2, replace it with the name "Difference."

Excel Output

Figure 7.3 also gives the Excel output. The computed t value is 3.865, which exceeds the 1% critical value of 2.539. We therefore reject H_0. Additionally, we find that the P-value is 0.00052.

Matched Pairs t Test—Using the ToolPak

Excel also provides an **Analysis** tool for a matched pairs t test. However, *it cannot be used with summarized data*, while the direct approach can.

1. Open a new workbook and enter the unscented and scented values with their labels in cells A3:A23 and B3:B23, respectively, as in Figure 7.4.

2. In cell C3 enter the label "Difference." In cell C4 type the formula $= A4 - B4$. Then select C4 and fill to cells C4:C23.

3. Choose **Tools – Data Analysis** from the Menu Bar, and then check the box ***t*-Test:Paired Two Sample for Means**. Click OK.

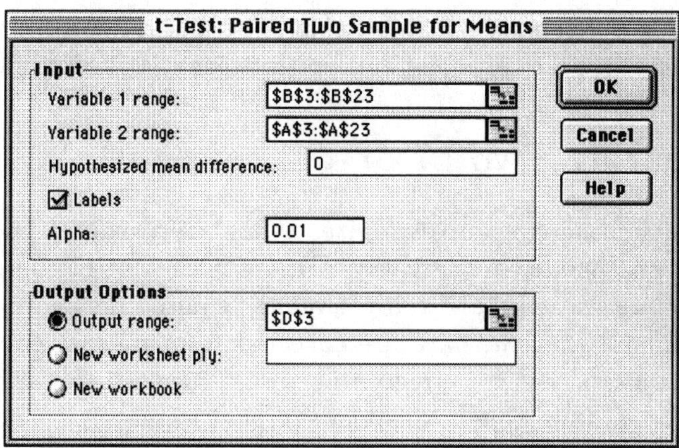

Figure 7.5: Paired t Test—Dialog Box

4. Complete the next dialog box shown in Figure 7.5. **Caution** is required in determining which scores are entered for **Variable 1 range** and which scores are entered for **Variable 2 range**. Since we are taking posttest−pretest ($\mu_1 - \mu_1$) scores and $\mu \equiv \mu_1 - \mu_1$ with $H_a : \mu > 0$, enter the posttest score range B3:B23 for Variable 1 and the pretest range A3:A23 for Variable 2.

Excel Output

The output appears in Figure 7.4 in the range D3:F16 beginning with cell D3, as specified in Figure 7.5. Individual sample means are given as well as the test statistic value t in cell E10. One-sided and two-sided critical values are provided in E14 and E16, as well as corresponding P-values in E13 and E15. Because $H_0 : \mu > 0$ and $t = 3.865$, we reject at level $\alpha = 0.01$. The P-value is 0.00052.

The entry $P(T <= t)$ one-tail in E13 is not $P(t(19) \leq 3.865)$, as the notation would suggest. Rather $P(T <= t)$ represents the tail area relative to t (so it is a lower tail if t is negative and an upper tail if t is positive). It is therefore not always a P-value.

Confidence Interval for Paired Data

The **Analysis ToolPak** does not provide a confidence interval directly, but it provides the information needed to carry out the calculations using the template in Section 7.1 applied to the differences. Again, adapt the workbook shown in Figure 7.1 for Example 7.1. In place of the "Data" range referenced in cells B7, B8, and B10 in Figure 7.1, point to the cells you have named "Difference," either with the direct approach or using the ToolPak. Excel will take the values from one workbook and use them in another. This is a reminder that data need not be on the same sheet or even the same workbook as your analysis.

Exercise. For Example 7.3, show that a 90% confidence interval for the difference in posttest mean and pretest mean is (1.382, 3.618).

7.2 Comparing Two Means

Independent simple random samples of sizes n_1 and n_2 are obtained from populations with means μ_1 and μ_2, respectively. We are interested in comparing μ_1 and μ_2. The appropriate statistics and critical values required depend on the assumptions made. Suppose that the data are collected from normal populations with standard deviations σ_1 and σ_2. Denote by $\bar{x}_1, s_1^2, \bar{x}_2$, and s_2^2 the corresponding summary statistics (the sample means and sample variances).

Inference is usually based on a two-sample statistic of the form

$$\frac{(\bar{x}_1 - \bar{x}_2) - (\mu_1 - \mu_2)}{\text{SE}} \tag{7.1}$$

where SE represents the standard error of the numerator.

If the underlying populations are normal with known population standard deviations σ_1 and σ_2, then we use

$$z = \frac{(\bar{x}_1 - \bar{x}_2) - (\mu_1 - \mu_2)}{\sqrt{\frac{\sigma_1^2}{n_1} + \frac{\sigma_2^2}{n_2}}}$$

which has a standard normal distribution.

If, however, σ_1 and σ_2 are unknown, then the appropriate ratio is

$$t = \frac{(\bar{x}_1 - \bar{x}_2) - (\mu_1 - \mu_2)}{\sqrt{\frac{s_1^2}{n_1} + \frac{s_2^2}{n_2}}} \tag{7.2}$$

which is called a two-sample t statistic.

Actually, the distribution of the statistic in (7.2) depends on σ_1 and σ_2 and does not have an exact t distribution. Nonetheless, it is used with t critical values in inference in one of two ways, each involving a computed value for degrees of freedom ν associated with the denominator of (7.1) to provide an approximate t statistic. The two options follow.

1. Use a value for ν given by

$$\nu = \frac{\left(\frac{s_1^2}{n_1} + \frac{s_2^2}{n_2}\right)^2}{\frac{1}{n_1-1}\left(\frac{s_1^2}{n_1}\right)^2 + \frac{1}{n_2-1}\left(\frac{s_2^2}{n_2}\right)^2} \tag{7.3}$$

2. Use

$$\nu = \min\{n_1 - 1, n_2 - 1\} \tag{7.4}$$

Approximation (7.4) is conservative in the sense of providing a larger margin of error and is recommended in the text when doing calculations without the aid of software. Approximation (7.3) is considered to provide a quite accurate approximation to the actual distribution.

Excel, like most statistical software, uses the value for ν given in (7.3) and provides three tools in the **Analysis ToolPak** for analyzing independent samples from normal populations: z test, t test (unequal variances), and pooled t test (equal variances). These involve different sets of assumptions and consequently different forms for the SE. Each tool provides a similar dialog box for user input, parameters, and the range for the actual raw data.

However, sometimes only summary statistics are available and the **Analysis ToolPak** cannot be used; instead, direct calculations are required.

Two-Sample z Statistic

For normal populations with known standard deviations, the "standard score"

$$z = \frac{(\bar{x}_1 - \bar{x}_2) - (\mu_1 - \mu_2)}{\sqrt{\frac{\sigma_1^2}{n_1} + \frac{\sigma_2^2}{n_2}}}$$

has a standard normal distribution. This ratio is also used for "large sample" procedures with σ_1 and σ_2 unknown and replaced by the sample standard deviations s_1 and s_2. Normality of the underlying population is not mandatory for large sample tests, and the corresponding confidence intervals and significance tests have approximately the specified values.

Two-Sample z Confidence Interval

Example 7.4. (Exercise 7.66, page 562 in the text.) The Johns Hopkins Regional Talent Searches give the Scholastic Aptitude Tests to 13-year-olds. In all, 19883 males and 19937 females took the tests between 1980 and 1982. The mean scores of males and females on the verbal test are nearly equal, but there is a clear difference between the sexes on the mathematics test. The reason for this difference is not understood. Here are the data (from a news article in *Science*, **224** (1983), 1029–1031).

Group	$\bar{x}$	s
Males	416	87
Females	386	74

Give a 99% confidence interval for the difference between the mean score for males and the mean score for females in the population that Johns Hopkins searches.

	A	B	C	D
1	*Two-Sample Z Confidence Interval*			
2				
3	Summary Statistics and User Input			
4	Group	n	xbar	sigma
5	Males	19883	416	87
6	Females	19937	386	74
7				
8	conf	0.99		
9				
10				
11	SE	0.810	=SQRT(D5^2/B5+D6^2/B6)	
12	z	2.576	=NORMSINV(0.5+conf/2)	
13	ME	2.085	=z*SE	
14				
15	Confidence Limits			
16				
17	lower	27.91	=(C5-C6)-ME	
18	upper	32.09	=(C5-C6)+ME	

Figure 7.6: Large Two-Sample z Confidence Interval

Solution. Figure 7.6 provides both formulas (column C) to derive the required confidence interval and the values obtained (column B). The user inputs are $\bar{x}_1$, $\bar{x}_2$, σ_1, σ_2, n_1, n_2, and the confidence level C. We have **Named** the ranges to refer to "conf," "SE," "z," and "ME." We can read off from cells B17:B18 that a 99% confidence interval is $(27.91, 32.09)$.

Two-Sample z Test

Following is a large-sample z test for a difference in means.

Example 7.5. (Exercise 7.61, page 560 in the text.) A bank compares two proposals to increase the amount its credit card customers charge on their cards. Proposal A offers to eliminate the annual fee for customers who charge \$2400 or more during the year. Proposal B offers a small percent of the total amount charged as a cash rebate at the end of the year. The bank offers each proposal to an SRS of its existing credit card customers. At the end of the year, the total amount charged by each customer is recorded. Here are the summary statistics.

Group	n	$\bar{x}$	s
A	150	1987	392
B	150	2056	413

Do the data show a significant difference between the amounts charged by customers offered the two plans? Give the null and alternative hypotheses and calculate the two-sample z statistic. State your conclusion.

Solution. The hypothesis to be tested is

$$\begin{aligned} H_0 : \mu_1 &= \mu_2 \\ H_a : \mu_1 &\neq \mu_2 \end{aligned}$$

where μ_1 and μ_2 are the hypothesized mean amounts charged by all bank customers were the bank to implement Proposals A and B, respectively. At the level $\alpha = 0.05$, the decision rule is to reject H_0 if the computed z value

$$z = \frac{\bar{x}_1 - \bar{x}_2}{\sqrt{\frac{\sigma_1^2}{n_1} + \frac{\sigma_2^2}{n_2}}}$$

satisfies $|z| > 1.96$. The P-value will be given by

$$P\text{-value } = 2\left(1 - \Phi(|z|)\right)$$

where $\Phi(z)$ is the cumulative $N(0,1)$ distribution function. The population standard deviations are not provided, but in view of the large sample sizes, we can use s_1 and s_2, the sample standard deviations, in place of σ_1 and σ_2. These calculations can readily be made by a hand calculator.

	A	B	C	D	F
1		*Large Two-Sample Z Test*			
2					
3		Summary Statistics and User Input			
4	Group	n	xbar	s	
5	A	150	1987	392	
6	B	150	2056	413	
7					
8	SE	46.493	=SQRT(D5^2/B5+D6^2/B6)		
9	z	-1.484	=((C5-C6)-null)/SE		
10	alpha	0.05			
11	null	0			
12	Lower Test				
13	lower_z		=NORMSINV(alpha)		
14	Decision		=IF(z<lower_z,"Reject H0","Do Not Reject H0")		
15	Pvalue		=NORMSDIST(z)		
16	Upper Test				
17	upper_z		=-NORMSINV(alpha)		
18	Decision		=IF(z>upper_z,"Reject H0","Do Not Reject H0")		
19	Pvalue		=1-NORMSDIST(z)		
20	Two-Sided Test				
21	two_z	1.960	=ABS(NORMSINV(alpha/2))		
22	Decision	Do Not Reject H0	=IF(ABS(z)>two_z,"Reject H0","Do Not Reject H0")		
23	Pvalue	0.1378	=2*(1-NORMSDIST(ABS(z)))		

Figure 7.7: Large Two-Sample z Test

While the full power of Excel comes from dealing with large data sets, even this simple example can illustrate the use of a spreadsheet to evaluate formulas within equations. Figure 7.7 is an Excel workbook containing a template for this type of problem showing the Excel formulas required. The calculations are carried out in column B and the corresponding formulas are displayed in the adjacent cells in column C. The problem at hand is a two-sided test, but formulas are additionally provided for both upper and lower tests, only one of which should be used at any time. For purposes of clarity, we have used **Named Ranges** rather than cell references to refer to α = alpha in cell B10, to z in B9, to the standard error

$$\text{SE} = \sqrt{\frac{s_1^2}{n_1} + \frac{s_2^2}{n_2}}$$

in B8, and to the critical values lower_z, upper_z, and two_z for the three types of alternative hypotheses. Remember to name your ranges when copying and adapting this template to your own workbook.

The Excel output appears in Figure 7.7 in cells B9, and B21:B23. The computed z statistic is -1.484 in B9 which does not exceed in absolute value the two-sided 5% critical value of 1.960 in B21. The conclusion "Do Not Reject H_0" appears in B22, and the P-value, 0.1378, if also desired, is printed by Excel in B23.

Two-Sample z Inference Using the ToolPak

While the aforementioned templates have been designed for summarized data, they can easily be modified for use with raw data, for instance, by entering = `AVERAGE`(Range) in place of the *value* of the sample mean, where Range is the cell range for the sample. Since population standard deviations are seldom known but instead are replaced by their sample estimates in most applications, if you are dealing with large raw data sets whose standard deviations have not yet been calculated, it is simplest to then let Excel do the calculation. So enter = `STDEV`(Range) where you would enter the value for the standard deviation and then proceed as before. Excel computes the standard deviation from the data and inserts it where required in the formulas.

There is a two-sample z test option included in the **Analysis ToolPak**. But for large sample sizes, the results of two-sample t and z options are virtually the same and the sequence of steps in the two tools are identical. Therefore, as we will discuss the two-sample t ToolPak in detail in the next section, we will not illustrate its z counterpart.

Two-Sample t Procedures

Using Summarized Data

Suppose only $\bar{x}_1, s_1^2, \bar{x}_2$, and s_2^2, not the raw data, are available. We will use Excel to calculate (7.2) and use it for a test of significance and a confidence interval.

> **Example 7.6.** (Example 7.14, page 542 in the text.) An educator believes that new directed reading activities in the classroom will help

elementary school pupils improve some aspects of their reading ability. She arranges for a third-grade class of 21 students to take part in these activities for an eight-week period. A control classroom of 23 third graders follows the same curriculum with the activities. At the end of the eight weeks, all students are given a Degree of Reading Power (DRP) test, which measures the aspects of reading ability that the treatment is designed to improve. The summary statistics are

Group	n	$\bar{x}$	s
Treatment	21	51.48	11.01
Control	23	41.52	17.15

Carry out the significance test at the level $\alpha = 0.05$.

	A	B	C	D	F	G
1	Two-Sample T Test Summarized Data					
2						
3	Summary Statistics and User Input					
4	Group	n	xbar	s		
5	Treatment	21	51.48	11.01		
6	Control	23	41.52	17.15		
7						
8	SE	4.308	=SQRT(D5^2/B5+D6^2/B6)			
9	t	2.312	=((C5-C6)-null)/SE			
10	alpha	0.05				
11	null	0				
12	mindf	20	=MIN(B5-1,B6-1)			
13	numerator	344.485	=POWER((D5^2/B5+D6^2/B6),2)			
14	denominator	9.099	=(D5^4/B5^2)/(B5-1)+(D6^4/B6^2)/(B6-1)			
15	df	38	=1+INT(B13/B14)			
16	Lower Test					
17	lower_t		=-TINV(2*alpha, df)			
18	Decision		=IF(t<lower_t,"Reject HO", "Do Not Reject HO")			
19	Pvalue		=IF(t<0, TDIST(ABS(t),df,1), 1-TDIST(t,df,1))			
20	Upper Test					
21	upper_t	1.686	=TINV(2*alpha, df)			
22	Decision	Reject HO	=IF(t>upper_t,"Reject HO","Do Not Reject HO")			
23	Pvalue	0.0131	=IF(t>0, TDIST(t,df,1), 1-TDIST(ABS(t),df,1))			
24	Two-Sided Test					
25	two_t		=TINV(alpha, df)			
26	Decision		=IF(ABS(t)>two_t,"Reject HO","Do Not Reject HO"))			
27	Pvalue		=TDIST(ABS(t), df, 2)			

Figure 7.8: Two-Sample t Test—Summarized Data

Solution. Because we hope to show that the treatment (group 1) is better than the control (group 2), the hypotheses are

$$\begin{aligned} H_0 : \mu_1 &= \mu_2 \\ H_a : \mu_1 &> \mu_2 \end{aligned}$$

At the level $\alpha = 0.05$, reject H_0 if the computed t value

$$t = \frac{\bar{x}_1 - \bar{x}_2}{\sqrt{\frac{s_1^2}{n_1} + \frac{s_2^2}{n_2}}}$$

satisfies $|t| > t^*$, where t^* is the upper $\alpha/2 = 0.025$ critical value of a t distribution.

As in Example 7.5, we have provided an Excel workbook, Figure 7.8, containing a template for this type of problem, where the calculations are carried out in column B while the actual formulas in the cells behind column B are given in the adjacent cells in column C. We are dealing with an upper-level test, but formulas are also provided for lower and two-sided alternatives.

Cell B8 gives the standard error (the denominator of (7.2)). Cell B12 gives the conservative degrees of freedom, 20, for the value of ν given by (7.4), while cells B13:B15 calculate the degrees of freedom, 38, corresponding to the value of ν in (7.3). The calculated t statistic in cell D9 is 2.312. Using the more accurate degrees of freedom, 38, the corresponding critical t value of 1.686 at level .05 is presented in cell B21. We therefore reject H_0 at level $\alpha = 0.05$. Observe that the calculated P-value is given in cell B23 in Figure 7.8 as 0.0131.

	A	B	C	D	E
1		*Two-Sample T CI Summarized Data*			
2					
3		Summary Statistics and User Input			
4	Group	n	xbar	s	
5	Treatment	21	51.48	11.01	
6	Control	23	41.52	17.15	
7					
8	conf	0.95	0.95		
9	mindf	20	=MIN(C5-1,C6-1)		
10	numerator	344.30	=POWER((E5^2/C5+E6^2/C6),2)		
11	denominator	9.10	=(E5^4/C5^2)/(C5-1)+(E6^4/C6^2)/(C6-1)		
12	df	38	=1+INT(C10/C11)		
13	Calculations				
14	SE	4.308	=SQRT(E5^2/C5+E6^2/C6)		
15	crit_t	2.024	=TINV(1-conf, df)		
16	ME	8.720	=crit_t*SE		
17	lower	1.23	=(D5-D6)-ME		
18	upper	18.67	=(D5-D6)+ME		

Figure 7.9: Two-Sample t Confidence Interval

Example 7.7. Find a 95% confidence interval for the mean improvement in the entire population of third graders in Example 7.6.

Solution. We modify the workbook for the z procedure given in Figure 7.6 to derive a two-sample t confidence interval. The confidence interval is given by

$$\bar{x}_1 - \bar{x}_2 \pm t^* \sqrt{\frac{s_1^2}{n_1} + \frac{s_2^2}{n_2}}$$

where t^* is a critical t-value on an appropriate number of degrees of freedom. The formulas required and values are shown in Figure 7.9. We find that a 95% confidence interval based on 38 degrees of freedom is (1.23, 18.67).

Using the Analysis ToolPak

As mentioned, Excel provides three tools in the **Analysis ToolPak** for comparing means from two populations based on independent samples. These are the two-sample t test, pooled two-sample t test, and two-sample z test. These tools provide dialog boxes in which the user locates the data and decides on the type of analysis desired. For large sample sizes, the results of two-sample t and z options are virtually the same and the sequence of steps in the two tools are identical. Therefore, as we will be discussing the two-sample t ToolPak in detail, we will not illustrate its z counterpart. Besides, there is a bug in the two-sample z ToolPak that outputs an incorrect two-sided P-value; moreover, there is no built-in confidence interval procedure. These considerations limit the usefulness of the two-sample z ToolPak, and we do not recommend using it.

We illustrate use of the ToolPak with Examples 7.6 and 7.7 not only for comparison purposes but also because the full data set appears as Table 7.2 in the text and may be readily referenced.

> **Example 7.8.** Redo Examples 7.6 and 7.7 with the raw data using the **Analysis ToolPak.**

Solution

1. Open a new workbook and enter the data from Table 7.2 on page 543 of the text. Insert the treatment group in cells A4:A24 and the control group in Cells B4:B26. Enter the label "Treatment" in A3 and the label "Control" in B3. (Refer to Figure 7.12, which also shows the output.)

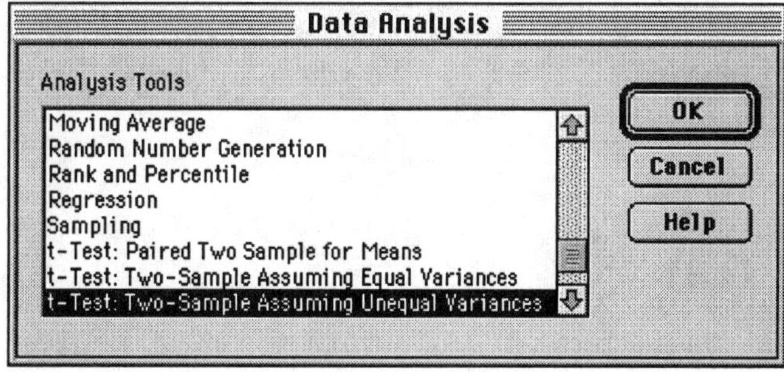

Figure 7.10: Two-Sample t Test Analysis Tools

2. From the Menu Bar choose **Tools – Data Analysis** and select ***t*-Test: Two-Sample Assuming Unequal Variances** from the list of selections (Figure 7.10). Click OK (equivalently, double-click your selection).

3. A dialog box (Figure 7.11) appears. Complete as shown. **Variable 1 range** refers to the cell addresses of the sample you have designated by subscript

1, in this case the treatment group. Type A3:A24 in its text area (with the flashing vertical I-beam). Alternatively, you can point to the data by clicking on cell A3 and dragging to the end of the treatment data, cell A24. The values \$A\$3:\$A\$24 will appear in the text area of the dialog box. Similarly, enter the range B3:B26 for the control group in the **Variable 2 range.**

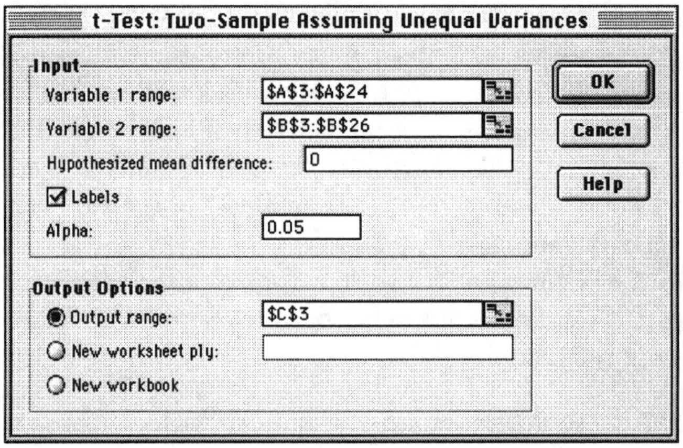

t-Test: Two-Sample Assuming Unequal Variances

Input
Variable 1 range: $A$3:$A$24
Variable 2 range: $B$3:$B$26
Hypothesized mean difference: 0
Labels
Alpha: 0.05

Output Options
Output range: $C$3
New worksheet ply:
New workbook

OK
Cancel
Help

Figure 7.11: Two-Sample t Test Dialog Box

4. The **Hypothesized mean difference** refers to the null value, which is 0 here. Check the **Labels** box, because your ranges included the labels for the two groups. The level of significance **Alpha** is the default 0.05. We will be placing the output in the same workbook as the data, so check the radio button **Output range** and type C3 in the text area. Finally, click OK. The output will appear in a block of cells whose upper left corner is cell C3 in Figure 7.12.

Excel Output

The output appears in the range C3:E15 in Figure 7.12. The range C18:E23 is not part of the output but is the result of additional formulas we have entered to give confidence intervals (discussed next). From cells D6:E6, we see that the sample means for treatment and control groups are 51.476 and 41.522, while the sample variances are 121.162 and 294.079, respectively. The degrees of freedom are 38 (Excel rounds up to the nearest integer). The computed t statistic in cell D11 is 2.311, while the one-sided critical t^* value on 38 degrees of freedom at the 5% level in cell D13 is 1.686. Since the computed t exceeds t^*, we reject the null hypothesis and conclude that there is strong evidence that the directed reading activities help elementary school pupils improve some aspects of their reading ability.

Excel provides P-values in cells D12 and D14. Here the P-value is 0.0132. As before, the entry "$P(T <= t)$ one-tail" in C12 needs some explanation. It

	A	B	C	D	E	F
1			*Two-Sample t Test*			
2						
3	Treatment	Control	t-Test: Two-Sample Assuming Unequal Variances			
4	24	10				
5	33	17		*Treatment*	*Control*	
6	43	19	Mean	51.476	41.522	
7	43	20	Variance	121.162	294.079	
8	43	26	Observations	21	23	
9	44	28	Hypothesized Mean Difference	0		
10	46	33	df	38		
11	49	37	t Stat	2.311		
12	49	37	P(T<=t) one-tail	0.0132		
13	52	41	t Critical one-tail	1.686		
14	53	42	P(T<=t) two-tail	0.0264		
15	54	42	t Critical two-tail	2.024		
16	56	42				
17	57	43				
18	57	46	Mean Difference	9.954	=D6-E6	
19	58	48	SE	4.308	=SQRT(D7/D8+E7/E8)	
20	59	53	t	2.024	=TINV(0.05, D10)	
21	61	54	ME	8.72	=D20*D19	
22	62	55	lower	1.23	=D18-D21	
23	67	55	upper	18.67	=D18+D21	
24	71	60				
25		62				
26		85				

Figure 7.12: Two-Sample t Test ToolPak Output

is meant to be a one-tailed P-value, which depends on the calculated t Stat, the computed t statistic. If t Stat < 0, then $P(T <= t)$ one-tail is in fact the lower tail corresponding to the area to the left of t Stat under a t density curve. But if t is positive, then $P(T <= t)$ one-tail is the area to the right of t Stat. It is therefore not always the P-value. For instance, if the test to be carried out were

$$\begin{aligned} H_0 : \mu_1 &= \mu_2 \\ H_a : \mu_1 &< \mu_2 \end{aligned}$$

then the P-value would be $1 - 0.0132 = 0.9868$ rather than 0.0121. $P(T <= t)$ two-tail is the correct P-value for a two-tailed test.

Confidence Intervals

The **ToolPak** does not print a confidence interval directly, but the output provides enough information to carry out the calculations. Details are given in cells C18:E23 of Figure 7.12 and are also shown isolated in Figure 7.13. The cells in column E of this block show the formulas that are the entries behind the cells in column D and whose values are evaluated and printed in the workbook by Excel. These formulas are the Excel equivalents of the formula

$$\bar{x}_1 - \bar{x}_2 \pm t^* \sqrt{\frac{s_1^2}{n_1} + \frac{s_2^2}{n_2}}$$

The information needed—the sample means, sample variances, sample sizes, and the critical t^* values—is part of the **ToolPak** output and is referenced in cells C18:E23. As before (Figure 7.9, from Example 7.7), the 95% confidence interval is (1.23, 18.67).

	C	D	E	F
18	Mean Difference	9.954	=D6-E6	
19	SE	4.308	=SQRT(D7/D8+E7/E8)	
20	t	2.024	=TINV(0.05, D10)	
21	ME	8.72	=D20*D19	
22	lower	1.23	=D18-D21	
23	upper	18.67	=D18+D21	

Figure 7.13: Two-Sample t Confidence Interval Formulas

The Pooled Two-Sample t Procedures

When the two populations are believed to be normal with the same variance, it is more common to use a pooled two-sample t based on an exact t distribution. The procedures are based on the statistic

$$t = \frac{(\bar{x}_1 - \bar{x}_2) - (\mu_1 - \mu_2)}{s_p\sqrt{\frac{1}{n_1} + \frac{1}{n_2}}} \tag{7.5}$$

where $s_p^2 = \frac{(n_1-1)s_1^2+(n_2-1)s_2^2}{n_1+n_2-2}$ is called the pooled sample variance. This statistic is known to have an exact t distribution on $\nu = n_1 + n_2 - 2$ degrees of freedom. The previous two-sample t analyses carry over with the obvious modifications for the degrees of freedom and use of the denominator in (7.5) in place of the denominator in (7.2).

Example 7.9. (Examples 7.19, 7.20, and 7.21, pages 551–554 in the text.) Does increasing the amount of calcium in our diet reduce blood pressure? A randomized comparative experiment gave one group of 10 black men a calcium supplement for 12 weeks. The control group of 11 black men received a placebo that appeared identical.

(a) Test the hypothesis that calcium lowers blood pressure more than a placebo by testing

$$\begin{aligned} H_0 : \mu_1 &= \mu_2 \\ H_a : \mu_1 &> \mu_2 \end{aligned}$$

at level $\alpha = 0.05$.

(b) Estimate the effect of calcium supplementation by computing a 90% confidence interval for the difference in population means $\mu_1 - \mu_2$.

Solution

1. Enter the data and labels in A3:A13 (Calcium) and B3:B14 (Placebo) of a workbook (Figure 7.14).

2. From the Menu Bar, choose **Tools – Data Analysis** and select **t-Test: Two-Sample Assuming Equal Variances** from the list of selections. (Refer to the dialog box in Figure 7.10.) Click OK.

3. Complete the next dialog box, which is similar to Figure 7.11, exactly as you did for the unequal variances case.

Excel Output

The output appears in the range C1:E14 in Figure 7.14, and we see that

$$\begin{aligned} \bar{x}_1 &= 5.000 & s_1^2 &= 76.444 \\ \bar{x}_2 &= -0.273 & s_2^2 &= 34.818 \\ s_p^2 &= 54.536 \end{aligned}$$

The computed pooled t statistic on 19 degrees of freedom is $t = 1.634$ (cell D12) while the $\alpha = 0.05$ upper critical value is $t^* = 1.729$ (cell D14). We conclude that although the calcium supplement appears to lower blood pressure, the difference between the calcium group and the placebo group is not significant at the 5% level. The P-value 0.0593 is shown in cell D13.

	A	B	C	D	E	F	G
1			*Pooled Two-Sample t Test*				
2							
3	Calcium	Placebo	t-Test: Two-Sample Assuming Equal Variances				
4	7	-1					
5	-4	12		*Calcium*	*Placebo*		
6	18	-1	Mean	5.000	-0.273		
7	17	-3	Variance	76.444	34.818		
8	-3	3	Observations	10	11		
9	-5	-5	Pooled Variance	54.536			
10	1	5	Hypothesized Mean Difference	0			
11	10	2	df	19			
12	11	-11	t Stat	1.634			
13	-2	-1	P(T<=t) one-tail	0.059			
14		-3	t Critical one-tail	1.729			
15			P(T<=t) two-tail	0.119			
16			t Critical two-tail	2.093			
17							
18			Mean Difference	5.273	=D6-E6		
19			SE	3.227	=SQRT(D9)*SQRT((1/10 + 1/11))		
20			t	1.729	=TINV(0.1, D11)		
21			ME	5.579	=D20*D19		
22			lower	-0.307	=D18-D21		
23			upper	10.852	=D18+D21		

Figure 7.14: Two-Sample t ToolPak Output

Confidence Intervals

In order to supplement the **Analysis ToolPak** output, which does not provide a confidence interval, we have also included in C18:D23 of Figure 7.14 formulas and output for the confidence intervals. These rely on the summary statistics produced by the ToolPak and are the Excel equivalents of the formula

$$\bar{x}_1 - \bar{x}_2 \pm t^* s_p \sqrt{\frac{1}{n_1} + \frac{1}{n_2}}$$

From D22:D23 we read a 90% confidence interval $(-0.307, 10.852)$.

7.3 Optional Topics in Comparing Distributions

Inference for Population Spread

Suppose that s_1^2 and s_2^2 are the sample variances of independent simple random samples of sizes n_1 and n_2 taken from normal populations $N(\mu_1, \sigma_1)$ and $N(\mu_2, \sigma_2)$, respectively. Then the ratio

$$F = \frac{s_1^2/\sigma_1^2}{s_2^2/\sigma_2^2}$$

has a known sampling distribution that does not depend on $\{\mu_1, \mu_2, \sigma_1, \sigma_2\}$, but only on the sample sizes. It has an F distribution on $n_1 - 1$ and $n_2 - 1$ degrees of freedom for the numerator and the denominator, respectively. The ratio on the right side of the equation is only one manifestation of the F distribution, which is also used in analysis of variance and regression.

The F Test for Equality of Spread

In this section the context is comparison of σ_1 and σ_2. It turns out for mathematical reasons that the appropriate parameter for testing the null hypothesis

$$H_0 : \sigma_1 = \sigma_2$$

is the ratio $\frac{\sigma_1}{\sigma_2}$ (equivalently $\frac{\sigma_1^2}{\sigma_2^2}$) rather than the difference, which we used for comparing means.

> **Example 7.10.** (Example 7.22, page 569 in the text.) Does increasing the amount of calcium in our diet reduce blood pressure? Examination of a large sample of people revealed a relationship between calcium intake and blood pressure. A randomized comparative experiment gave one group of 10 black men a calcium supplement for 12 weeks. The control group of 11 black men received a placebo that appeared identical. Table 7.2 gives the seated systolic (heart contracted)

Table 7.2: Seated Systolic Blood Pressure

Calcium group			Placebo group		
Begin	End	Decrease	Begin	End	Decrease
107	100	7	12	124	−1
110	114	−4	109	97	12
123	105	18	112	113	−1
129	112	17	102	105	−3
112	115	−3	98	95	3
111	116	−5	114	119	−5
107	106	1	119	114	5
112	102	10	112	114	2
136	125	11	110	121	−11
102	104	−2	117	118	−1
			130	133	−3

blood pressure for all subjects at the beginning and end of the 12-week period, in millimeters of mercury.

The analysis employed a pooled two-sample t test, which required assumption of equal population variances. Using level $\alpha = 0.05$ test

$$\begin{aligned} H_0 : \sigma_1 &= \sigma_2 \\ H_a : \sigma_1 &\neq \sigma_2 \end{aligned}$$

	A	B	C	D	E	F
1	*F Test for Equality of Variances*					
2	Differences in Systolic BP					
3	Calcium	Placebo				
4	7	-1		F-Test Two-Sample for Variances		
5	-4	12				
6	18	-1			*Calcium*	*Placebo*
7	17	-3		Mean	5.00	-0.27
8	-3	3		Variance	76.444	34.818
9	-5	-5		Observations	10	11
10	1	5		df	9	10
11	10	2		F	2.1955	
12	11	-11		P(F<=f) one-tail	0.1182	
13	-2	-1		F Critical one-tail	3.7790	
14		-3				

Figure 7.15: F Test Data and Output

Solution. We will use the F test in the **Analysis ToolPak**.

1. Enter the data from Table 7.2 in columns A and B of a workbook (Figure 7.15).

2. From the Menu Bar choose **Data Analysis** – ***F*-Test Two-Sample for Variances** and complete the dialog box of Figure 7.16. Notice that we have inserted not the specified level of significance $\alpha = 0.05$ in this box but rather half the value, 0.025, to reflect the fact that our test is two-sided while the cells E12:E13 give the P-value and the critical value for a one-sided upper-tailed test.

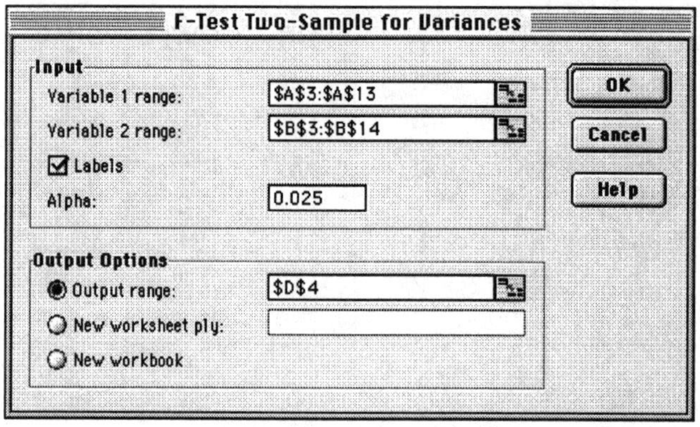

Figure 7.16: F Test Dialog Box

Caveat

This tool requires that the larger of the two sample variances be in the numerator, so repeat this procedure by reversing the variables if the output shows that the variance of the data in **Variable 1 range** in Figure 7.16 is less than that in **Variable 2 range** (which is not the case here).

Excel Output

The output appears in cells D4:F13, as shown in Figure 7.15. The computed value of F under H_0

$$F = \frac{s_1^2}{s_2^2} = 2.1955$$

appears in E11. (Note that $s_1^2 > s_2^2$ as required.) The critical F value is 3.779 and therefore the data are not significant at the 5% level.

We can obtain from cell E12 a one-sided P-value, which we need to double and find that the P-value $= 0.236$.

The F Distribution Function

This a good place to record the syntax for the F distribution. Suppose that F is a random variable having an F distribution with degrees of freedom ν_1 for the numerator and ν_2 for the denominator. Then for any $x > 0$,

$$\texttt{FDIST}(x, \nu_1, \nu_2) = P(F > x)$$

while for any $0 < p < 1$, the upper p critical value is obtained from the inverse

$$P\left(F > \texttt{FINV}(p, \nu_1, \nu_2)\right) = p$$

Chapter 8

Inference for Proportions

In this chapter we discuss data representing the counts or proportions of outcomes occurring in different categories. The methods of this chapter answer questions such as the following.

- What is the prevalence of frequent binge drinking among U.S. college students? A survey designed to answer this question sampled 17592 students on 140 campuses. Are there gender differences in this behavior?

- Does taking aspirin regularly help prevent heart attacks? A double-blind randomized comparative experiment assigned 11000 male doctors to take aspirin and another 11000 to take a placebo. After 5 years, 104 of the aspirin group and 189 of the control group had died of heart attacks. Is this difference large enough to convince us that aspirin works?

8.1 Inference for a Single Proportion

To estimate the proportion p of some characteristic in a population, it is common to take an SRS of size n and count X = the number in the sample possessing the characteristic. For large n, the distribution of X is approximately binomial $B(n, p)$, and by the central limit theorem the sample proportion

$$\hat{p} \text{ is approximately } N\left(p, \sqrt{\frac{p(1-p)}{n}}\right)$$

Inference is then based on the procedures for estimating a normal mean discussed in Chapter 6.

Confidence Intervals

The standard error of $\hat{p}$ is

$$\mathrm{SE}_{\hat{p}} = \sqrt{\frac{\hat{p}(1-\hat{p})}{n}}$$

where we have replaced p with $\hat{p}$ in the expression for the standard deviation of $\hat{p}$. Therefore a large-sample level C confidence interval for p is given as

$$\hat{p} \pm z^*\mathrm{SE}_{\hat{p}}$$

where z^* is the upper $(1-C)/2$ standard normal critical value.

Example 8.1. (Example 8.1, page 587 in the text.) Alcohol abuse has been described by college presidents as the number one problem on campus, and it is an important cause of death in young adults. How common is it? A survey of 17096 students in four-year colleges collected information on drinking behavior and alcohol-related problems. The researchers defined *frequent binge drinking* as having five or more drinks in a row, three or more times in the past two weeks. According to this definition, 3314 students were classified as frequent binge drinkers. The proportion of drinkers is

$$\hat{p} = \frac{3314}{17096} = 0.194$$

Find a 95% confidence interval for the proportion of binge drinkers among all four-year college students.

	A	B	C
1	*Confidence Interval for a Population Proportion*		
2			
3		Values	Formulas
4	User Input		
5	conf	0.95	0.95
6	Summary Statistics		
7	n	17096	
8	X	3314	
9	Calculations		
10	p_hat	0.1938	=X/n
11	SE	0.0030	=SQRT(p_hat*(1-p_hat)/n)
12	z	1.96	=NORMSINV(0.5+conf/2)
13	ME	0.006	=z*SE
14	Confidence Limits		
15	lower	0.188	=p_hat-ME
16	upper	0.200	=p_hat+ME

Figure 8.1: Confidence Interval for a Population Proportion

Solution. Figure 8.1 shows the Excel formulas required for the calculation (in column C), together with the corresponding values obtained when these formulas

are entered in column B on your workbook. The formulas parallel those for the confidence interval for a normal mean. From cells B15:B16 we find that a 95% confidence interval is (0.188, 0.200).
Note: We remind you that the formulas in Figure 8.1 require **Named Ranges** to refer to the variables by their names, or else the cell references must be used.

Significance Tests

For testing the null hypothesis

$$H_0 : p = p_0$$

we use the test statistic

$$z = \frac{\hat{p} - p_0}{\sqrt{\frac{p_0(1-p_0)}{n}}}$$

for a large-sample procedure. For example, if the alternative is two-sided

$$H_a : p \neq p_0$$

then we reject H_0 at level α if

$$|z| > z^*$$

where z^* is the upper $\frac{\alpha}{2}$ standard normal critical value. Furthermore the P-value is $2P(Z > |z|)$, where Z is $N(0, 1)$.

> **Example 8.2.** (Example 8.2, page 589 in the text.) The French naturalist Count Buffon once tossed a coin 4040 times and obtained 2048 heads. To assess whether the data provide evidence that the coin was not balanced, we test
>
> $$H_0 : p = 0.5$$
> $$H_a : p \neq 0.5$$
>
> where p is the probability that Buffon's coin lands heads.

Solution. Figure 8.2 gives the required formulas in column C for lower, upper, and two-sided tests. Column B contains the data and calculations. Cell B11 gives $\hat{p} = 0.5069$, cell B12 gives the standard error SE $= 0.0079$, and cell B13 gives the calculated value of $z = 0.881$. Our problem is two-sided, so only the values in rows 22–25 of the template are relevant. The P-value is 0.378, and we therefore do not reject H_0.

These calculations are straightforward, and Figure 8.2 merely provides a systematic way of carrying them out. Nothing so elaborate is really needed for calculations that could be done with a hand calculator.

	A	B	C
1			*Significance Test for a Population Proportion*
2			
3	User Input		
4	p0	0.50	
5	alpha	0.01	
6			
7	Summary Statistics		
8	n	4040	
9	X	2048	
10	Calculations		
11	p_hat	0.5069	=X/n
12	SE	0.0079	=SQRT(p0*(1-p0)/n)
13	z	0.881	=(p_hat-p0)/SE
14	Lower Test		
15	lower_z		=NORMSINV(alpha)
16	Decision		=IF(z<lower_z,"Reject H0","Do Not Reject H0")
17	Pvalue		=NORMSDIST(z)
18	Upper Test		
19	upper_z		=-NORMSINV(alpha)
20	Decision		=IF(z>upper_z,"Reject H0","Do Not Reject H0")
21	Pvalue		=1-NORMSDIST(z)
22	Two-Sided Test		
23	two_z	2.576	=ABS(NORMSINV(alpha/2))
24	Decision	Do Not Reject H0	=IF(ABS(z)>two_z,"Reject H0","Do Not Reject H0")
25	Pvalue	0.378	=2*(1-NORMSDIST(ABS(z)))

Figure 8.2: Significance Test for a Population Proportion

Exercise. (Adapted from Example 8.4, page 591 in the text.) Simulate 4040 tosses of a fair coin. Repeat the simulation 100 times and plot the lower and upper end points of the 100 confidence intervals on a graph with the constant line $p = 0.5$. How many intervals contain 0.5?

8.2 Comparing Two Proportions

Large-sample inference procedures for comparing the proportions p_1 and p_2 in two populations based on independent SRS of sizes n_1 and n_2, respectively, are also based on the normal approximation. The natural estimate $D = \hat{p}_1 - \hat{p}_2$ of the difference in proportions $p_1 - p_2$ is approximately normal with mean $p_1 - p_2$ and standard deviation

$$\sigma = \sqrt{\frac{p_1(1-p_1)}{n_1} + \frac{p_2(1-p_2)}{n_2}}$$

Confidence Intervals

We must replace the unknown parameters p_1 and p_2 by their estimates $\hat{p}_1$ and $\hat{p}_2$ to obtain an estimated standard error

$$\mathrm{SE}_D = \sqrt{\frac{\hat{p}_1(1-\hat{p}_1)}{n_1} + \frac{\hat{p}_2(1-\hat{p}_2)}{n_2}}$$

and an approximate level C confidence interval

$$\hat{p}_1 - \hat{p}_2 \pm z^*\mathrm{SE}_D$$

where z^* is the upper $(1 - C)/2$ standard normal critical value.

> **Example 8.3.** (Example 8.8, page 603 in the text.) In the binge drinking study, Example 8.1, data were also summarized by gender. In this table the $\hat{p}$ column gives the sample proportions of frequent binge drinkers. The last line gives the totals used in Example 8.1. Find a 95% confidence interval for the difference between the proportions of men and women who are frequent binge drinkers.

Solution. Table 8.1 conveniently summarizes the data. The corresponding Excel workbook is in Figure 8.3. Results of the calculations are given in column B with the corresponding formulas in the adjacent column C. Based on this data set we are 95% confident that the difference in proportions between men and women is in the range (0.045, 0.069).

Table 8.1: Frequent Binge Drinkers

Population	n	X	$\hat{p} = X/n$
1 (men)	7180	1630	0.227
2 (women)	9916	1684	0.170
Total	17096	3314	0.194

	A	B	C	D	E
1	*Confidence Interval for the Difference in Proportions*				
2					
3		Summary Statistics and User Input			
4	Group	n	X	p_hat=X/n	
5	Men	7180	1630	0.2270	
6	Women	9916	1684	0.1698	
7					
8	conf	0.95			
9					
10	SE	0.00622	=SQRT((D5*(1-D5)/B5)+(D6*(1-D6)/B6))		
11	z	1.960	=NORMSINV(0.5+conf/2)		
12	ME	0.012	=z*SE		
13					
14	Confidence Limits				
15					
16	lower	0.045	=(D5-D6)-ME		
17	upper	0.069	=(D5-D6)+ME		

Figure 8.3: Confidence Interval—Difference in Proportions

Significance Tests

The null hypothesis

$$H_0 : p_1 = p_2$$

is tested using the statistic

$$z = \frac{\hat{p}_1 - \hat{p}_2}{\mathrm{SE}_{Dp}}$$

where SE_{Dp} is the estimated standard deviation based on the pooled estimate

$$\hat{p} = \frac{x_1 + x_2}{n_1 + n_2}$$

of the common value $p \equiv p_1 = p_2$ of the population proportions. Here x_1 and x_2 are the number of counts possessing the characteristic being counted in sample 1 and sample 2, respectively, and

$$\mathrm{SE}_{Dp} = \sqrt{\hat{p}(1 - \hat{p})\left(\frac{1}{n_1} + \frac{1}{n_2}\right)}$$

The decision rules based on z are then analogous to those in Section 8.1. For example, if the alternative hypothesis is

$$H_a : p_1 > p_2$$

then we reject at level α if

$$z > z^*$$

where z^* is the upper α standard normal critical value. Furthermore, the P-value is $P(Z > z)$, where Z is $N(0, 1)$.

Example 8.4. (Exercise 8.36, page 611 in the text.) In the 1996 regular baseball season, the World Series Champion New York Yankees played 80 games at home and 82 games away. They won 49 of their home games and 43 of the games played away. We can consider these games as samples from hypothetically large populations of games played at home and away.
(a) Most people think it is easier to win at home than away. Formulate null and alternative hypotheses to examine this idea.
(b) Compute the z statistic and its P-value. What conclusion do you draw?

Solution

(a) Let

$$p_1 = \text{probability that the Yankees win a home game}$$

$$p_2 = \text{probability that the Yankees win an away game}$$

Since we are trying to "prove" that it is easier to win at home than away, the appropriate significance test should be

$$H_0 : p_1 = p_2$$
$$H_a : p_1 > p_2$$

	A	B	C	D	E
1		*Significance Test for the Difference in Proportions*			
2					
3	Summary Statistics and User Input				
4	Group	n	X	p_hat	
5	home	80	49	0.613	
6	away	82	43	0.524	
7					
8	null	0	Calculations		
9	alpha	0.05	pooled_p	0.568	=(C5+C6)/(B5+B6)
10	alternate	upper	SE	0.078	=SQRT(pooled_p*(1-pooled_p)*(1/B5 + 1/B6))
11			z	1.132	=((D5-D6)-null)/SE
12	Lower Test				
13	lower_z		=NORMSINV(alpha)		
14	Decision		=IF(z<lower_z,"Reject H0","Do Not Reject H0")		
15	Pvalue		=NORMSDIST(z)		
16	Upper Test				
17	upper_z	1.645	=-NORMSINV(alpha)		
18	Decision	Do Not Reject H0	=IF(z>upper_z,"Reject H0","Do Not Reject H0")		
19	Pvalue	0.129	=1-NORMSDIST(z)		
20	Two-Sided Test				
21	two_z		=ABS(NORMSINV(alpha/2))		
22	Decision		=IF(ABS(z)>two_z,"Reject H0","Do Not Reject H0")		
23	Pvalue		=2*(1-NORMSDIST(ABS(z)))		

Figure 8.4: Significance Test—Difference in Proportions

(b) Figure 8.4 is a template that provides all the formulas required (located in the relevant cells in Columns C and E). Although this is an upper test, we have provided the formulas for lower and two-tailed tests.

The sample proportion of games won at home is (from cell D5)

$$\hat{p}_1 = \frac{49}{80} = 0.613$$

and the proportion of games won away is (from cell D6)

$$\hat{p}_2 = \frac{43}{82} = 0.524$$

Cell D9 gives the pooled sample proportion

$$\hat{p} = \frac{49 + 43}{80 + 82} = 0.568$$

The z test statistic in cell D11 is 1.132, which is not significant (cell B18), and the P-value in cell B19 is 0.129. There is no evidence of a difference between home and away games.

Chapter 9

Inference for Two-Way Tables

In Example 8.3 we were interested in comparing two populations (male, female) with respect to one response variable, "frequent binge drinker." The response variable had two values, "yes" or "no." A test was carried out of the null hypothesis

$$H_0 : p_1 = p_2 \qquad (9.1)$$

where p_1 and p_2 are the proportions in the respective two populations who are binge drinkers.

There is another way to view and display the data. We may consider measuring two variables, gender and binge category, on the 17096 students who were surveyed. With two possible levels for each variable, there are four combinations of measurements, and we could display the results as a frequency table or a bar graph involving four categories. But this would cause us to lose track of how the four categories are related to the levels and the variables (essentially the "geometry" of the design). It is more natural to display the data in a different (though equivalent) way than was presented in Table 8.1 by using a "two-dimensional" frequency table, shown here in Table 9.1.

Table 9.1: Binge Drinkers—2 × 2 Table

	Gender		
Frequent binge drinker	Men	Women	Total
Yes	1630	1684	3314
No	5550	8232	13782
Total	7180	9916	17096

This presentation of the data shows that there are two variables of interest, one that might be considered as an explanatory variable (gender) and the other as the response variable. The significance test of the null hypothesis in (9.1) judges whether there is a relationship between the two variables. If $p_1 = p_2$, then there is no relationship.

We wish to generalize this test to the situation in which there are more than two populations of interest or where the response variable can take more than two values.

A table, such as Table 9.1, showing data collected on two categorical variables having r rows and c columns of values for each of the two variables is called an $r \times c$ table. In this chapter we discuss a technique based on the chi-square distribution for deciding if there is a relationship between two categorical variables.

9.1 Inference for Two-Way Tables

Example 9.1. (Examples 9.1–9.6, pages 624–632 in the text.) Do men and women participate in sports for the same reasons? One goal for sports participants is social comparison – the desire to win or to do better than other people. Another is mastery – the desire to improve one's skills or to try one's best. A study on why students participate in sports collected data from 67 male and 67 female undergraduates at a large university. Each student was classified into one of four categories based on his or her responses to a questionnaire about sports goals. The four categories were high social comparison-high mastery (HSC-HM), high social comparison-low mastery (HSC-LM), low social comparison-high mastery (LSC-HM), and low social comparison-low mastery (LSC-LM). One purpose of the study was to compare the goals of male and female students. The data are displayed in a two-way table (Table 9.2). The entries in this table are the observed, or sample, counts. For example, there are 14 females in the high social comparison-high mastery group. Determine whether there is an association between gender and goal.

Table 9.2: Observed Counts for Sports Goals

	Sex		
Goal	Female	Male	Total
HSC-HM	14	31	45
HSC-LM	7	18	25
LSC-HM	21	5	26
LSC-LM	25	13	38
Total	67	67	134

Example 9.1 is a direct generalization of Example 8.3. One variable, with

categories displayed across the top of Table 9.1, is also gender, the explanatory variable. But now the second variable has four levels (as opposed to the two levels in Example 8.3) and the objective is to determine whether the proportions in the population for each level of the second variable are the same in males as they are in females. Another possible way of expressing this is to say that we are asking whether the cross classifications into the two variables are independent of each other. Which interpretation is appropriate depends on how the data are collected, essentially whether with fixed or random marginal totals, but the methodology is the same in either case, as discussed in the text.

Each combination of row and column levels defines a cell, and we call a table, such as Table 9.2, with (more generally) r rows and c columns an $r \times c$ contingency table. The two columns labeled "Female" and "Male" represent the results of independent samples from the respective populations.

Suppose that

$$\begin{aligned} \mathbf{p}_1 &= (p_{11}, p_{12}, p_{13}, p_{14}) \\ \mathbf{p}_2 &= (p_{21}, p_{22}, p_{23}, p_{24}) \end{aligned}$$

represent the population proportions of all female (respectively male) undergraduates who are in the four goal levels. Thus, for instance, p_{12} = the population proportion among all females whose goal rating is HSC-LM, the second row in Table 9.2.

The null hypothesis of interest is

$$H_0 : \mathbf{p}_1 = \mathbf{p}_2$$

meaning equality of the vector of proportions, thereby generalizing (9.1). However, this methodology is not limited to $c = 2$, which is the example here. With c columns we would be testing

$$H_0 : \mathbf{p}_1 = \mathbf{p}_2 = \cdots = \mathbf{p}_c$$

for corresponding population proportion vectors $\mathbf{p}_i$.

Some preparation is needed before we can use the Excel function `CHIINV`, which returns the P-value for this significance test. We need to first create a table of expected cell counts.

Expected Cell Counts

Consider row 1 in Table 9.2, labeled HSC-HM, in which there are 14 counts under Female and 31 counts under Male for a total sum of row counts equal to 45.

Under the null hypothesis, we may pool the data for Male and Female with regard to the variable "Goal." Since the total number of subjects is 134, we get a pooled estimate

$$\frac{14 + 31}{67 + 67} = \frac{45}{134} = 0.3358$$

for the common value $p_{11} = p_{21}$.

The column total for "Female" is 67. Therefore the number of counts in cell "Female $\times$ HSC-HM" is a binomial random variable on 67 trials with "success probability" $\frac{45}{134}$. Under H_0 we estimate this with the usual binomial mean

$$67 \times \frac{45}{134} = \frac{45 \times 67}{134} = 22.5$$

leading to the useful mnemonic

$$\text{expected count} = \frac{\text{row total} \times \text{column total}}{\text{table total}}$$

Expected counts need to be computed for all cells. These calculations need to be hand-coded in the Excel workbook, and we now show how to do this. Copy the observed counts to an Excel workbook. This is shown in Figure 9.1 where the data appear in block A4:C8 (including labels).

	A	B	C	D	E	F	G	H	I
1				*Computing the Expected Table*					
2									
3		Observed Counts					Formulas for Expected Counts		
4		Female	Male	Row Total			Female	Male	Row Total
5	HSC-HM	14	31	45		HSC-HM	=B$9*$D5/$D$9	=C$9*$D5/$D$9	=SUM(B14:C14)
6	HSC-LM	7	18	25		HSC-LM	=B$9*$D6/$D$9	=C$9*$D6/$D$9	=SUM(B15:C15)
7	LSC-HM	21	5	26		LSC-HM	=B$9*$D7/$D$9	=C$9*$D7/$D$9	=SUM(B16:C16)
8	LSC-LM	25	13	38		LSC-LM	=B$9*$D8/$D$9	=C$9*$D8/$D$9	=SUM(B17:C17)
9	Column Total	67	67	134		Column Total	=SUM(B14:B17)	=SUM(C14:C17)	=SUM(D14:D17)
10									
11									
12		Expected Counts							
13		Female	Male	Row Total		Chi-Square Test			
14	HSC-HM	22.5	22.5	45		P-value:	1.622E-05		
15	HSC-LM	12.5	12.5	25			=CHITEST(B5:C8,B14:C17)		
16	LSC-HM	13	13	26					
17	LSC-LM	19	19	38					
18	Column Total	67	67	134					

Figure 9.1: Expected Cell Counts and P-value

1. For the row totals, type the formula = SUM(B5:C5) in cell D5 and press Enter.
2. Select cell D5, then click the fill handle and fill column D to cell D8.
3. Enter the formula = SUM(B5:B8) in cell B9 and fill row 9 to cell D9 to produce the row totals and the overall table total.

Next we produce the table of expected counts. Again refer to Figure 9.1.

1. Copy the entire observed table block A3:D9 to another location, say A12:D18, and change the label "Observed Counts" to "Expected Counts."

2. Select cell B14 and enter the formula = B$9*$D5/$D$9, which involves both absolute and relative cell reference. Fill the formula to the other cells in the block B14:C17 by first filling from cell B14 to C14 and then from block B14:C14 to B17:C17. This fills the table with expected counts. We require an absolute reference $9 since the column totals are always in row 9. We require an absolute reference $D because the row totals are always in column D. We don't want these to change when the formula is filled to all the cells in the expected table, so absolute references are mandated as shown.

3. Finally obtain the marginal sums as before.

Figure 9.1 shows the formulas in block F4:I9 that are behind the cells of expected counts in block A13:D18.

The Chi-Square Test

The statistic

$$X^2 = \sum \frac{(\text{observed count} - \text{expected count})^2}{\text{expected count}}$$

where the sum is taken over counts in all the cells, was introduced by Karl Pearson in 1900 to measure how well the model H_0 fits the data. Under H_0 it has a sampling distribution that is approximated by a one-parameter family of distributions known as chi-square and denoted by the symbol χ^2. The parameter is called the degrees of freedom ν, and for an $r \times c$ contingency table it is known that $\nu = (r-1)(c-1)$. If H_0 is true, then X^2 should be "small," while if H_0 is false then X^2 should be "large." This leads to the criterion

$$\text{chi-square test:} \quad \text{Reject } H_0 \text{ if } X^2 > \chi^2$$

where χ^2 is the upper critical α value of a chi-square distribution on $(r-1)(c-1)$ degrees of freedom.

The P-Value

The Excel function CHITEST calculates the P-value associated with the Pearson X^2 statistic. The syntax is

= CHITEST(actual_range, expected_range)

where "actual_range" refers to the observed table of counts B5:C8 and where "expected_range" refers to the expected table of counts B14:C17. Refer to block F14:G15 of Figure 9.1 where we have given the formula and its value. The P-value is 0.000016, so we reject H_0 and conclude that there is an association between gender and sports goals.

9.2 Formulas and Models for Two-Way Tables

Computing the Chi-Square Statistic

Although not required for the function CHITEST, it is instructive to do the calculation of X^2 and then carry out the test based on value of X^2. We have shown this in block F3:H13 on the right half of Figure 9.2. For convenience the required formulas are displayed in Figure 9.3, and we present the details.

	A	B	C	D	E	F	G	H
1				*Computing Chi Square*				
2								
3		Observed Counts					Chi-Square Cell Values	
4		Female	Male	Row Total			Female	Male
5	HSC-HM	14	31	45		HSC-HM	3.2111	3.2111
6	HSC-LM	7	18	25		HSC-LM	2.4200	2.4200
7	LSC-HM	21	5	26		LSC-HM	4.9231	4.9231
8	LSC-LM	25	13	38		LSC-LM	1.8947	1.8947
9	Column Total	67	67	134				
10						Chi-Square	24.898	
11		Expected Counts				Critical 5%	7.815	
12		Female	Male	Row Total				
13	HSC-HM	22.5	22.5	45		Decision:	Reject H0	
14	HSC-LM	12.5	12.5	25				
15	LSC-HM	13	13	26				
16	LSC-LM	19	19	38				
17	Column Total	67	67	134				

Figure 9.2: Computing the Value of X^2—Values

	A	B	C	D	E	F	G	H
1				*Computing Chi Square*				
2								
3		Observed Counts					Chi-Square Cell Values	
4		Female	Male	Row Total			Female	Male
5	HSC-HM	14	31	45		HSC-HM	=(B5-B13)^2/B13	=(C5-C13)^2/C13
6	HSC-LM	7	18	25		HSC-LM	=(B6-B14)^2/B14	=(C6-C14)^2/C14
7	LSC-HM	21	5	26		LSC-HM	=(B7-B15)^2/B15	=(C7-C15)^2/C15
8	LSC-LM	25	13	38		LSC-LM	=(B8-B16)^2/B16	=(C8-C16)^2/C16
9	Column Total	67	67	134				
10						Chi-Square	=SUM(G5:H8)	
11		Expected Counts				Critical 5%	=CHIINV(0.05, 3)	
12		Female	Male	Row Total				
13	HSC-HM	22.5	22.5	45		Decision:	=IF(G10>G11, "Reject H0", "Do Not Reject H0")	
14	HSC-LM	12.5	12.5	25				
15	LSC-HM	13	13	26				
16	LSC-LM	19	19	38				
17	Column Total	67	67	134				

Figure 9.3: Computing the Value of X^2—Formulas

1. Copy the block A4:C8 to a convenient location, shown here copied to cells F4:H8. Change the label "Response" to "Chi-Square Cell Values."

2. The equation for X^2 is

$$X^2 = \sum \frac{(\text{observed count} - \text{expected count})^2}{\text{expected count}}$$

which we translate into Excel by the formula = (B5-B13)ˆ2/B13 entered in cell G5 and then filled to the block G5:H8. See Figure 9.3.

3. Sum all six cell entries by entering = `SUM`(G5:H8) in cell G10. We get the value

$$X^2 = 10.500$$

The critical 5% χ^2 value is given by = `CHIINV`(0.05,3) in cell G11 and is

$$\chi^2_{.05} = 5.991$$

We therefore reject H_0, which is the same conclusion drawn using P-value.

You may arrange for the decision to appear on your workbook using the formula = `IF`($G10 > G11$, "Reject H0", "Do Not Reject H0"), which we have entered in cell G13.

Chapter 10

Inference for Regression

We pointed out in Chapter 2 that there are three procedures in Excel for regression analysis. These are complementary for the most part. We have already used the **Insert Trendline** command to fit and draw the least-squares regression line. In this chapter we derive additional information about the regression model with the **Regression** tool in the ToolPak. We also introduce some relevant Excel functions.

10.1 Simple Linear Regression

The general regression model for n pairs of observation (x_i, y_i) is

$$DATA = FIT + RESIDUAL$$

which is expressed mathematically as

$$y_i = \beta_0 + \beta_1 x_i + \varepsilon_i$$

The function

$$\mu_y = \beta_0 + \beta_1 x$$

is called the population (true) regression curve of y on x, here taken to be linear. It represents the mean response of y as a function of x. The quantities $\{\varepsilon_i\}$ are assumed to be independent normally distributed random variables with a common mean 0 and common standard deviation σ. Therefore there are three unknown parameters $(\beta_0, \beta_1, \sigma)$.

The parameters β_0, β_1 are estimated by the method of least-squares (described in Chapter 2) by the values b_0, b_1, respectively, where

$$\begin{aligned} b_0 &= \bar{y} - b_1 \bar{x} \\ b_1 &= r\,\frac{s_y}{s_x} \end{aligned}$$

Recall that s_x and s_y are the sample standard deviations of the x and y data sets respectively.

The least-squares regression line, which estimates the population regression line, is given by the equation

$$\hat{y} = b_0 + b_1 x$$

Algebra shows that b_0 and b_1 are unbiased estimators of β_0 and β_1. They are random variables having sampling distributions:

(i) b_0 is normal with

$$\begin{aligned} \text{mean} &= \mu_{b_0} = \beta_0 \\ \text{standard deviation} &= \sigma_{b_0} = \sigma\sqrt{\frac{1}{n} + \frac{\bar{x}^2}{\sum_{i=1}^{n}(x_i - \bar{x})^2}} \end{aligned}$$

(ii) b_1 is normal with

$$\begin{aligned} \text{mean} &= \mu_{b_1} = \beta_1 \\ \text{standard deviation} &= \sigma_{b_1} = \frac{\sigma}{\sqrt{\sum_{i=1}^{n}(x_i - \bar{x})^2}} \end{aligned}$$

(iii) In a manner entirely analogous to the case of estimating the mean of a data set, we obtain an estimate of σ^2 for inference purposes. This estimate is provided by

$$s^2 = \frac{\sum_{i=1}^{n} e_i^2}{n-2}$$

where

$$e_i = y_i - \hat{y}_i$$

is called the residual at x_i and is an estimate of the sampling error $\{\varepsilon_i\}$. The denominator $(n-2)$ gives an unbiased estimate because

$$\frac{(n-2)s^2}{\sigma^2} \quad \text{is chi-squared on } (n-2) \; df$$

Regression analysis thus consists of estimating β_0, β_1, and σ^2, and making inferences about them.

The Regression Tool

Example 10.1. (Examples 10.1–10.5, pages 664–677 in the text.) Figure 10.1 shows data from a sample of 92 males aged 20 to 29 relating skinfold thickness to body density. Body density is a proxy for fat content, a variable having medical importance. In practice, fat content is found by measuring body density, the weight per unit volume of the body. High fat content corresponds to low body density. Body density is hard to measure directly – the standard method requires that

	A	B	C	D	E	F	G	H	I	J	K	L	M	N	O
1							*Complete Data Set for Example 10.1*								
2															
3	LSKIN	1.27	1.56	1.45	1.52	1.51	1.51	1.5	1.62	1.5	1.75	1.43	1.81	1.6	
4	DEN	1.093	1.063	1.078	1.056	1.073	1.071	1.076	1.047	1.089	1.053	1.057	1.051	1.074	
5															
6	LSKIN	1.49	1.29	1.52	1.83	1.58	1.7	1.59	2.02	1.84	1.87	1.83	1.36	1.47	
7	DEN	1.07	1.081	1.064	1.037	1.06	1.065	1.058	1.042	1.045	1.026	1.046	1.063	1.067	
8															
9	LSKIN	1.84	1.46	1.74	1.52	1.82	1.53	1.54	1.61	1.22	1.44	1.18	1.35	1.52	
10	DEN	1.05	1.063	1.06	1.078	1.055	1.059	1.076	1.06	1.086	1.083	1.084	1.082	1.071	
11															
12	LSKIN	1.64	1.73	1.38	1.3	1.73	1.57	1.74	1.41	1.53	1.54	1.83	2.08	1.91	
13	DEN	1.056	1.046	1.096	1.074	1.054	1.063	1.058	1.076	1.066	1.076	1.058	1.039	1.029	
14															
15	LSKIN	1.47	1.67	1.89	1.84	1.38	1.5	1.49	1.57	1.5	1.34	1.52	1.74	1.24	
16	DEN	1.07	1.061	1.034	1.05	1.063	1.071	1.058	1.057	1.092	1.07	1.054	1.053	1.084	
17															
18	LSKIN	1.43	2.01	1.75	1.65	1.79	1.32	1.64	1.95	1.2	1.36	1	1.73	1.56	
19	DEN	1.079	1.03	1.041	1.06	1.072	1.071	1.056	1.037	1.083	1.069	1.096	1.057	1.063	
20															
21	LSKIN	1.52	1.53	1.86	1.85	1.62	1.39	1.52	1.69	1.27	1.21	1.58	1.1	1.53	1.46
22	DEN	1.08	1.068	1.05	1.043	1.056	1.086	1.06	1.047	1.082	1.093	1.059	1.093	1.067	1.073

Figure 10.1: Skinfold Thickness Data

subjects be weighed under water. For this reason, scientists have sought variables that are easier to measure and that can be used to predict body density. Research suggests that *skinfold thickness* can accurately predict body density. To measure skinfold thickness, pinch a fold of skin between calipers at four body locations to determine the thickness, and add the four thicknesses. There is a linear relationship between body density and the logarithm of the skinfold thickness measure. The explanatory variable x is the log of the sum of the skinfold measures, denoted by LSKIN, and the response variable y is body density denoted by DEN. Density is measured in 10^3 kg/m^3 and skinfold thickness in mm. The skinfolds were measured at the biceps, triceps, subscapular and suprailiac areas. Analyse the data using the regression tool.

(a) Plot the data and confirm that a straight-line fit is appropriate.

(b) Fit the least-squares regression line to the data.

(c) Find the standard error.

(d) Give 95% confidence intervals for the parameters β_1 and β_0.

(e) Test the null hypothesis $H_0 : \beta_1 = 0$.

(f) Compute r^2. Examine the residuals for any deviation from a straight line.

Using the Regression Tool

1. Enter the 92 data pairs (including labels) from Figure 10.1 into columns,

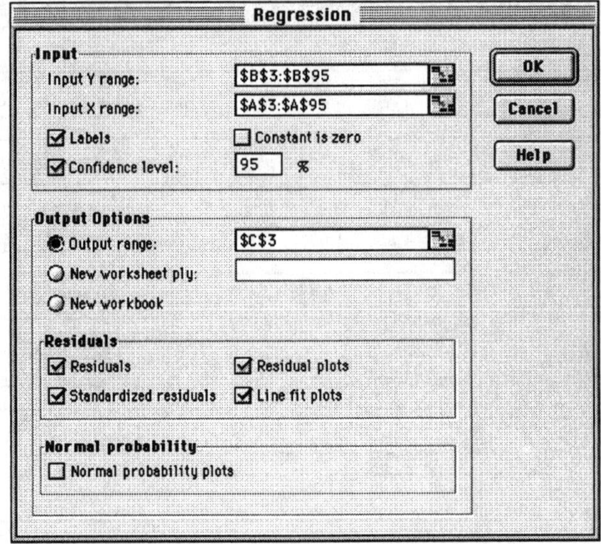

Figure 10.2: Regression Tool Dialog Box

with the independent variable x to the left of the dependent variable y, say in columns A3:A95 and B3:B95, with the first row reserved for labels. (Excel requires this kind of data in contiguous regions.) Refer to Figure 10.3 later in this section; it shows the **Regression** tool output and also part of columns A and B.

2. From the Menu Bar choose **Tools – Data Analysis** and select **Regression** from the tools listed in the **Data Analysis** dialog box. Click OK to display the **Regression** dialog box (Figure 10.2).

3. Type the cell references B3:B95 in Figure 10.2 (or point and drag over the data in column B) for **Input Y range.** Do the same with respect to Column A for **Input X range.** Check the box **Labels**, leave **Constant is zero** clear because we are not forcing the line through the origin, and check the box **Confidence level** and insert "95." Under **Output options** select the radio button **Output range** and enter C3 to locate the upper left corner where the output will appear in the same workbook.

   Check the following boxes:

   Residuals. To obtain predicted or fitted values $\hat{y}$ and their residuals.

   Residual Plots. For a scatterplot of the residuals against their x values.

   Standardized Residuals. To obtain residuals divided by their standard error (useful to identify outliers).

   Line Fit Plots. To obtain a scatterplot of y against x.

Do not check Normal Probability Plots.

Note: In making these selections, either use the Tab key to move from option to option or use the mouse. Click OK.

Excel Output

The output is separated into six regions: Regression Statistics, ANOVA table, statistics about the regression line parameters, residual output, scatterplot with fitted line, and residual plot. We have reproduced a portion of the output in Figure 10.3. (Part of the output also appears in the text at the top of Figure 10.5 page 669.) We next interpret the output in each of these regions.

	A	B	C	D	E	F	G	H	I
1				*Regression Tool Example - Skinfold Thickness*					
2									
3	LSKIN	DEN	SUMMARY OUTPUT						
4	1.27	1.093							
5	1.56	1.063	*Regression Statistics*						
6	1.45	1.078	Multiple R	0.8488					
7	1.52	1.056	R Square	0.7204					
8	1.51	1.073	Adjusted R Square	0.7173					
9	1.51	1.071	Standard Error	0.0085					
10	1.5	1.076	Observations	92					
11	1.62	1.047							
12	1.5	1.089	ANOVA						
13	1.75	1.053		*df*	*SS*	*MS*	*F*	*Significance F*	
14	1.43	1.057	Regression	1	0.01691	0.0169	231.894	1.221E-26	
15	1.81	1.051	Residual	90	0.00656	0.0001			
16	1.6	1.074	Total	91	0.02347				
17	1.49	1.07							
18	1.29	1.081		*Coefficients*	*Standard Error*	*t Stat*	*P-value*	*Lower 95%*	*Upper 95%*
19	1.52	1.064	Intercept	1.1630	0.0066	177.296	0.0000	1.1500	1.1760
20	1.83	1.037	LSKIN	-0.0631	0.0041	-15.228	0.0000	-0.0714	-0.0549
21	1.58	1.06							

Figure 10.3: Data and Regression Tool Output

Parameter Estimates and Inference

Rows 19 and 20 of the output provide statistics for the regression line parameters. From cell D19 we read $b_0 = 1.163$, and from cell D20 we read $b_1 = -0.063$. The regression line is therefore

$$\hat{y} = 1.163 - 0.063x$$

Significance Tests

The test

$$H_0 : \beta_1 = 0 \qquad \text{vs.} \qquad H_a : \beta_1 \neq 0$$

is useful in assessing whether a simple model of a straight line through the origin provides an equally good fit. The appropriate test statistic

$$t = \frac{b_1}{\mathrm{SE}_{b_1}}$$

where

$$\mathrm{SE}_{b_1} = \frac{s}{\sqrt{\sum_{i=1}^{n}(x_i - \bar{x})^2}}$$

is the standard error of b_1 which is obtained from the standard deviation of b_1 by replacing the unknown σ with s. From E20 in Figure 10.3 we read $\mathrm{SE}_{b_1} = 0.0041$, and from F20 we have the t statistic, denoted as t stat,

$$t = \frac{b_1}{\mathrm{SE}_{b_1}} = \frac{-0.0631}{0.0041} = -15.228.$$

The corresponding two sided P-value appears in cell G20,

$$P\text{-value} = 0.0000$$

Similarly, for testing

$$H_0 : \beta_0 = 0 \qquad \text{vs.} \qquad H_a : \beta_0 \neq 0$$

the appropriate test statistic is

$$t = \frac{b_0}{\mathrm{SE}_{b_0}}$$

where

$$\mathrm{SE}_{b_0} = s\sqrt{\frac{1}{n} + \frac{\bar{x}^2}{\sum_{i=1}^{n}(x_i - \bar{x})^2}}$$

estimates the true standard deviation σ_{b_0} of b_0. From D19:G19 we read the relevant estimates and the computed test statistic

$$t = \frac{b_0}{\mathrm{SE}_{b_0}} = \frac{1.1630}{0.0066} = 177.296$$

together with the two-sided

$$P\text{-value} = 0.0000$$

Significance tests for other null values can be carried out using the templates developed in Chapter 7, which required only the summary statistics. The main change is to use $n - 2$ instead of $n - 1$ for the degrees of freedom.

Confidence Intervals

We used the default of 95% for the confidence level when we completed the **Regression** tool dialog box. The lower and upper 95% confidence limits appear in H19:I20 in Figure 10.3. Thus 95% confidence intervals are

$$\begin{array}{ll} \text{for } \beta_0 & (1.150, 1.176) \\ \text{for } \beta_1 & (-0.0714, -0.0549) \end{array}$$

Scatterplot

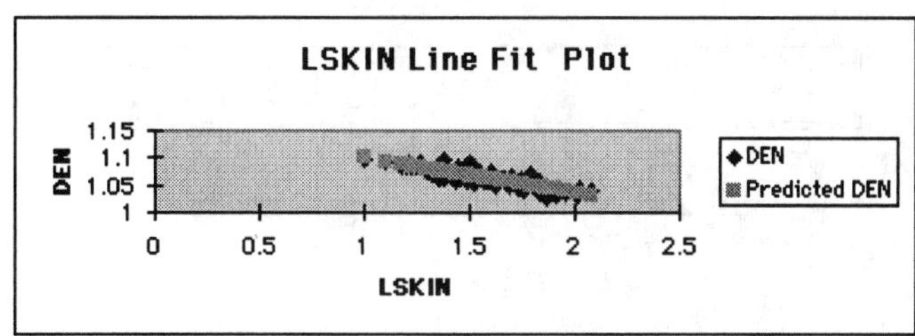

Figure 10.4: Default Scatterplot

Figure 10.4 shows one of the scatterplots produced, DEN against LSKIN. By default Excel uses markers even for the predicted values and the scales need to be changed to produce a more useful graph.

Changing Markers

We have modified the Excel scatterplot by enlarging it to make it more readable, changing the scale of the X and Y axes, replacing the diamond-shaped data plotting markers with circles, and replacing the square-shaped predicted values with a line.

1. To resize the **Chart Area** activate the Chart and drag one or more of the handles to the desired size.

2. To resize the **Plot Area** click within the plot area and drag one or more handles to the desired size.

3. To change the scale on the X axis activate the Chart and double-click the X axis. (Equivalently, click the X axis once and choose **Format – Selected Axis...** from the Menu Bar.) Then under the **Scale** tab change **Minimum** to 1 and **Maximum** to 2.1. Similarly edit the Y-axis and under the **Scale** tab set **Minimum** to 1.02 and **Maximum** to 1.11.

4. To change the data markers first activate the Chart and select one of the data points. From the Menu Bar choose **Format – Selected Data Series...**, click the **Patterns** tab, and select a **Custom Marker** (Figure 10.5).

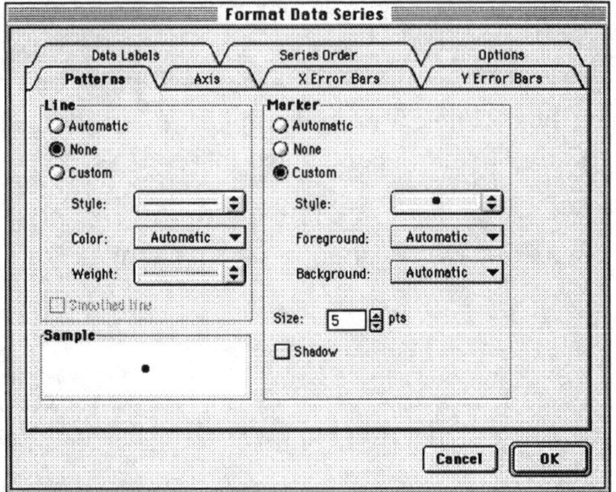

Figure 10.5: Formatting Markers

Changing Predicted Markers to a Line

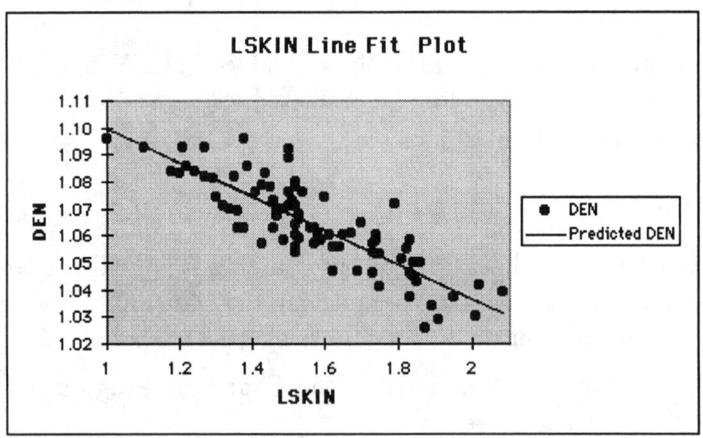

Figure 10.6: Regression Line and Enhanced Scatterplot

The other important enhancement to the markers is to change the predicted ones in the default scatterplot into a line. This is a useful enhancement even in other contexts and we separate its description here. The steps are as follows.

1. Activate the Chart and select one of the predicted markers. Choose **Format-Selected Data Series...** from the Menu Bar .

2. In the **Format Data Series** dialog box, click the **Patterns** tab. Select radio button **Custom** for **Line**, and then pick a Color. Finally, select **None**

for **Marker.** The markers disappear and are replaced by the regression line (Figure 10.6).

Residual Plot

The enhanced residual plot appears in Figure 10.7. The residuals show a random scatter about the X axis indicating that the straight-line fit is appropriate.

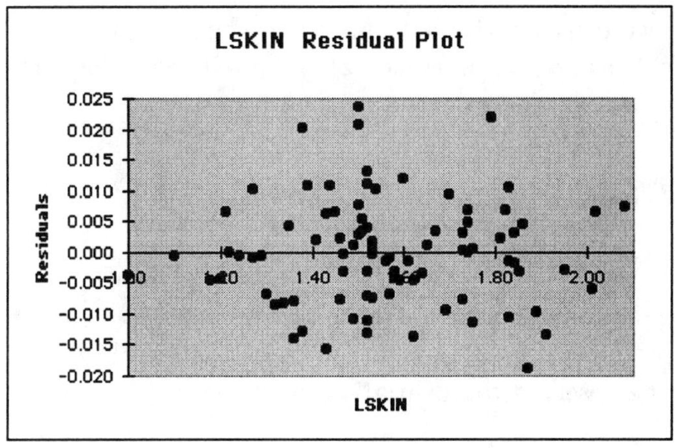

Figure 10.7: Residual Plot

	C	D	E	F
24	RESIDUAL OUTPUT			
25				
26	*Observation*	*Predicted DEN*	*Residuals*	*Standard Residuals*
27	1	1.082837953	0.010162047	1.196664361
28	2	1.064533444	-0.00153344	-0.18057562
29	3	1.071476534	0.006523466	0.76819163
30	4	1.067058204	-0.0110582	-1.302194137
31	5	1.067689394	0.005310606	0.625367386
32	6	1.067689394	0.003310606	0.389850996
33	7	1.068320584	0.007679416	0.904314182
34	8	1.060746305	-0.0137463	-1.618740007
35	9	1.068320584	0.020679416	2.435170714
36	10	1.052540835	0.000459165	0.054070411
37	11	1.072738913	-0.01573891	-1.853386038
38	12	1.048753696	0.002246304	0.264520751
39	13	1.062008684	0.011991316	1.412075676
40	14	1.068951774	0.001048226	0.123437225
41	15	1.081575573	-0.00057557	-0.0677784
42	16	1.067058204	-0.0030582	-0.360128579
43	17	1.047491316	-0.01049132	-1.235438399
44	18	1.063271064	-0.00327106	-0.385194628

Figure 10.8: Residual Output

Residual Output

Figure 10.8 shows a portion of the residual output with all predicted values, residuals and standardized residuals. With this at hand, other diagnostic scatterplots can be quickly obtained, in addition to the default output—for example, residuals versus x-variable.

10.2 More Detail about Simple Linear Regression

In this section we consider the remaining two elements of the output from the **Regression** tool, as well as inference about predicted values and use of Excel regression functions

ANOVA F and Regression Statistics

ANOVA Table

The ANOVA approach uses an F-test to determine whether a substantially better fit is obtained by the regression model than by a model with $\beta_1 = 0$. The ANOVA output breaks the observed total variation in the data

$$\text{SST} = \sum_{i=1}^{n}(y_i - \bar{y})^2$$

into two components, residual or error sum of squares

$$\text{SSE} = \sum_{i=1}^{n}(y_i - \hat{y}_i)^2$$

and a model sum of squares

$$\text{SSM} = \sum_{i=1}^{n}(\hat{y}_i - \bar{y})^2$$

connected by the identity

$$\text{SST} = \text{SSE} + \text{SSM}$$

The test criterion is based on how much smaller the residual sum of squares is under each fit and is based on the F ratio

$$F = \frac{\text{MSM}}{\text{MSE}}$$

where $\text{MSM} = \frac{\text{SSM}}{1}$ is the mean square for the model fit while $\text{MSE} = \frac{\text{SSE}}{n-2}$ is the mean square for error. Under the null hypothesis that $\beta_1 = 0$, this ratio has an F distribution with 1 degree of freedom in the numerator and $n - 2$ degrees of freedom in the denominator.

	C	D	E	F	G	H
12	ANOVA					
13		*df*	*SS*	*MS*	*F*	*Significance F*
14	Regression	1	0.01691	0.0169	231.894	1.2213E-26
15	Residual	90	0.00656	0.0001		
16	Total	91	0.02347			

Figure 10.9: ANOVA Output

All the above calculations appear in rows 13–16 in Figure 10.9. This is an advanced topic, and the reader is referred to the text for a more complete discussion. The only point we add here is that the F test is identical to the earlier two-sided test for $H_0 : \beta_1 = 0$ vs. $H_a : \beta_1 \neq 0$ and the F statistic 231.894 in cell G14 is always the square of the t stat $= -15.228$ appearing in cell F20 in Figure 10.3 in this context.

Regression Statistics

	C	D
5	*Regression Statistics*	
6	Multiple R	0.8488
7	R Square	0.7204
8	Adjusted R Square	0.7173
9	Standard Error	0.0085
10	Observations	92

Figure 10.10: Regression Statistics Output

This is the last component of the output from Figure 10.3 which we discuss briefly and isolate in Figure 10.10, which gives, for instance, the coefficient of determination $r^2 = 0.720$ and the standard error $s = 0.00854$, in addition to more advanced statistics such as the Adjusted R Square used in multiple regression.

Inference about Predictions

In the preceding section we drew inferences on the parameters β_0 and β_1 in the regression line $\mu_y = \beta_0 + \beta_1 x$. Here we examine the mean response and the prediction of a single outcome, both at a specified value x^* of the explanatory variable.

- To estimate the mean response, we use a confidence interval for the parameter μ_y based on the point estimate

  $$\hat{\mu_y} \equiv \hat{y} = b_0 + b_1 x$$

  A level C confidence interval for μ_y is given by

  $$\hat{y} \pm t^* \mathrm{SE}_{\hat{\mu}} \qquad (10.1)$$

where the standard error is given by

$$\text{SE}_{\hat{\mu}} = s\sqrt{\frac{1}{n} + \frac{(x^* - \bar{x})^2}{\sum_{i=1}^{n}(x_i - \bar{x})^2}}$$

- To predict a single observation at x^*, a level C prediction interval given by

$$\hat{y} \pm t^* \text{SE}_{\hat{y}} \tag{10.2}$$

where the appropriate standard error is now given by

$$\text{SE}_{\hat{y}} = s\sqrt{1 + \frac{1}{n} + \frac{(x^* - \bar{x})^2}{\sum_{i=1}^{n}(x_i - \bar{x})^2}}$$

In each case t^* is the upper $(1 - C)/2$ critical value of the Student t-distribution with $n - 2$ degrees of freedom.

Unfortunately Excel's Regression tool does not provide either of these intervals and (10.1) and (10.2) need to be constructed using Excel functions and appropriate cell references.

There are several ways to proceed. The Regression tool gives, inter alia, b_0, b_1, and s from which $\hat{y}$ and $\text{SE}_{\hat{\mu}}$ can be determined. We illustrate with Example 10.2. An alternative to the Regression tool uses Excel regression functions. This is the third method mentioned in Chapter 2 for regression analysis and is taken up in detail at the end of this section.

Example 10.2. (Based on Examples 10.8–10.14, pages 686–691, and Exercise 10.23, page 702 in the text.) Like many other businesses, technological advances and new methods have produced dramatic results in agriculture. Cells A3:B7 in Figure 10.11 give the data on the average yield in bushels per acre of U.S. corn for selected years. Here, year is the explanatory variable x, and yield is the response variable y. A scatterplot suggests that we can use linear regression to model the relationship between yield and time.

(a) Assuming that the linear pattern of increasing corn yield continues into the future, estimate the yield in the year 2006.

(b) Give a 95% prediction interval for the yield in the year 2006 which quantifies the uncertainty in this estimate.

(c) Give a 95% confidence interval for the mean yield in the year 1990.

Solution

1. Enter the four pairs of observations from the table in Example 10.8 on page 686 in the text into columns A and B of a workbook (Figure 10.11).

	A	B	C	D	E	K
1			*Confidence and Prediction Intervals*			
2						
3	Year	Yield	SUMMARY OUTPUT			
4	1966	73.1				
5	1976	88.0	*Regression Statistics*			
6	1986	119.4	Multiple R	0.976		
7	1996	127.1	R Square	0.952		
8			Adjusted R Square	0.928		
9			Standard Error	6.847		
10			Observations	4		
11						
12			ANOVA			
13				*df*	*SS*	
14			Regression	1	1870.18	
15			Residual	2	93.76	
16			Total	3	1963.94	
17						
18				*Coefficients*	*Standard Error*	
19			Intercept	-3729.35	606.60	
20			Year	1.93	0.31	
21						
22	Prediction Interval for a Future Observation					
23	Fit	150.25	=D19+D20*2006			
24	StDev Fit	10.83	=D9*SQRT(1+1/4 + (2006-AVERAGE(A4:A7))^2/(4*VARP(A4:A7)))			
25	PI lower limit	103.67	=B23-TINV(0.05,2)*B24			
26	PI upper limit	196.83	=B23+TINV(0.05,2)*B24			
27						
28	Confidence Interval for Mean Response					
29	Fit	119.31	=D19+D20*1990			
30	StDev Fit	4.39	=D9*SQRT(1/4 + (1990 -AVERAGE(A4:A7)^2/(4*VARP(A4:A7)))			
31	CI lower limit	100.40	=B29-TINV(0.05,2)*B30			
32	CI upper limit	138.22	=B28+TINV(0.05,2)*B30			

Figure 10.11: Confidence and Prediction Intervals

2. Following the instructions given in Example 10.1, invoke the **Regression** tool, completing the dialog box without checking the boxes for residuals, and output the results beginning in cell C3. Figure 10.11 also shows the output. The value of b_0 is in cell D19, b_1 is in D20, and s is in D9. The regression line is (also at top of page 688 in the text)

$$\hat{y} = -3729.35 + 1.93x$$

3. In a blank cell, say B23 in Figure 10.11, type the formula

$$= D19 + D20 * 2006$$

giving a predicted value $\hat{y} = 150.25$ (top of page 691 in the text). In cell B24 enter the formula

$$= D9 * \texttt{SQRT}(1 + 1/4 + (2006 - \texttt{AVERAGE}(A4 : A7))\hat{}2/(4 * \texttt{VARP}(A4 : A7)))$$

giving $\text{SE}_{\hat{y}} = 10.83$. In B25, enter the formula

$$= B23 - \texttt{TINV}(0.05, 2) * B24$$

and in B26, enter the formula

$$= B23 + \texttt{TINV}(0.05, 2) * B24$$

(For convenience we have shown these formulas on the workbook.) You can read the lower and upper 95% prediction interval $(103.67, 196.83)$ in B25:B26 (as well as in the middle of page 691 in the text).

4. Entirely analgous steps are used to derive a confidence interval for the mean yield in 1990, the only difference being in the formula for the standard error SE_{μ}. Again, all formulas are shown in Figure 10.11. The required 95% confidence interval is (100.40, 138.22).

Regression Functions

FORECAST

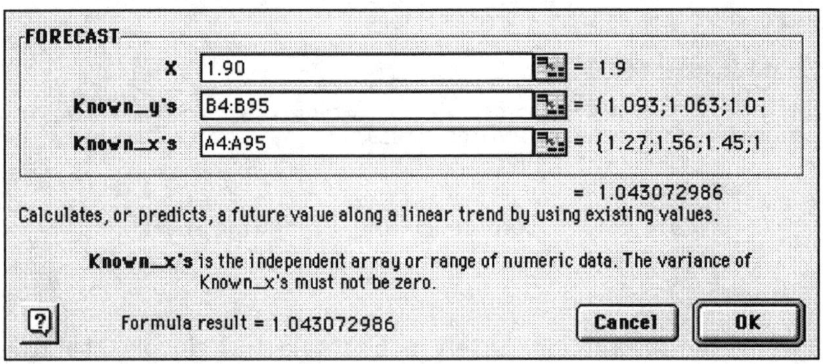

Figure 10.12: The FORECAST Dialog Box

A third method for regression analysis uses Excel functions. Suppose we wish to predict body density when the skinfold thickness explanatory variable is 1.90. Without having to derive the regression line and calculating

$$\hat{y} = 1.163 - 0.0631(1.90) = 1.043$$

we could use the FORECAST function, whose syntax is

$$\texttt{FORECAST}(x, \text{Known\_}y\text{'s}, \text{Known\_}x\text{'s}).$$

Here "Known_x's" refers to the set $\{x_i\}$ while "Known_y's" refers to the $\{y_i\}$. The function can be called from the **Formula Palette** or the **Function Wizard**, in which case you should select Statistical for Function category, FORECAST for Function name, and insert the required parameters in the dialog box to obtain the predicted value (Figure 10.12).

TREND

A more general method for obtaining predicted values is **TREND** with syntax

$$\texttt{TREND}(\text{Known\_}y\text{'s, Known\_}x\text{'s, New\_}x\text{'s, Const})$$

The parameter "New_x's" is the range of x-values for which predictions are desired. The parameter "Const" determines whether the regression line is forced through the origin ($\beta_0 = 0$). Use the value 1 for general β_0.

Example 10.3. (Example 10.2 continued.) Predict corn yields in the years 2000, 2001, 2002, 2003, and 2004.

D4 = {=TREND(B4:B7,A4:A7,C4:C8)}

	A	B	C	D	E
1			*Using the TREND Function*		
2					
3	Year	Yield	x	predicted	
4	1966	73.1	2000	138.646	
5	1976	88.0	2001	140.58	
6	1986	119.4	2002	142.514	
7	1996	127.1	2003	144.448	
8			2004	146.382	
9					

Figure 10.13: The **TREND** Function

Solution

1. Select a region of cells D4:D8 (Figure 10.13) for the output.
2. Click the **Formula Palette**, select the TREND function, and enter the values B4:B7, A4:A7, and C4:C8 into the argument fields "Known_y's," "Known_x's," and "New_x's," respectively. Leave the "Const" field blank. Click OK.
3. Click the mouse pointer in the **Formula Bar**, hold down the **Shift** and **Control** keys (either **Macintosh** or **Windows**), and press Enter to **Array-Enter** the formula. The formula will appear surrounded by braces in the Formula Bar, indicating that it has been array-entered, and the predicted values {138.648, 140.580, 142.514, 144.448, 146.382} will appear in the output range (Figure 10.13).

LINEST

The last regression function described here is LINEST, which returns the estimated least-squares line coefficients, their standard errors, r^2, s, the computed F statistic with its degrees of freedom, and the regression and error sums of squares. The syntax is

$$\texttt{LINEST}(x, \text{Known\_}y\text{'s, known\_}x\text{'s, Const, Stats})$$

C4 = {=LINEST(B4:B7,A4:A7,,TRUE)}

	A	B	C	D
1			*Using the LINEST Function*	
2				
3	Year	Yield		
4	1966	73.1	1.93400	-3729.35400
5	1976	88.0	0.30621	606.60338
6	1986	119.4	0.9523	6.8470
7	1996	127.1	39.892	2
8			1870.178	93.76200
9				
10			b_1	b_0
11			SE_b_1	SE_b_1
12			R Square	s
13			F	df
14			SSM	SSE

Figure 10.14: The `LINEST` Function

If the field "Stats" is set to false, then only the regression coefficients are output; true will produce an entire block of output. The default is false.

> **Example 10.4.** (Example 10.2 continued.) Use `LINEST` to perform a regression analysis of corn yield versus time for the data in Figure 10.11.

Solution

1. Select a block of cells C4:D8 with two columns and five rows. (For multiple regression you require a block in which the number of colunns equals the number of independent variables plus one.)

2. Click the **Formula Palette**, select the `LINEST` function, and enter the values B4:B7 and A4:A7 into the argument fields "Known_y's" and "Known_x's," respectively. Leave the "Const" field blank and type true in the "Stats" field. Click OK.

3. The formula = `LINEST`(B4:B7,A4:A7) will be visible in the Formula Bar, but only the value 1.93400 in cell C4 will appear in your sheet. With the block C4:D8 still selected, click the mouse pointer in the **Formula Bar**, as you did for the `TREND` function, hold down the **Shift** and **Control** keys, and press Enter to **Array-Enter** the formula. The formula will appear surrounded by braces in the Formula Bar, indicating that it has been array-entered, and the output will appear in C4:C8 (Figure 10.14)

In cells C10:D14 in Figure 10.14, we have described the contents of the values in C4:D8. These may be compared with the output from the **Regression** tool for this data set, shown in Figure 10.11. For instance, cell D8 in Figure 10.14 contains the value 93.76200. This represents the error sum of squares SSE, as seen from the corresponding description in D14, which is identical to the output in cell E15 of the **Regression** tool output in Figure 10.11.

Note: Both `TREND` and `LINEST` can be used with multiple regression.

Chapter 11

Multiple Regression

Multiple regression extends simple linear regression by fitting surfaces to data involving two or more explanatory variables $\{x_1, \ldots, x_p\}$ used to predict a response variable y. For example,

> We want to predict the college grade point average of newly admitted students. We have data on their high school grades in several subjects and their scores on the two parts of the Scholastic Aptitude Test (SAT). How well can we predict college grades from this information? Do high school grades or SAT scores predict college grades more accurately?

Multiple Linear Regression Model

There are n sets of observations

$$(x_{i1}, x_{i2}, \ldots, x_{ip}, y_i) \qquad 1 \leq i \leq n$$

where y_i is the response and $(x_{i1}, \ldots, x_{ip})$ and the values of the p explanatory variables measured on the ith subject.

As in Chapter 10, the model is

$$DATA = FIT + RESIDUAL$$

expressed mathematically as

$$y_i = \beta_0 + \beta_1 x_{i1} + \beta_2 x_{i2} + \cdots + \beta_p x_{ip} + \varepsilon_i$$

The FIT portion involving the $\{\beta_i\}$ is a linear model and expresses the population regression equation

$$\mu_y = \beta_0 + \beta_1 x_1 + \cdots + \beta_p x_p$$

which is the mean of the response variable for explanatory variables $(x_1, \ldots, x_p)$. The $RESIDUAL$ component represents the variation in the observations and the

$\{\varepsilon_i\}$ are assumed to be independent and identically distributed $N(0, \sigma)$ random variables.

The regression coefficient β_i measures how much the response changes if x_i changes one unit, keeping all other explanatory variables fixed.

The parameters are estimated by the Principle of Least-Squares described in Section 10.1, and the estimates of $\{\beta_1, \ldots, \beta_p\}$ are denoted by

$$\{b_1, \ldots, b_p\}$$

giving a predicted response of

$$\hat{y}_i = b_0 + b_1 x_{i1} + b_2 x_{i2} + \cdots + b_p x_{ip}$$

The ith residual is

$$e_i = y_i - \hat{y}_i$$

and is used to estimate the population variance using

$$s^2 = \frac{\sum_{i=1}^{n} e_i^2}{n - p - 1}$$

Excel uses the **Regression** tool for multiple regression. The data is entered into a workbook one row for each observation so that the columns correspond to the explanatory variables, followed by the response variable (all in adjacent columns).

As in Chapter 10, the output is separated into six regions: regression statistics, ANOVA table, statistics about parameters, residuals, scatterplots, and residual plots.

The statistics about the parameters allow you to

- construct confidence intervals of the form

  $$b_j \pm t^* \mathrm{SE}_{b_j}$$

  for the parameter β_j where SE_{b_j} is the standard error of b_j

- test the null hypothesis

  $$H_0 : \beta_j = 0$$

  using the test statistic

  $$t = \frac{b_j}{\mathrm{SE}_{b_j}}$$

The ANOVA table provides an F statistic for testing

$$H_0 : \beta_1 = \beta_2 = \cdots = \beta_p = 0$$
$$H_a : \text{ at least one of the } \beta_j \text{ is not } 0$$

and gives the estimate s^2.

The various plots can be used for diagnostic purposes and the regression statistics include the squared multiple correlation coefficient.

Case Study

Example 11.1. (Example 11.1, page 712 in the text; data also on the Student CD-ROM.) The Computer Science department of a large university was interested in understanding why a large proportion of their first-year students failed to graduate as computer science majors. An examination of records from the registrar indicated that most of the attrition occurred during the first three semesters. Therefore, they decided to study all first-year students entering their program in a particular year and to follow their progress for the first three semesters.

The variables studied included the grade point average after three semesters and a collection of variables that would be available as students entered their program. The purpose of the study was to attempt to predict success in the early university years. One measure of success was the cumulative grade point average (GPA) after three semesters. Among the explanatory variables recorded at the time the students enrolled in the university were high school grades in mathematics (HSM), science (HSS), and English (HSE), coded on a scale from 1 to 10.

Also recorded were SATM (SAT Math score) and SATV (SAT Verbal score). The first 20 cases in the full data set are shown in Figure 11.1 in columns A:G.

	A	B	C	D	E	F	G
1	*Multiple Regression Case Study*						
2							
3	Student	HSM	HSS	HSE	SATM	SATV	GPA
4	1	10	10	10	670	600	3.32
5	2	6	8	5	700	640	2.26
6	3	8	6	8	640	530	2.35
7	4	9	10	7	670	600	2.08
8	5	8	9	8	540	580	3.38
9	6	10	8	8	760	630	3.29
10	7	8	8	7	600	400	3.21
11	8	3	7	6	460	530	2.00
12	9	9	10	8	670	450	3.18
13	10	7	7	6	570	480	2.34
14	11	9	10	6	491	488	3.08
15	12	5	9	7	600	600	3.34
16	13	6	8	8	510	530	1.40
17	14	10	9	9	750	610	1.43
18	15	8	9	6	650	460	2.48
19	16	10	10	9	720	630	3.73
20	17	10	10	9	760	500	3.8
21	18	9	9	8	800	610	4.00
22	19	9	6	5	640	670	2.00
23	20	9	10	9	750	700	3.74

Figure 11.1: GPA Data Set

Carry out a multiple regression analysis on these data including only the first three explanatory variables HSM, SSS, HSE.

Preliminary Exploratory Data Analysis

In this complex setting with several explanatory variables, it is important to use some of the descriptive tools described in Chapters 1 and 2 as preliminary steps to first obtain numerical and graphical summaries. These are shown in Figures 11.2, 11.3, and 11.4 and are obtained as follows.

1. Referring to Figure 11.2, which is part of the same workbook as Figure 11.1, label the cells as shown in I3:N3 and H4:H7.

	H	I	J	K	L	M	N
3		HSM	HSS	HSE	SATM	SATV	GPA
4	mean	8.3214	8.0893	8.0938	595.29	504.55	2.6352
5	Std_Dev	1.6387	1.6997	1.5079	86.40	92.61	0.7794
6	minimum	2	3	3	300	285	0.12
7	maximum	10	10	10	800	760	4

Figure 11.2: Summary Statistics

2. Select H4 and enter = AVERAGE(B4:B227). In our workbook, B4:B227 is the range for all the HSM scores. Use whatever range is appropriate in your own workbook. In I5 enter = STDEV(B4:B227), in I6 enter = MIN(B4:B227), and in I7 enter = MAX(B4:B227).

3. Select I4:I7, click the fill handle in the lower right corner of I7, and fill to the range I4:N7.

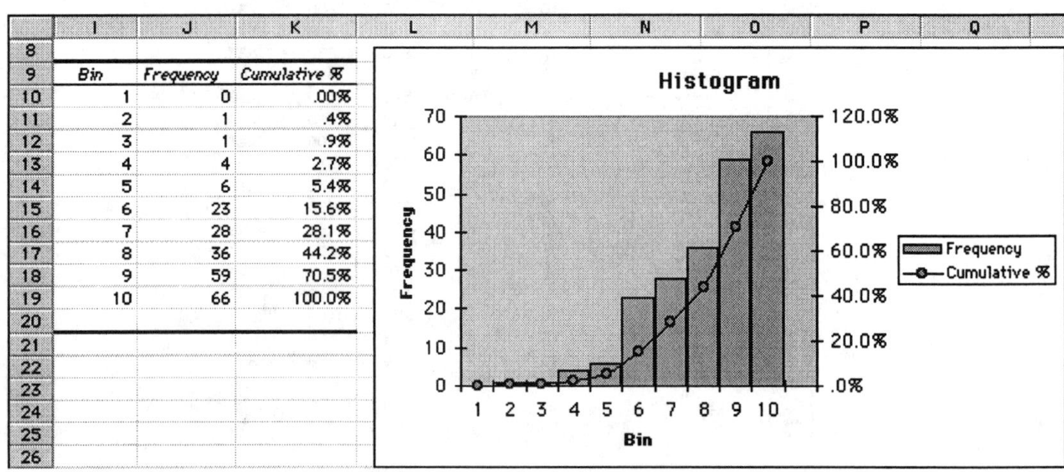

	I	J	K
9	Bin	Frequency	Cumulative %
10	1	0	.00%
11	2	1	.4%
12	3	1	.9%
13	4	4	2.7%
14	5	6	5.4%
15	6	23	15.6%
16	7	28	28.1%
17	8	36	44.2%
18	9	59	70.5%
19	10	66	100.0%

Figure 11.3: Frequency Table and Raw Histogram

Next, we choose **Tools – Data Analysis – Histogram** from the Menu Bar to produce frequency tables and histograms for the explanatory variables (refer to

Chapter 1). Figure 11.3 shows a table and a histogram for the HSM scores. The histogram distribution is strongly skewed to the right.

We can also examine pairwise correlations for all variables by choosing **Tools – Data Analysis – Correlation** from the Menu Bar, as in Section 2.2. The correlation matrix appears in Figure 11.4.

	H	I	J	K	L	M	N
28		*HSM*	*HSS*	*HSE*	*SATM*	*SATV*	*GPA*
29	HSM	1					
30	HSS	0.5757	1				
31	HSE	0.4469	0.5794	1			
32	SATM	0.4535	0.2405	0.1083	1		
33	SATV	0.2211	0.2617	0.2437	0.4639	1	
34	GPA	0.4365	0.3294	0.2890	0.2517	0.1145	1

Figure 11.4: Pairwise Correlations

Estimation of Parameters and ANOVA Table

Using the Regression Tool

Having explored some distributional aspects of the explanatory and response variables, we are ready to run a multiple regression. We use the same dialog box as in the one-variable regression that appears after we choose **Tools – Data Analysis – Regression** from the Menu Bar, so we will not repeat the explicit steps given in Chapter 10 but refer the reader to that chapter.

Excel Output

Figure 11.5 shows a portion of the multiple regression output for a model involving only the three predictor variables HSM, HSS, HSE. The ANOVA F statistic is given in cell W12 as 18.861. Cells T12:T13 show that the numerator degrees of freedom are DFM $= p = 3$ and the error (residual) degrees of freedom are DFE $= n - p - 1 = 224 - 3 - 1 = 220$. The P-value is 0.0000 in cell X12, and therefore we reject

$$H_0 : \beta_1 = \beta_2 = \beta_3 = 0$$

and conclude that at least one of the three regression coefficients is not 0. Cell T7 is the estimate for standard error

$$s = 0.700$$

and in cell T5 we see that the squared multiple correlation is

$$R^2 = 0.205$$

which indicates that 20.5% of the observed variation in the GPA scores $\{y_i\}$ is accounted for by the linear regression on these three high school scores. At the bottom of Figure 11.5, we find in cells T17:T20 the least-squares estimates for $\beta_0, \beta_1, \beta_2$, and β_3, respectively. Thus the fitted regression equation is

$$\widehat{\text{GPA}} = 0.590 + 0.169 \text{ HSM } + 0.034 \text{ HSS } + 0.045 \text{ HSE}$$

Compare with the same equation shown in the middle of page 723 in the text.

In the same portion of the output we can find confidence intervals (X17:Y20) and values of the t-test statistics (V17:V20) as well as P-values (W17:W20). We conclude that only HSM is a significant explanatory variable.

	S	T	U	V	W	X	Y
1	SUMMARY OUTPUT						
2							
3	*Regression Statistics*						
4	Multiple R	0.452					
5	R Square	0.205					
6	Adjusted R Square	0.194					
7	Standard Error	0.700					
8	Observations	224					
9							
10	ANOVA						
11		*df*	*SS*	*MS*	*F*	*Significance F*	
12	Regression	3	27.712	9.237	18.861	0.0000	
13	Residual	220	107.750	0.490			
14	Total	223	135.463				
15							
16		*Coefficients*	*Standard Error*	*t Stat*	*P-value*	*Lower 95%*	*Upper 95%*
17	Intercept	0.590	0.294	2.005	0.046	0.010	1.170
18	HSM	0.169	0.035	4.749	0.000	0.099	0.239
19	HSS	0.034	0.038	0.914	0.362	-0.040	0.108
20	HSE	0.045	0.039	1.166	0.245	-0.031	0.121

Figure 11.5: Regression Tool Output

Interpretation

Some of the numerical results appear contradictory. The value $R^2 = 0.205$ is small, indicating that the model does not explain much of the variation. Yet the small P-value for the test $H_0 : \beta_1 = 0$ against $H_a : \beta_1 \neq 0$ suggests that the HSM score is significant.

Moreover, if we ran simple regressions of GPA against each of HSS and HSE we would find that the corresponding explanatory variables taken individually are significant, while taken all three together, they are not.

A partial explanation for this may be found in the relatively high correlations between HSM and HSS, HSE as shown in Figure 11.4. This means that there is overlap in predictive information contained in these variables.

Residuals

Residual plots are useful diagnostic aids for checking not merely the linearity but, for instance, the assumption of constant variance in a model. Excel provides plots of the residuals against each of the explanatory variables. These plots appear to the right of the numerical summaries and are omitted here.

Chapter 12

One-Way Analysis of Variance

Analysis of variance (ANOVA) is a technique for comparing the means of two or more populations. It is a direct generalization of the two-sample t test described in Chapter 7. In particular, the F statistic in ANOVA, when there are two populations, is precisely the square of Student's t, and the ANOVA F test is then identical to the Student two-sample t procedure.

As the name suggests, ANOVA consists of separating the variability in a data set into two components and judging whether a fit that assumes equal population means is substantially better than a fit in which all means are assumed to be the same. This is achieved by comparing the residual variation following the model fit in both cases by a ratio called an F, which may be viewed as a signal-to-noise ratio. (See also the discussion in Chapter 10 on the ANOVA output from the **Regression** tool.)

The ANOVA Model

Suppose there are I normal populations labeled $1 \leq i \leq I$ and that independent random samples $\{x_{ij} : 1 \leq j \leq n_i\}$ of size $n_i \geq 1$, $1 \leq i \leq I$ are taken from each. The ANOVA model is

$$x_{ij} = \mu_i + \varepsilon_{ij} \qquad 1 \leq i \leq I,\ 1 \leq j \leq n_i$$

where $\{\varepsilon_{ij}\}$ are independent $N(0, \sigma)$ random variables. The populations thus have means μ_i and a common standard deviation σ, and we express this as

$$DATA = FIT + RESIDUAL$$

The one-way ANOVA significance test is

$$H_0 : \mu_1 = \mu_2 = \cdots = \mu_I$$
$$H_a : \text{ not all of the } \mu_i \text{ are equal}$$

Under H_0 we estimate the common value μ with the overall mean

$$\bar{x} = \sum_{i=1}^{I} \sum_{j=1}^{n_i} x_{ij}$$

What is left over $x_{ij} - \bar{x}$ is called the residual under H_0, and then the total residual variation in the data is given by

$$\text{SST} = \sum_{i=1}^{I} \sum_{j=1}^{n_i} (x_{ij} - \bar{x})^2$$

called the **total sum of squares.**

If we do not assume H_0 is true, then we should estimate each individual μ_i by the corresponding sample mean $\bar{x}_i$. The remainder $x_{ij} - \bar{x}_i$ is then the residual under an unrestricted model, and the total residual variation in the data is given by

$$\text{SSE} = \sum_{i=1}^{I} \sum_{j=1}^{n_i} (x_{ij} - \bar{x}_i)^2$$

called the **error** (or within groups) **sum of squares.** Intuitively, a better fit always occurs with the unrestricted model. We can show with a little algebra that the difference SSG = SST − SSE is positive and can also be expressed as

$$\text{SSG} = \sum_{i=1}^{I} \sum_{j=1}^{n_i} (\bar{x}_i - \bar{x})^2$$

called the **between groups sum of squares.** Remarkably, it turns out that

$$\text{SST} = \text{SSG} + \text{SSE}$$

which is the key to the partition of the variation.

The magnitude of SSG measures the improvement in the fit as measured by the residual sum of squares. In order to reject H_0, the improvement must be significantly beyond what might be expected due to chance, and one is led to consider the ratio $\frac{\text{SSG}}{\text{SSE}}$. Define the mean squares

$$\text{MSG} = \frac{\text{SSG}}{I - 1} \qquad \text{MSE} = \frac{\text{SSE}}{N - I}$$

where $N = \sum_{i=1}^{I} n_i$, and form the ratio

$$F = \frac{\text{MSG}}{\text{MSE}}$$

known to have an F distribution with $I - 1$ degrees of freedom for the numerator and $N - I$ degrees of freedom for the denominator (denoted by $F(I - 1, N - I)$).

The decision rule is

$$\text{Reject } H_0 \text{ at level } \alpha \text{ if } F > F^*$$

where F^* is the upper α critical value of an $F(I-1, N-I)$ distribution, that is, F^* satisfies

$$P\left(F(I-1, N-I) > F^*\right) = \alpha$$

We observe that MSE can also be expressed as

$$\text{SSE} = \sum_{i=1}^{I} (n_i - 1)s_i^2$$

where $\{s_i\}$ are the sample variances and, consequently, MSE is a pooled sample variance

$$s_p^2 \equiv \text{MSE} = \frac{\sum_{i=1}^{I}(n_i - 1)s_i^2}{\sum_{i=1}^{I}(n_i - 1)}$$

and therefore an unbiased estimate of σ^2.

Finally, we define the ANOVA coefficient of determination

$$R^2 = \frac{\text{SSG}}{\text{SST}}$$

as the fraction of the total variance "explained by model H_0."

Testing Hypotheses in a One-Way ANOVA

We illustrate the implementation of the preceding discussion with the following worked exercise.

> **Example 12.1.** (Exercise 12.11, page 782 in the text.) The presence of harmful insects in farm fields is detected by erecting boards covered with a sticky material and then examining the insects trapped on the board. To investigate which colors are most attractive to cereal leaf beetles, researchers placed six boards of each of four colors in a field of oats. Table 12.1 gives data on the number of cereal leaf beetles trapped. (Modified from M. C. Wilson and R. E. Shade, "Relative attractiveness of various luminescent colors to the cereal leaf beetle and the meadow spittlebug," *Journal of Economic Entomology*, **60** (1967), 578–580.)
> (a) Make a table of means and standard deviations for the four colors, and plot the data and the means.
> (b) State H_0 and H_a for an ANOVA on these data, and explain in words what ANOVA tests in this setting.
> (c) Using Excel, run the ANOVA. What are the F statistic and its P-value? State the values of s_p and R^2. What do you conclude?

Table 12.1: Luminescent Colors and Insect Attractiveness

Color	Insects trapped					
Lemon yellow	45	59	48	46	38	47
White	21	12	14	17	13	17
Green	37	32	15	25	39	41
Blue	16	11	20	21	14	7

Plotting the Data and the Sample Means

The first step in ANOVA is usually exploratory, where the data and means are plotted. Such a plot will help to visually confirm or dispel the assumption of equal variances and to indicate possible outliers or skewness in the data that might call into question use of this technique.

The step is easily carried out in Excel using the **ChartWizard**, which produces side-by-side displays of the samples $\{x_{ij}\}$, their sample means $\{\bar{x}_i\}$, and the overall mean $\bar{x}$, a good preliminary display of ANOVA data.

Means and Standard Deviations

The mean and standard deviations are evaluated with the Excel function `AVERAGE` and `STDEV`. Refer to Figure 12.1 throughout.

1. Enter the data and labels in B1:E7 of a workbook.

2. Enter the label "mean" in A8 and then the formula = `AVERAGE`(B2:B7) in B8. The value 47.17 appears, which is the sample mean of the "Yellow" observations. Select cell B8 and, using the fill handle, drag to E8, filling the cells with the means for the other colors.

3. Enter the label "stdev" in A9 and the formula = `STDEV`(B2:B7) in B9. Then select B9 and drag the fill handle to E9. Now the standard deviations of all the samples appear.

Plotting

The workbook needs to be prepared for the **ChartWizard** by relocating the data, coding the samples, relocating the sample means, and entering the overall mean. Complete columns G, H, I, J as indicated.

Since the sample means have already been calculated in B8:E8, an efficient way to relocate them to I26:I29 is to select B8:E8, choose **Edit** – **Copy** from the Menu Bar, then select cell I26 and choose **Edit** – **Paste Special** from the Menu Bar. Complete the **Paste Special** dialog box as in Figure 12.2. The radio button for **values** needs to be selected, because the contents of B8:E8 are formulas, not

	A	B	C	D	E	F	G	H	I	J
1		Yellow	White	Green	Blue		Color	Insects	Means	Overall
2		45	21	37	16		1	45		
3		59	12	32	11		1	59		
4		48	14	15	20		1	48		
5		46	17	25	21		1	46		
6		38	13	38	14		1	38		
7		47	17	41	7		1	47		
8	mean	47.17	15.67	31.33	14.83		2	21		
9	stdev	6.795	3.327	9.771	5.345		2	12		
10							2	14		
11							2	17		
12							2	13		
13							2	17		
14							3	37		
15							3	32		
16							3	15		
17							3	25		
18							3	38		
19							3	41		
20							4	16		
21							4	11		
22							4	20		
23							4	21		
24							4	14		
25							4	7		
26							1		47.17	27.25
27							2		15.67	27.25
28							3		31.33	27.25
29							4		14.83	27.25

Figure 12.1: Preparing the Data for the One-Way ANOVA Tool

values, and a straight **Edit** – **Copy** will change the relative cell references. The check box **Transpose** merely converts a row selection into a column. Finally, the common value 27.25 in J26:J29 is the overall mean obtained from the Excel formula = `AVERAGE`(B2:E7).

Users of Excel 5/95

Begin as always by clicking the **ChartWizard** button.

- In Step 1 of 5 enter G1:J29 for the range. This will produce a simultaneous scatterplot with all three variables—data, sample means, overall mean—plotted on the y-axis (with different markers) against the colors (column G) on the x-axis.

- In Step 2 select **XY (Scatter)**.

- In Step 3 select Format **1**.

- In Step 4 use **Columns** for Data Series in, First "1" Column for X Data, and First "1" Row for Legend Text.

- In Step 5 select **Yes** for Add a Legend?, type "Plot of the Data and the Means" for the Chart Title, type "Color" for Axis Title Category (X), and type "Insects Trapped" for Axis Title Value (Y). Click Finish.

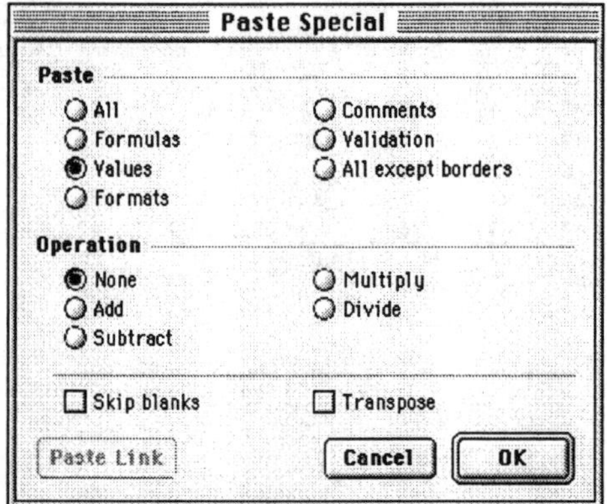

Figure 12.2: Paste Special

Users of Excel 97/98

- In Step 1 select **XY (Scatter)** for Chart Type and the upper left Chart sub-type **Scatter**.

- In Step 2 under under the **Data Range** tab enter G1:J29 (which will produce a simultaneous scatterplot with all three variables – data, sample means, and overall mean – plotted on the y-axis (with different markers) against the colors (column G) on the x-axis, and select the Series radio button for **Columns**.

- In Step 3 under the **Titles** tab, type "Plot of the Data and the Means" as the Chart Title, "Color" as Value (X) Axis, and "Insects Trapped" as Value (Y) Axis. Under the **Axes** tab, both check boxes should be selected. Under the **Gridlines** tab clear all check boxes. Under the **Legend** tab check the **Show legend** and locate it to the right. Finally, under the **Data Labels** tab select the radio button **None**.

- In Step 4 embed the graph in the current workbook by selecting the radio button **As object in**. Click Finish.

A scatterplot like the one shown in Figure 12.3 appears with distinct markers representing the individual observations, the sample means, and the overall mean. Editing will enhance the usefulness of this plot.

Editing the Plot

We edit Figure 12.3 to more clearly emphasize the observations, the sample means, and the overall mean. The result resembles side-by-side boxplots.

1. Activate the chart for editing.

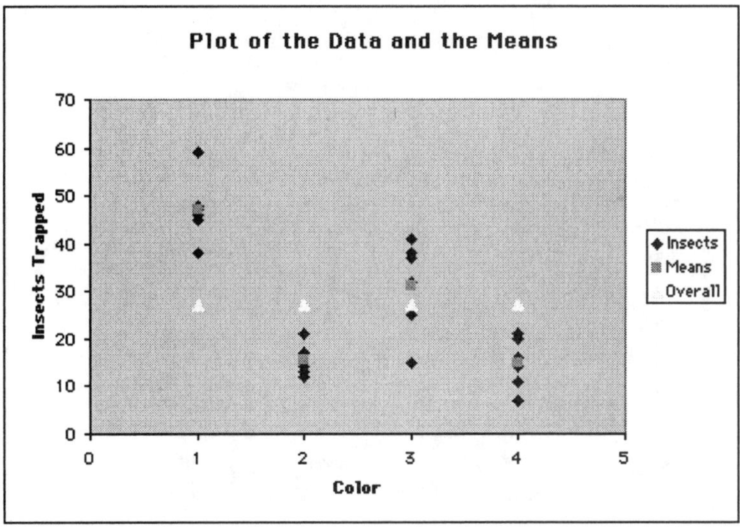

Figure 12.3: Default Plot—Data and Means

2. Click one of the markers representing a sample mean (the entire series will then appear highlighted). From the Menu Bar, choose **Format – Selected Data Series....** In the dialog box under the **Patterns** tab, select radio buttons **Automatic** for **Line** and **None** for **Marker**. Click OK. The markers for sample means are now replaced by a connecting polygonal line.

3. Repeat this step after selecting a marker for the overall mean. The four markers are replaced by a horizontal line.

4. Click a marker for the observations and change the Style of the marker to a square.

5. Select the horizontal axis, and from the Menu Bar choose **Format – Selected Axis....** In the dialog box click the **Scale** tab. Then type "1" for **Minimum**, "4" for **Maximum**, "1" for **Major unit** and "1" for **Minor unit**. Then click the **Patterns** tab, click **None** for **Tick Mark Labels**, and click **None** for **Major** and **Minor** Tick Mark Type. Click OK.

6. Activate the **Plot Area** and change the background color to white using the color swatch. Click OK.

7. Type the word "Yellow" in the **Formula Bar** and press enter. The word "Yellow" appears in a grey shaded text rectangle. With the cursor, drag it to the location shown in Figure 12.4. Click outside the text rectangle to deselect it. Now repeat for the other three labels "White," "Green," and "Blue."

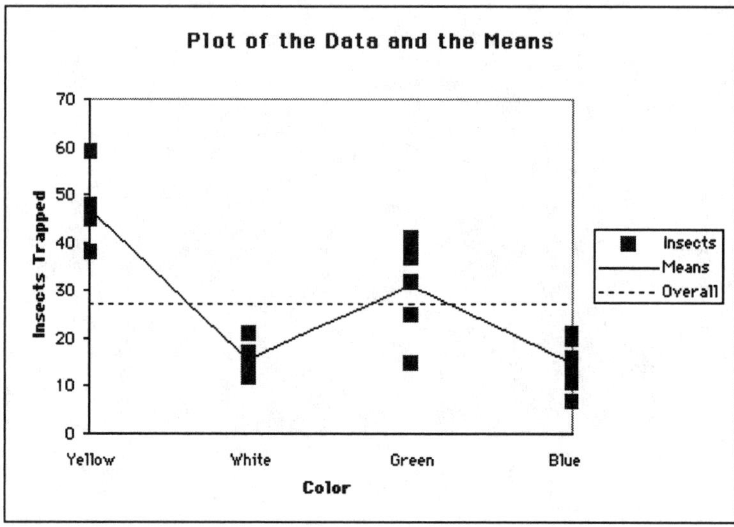

Figure 12.4: Enhanced Plot—Data and Means

The result of this enhancement is Figure 12.4, which immediately shows the dominant features of the data set much more aggressively than the numerical counterpart in cells A1:E9 of Figure 12.1.

Using the One-Way ANOVA Tool

1. From the Menu Bar choose **Tools** – **Data Analysis** and select **Anova: Single Factor** from the tools listed in the **Data Analysis** dialog box. Click OK to display the Anova: Single Factor dialog box (Figure 12.5).

2. Type the cell references B1:E7 (or point and drag on the workbook) for **Input range**. Select the radio button Grouped By: **Columns**. Check the box **Labels in first row** and use "0.05" for Alpha. Enter K1 for **Output range**. Click OK.

Excel Output—ANOVA Table

The output from the **Anova: Single Factor** tool appears in Figure 12.6, from which we can read off all the variables described earlier in this chapter. The first half of the output provides summary statistics for the four samples. Had we not desired a plot we would have read the sample means and standard deviations directly from this output.

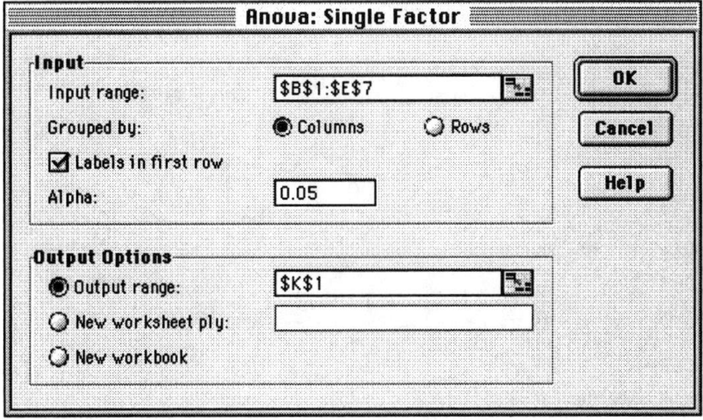

Figure 12.5: Single-Factor ANOVA Dialog Box

	K	L	M	N	O	P	Q
1	Anova: Single Factor						
2							
3	SUMMARY						
4	*Groups*	*Count*	*Sum*	*Average*	*Variance*		
5	Yellow	6	283	47.167	46.167		
6	White	6	94	15.667	11.067		
7	Green	6	188	31.333	95.467		
8	Blue	6	89	14.833	28.567		
9							
10							
11	ANOVA						
12	*Source of Variation*	*SS*	*df*	*MS*	*F*	*P-value*	*F crit*
13	Between Groups	4210.167	3	1403.389	30.97	1.03E-07	3.10E+00
14	Within Groups	906.333	20	45.317			
15							
16	Total	5116.5	23				

Figure 12.6: Single-Factor ANOVA Output

Source of Variation

Between Groups sum of squares refers to

$$\text{SSG} = 4210.167$$

Within Groups (error) sum of squares is

$$\text{SSE} = 906.333$$

The Total sum of squares is

$$\text{SST} = 5116.5$$

Further, we read the degrees of freedom (*df*)

$$\text{DFG} = 3 \quad \text{DFE} = 20 \quad \text{DFT} = 23$$

Consequently, the mean squares (*MS*) are

$$\begin{aligned} \text{MSG} &= \frac{4210.167}{3} = 1403.389 \\ \text{MSE} &= \frac{906.333}{20} = 45.317 \end{aligned}$$

and the calculated F statistic is

$$F = \frac{\text{MSG}}{\text{MSE}} = \frac{1403.389}{45.317} = 30.97$$

The 5% critical F value (F *crit*) is

$$F^* = 3.098$$

The P-value is 0.00000.

We conclude that for the significance test

$$\begin{aligned} &H_0 : \mu_1 = \mu_2 = \mu_3 = \mu_4 \\ &H_a : \text{ not all of the } \mu_i \text{ are equal} \end{aligned}$$

we reject H_0 at the 5% level (in fact at *any* reasonable level) in view of the small P-value.

We observe that the pooled estimate of variance is

$$s_p^2 = \text{ MSE } = 45.17$$

which we note is also (because of the equal sample sizes) the average of the individual sample variances.

Finally, although the coefficient of determination is not provided, we can do the arithmetic to derive

$$R^2 = \frac{\text{SSG}}{\text{SST}} = \frac{4210.167}{5116.5} = 0.823$$

Chapter 13

Two-Way Analysis of Variance

In a one-way ANOVA, independent samples are taken from I populations each differing with respect to one categorical variable, the population mean. We can view this variable as representing the levels of a particular factor as discussed in Chapter 3 of the text. In this setting the one-way ANOVA then describes the analysis of a **completely randomized design**.

Another design discussed in Chapter 3 was a **randomized block design**. We recall that a block is a group of experimental units similar in some way that is expected to influence the response. The paired two-sample t procedure is an example of a block design analysis.

Two-way ANOVA may then be viewed as a generalization of the paired t test when there are more than two populations. The approach parallels one-way ANOVA by partitioning the total variation into components that can be interpreted as representing the contributions of factor effects or block effects.

However, the analysis of a randomized block design also applies when there are two or more factors whose effect is of interest, where one is not necessarily a blocking variable and where there are multiple observations (replications) for each combination of factor levels. That is the model we consider here.

The Two-Way ANOVA Model

The data comprise samples of size n_{ij} for $I \times J$ treatments representing the combinations of I levels of factor A and J levels of factor B. The data are represented by $\{x_{ijk} : 1 \leq i \leq I,\ 1 \leq j \leq J,\ 1 \leq k \leq n_{ij}\}$, where x_{ijk} represents the kth observation of treatment combination (i, j). The model states

$$x_{ijk} = \mu_{ij} + \varepsilon_{ijk} \quad 1 \leq I,\ 1 \leq j \leq J,\ 1 \leq k \leq n_{ij}$$

where $\{\varepsilon_{ijk}\}$ are independent $N(0, \sigma)$ random variables. Note the assumption of common variance, as with the one-way model.

First, we view the model as a one-way ANOVA for $I \times J$ populations (treatments).

$$DATA = FIT + RESIDUAL$$

In particular, we can estimate an overall mean μ by

$$\bar{x} = \frac{1}{N}\sum_{i=1}^{I}\sum_{j=1}^{J}\sum_{k=1}^{n_{ij}} x_{ijk}$$

and then produce a corresponding **total sum of squares** where N is the overall total number of observations.

$$\text{SST} = \sum_{i=1}^{I}\sum_{j=1}^{J}\sum_{k=1}^{n_{ij}}(x_{ijk} - \bar{x})^2$$

Likewise, we can estimate the residual or error variation using the sample means for each (i, j) combination of $I \times J$ treatments

$$\bar{x}_{ij} = \frac{1}{n_{ij}}\sum_{k=1}^{n_{ij}} x_{ijk}$$

and then we have the error (residual) sum of squares

$$\text{SSE} = \sum_{i=1}^{I}\sum_{j=1}^{J}\sum_{k=1}^{n_{ij}}(x_{ijk} - \bar{x}_{ij})^2$$

As with the one-way ANOVA this formula can be reexpressed as

$$\text{SSE} = \sum_{i=1}^{I}\sum_{j=1}^{J}(n_{ij} - 1)s_{ij}^2$$

where s_{ij}^2 is the sample variance for the (i, j) combination, which leads to a pooled estimate for the common variance σ^2

$$s_p^2 = \frac{\text{SSE}}{\sum_{i=1}^{I}\sum_{j=1}^{J}(n_{ij} - 1)} = \frac{\text{SSE}}{N - IJ}$$

The corresponding between groups sum of squares is here denoted as (with a change in notation, SSM replacing SSG)

$$\text{SSM} = \sum_{i=1}^{I}\sum_{j=1}^{J}\sum_{k=1}^{n_{ij}}(\bar{x}_{ij} - \bar{x})^2$$

To this stage, the analysis is for a one-way ANOVA. But because of the way the data were collected (the design), it is then possible to partition SSM further into additional sums of squares that can be identified as arising from model parameters. We therefore prescribe μ_{ij} using a **linear model**

$$\mu_{ij} = \mu + \alpha_i + \beta_j + \gamma_{ij}$$

where μ is an overall mean, α_i represents an effect due to level i of factor A, β_j represents an effect due to level j of factor B (both of these are called **main effects**), and γ_{ij} represents an **interaction** effect between factor A and factor B. We say that two factors interact if the difference in mean response for two levels of one factor is not constant across levels of the second factor.

Excel provides two versions of **two-way ANOVA** in the **Analysis ToolPak**. The first is where $n_{ij} = 1$ for all (i, j) (in which case γ_{ij} cannot be estimated). The other is where $n_{ij} \equiv n \geq 2$, so there is replication of observations at each (i, j) level but the number of replications is the same. This is called a **balanced** design. With unequal sample sizes, the ANOVA formulas become more complex and the factor effect components (sums of squares) are no longer orthogonal (they don't add up). For this reason, we now limit the discussion to balanced designs with the same number of observations $n_{ij} \equiv n$ per treatment.

The parameters in the representation of μ_{ij}

$$\mu_{ij} = \mu + \alpha_i + \beta_j + \gamma_{ij}$$

are not uniquely determined, because we can add and subtract constants on the right-hand side, changing the parameters but maintaining equality. It is necessary to impose constraints for uniqueness, namely,

$$\sum_{i=1}^{I} \alpha_i = 0, \quad \sum_{j=1}^{J} \beta_j = 0, \quad \sum_{i=1}^{I} \gamma_{ij} = 0, \quad \sum_{j=1}^{I} \gamma_{ij} = 0$$

We can then **interpret** α_i as the contribution or deviation of level i of factor A from a baseline (of 0 in view of $\sum_{i=1}^{I} \alpha_i = 0$). Likewise, β_j is the contribution of level j of factor B and γ_{ij} is any possible interaction of combination (i, j).

If we fix attention on level i of factor A and average over all levels of factor B, we can estimate $\mu + \alpha_i$ using the sample mean of all observations with the same value i,

$$\frac{1}{Jn} \sum_{j=1}^{J} \sum_{k=1}^{n} x_{ijk} = \bar{x}_{i\bullet}$$

We can also estimate $\mu + \beta_j$ using

$$\frac{1}{In} \sum_{i=1}^{I} \sum_{k=1}^{n} x_{ijk} = \bar{x}_{\bullet j}$$

We then take differences and find the natural estimates (recalling that $\bar{x}$ is the overall mean):

$$\begin{aligned} \hat{\alpha}_i &\equiv \bar{x}_{i\bullet} - \bar{x} && \text{estimates } \alpha_i \\ \hat{\beta}_i &\equiv \bar{x}_{\bullet j} - \bar{x} && \text{estimates } \beta_j \\ \hat{\gamma}_{ij} &\equiv \bar{x}_{ij} - \bar{x}_{i\bullet} - \bar{x}_{\bullet j} + \bar{x} && \text{estimates } \gamma_{ij} \end{aligned}$$

Then a minor miracle occurs. Each of these estimates contributes to a marvelous sum of squares decomposition; namely, if we define

$$\begin{aligned} \text{SSA} &= \sum_{i=1}^{I} \sum_{j=1}^{J} \sum_{k=1}^{n} \hat{\alpha}_i^2 \\ \text{SSB} &= \sum_{i=1}^{I} \sum_{j=1}^{J} \sum_{k=1}^{n} \hat{\beta}_j^2 \\ \text{SSAB} &= \sum_{i=1}^{I} \sum_{j=1}^{J} \sum_{k=1}^{n} \hat{\gamma}_{ij}^2 \end{aligned}$$

then

$$\text{SSM} = \text{SSA} + \text{SSB} + \text{SSAB}$$

Each term on the right carries a degrees of freedom

$$\begin{aligned} \text{DFA} &= I - 1 \\ \text{DFB} &= J - 1 \\ \text{DFAB} &= (I-1)(J-1) \end{aligned}$$

giving corresponding mean squares and F ratios as in one-way ANOVA. For instance,

$$\begin{aligned} \text{MSA} &= \frac{\text{SSA}}{I-1} \\ F &= \frac{\text{MSA}}{\text{MSE}} \end{aligned}$$

and this particular F ratio is used to test

$$H_{0A}: \alpha_i = 0, \quad 1 \le i \le I$$
$$H_{aA}: \text{ at least two } \alpha_i \text{ are not zero}$$

The decision rule is that if

$$F = \frac{\text{MSA}}{\text{MSE}} \text{ exceeds the critical value } F^*(I-1, N-IJ)$$

then we reject H_{0A} and conclude that there is a difference among the means for the levels of factor A; that is, factor A is significant. An analogous analysis is carried out for factor B and then for interaction if neither factor A nor factor B is judged significant.

The ANOVA Tool

While the theory just described may seem intimidating, the practical implementation of the procedure could not be simpler once the data are properly recorded in the workbook. Excel outputs summary statistics, sums of squares, mean squares, F ratios, critical F values, and P-values in an extensive ANOVA table.

Table 13.1: Modulus of Elasticity of Wood Flakes

	Size of flakes	
Species	$S1$	$S2$
Aspen	308	278
	428	398
	426	331
Birch	214	534
	433	512
	231	320
Maple	272	158
	376	503
	322	220

Example 13.1. (Exercise 13.11, page 818 in the text.) A large research project studied the physical properties of wood materials constructed by bonding together small flakes of wood. Different species of trees were used, and the flakes were made in different sizes. One of the physical properties measured was the tension modulus of elasticity in the direction perpendicular to the alignment of the flakes, in pounds per square inch (psi). Some of the data are given in Table 13.1. The sizes of the flakes are $S1 = 0.015$ inches by 2 inches and $S2 = 0.025$ inches by 2 inches. (Data provided by Michael Hunt and Bob Lattanzi of the Purdue University Forestry Department.)
(a) Compute means and standard deviations for the three observations in each species–size group. Find the marginal mean for each species and for each size of flakes. Display the means and marginal means in a table.
(b) Plot the means of the six groups. Put species on the x-axis and modulus of elasticity on the y-axis. For each size connect the three points corresponding to the different species. Describe the patterns you see. Do the species appear to be different? What about the sizes? Does there appear to be an interaction?
(c) Run a two-way ANOVA on these data. Summarize the results of

the significance tests. What do these results say about the impressions that you described in part (b) of this example?

Using the Two-Way ANOVA Tool

We refer to species as Factor A (3 levels: Aspen, Birch, Maple) and size of flakes as Factor B (2 levels: $S1$, $S2$). There are 3 replications per treatment so $n = 3$ for a total of $N = I \times J \times n = 3 \times 2 \times 3 = 18$ observations.

	A	B	C
1	Two-Way ANOVA		
2			
3	Species	Size of Flakes	
4		S1	S2
5	Aspen	308	278
6		428	398
7		426	331
8	Birch	214	534
9		433	512
10		231	320
11	Maple	272	158
12		376	503
13		322	220

Figure 13.1: Preparing Data for Two-Way ANOVA Tool

1. Enter the data exactly as shown in Figure 13.1. Excel will balk if the data are not laid out correctly.

2. From the Menu Bar, choose **Tools – Data Analysis – Anova: Two-Factor With Replication** and complete the dialog box with entries as shown in Figure 13.2. The output appears in cells D1:J36.

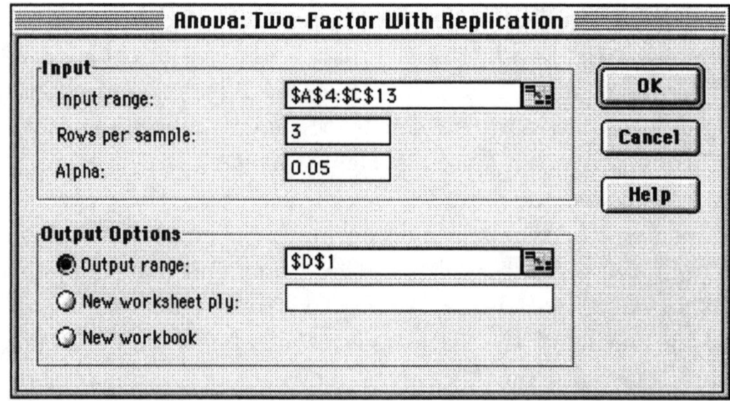

Figure 13.2: Two-Way ANOVA Dialog Box

	D	E	F	G	H	I	J
1	Anova: Two-Factor With Replication						
2							
3	SUMMARY	S1	S2	Total			
4	*Aspen*						
5	Count	3	3	6			
6	Sum	1162	1007	2169			
7	Average	387.33	335.67	361.5			
8	Variance	4721.33	3616.33	4135.9			
9							
10	*Birch*						
11	Count	3	3	6			
12	Sum	878	1366	2244			
13	Average	292.67	455.33	374			
14	Variance	14842.33	13857.33	19418			
15							
16	*Maple*						
17	Count	3	3	6			
18	Sum	970	881	1851			
19	Average	323.33	293.67	308.5			
20	Variance	2705.33	33826.33	14876.7			
21							
22	*Total*						
23	Count	9	9				
24	Sum	3010	3254				
25	Average	334.44	361.56				
26	Variance	7317.03	18102.53				
27							
28							
29	ANOVA						
30	*Source of Variation*	*SS*	*df*	*MS*	*F*	*P-value*	*F crit*
31	Sample	14511	2	7255.5	0.592	0.569	3.885
32	Columns	3307.556	1	3307.556	0.270	0.613	4.747
33	Interaction	41707.444	2	20853.722	1.701	0.224	3.885
34	Within	147138	12	12261.5			
35							
36	Total	206664	17				

Figure 13.3: Two-Way ANOVA Output

Excel Output—Two-Way ANOVA

There are two components to the output Figure 13.3. The top portion provides summary statistics such as sample means and variances, which are useful in plotting. The lower half is the ANOVA table.

Summary

Referring to Figure 13.3, we can read in cells E5:E8 that there were 3 observations for the treatment (Aspen $\times$ $S1$), the average was 387.33 (cell E7), and the sample variance was 4721.33 (cell E8). In the notation used earlier, this indicates that

$$\bar{x}_{11} = 387.33$$
$$s^2_{11} = 4721.33$$

We can read off corresponding means and variances for all treatment combinations.

The column on the right labeled "Total" (G3:G20) provides summaries for each level of factor A (summed or averaged over all levels of factor B). Thus, for instance, from G11:G14 we see that the average for Birch (cell G13) is

$$\bar{x}_{2\bullet} = 374$$

The corresponding summaries for factor B are in the row block headed by the label Total in D22:D26. For example, the average for $S2$ (cell F25) is

$$\bar{x}_{\bullet 2} = 361.56$$

The ANOVA Table for Two-Way ANOVA

ANOVA Table

There are three separate possible significance tests: Factor A, Factor B, and Interaction. Block D29:J36 presents the ANOVA table listing the four sources of variation (Sample = Factor A, Column = Factor B, Interaction, Within = Error) and the Total in D31:D36. The next columns have the corresponding sums of squares, degrees of freedom, mean square, computed F, P-value, and critical F^* values. We can immediately read off the conclusions.

First, there does not appear to be any interaction, so we then examine the two factors. For Factor A, since F does not exceed $F^*(2, 12)$ we conclude that the species do not appear to be different. For Factor B, we conclude that the size of flakes is not a significant factor.

Profile (Interaction) Plot

A profile plot is a simple graphical diagnostic tool for displaying the numerical summaries in the Excel output. It is handy for seeing possible interaction visually. A profile plot is a graph of the sample means of the levels of one factor (on the y-axis) plotted against the levels of the other factor on the x-axis. The following steps show how to use the **ChartWizard** and the summary output to produce a profile plot. We will plot species along the horizontal axis.

1. Enter the data as in the top half of Figure 13.4. (The bottom half is used for reversing the roles of the factors when plotting.) The means to be entered can be copied from the summary of the Excel output using the **Edit – Paste Special** to transpose the row output of means in Figure 13.3 to columns.

2. Click the **ChartWizard** button.

3. **Users of Excel 5/95**

    - In Step 1 enter the range K3:O6, in Step 2 select **Line** chart, in Step 3 select Format **1**, in Step 4 select Data Series in **Rows**, Row "1" for

	K	L	M	N	O
1	Profile Plot				
2	of Means		*Species*		
3			Aspen	Birch	Maple
4	S1		387.33	292.67	323.33
5	S2		335.67	455.33	293.67
6					
7					
8	Profile Plot				
9	of Means		*Size of Flakes*		
10			S1	S2	
11	Aspen		387.33	335.67	
12	Birch		292.67	455.33	
13	Maple		323.33	293.67	

Figure 13.4: Interactions—Data

Category(X) Labels, Use Column "1" for Legend Text, and finally in Step 5 select **Yes** for Add a Legend? and type the name for the chart and the axis titles.

Users of Excel 97/98

- In Step 1 select **Line** for Chart Type and the upper left Chart subtype **Line**. In Step 2 under the **Data Range** tab, enter K3:O6 and select the Series radio button for **Rows**. In Step 3 enter the titles under the **Titles** tab, remove gridlines under the **Gridlines** tab, check **Show legend** under the **Legend** tab, and under the **Data Labels** tab select the radio button **None**. In Step 4 click Finish.

4. Finally, make any additional editing changes so that the resulting chart appears similar to Figure 13.5.

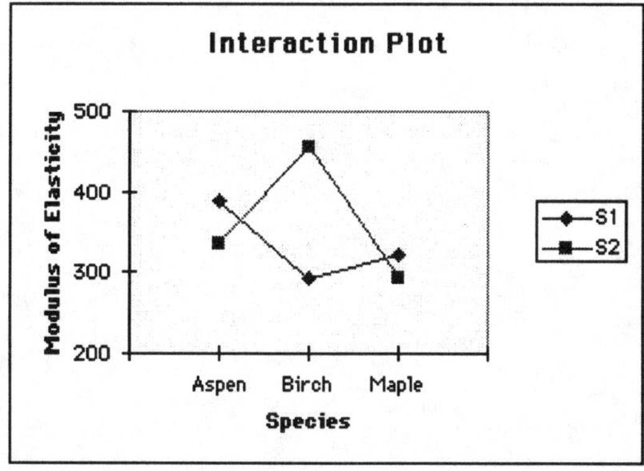

Figure 13.5: Interaction Plot—Species

Exercise. By following the same steps with the bottom portion of Figure 13.4, produce the interaction plot shown in Figure 13.6 with Size of Flakes on the horizontal axis.

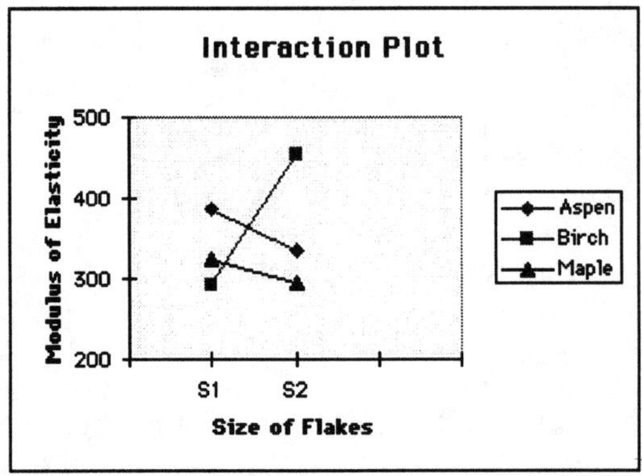

Figure 13.6: Interaction Plot—Flakes

Output

The interaction plots show lines that cross each other. This pattern indicates that interaction is present. However, the F test for ANOVA did not indicate this interaction. The apparent contradiction results because the estimated standard deviation (s_p^2 from cell G34)

$$s_p = 110.73$$

is so large that such a crossover in means can be explained as random variation. This example indicates that some caution needs to be exercised in interpreting interaction plots.

Chapter 14

Nonparametric Tests

Nonparametric procedures are based on the ranks of observations and replace assumptions of normality with less stringent assumptions, such as symmetry and continuity of distributions. Excel does not provide nonparametric tests. Nonetheless, these tests may readily be developed.

14.1 The Wilcoxon Rank Sum Test

The Wilcoxon rank sum test is a procedure for comparing independent samples from two populations. Suppose

$$(x_{11}, x_{12}, \ldots, x_{1n_1}), (x_{21}, x_{22}, \ldots, x_{2n_2})$$

are the two samples. Generally sample 1, the x_{1j} observations, might be the control group, while sample 2, the x_{2j} observations, might be the treatment group.

We wish to test whether both samples can be assumed to arise from identical populations or whether the populations are shifted by a constant. The precise assumption is that if F_i represents the cumulative distribution function of the the x_i-observations, $1 \leq i \leq 2$, then

$$F_2(t) = F_1(t + \Delta) \qquad \text{for all real } t$$

where Δ is a constant representing an unknown shift in the two distributions. If Δ is positive, then the second sample would contain systematically higher values than the first sample.

The null hypothesis is thus

$$H_0 : \mu_1 - \mu_2 = 0$$

where μ_1 and μ_2 are the two population means (or medians).

It is also possible to test a nonzero difference

$$H_0 : \mu_1 - \mu_2 = \Delta_0$$

by subtracting Δ_0 from each sample 1 observation and applying the first test for a zero difference in means to

$$(x_{11} - \Delta_0, x_{12} - \Delta_0, \ldots, x_{1n_1} - \Delta_0), (x_{21}, x_{22}, \ldots, x_{2n_2})$$

Description of the Procedure

1. Rank all the observations from smallest to largest into one list with $N = n_1 + n_2$ observations.

2. Let r_j be the rank of the observation x_{1j}.

3. Set
$$W = \sum_{j=1}^{n_1} r_j$$
which is the sum of the ranks associated with the data from sample 1.

If H_0 is true, then each overall ranking of the N combined observations would have the same probability, but if, for instance,

$$H_a : \mu_1 - \mu_2 > 0$$

then the ranks of sample 1, contributing to W and arising from the population with the larger mean under H_a, would be larger than expected under H_0, leading to an observed W value above the mean. The Wilcoxon rank sum test therefore rejects H_0 if W is beyond some reasonable value. This test is sometimes called the Mann-Whitney test because it was originally derived as an alternative, but equivalent formulation by H. B. Mann and D. R. Whitney in 1947.

The exact distribution of W has been tabulated, but we will base our procedure on the normal approximation. It can be shown that if H_0 is true, then

$$\mu_W = \text{mean of } W = \frac{n_1(n_1 + n_2 + 1)}{2}$$

$$\sigma_w = \text{standard deviation of } W = \sqrt{\frac{n_1 n_2 (n_1 + n_2 + 1)}{12}}$$

Calculate

$$z = \frac{W - \mu_W}{\sigma_W}$$

For a fixed level α test:

if $H_a : \mu_1 - \mu_2 > 0$	reject H_0 if $z > z^*$
if $H_a : \mu_1 - \mu_2 < 0$	reject H_0 if $z < -z^*$
if $H_a : \mu_1 - \mu_2 \neq 0$	reject H_0 if $\lvert z \rvert > z^*$

where z^* represents the corresponding upper critical value of a standard normal distribution.

The Wilcoxon Rank Sum Test in Practice

Example 14.1. (Examples 14.1–14.5, pages 4-10 in the additional chapters in the Student CD-ROM.) Does the presence of small numbers of weeds reduce the yield of corn? Lamb's-quarter is a common weed in corn fields. A researcher planted corn at the same rate in eight small plots of ground, then weeded the corn rows by hand to allow no weeds in four randomly selected plots and exactly three lamb's-quarter plants per meter of row in the other four plots. Here are the yields of corn (bushels per acre) in each of the plots.

Weeds per meter	Yield (bu/acre)			
0	166.7	172.2	165.0	176.9
3	158.6	176.4	153.1	156.0

Normal quantile plots suggest that the data may be right-skewed. The samples are too small to assess normality adequately or to rely on the robustness of the two-sample t test. We may prefer to use a test that does not require normality. Carry out the significance test

$$H_0: \text{ no difference in distribution of yields}$$
$$H_a: \text{ yields are systematically higher in weed-free plots}$$

Solution. Figure 14.1 shows the Excel formulas required and the corresponding values taken when applied to the above data set.

1. Enter the labels "Sample 1," "Sample 2," "Population," "Combined," and "Rank" in cells A3:E3.

2. Record the observations in Sample 1 in cells A4:A7 and copy them to D4:D7. Record the observations in Sample 2 in cells B4:B7 and copy them to D8:D11. Enter the value 1 in C4:C7, the value 2 in C8:C11, and finally the values $\{1, 2, 3, 4, 5, 6, 7, 8\}$ in E4:E11.

3. **Name** the ranges for Sample 1, Sample 2, Population, Combined, and Rank.

4. **Rank** the combined data set that has been copied into D4:D11, and carry the corresponding sample numbers (Sample 1 = 0 weeds per meter, Sample 2 = 3 weeds per meter) back to C4:C11 as follows. Select C3:D11 and from the Menu Bar choose **Data – Sort** to bring up the **Sort** dialog box. Select the radio buttons for "Ascending" and for "Header row". Then click OK. This ranks the combined sample in D4:D11 in increasing order and also carries the corresponding sample labels in C4:C11. Figure 14.1 shows the results of the sort.

	A	B	C	D	E	F
1			*Wilcoxon Rank Sum Test*			*Test*
2						
3	Sample1	Sample2	Population	Combined	Rank	
4	166.7	158.6	2	153.1	1	
5	172.2	176.4	2	156.0	2	
6	165.0	153.1	2	158.6	3	
7	176.9	156.0	1	165.0	4	
8			1	166.7	5	
9			1	172.2	6	
10			2	176.4	7	
11			1	176.9	8	
12						
13	User Input					
14	alpha	0.05				
15	Summary Statistics					
16	n_1	4		=COUNT(Sample1)		
17	n_2	4		=COUNT(Sample2)		
18	Calculations					
19	W	23		=SUMIF(Population, "=1", Rank)		
20	mu	18		=n_1*(n_1+n_2+1)/2		
21	sigma	3.464		=SQRT(n_1*n_2*(n_1+n_2+1)/12)		
22	z	1.443		=(W-mu)/sigma		
23	Lower Test					
24	lower_z			=NORMSINV(alpha)		
25	Decision			=IF(z<lower_z,"Reject H0","Do Not Reject H0")		
26	Pvalue			=NORMSDIST(z)		
27	Upper Test					
28	upper_z	1.645		=-NORMSINV(alpha)		
29	Decision	Do Not Reject H0		=IF(z>upper_z,"Reject H0","Do Not Reject H0")		
30	Pvalue	0.0745		=1-NORMSDIST(z)		
31	Two-Sided Test					
32	two_z			=ABS(NORMSINV(alpha/2))		
33	Decision			=IF(ABS(z)>two_z,"Reject H0","Do Not Reject H0")		
34	Pvalue			=2*(1-NORMSDIST(ABS(z)))		

Figure 14.1: Wilcoxon Rank Sum Test—Formulas and Values

5. Next we carry out calculations akin to those in Chapter 6 (see Figure 6.2) for a one-sample test of a normal mean. Enter the labels as shown in cells A13:A34 on Figure 14.1, and **Name** the corresponding ranges in the respective column B cells to be able to refer to n_1, n_2, W, mu, sigma, z, lower_z, upper_z, and two_z by name in the ensuing formulas shown. The formulas to be entered in column B are presented in column D, and the values taken when the formulas are applied to this data (that is, what you will see in your workbook) are shown in the corresponding cells in column B. For instance, enter the formula from cell D19

$$= \texttt{SUMIF}(\text{Population, ``}= 1\text{'', Rank})$$

into cell B19. This Excel function adds those cells under Rank whose corresponding Population is 1. In other words, the function adds the ranks corresponding to observations from Sample 1. In cell B19 appears the answer 23, the value of the Wilcoxon rank sum statistic. Formulas for all three types of alternate hypotheses are provided in Figure 14.1, but the values are shown only for the particular alternative in this problem (upper test).

When using this template, remember to use only the appropriate cells for the problem at hand.

Interpreting the Results

We read off

$$\begin{aligned} W &= 23 \quad \text{(cell B19)} \\ \mu_W &= 18 \quad \text{(cell B20)} \\ \sigma_W &= 3.464 \quad \text{(cell B21)} \\ \text{z statistic} &= 1.443 \quad \text{(cell B22)} \end{aligned}$$

The alternate hypothesis is

$$H_a : \mu_1 - \mu_2 > 0$$

so that rows 27–30 are appropriate (rows 23–26 for a lower-tailed test and rows 31–34 for a two-tailed test). We find that

$$\begin{aligned} \text{upper critical value } z^* &= 1.645 \quad \text{(cell B28)} \\ \text{decision rule} &= \text{“Do not reject”} \quad \text{(cell B29)} \\ P\text{-value} &= 0.0745 \quad \text{(cell B30)} \end{aligned}$$

We conclude that the data are not significant at the nominal 5% level of significance.

Continuity Correction for the Normal Approximation

A more accurate P-value is obtained by applying the continuity correction, which adjusts for the fact that a continuous distribution (the normal) is being used to approximate a discrete distribution W. If we use a correction of 0.5, then the upper-tailed test statistic becomes

$$z = \frac{W - 0.5 - \mu_W}{\sigma_W}$$

and this leads to

$$z = 1.299$$

and

$$P\text{-value} = 1 - \Phi(1.299) = 0.0970$$

Note: These more accurate values appear in Examples 14.4 and 14.5 on page 8 in the Student CD-ROM.

Ties

Theoretically, the assumption of a continuous distribution ensures that all $n_1 + n_2$ observed values will be different. In practice, ties are sometimes observed. The common practice is to average the ranks for the tied observations and carry on as above with a change in the standard deviation. Use

$$\sigma_W^2 = \frac{n_1 n_2}{12}\left(n_1 + n_2 + 1 - \frac{\sum_{i=1}^{G} t_i(t_i^2 - 1)}{(n_1 + n_2)(n_1 + n_2 - 1)}\right)$$

where G is the number of tied groups and t_i is the number of tied observations in the ith tied group. Unless G is large, the adjustment in the formula for the variance makes little difference.

14.2 The Wilcoxon Signed Rank Test

The Wilcoxon signed rank test is a nonparametric version of the one-sample procedures based on the assumption of normal population discussed in Chapters 6 and 7. The key assumption is that the data arise from a population symmetric about its mean.

For this reason one of its most useful applications is in a matched-pairs setting with n pairs (x_{1i}, x_{2i}) of observations, where it is natural to assume that the populations from which the pairs are taken differ only by a shift in the mean (that is, the population distribution shapes are otherwise the same). The differences then satisfy the requirement of symmetry under the null hypothesis of equality of means.

Description of the Matched-Pairs Procedure

The data consist of n pairs (x_{1i}, x_{2i}) of observations. The $\{x_{1i}\}$ are a sample from a population with mean μ_1 and the $\{x_{2i}\}$ are a sample from a population with mean μ_2. The null hypothesis is

$$H_0 : \mu_1 - \mu_2 = 0$$

1. Form the absolute differences $|d_j|$, where $d_j = x_{1j} - x_{2j}$.
2. Let r_j be the rank of $|d_j|$ in the joint ranking of the $\{|d_j|\}$, from smallest to largest.
3. Form the sum of the positive signed ranks

$$W^+ = \sum r_j$$

where the sum is taken over all ranks r_j for which the corresponding difference d_j is positive.

The Wilcoxon signed rank procedure rejects H_0 if W^+ is beyond some reasonable value, in particular for values of W^+ that are too large or too small.

As with the Wilcoxon rank sum statistic W, there exist tables of the exact distribution of W^+, but we will base our procedure on the normal approximation. When H_0 is true the mean and the standard deviation of W^+ are given by

$$\mu_{W+} = \frac{n(n+1)}{4}$$

$$\sigma_{W+} = \sqrt{\frac{n(n+1)(2n+1)}{24}}$$

We then calculate

$$z = \frac{W^+ - \mu_{W+}}{\sigma_{W+}}$$

and for a fixed level α test:

$$\begin{array}{ll} \text{if } H_a : \mu_1 - \mu_2 > 0 & \text{reject } H_0 \text{ if } z > z^* \\ \text{if } H_a : \mu_1 - \mu_2 < 0 & \text{reject } H_0 \text{ if } z < -z^* \\ \text{if } H_a : \mu_1 - \mu_2 \neq 0 & \text{reject } H_0 \text{ if } |z| > z^* \end{array}$$

The Wilcoxon Signed Rank Test in Practice

Example 14.2. (Examples 14.8–14.10, pages 19–20 in the Student CD-ROM.) A study of early childhood education asked kindergarten students to tell two fairy tales that had been read to them earlier in the week. The first tale had been read to them and the second had been read but also illustrated with pictures. An expert listened to a recording of the children and assigned a score for certain uses of language. Here are the data for five "low progress" readers in a pilot study:

Child	1	2	3	4	5
Story 2	0.77	0.49	0.66	0.28	0.38
Story 1	0.40	0.72	0.00	0.36	0.55
Difference	0.37	−0.23	0.66	−0.08	−0.17

We wonder if illustrations improve how the children retell a story. We would like to test the hypotheses

H_0 : scores have the same distribution for both stories

H_a : scores are systematically higher for story 2

Because this is a matched-pairs design, we base our inference on the differences. The matched-pairs t test gives $t = 0.635$ with a one-sided P-value of 0.280. As displays of the data suggest a mild lack of normality, carry out a nonparametric significance test.

Solution. Figure 14.2 shows the Excel formulas required and the corresponding values taken for the example data set. The calculations are similar to those in the previous section.

	A	B	C	D	E	F	G
1	*Wilcoxon Signed Rank Test - Matched Pairs*						
2							
3	Sample1	Sample2	Diff		Ranked_Diff	Ranked_Abs_Diff	Rank
4	0.77	0.40	0.37		-0.08	0.08	1
5	0.49	0.72	-0.23		-0.17	0.17	2
6	0.66	0.00	0.66		-0.23	0.23	3
7	0.28	0.36	-0.08		0.37	0.37	4
8	0.38	0.55	-0.17		0.66	0.66	5
9							
10							
11	User Input						
12	alpha	0.05					
13	Summary Statistics						
14	n	5		=COUNT(Sample1)			
15	Calculations						
16	Wplus	9		=SUMIF(Ranked_Diff, ">0", Rank)			
17	mu	7.5		=n*(n+1)/4			
18	sigma	3.708		=SQRT(n*(n+1)*(2*n+1)/24)			
19	z	0.405		=(Wplus-mu)/sigma			
20	Lower Test						
21	lower_z			=NORMSINV(alpha)			
22	Decision			=IF(z<lower_z,"Reject H0","Do Not Reject H0")			
23	Pvalue			=NORMSDIST(z)			
24	Upper Test						
25	upper_z	1.645		=-NORMSINV(alpha)			
26	Decision	Do Not Reject H0		=IF(z>upper_z,"Reject H0","Do Not Reject H0")			
27	Pvalue	0.3429		=1-NORMSDIST(z)			
28	Two-Sided Test						
29	two_z			=ABS(NORMSINV(alpha/2))			
30	Decision			=IF(ABS(z)>two_z,"Reject H0","Do Not Reject H0")			
31	Pvalue			=2*(1-NORMSDIST(ABS(z)))			

Figure 14.2: Wilcoxon Signed Rank—Matched Pairs

1. Enter the labels "Sample 1," "Sample 2," and "Diff" in cells A3:C3 and the labels "Ranked_Diff," "Ranked_Abs_Diff," and "Rank" in cells E3:G3.

2. Record the observations in Sample 1 in cells A4:A8 and the observations for Sample 2 in cells B4:B8. In cells C4:C8 record the difference between Sample 1 and Sample 2. Enter the values $\{1, 2, 3, 4, 5\}$ in G4:G8.

3. **Name** the ranges for the corresponding labels Sample 1, Sample 2, Ranked_Diff, Ranked_Abs_Diff, and Rank to include the respective cells in rows 4–8.

4. In a different part of the sheet, copy the values of the differences (not their formulas, if Excel calculated them) and then enter their absolute values using the Excel function `ABS()`, which gives the absolute value of its argument. As in step 4 of the previous section, **rank** the absolute values of the differences in increasing order and carry along the actual differences. Copy the results to cells E4:F8 as shown in Figure 14.2, showing the ranked absolute differences

in column F and the actual ranked differences in column E. We need the latter to recognize which absolute differences correspond to positive differences in calculating W^+.

5. The calculations required are shown in the lower portion of Figure 14.2, where we have included in column D the formulas that are to be entered in column B. The actual values taken by these formulas—the values that will appear on your workbook—are in column B. Refer to step 5 of the previous section for the analogous details, and remember to name all ranges used in the formulas shown.

Interpreting the Results

We read off

$$\begin{aligned} W^+ &= 9 && \text{(cell B16)} \\ \mu_{W^+} &= 7.5 && \text{(cell B17)} \\ \sigma_{W^+} &= 3.708 && \text{(cell B18)} \\ z\text{statistic} &= 0.405 && \text{(cell B19)} \end{aligned}$$

The alternate hypothesis requires an upper-tailed test for which

$$P\text{-value} = 0.3429 \qquad \text{(cell B27)}$$

We conclude that the data are not significant.

Continuity Correction for the Normal Approximation

As with the Wilcoxon rank sum test, a more accurate P-value is obtained with the continuity correction

$$z = \frac{W^+ - 0.5 - \mu_{W^+}}{\sigma_{W^+}}$$

and this leads to

$$z = 0.270$$

$$P\text{-value} = 1 - \Phi(0.270) = 0.3937$$

Note: These more accurate values are used in Example 14.10 on page 20 in the Student CD-ROM.

Ties and Zero Values

If there are zeros among the differences $\{d_i\}$, discard them and use for n the number of nonzero $\{d_i\}$. If there are any ties, then use the average rank for each set of tied observations and apply the procedure with variance

$$\sigma^2_{W+} = \frac{1}{24}\left(n(n+1)(2n+1) - \frac{\sum_{i=1}^{G} t_i(t_i^2-1)}{2}\right)$$

where G is the number of tied groups and t_i are the number of tied observations in the ith tied group.

14.3 The Kruskal-Wallis Test

In this section we generalize the Wilcoxon rank sum test to situations involving independent samples from I populations when the assumptions required for validity of the one-way ANOVA in Chapter 12 cannot be substantiated.

The data consist of $N = \sum_{i=1}^{I} n_i$ observations with $n_i \geq 1$ observations $\{x_{ij} : 1 \leq j \leq n_i\}$ taken from population i. The assumption replacing normality is

$$x_{ij} = \mu_i + \varepsilon_{ij} \quad 1 \leq i \leq I,\ 1 \leq j \leq n_i$$

where the errors $\{\varepsilon_{ij}\}$ are mutually independent with mean 0 and have the same *continuous* distribution. If we let $F(x)$ be the cumulative distribution function (c.d.f.) of a generic error term, this is assumption is tantamount to $F_i(x)$ being the c.d.f. of population i, where

$$F_i(x) \equiv F(x - \mu_i) \quad 1 \leq i \leq I$$

The significance test is

$$H_0 : \mu_1 = \mu_2 = \cdots = \mu_I$$
$$H_a : \text{not all of the } \mu_i \text{ are equal}$$

and the procedure generalizing the rank sum test is called the Kruskal-Wallis test.

Description of the Procedure

1. **Rank** all the observations jointly from smallest to largest.

2. Let r_{ij} be the rank of observation x_{ij}.

3. Set
$$R_i = \sum_{j=1}^{n_i} r_{ij}$$
which is the sum of the ranks associated with sample i.

Denote by

$$\bar{R}_i = \frac{1}{n_i} R_i$$

the average rank in sample i. If H_0 is true, then by symmetry the mean of any rank r_{ij} is $E(r_{ij}) = \frac{N+1}{2}$, which is the average of the integers $\{1, 2, \ldots, N\}$, and therefore $E[\bar{R}_i] = E\left[\frac{1}{n_i} R_i\right] = E\left[\frac{1}{n_i}\sum_{j=1}^{n_i} r_{ij}\right] = \frac{N+1}{2}$. Thus, we would expect the ranks to be uniformly intermingled among the I samples. But if H_0 is false, then some samples will tend to have many small ranks, while others will have many large ranks. Just as in ANOVA, we take the sum of squares of the differences between the average rank $\bar{R}_i$ of each sample and the overall average $\frac{N+1}{2}$ by computing

$$\frac{12}{N(N+1)} \sum_{i=1}^{I} n_i \left(\bar{R}_i - \frac{N+1}{2}\right)^2$$

which can be expressed equivalently as

$$H = \frac{12}{N(N+1)} \sum_{i=1}^{I} \frac{R_i^2}{n_i} - 3(N+1)$$

and called the Kruskal-Wallis statistic. We then reject H_0 for "large" values of H.

Tables of critical values exist for small values of the $\{n_i\}$, but it is customary to use a normal approximation, which provides an approximate sampling distribution:

H is approximately chi-square with $I - 1$ degrees of freedom.

Therefore the test is

$$\text{Reject } H_0 \text{ if } H > \chi^2$$

where χ^2 is the upper critical α value of a chi-square distribution on $I - 1$ degrees of freedom.

The Kruskal-Wallis Test in Practice

Example 14.3. (Examples 14.13 and 14.14, pages 25–29 in the Student CD-ROM.) Lamb's-quarter is a common weed that interferes with the growth of corn. A researcher planted corn at the same rate in 16 small plots of ground, then randomly assigned the plots to four groups. He weeded the plots by hand to allow a fixed number of lamb's-quarter plants to grow in each meter of corn row. These numbers were 0, 1, 3, and 9 in the four groups of plots. No other weeds were allowed to grow, and all plots received identical treatment, except for the weeds. Here are the yields of corn (bushels per acre) in each of the plots.

Weeds per meter	Corn yield	Weeds per meter	Corn yield	Weeds per meter	Corn yield	Weeds per meter	Corn yield
0	166.7	1	166.2	3	158.6	9	162.8
0	172.2	1	157.3	3	176.4	9	142.4
0	165.0	1	166.7	3	153.1	9	162.7
0	176.9	1	161.1	3	156.0	9	162.4

The summary statistics follow:

Weeds	n	Mean	Std Dev
0	4	170.200	5.422
1	4	162.825	4.469
3	4	161.025	10.498
9	4	157.575	10.118

The sample standard deviations do not satisfy our rule of thumb that for safe use of ANOVA the largest should not exceed twice the smallest. Normal quantile plots show that outliers are present in the yields for 3 and 9 weeds per meter. Use the Kruskal-Wallis procedure to test

$$H_0: \text{ yields have the same distribution in all groups}$$
$$H_a: \text{ yields are systematically higher in some groups than others.}$$

Solution. Figure 14.3 shows the Excel formulas required and the corresponding values taken when applied to the example data set.

As we have already described in detail the construction of the two earlier nonparametric procedures, we leave as an exercise the application of this workbook. Beginning in row 9, column C shows the formulas to be entered into the adjacent cells of column B, where the numerical evaluation of the formulas is shown.

	A	B	C	D	E	F	G	H
1			*Kruskal-Wallis Test*					
2								
3	Sample1	Sample2	Sample3	Sample4		Pop	Combined	Rank
4	166.7	166.2	158.6	162.8		4	142.4	1
5	172.2	157.3	176.4	142.4		3	153.1	2
6	165.0	166.7	153.1	162.7		3	156.0	3
7	176.9	161.1	156.0	162.4		2	157.3	4
8						3	158.6	5
9	R_1	52	=SUMIF(Pop, "=1", Rank)			2	161.1	6
10	R_2	34	=SUMIF(Pop, "=2", Rank)			4	162.4	7
11	R_3	25	=SUMIF(Pop, "=3", Rank)			4	162.7	8
12	R_4	25	=SUMIF(Pop, "=4", Rank)			4	162.8	9
13						1	165.0	10
14	n_1	4	=COUNT(A4:A7)			2	166.2	11
15	n_2	4	=COUNT(B4:B7)			1	166.7	12
16	n_3	4	=COUNT(C4:C7)			2	166.7	13
17	n_4	4	=COUNT(D4:D7)			1	172.2	14
18	N	16	=n_1+n_2+n_3+n_4			3	176.4	15
19						1	176.9	16
20		676	=R_1^2/n_1					
21		289	=R_2^2/n_2					
22		156.25	=R_3^2/n_3					
23		156.25	=R_4^2/n_4					
24								
25	H	5.3602941	=(12/(N*(N+1)))*SUM(B21:B24) - 3*(N+1)					
26	Critical 5%	7.815	=CHIINV(0.05,3)					
27	P-value:	1.47E-01	=CHIDIST(H,3)					

Figure 14.3: Kruskal-Wallis Test—Formulas and Values

Interpreting the Results

We read off

$$\begin{aligned} H &= 5.36 \quad \text{(cell B25)} \\ \chi^2 &= 7.815 \quad \text{(cell B26)} \\ P\text{-value} &= 0.147 \quad \text{(cell B27)} \end{aligned}$$

The data are not significant at the 5% level, meaning that there is no convincing evidence that more weeds decrease yield.

Ties

We have ignored a tie in the above calculation; the value 166.7 appears in Sample 1 and in Sample 2. For a more accurate calculation, we give to the value 166.7 the average rank 12.5. Then replace H with

$$H' = \frac{H}{1 - \sum_{i=1}^{G} \frac{t_i(t_i^3-1)}{N^3-N}}$$

where G is the number of tied groups and t_i are the number of tied observations in the ith tied group.

> **Exercise.** Replace the ranks of the two observations 166.7 in your workbook with common rank 12.5 and then calculate H' as
>
> $$H' = \frac{H}{1 - \frac{2(2^3-1)}{16^3-16}} = \frac{H}{.9965686}$$
>
> Show that the P-value becomes 0.134.

Note: This is the P-value given on page 29 in the Student CD-ROM, which is also adjusted for the presence of ties.

Chapter 15

Logistic Regression

Excel does not possess a built-in logistic regression tool. Logistic regression is a highly specialized topic, best used under the guidance of a statistician. However, it is still possible within Excel to obtain estimates of the parameters in simple logistic regression with little effort using weighted least squares.

Binomial Distributions and Odds

Recall from the discussion in Section 2.5 and Chapter 10 that regression refers to fitting models for the mean value of a response as a function of an explanatory variable. For n pairs of observations (x_i, y_i) the simple linear regression model is

$$y_i = \beta_0 + \beta_1 x_i + \varepsilon_i$$

with the errors $\{\varepsilon_i\}$ assumed independent $N(0, \sigma)$ and the regression function

$$\mu_y = \beta_0 + \beta_1 x_i$$

representing the mean of y_i as a function of x.

It is possible to fit models *other* than a straight line. If you examine Figure 2.14, you will observe that Excel can fit

$$\begin{aligned} \mu_y &= a + b \log x \quad \text{(logarithmic)} \\ \mu_y &= a x^b \quad \text{(power)} \\ \mu_y &= a e^{bx} \quad \text{(exponential)} \end{aligned}$$

as well as polynomial and moving average models. These nonlinear models are fitted using transformations which linearize the regression curve.

In some circumstances, the response variable is discrete, not continuous. An example of a discrete variable might be the number of cases of skin cancer in a metropolitan area. A special case of a discrete variable is a binary response, say $\{0, 1\}$, which leads to the logistic regression model. We restrict attention to binary response in the remainder of this chapter.

First, we will identify some major differences between regression with binary responses and regression with continuous responses. Assume, as is customary, that the errors have mean 0, which endows the regression curve

$$\mu_y = \beta_0 + \beta_1 x$$

with the usual meaning as the mean response. In the Bernoulli case, y_i can take only two values so that

$$\mu_y = P[y_i = 1] = \beta_0 + \beta_1 x_i = p_i$$

and the variance of the error term is therefore the variance of a Bernoulli random variable, $p_i(1 - p_i)$. Note:

- Errors cannot be normal.
- Errors have nonconstant variance.
- The mean response is constrained to lie within the interval [0,1] since it represents a probability.

Because of these differences, ordinary regression methodology is not appropriate.

The explanatory variables are $(x_1, x_2, \ldots, x_c)$ with n_i observations at the value x_i. The number of "successes" at x_i is denoted by s_i, which is a binomial $\text{Bin}(n_i, p_i)$ random variable on n_i trials and success probability $p_i \equiv p(x_i)$, which is a function of x_i. It is the function $p(x_i)$ that is of interest.

We have dealt with binary responses on two occasions in this text; in Chapter 8 we considered $c = 2$, while in Chapter 9 we dealt with $c \geq 2$. Here we also consider $c \geq 2$ but impose a model (the regression curve) relating all the probabilities p_i as a function of x_i.

A scatterplot of such data shows binary observations having y values of 0 and 1 and the interpretation of the regression curve as somehow passing near the data is lost because of the categorical (binary) nature of the response. However, the interpretation is partially regained if a scatterplot is made of the sample proportions $\hat{p}_i = \frac{s_i}{n_i}$ on the y-axis against their corresponding x_i on the x-axis (as shown in Figure 15.4 on page 44 of the Student CD-ROM).

We may then consider fitting a curve through the $(x_i, \hat{p}_i)$ pairs. In order to facilitate this approach, statisticians have introduced the logit transformation based on the odds. Define the odds ratio

$$\text{ODDS} = \frac{p}{1 - p}$$

and then take the natural logarithm of the odds to define the logit function:

$$\text{logit}(p) = \log(\text{ODDS}) = \log\left(\frac{p}{1 - p}\right)$$

There is a mathematical reason for using ODDS as the *natural* parameter that comes from the factorization of the likelihood function. These are advanced details, beyond the scope of this presentation.

The Logistic Regression Model

The **statistical model for logistic regression** posits a linear form

$$\text{logit}(p) = \beta_0 + \beta_1 x$$

that is equivalent to

$$p \equiv p(x) = \frac{1}{1 + e^{-(\beta_0 + \beta_1 x)}}$$

in terms of the original probability p (Figure 15.4 of the Student CD-ROM).

In the text, estimates for the parameters β_0, β_1 are presented based on the output from a specialized statistics package called SAS. Although Excel does not provide a logistic regression output, we can obtain estimates using the **spreadsheet** features of Excel and **weighted least-squares.**

Weighted Least-Squares

The least-squares criterion minimizes the sum of the squared residuals

$$\sum_{i=1}^{n} e_i^2 = \sum_{i=1}^{n} (y_i - \hat{y}_i)^2$$

where $\{y_i\}$ are the observed values and $\{\hat{y}_i\}$ are the fitted values where $\hat{y}_i = b_0 + b_1 x_i$. When the errors do not have a constant variance, it is more **efficient** (in the sense of producing estimates with smaller variance) to weight the residuals. It is intuitively reasonable to give a residual more weight (for accuracy) if its corresponding variance is smaller.

Let w_i be the weight assigned to an observation at x_i. Then w_i is inversely proportional to the variance σ_i^2 of the error ε_i and the criterion for weighted least-squares is

$$\text{minimize} \quad \sum_{i=1}^{n} w_i (y_i - b_0 - b_1 x_i)^2$$

The solution is obtained (as in Chapter 2) by differentiating with respect to b_0, b_1. We find

$$b_1 = \frac{\sum w_i x_i y_i - \frac{(\sum w_i x_i)(\sum w_i y_i)}{\sum w_i}}{\sum w_i x_i^2 - \sum w_i x_i}$$

$$b_0 = \frac{\sum w_i y_i - b_1 \sum w_i x_i}{\sum w_i}$$

When $w_i \equiv 1$ for all i, these two equations reduce to the equations obtained for finding the ordinary least-squares estimates.

These equations represent the quintessential spreadsheet operations, columns of numbers that are added and whose totals are then algebraically manipulated. Therefore, they are ideal for spreadsheet logic, and *it is fitting in the final chapter, dealing with a sophisticated statistical model for which Excel does not provide a built-in tool, that we can obtain a solution by setting up our own columns of variables.*

The variable y_i, which is appropriate in this setting, is

$$y_i = \log \frac{\hat{p}_i}{1 - \hat{p}_i}$$

and the weights are approximately $w_i = n_i p_i(1 - p_i)$. Since the $\{p_i\}$ are unknown, we employ instead

$$w_i = n_i \hat{p}_i(1 - \hat{p}_i)$$

We organize our workbook with columns for

$$x_i, y_i, w_i, w_i x_i, w_i y_i, w_i x_i^2, w_i y_i, w_i x_i y_i$$

which are then used to determine b_0, b_1 (Figure 15.1).

Fitting and Interpreting the Logistic Regression Model

Example 15.1. (Examples 15.7–15.8, pages 42–45 in the Student CD-ROM.) An experiment was designed to examine how well the insecticide rotenone kills aphids that feed on the chrysanthemum plant called *macrosiphoniella sanborni.* The explanatory variable is the concentration (in log of mg/l) of the insecticide. At each concentration, approximately 50 insects were exposed. Each insect was either killed or not killed. We summarize the data using the number killed. The response variable for logistic regression is the log odds of the proportion killed. Here are the data:

Concentration (log)	Number of insects	Number killed
0.96	50	6
1.33	48	16
1.63	46	24
2.04	49	42
2.32	50	44

Fit a logistic model

$$p_i = \frac{1}{1 + e^{-(\beta_0 + \beta_1 x_i)}}$$

to the probability that an aphid will be killed as a function of the concentration x_i.

	A	B	C	D	E	F	G	H	I	J	K
1					Logistic Regression by Weighted Least Squares						
2											
3					logit	weight			calculations		
4	x	n	s	p	y	w	w*x	w*y	w*x*y	w*x^2	w*y^2
5	0.96	50	6	0.120	-1.992	5.280	5.069	-10.520	-10.099	4.866	20.960
6	1.33	48	16	0.333	-0.693	10.667	14.187	-7.394	-9.833	18.868	5.125
7	1.63	46	24	0.522	0.087	11.478	18.710	0.999	1.628	30.497	0.087
8	2.04	49	42	0.857	1.792	6.000	12.240	10.751	21.931	24.970	19.262
9	2.32	50	44	0.880	1.992	5.280	12.250	10.520	24.406	28.419	20.960
10				sums		38.705	62.455	4.356	28.033	107.620	66.395
11											
12				=s/n	=LN(p/(1-p))	=n*p*(1-p)	=w*x	=w*y	=w*x*y	=w*x^2	=w*y^2
13											
14	b1=	3.070	=(I10-G10*H10/F10)/(J10-G10^2/F10)								
15	b0=	-4.841	=(H10-B14*G10)/F10								

Figure 15.1: Weighted Least-Squares Estimates for Logistic Regression

Solution. Figure 15.1 shows how we have set up the workbook. Use **Named Ranges** for the labels and variables shown in row 4. The formulas required for calculating the columns are shown in cells D12:K12. The weighted least-squares estimates are in B14:B15, and in C14:C15 we have given the formulas to be entered into B14:B15. (Your workbook will not have C14:C15.)

We find the weighted least-squares estimates to be

$$\begin{aligned} b_1 &= 3.070 \\ b_0 &= -4.841 \end{aligned}$$

These compare favorably with the SAS output presented in the text

$$\begin{aligned} b_1 &= 3.10 \\ b_0 &= -4.89 \end{aligned}$$

Exercise. (Examples 15.1–15.4, pages 36–41 in the Student CD-ROM.) In Chapter 8 we presented the results of a survey on binge drinking. The text discusses this example to illustrate logistic regression. Adapt the workbook developed in the previous section to show that the weighted least-squares estimates are

$$\begin{aligned} b_1 &= 0.362 \\ b_0 &= -1.587 \end{aligned}$$

These are identical to those in your text. Your results should look like those in Figure 15.2.

Exercise. Adapt the workbook shown in Figure 15.1 and the formula $=$ `CRITBINOM`$(n, p,$ `RAND`$())$ used in Section 5.2 to simulate data for a

	A	B	C	D	E	F	G	H	I	J	K
1					*Logistic Regression by Weighted Least Squares*						
2											
3					*logit*	*weight*			*calculations*		
4	x	n	s	p	y	w	w*x	w*y	w*x*y	w*x^2	w*y^2
5	0	9916	1684	0.170	-1.587	1398.012	0.000	-2218.445	0.000	0.000	3520.356
6	1	7180	1630	0.227	-1.225	1259.958	1259.958	-1543.723	-1543.723	1259.958	1891.398
7			sums			2657.970	1259.958	-3762.169	-1543.723	1259.958	5411.753
8											
9				=s/n	=LN(p/(1-p))	=n*p*(1-p)	=w*x	=w*y	=w*x*y	=w*x^2	=w*y^2
10											
11	b1=	0.362	=(I7-G7*H7/F7)/(J7-G7^2/F7)								
12	b0=	-1.587	=(H7-B11*G7)/F7								

Figure 15.2: Binge Drinking

logistic regression. Choose the same values $\{x_i\}$ and $\{n_i\}$ as in Example 15.1. Set the parameters to be

$$\begin{aligned} \beta_1 &= -5.00 \\ \beta_0 &= 3.00 \end{aligned}$$

and obtain estimates b_1, b_0.

Next, generate repeated samples by pressing the function key (F9) and constructing histograms of the values for b_0 and b_1 and comparing them with the true values. What can you conclude about the mean, bias, standard error, and confidence intervals? Construct scatterplots of the simulated values of (b_0, b_1).

Appendix—Excel 5/95

With respect to data analysis capabilities, the releases of Excel 97 (Windows) and Excel 98 (Macintosh) corrected some bugs in the Data Analysis Toolpak and changed the interface to the construction of charts and formulas. Components of some dialog boxes were juxtaposed or shifted (horizontally or vertically). Otherwise few substantive changes were made that affect use of Excel in statistics.

Excel 97/98 has been used as the basis in this book. Still, there are Excel 5/95 users. Also, a student may be exposed to one version at school, another at home, and possibly a third at work.

An instructor using this manual could make the adjustments for Excel 5/95. Nonetheless, in order to provide a ramp into Excel 5/95, we provide this appendix to cover all changes that are relevant, mainly material in Chapters 1 and 2. Remarkably, it is only in these opening chapters on data description and functions where a detailed separate exposition is required. Elsewhere in the book, where Excel 5/95 and Excel 97/98 differ, either because of dialog boxes or different pull-down menus, for instance, parallel step-by-step descriptions are indicated for each version. There are only a few techniques that require this, and in each case the difference requires but a few lines.

Figures in this Appendix were taken using Excel 5 on a Macintosh.

A.1 The ChartWizard

The **ChartWizard** is a step-by-step approach to creating informative graphs. In Excel 5/95 a sequence of five dialog boxes guides the user through the creation of a customized chart. The user provides details about the chart type, formatting, titles, legends, etc. The **ChartWizard** can be activated either from the button on the **Standard Toolbar** or by choosing **Insert – Chart** from the Menu Bar. If you embed the chart on the workbook the mouse pointer becomes a cross hair + with the image of a chart next to it. Click on the workbook to locate the upper left corner of your graph output to create a default size. Otherwise click and drag the cursor to the lower right corner to create the desired size chart.

We illustrate use of the ChartWizard with the same example (Example 1.1) from Chapter 1.

	A	B	C
1	*Marital Status*	*Count (millions)*	*Percent*
2	Never married	43.9	22.9
3	Married	116.7	60.9
4	Widowed	13.4	7
5	Divorced	17.6	9.2

Figure A.1: Marital Status of U.S. Adults

Example A.1. (Page 6 in the text.) Figure A.1 shows the marital status for all Americans age 18 and over. Create a bar graph.

Creating a Bar Chart

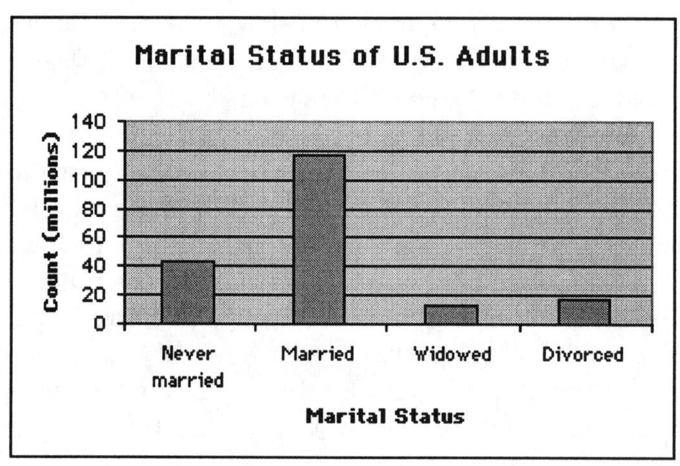

Figure A.2: Excel Bar Chart

The steps for creating a bar graph follow. For other types of graphical displays, make appropriate choices from the same sequence of dialog boxes. Figure A.2 is a bar graph, produced by Excel, that displays the same information as in Figure A.1. First enter the data and labels in cells A1:B7

Step 1. Select the cells where you have located the data, in this case cells A2:B5, and click on the **ChartWizard** on the **Standard Toolbar**. The pointer changes to a cross hair +. Now click in cell A6 (or some other cell) to locate the chart. (Alternatively you may first click on the **ChartWizard**, then click in cell A6 to locate the chart, and finally click and drag over the range A2:B5.) In either case, dialog box 1 of 5 appears (Figure A.3) with the selected range highlighted. Modify the range if necessary and then confirm by clicking the Next button.

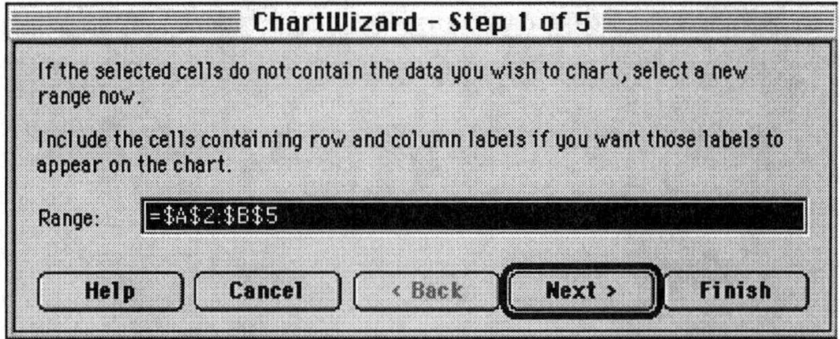

Figure A.3: ChartWizard—Step 1

Step 2. The **ChartWizard** (Figure A.4) displays the types of graphs that are available. Select **Column** and click Next.

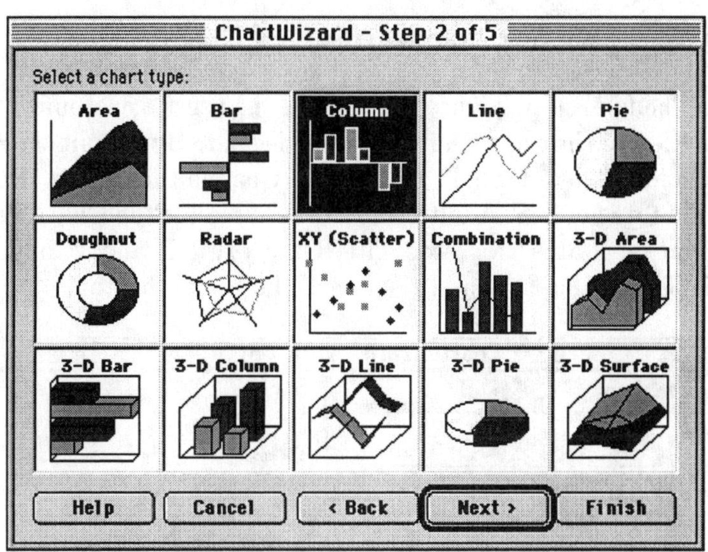

Figure A.4: ChartWizard—Step 2

Step 3. Various formatting options (Figure A.5) available for the chart type selected in Step 2 are presented. Select Format **6** and click Next.

Step 4. This dialog box (Figure A.6) presents the chart as it will appear by default. You may confirm or change its appearance. Because each variable is located in a column, select the radio button **Columns**. The first column of the data set will be located on the X axis. Therefore "1" should be in the text area for the Column(s) for Category(X) axis labels. We don't require a legend since only one variable is plotted. Thus "0" should be in the box for

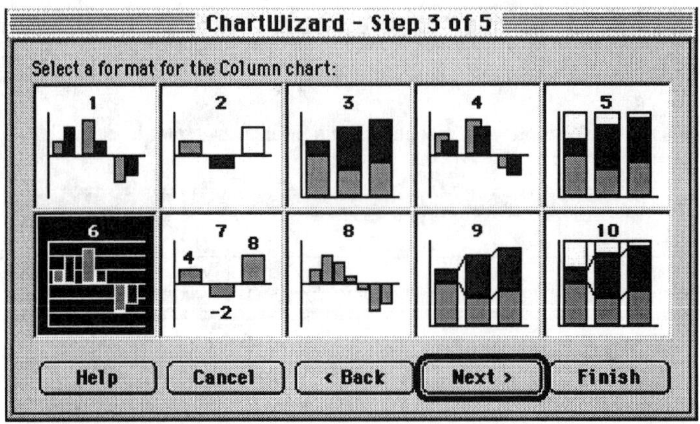

Figure A.5: ChartWizard—Step 3

Row(s) for Legend Text. All these choices are the defaults here. Click Next.

Step 5. The final step (Figure A.7) allows the user to customize the chart by adding titles for the chart, the axes, and a legend. By default the radio button for **Yes** under Add a legend? is selected. Change this to **No**. Enter the text "Marital Status of U.S. Adults" in the text area for the Chart Title. Finally, type "Marital Status" for Axis Title Category (X) and "Count (millions)" for Axis Title Value (Y), and then click the Finish button.

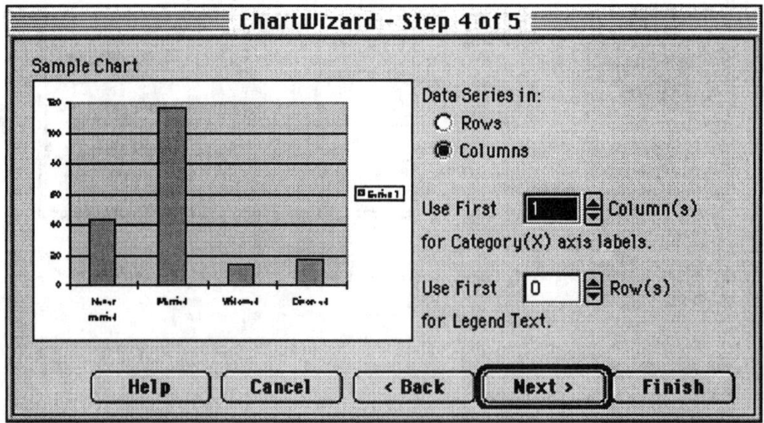

Figure A.6: ChartWizard—Step 4

The chart appears with **eight handles** (Figure A.8), indicating that it is selected. It can be resized by selecting a handle and then dragging the handle to the desired size. The chart can also be moved. Click the interior of the chart and drag to another location (holding the mouse button down). Click outside the chart to deselect.

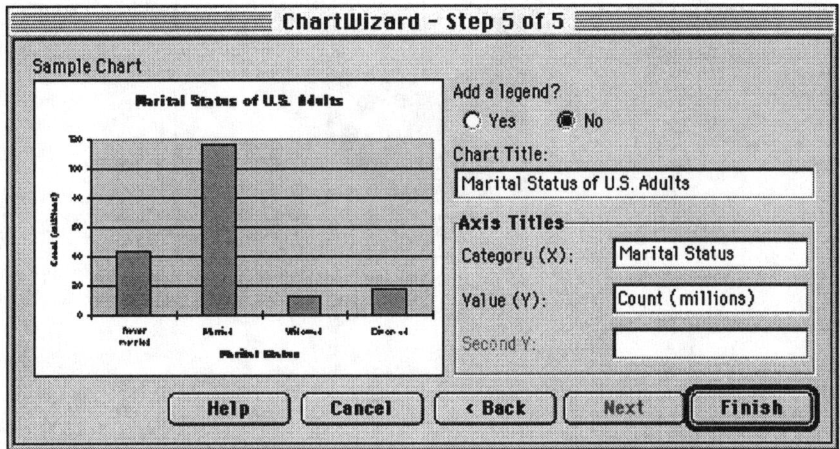

Figure A.7: ChartWizard—Step 5

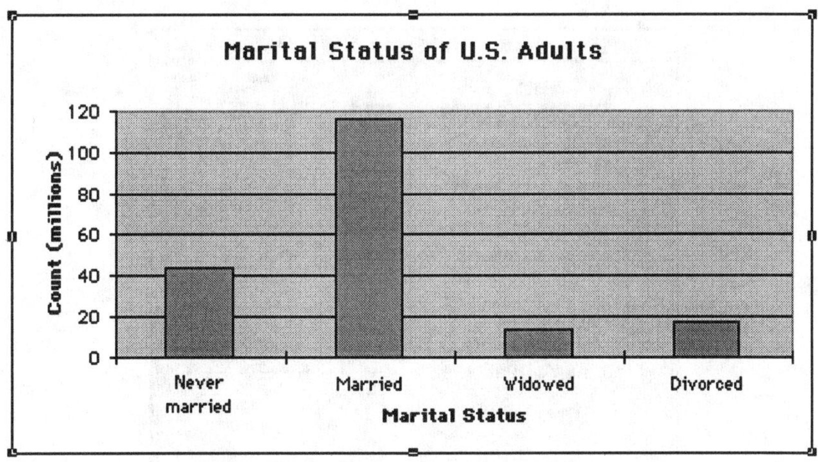

Figure A.8: Bar Chart Selected

Pie Charts

To produce a pie chart as in Figure A.9, select **Pie** in place of **Column** in Step 2 of the ChartWizard and follow the remaining steps.

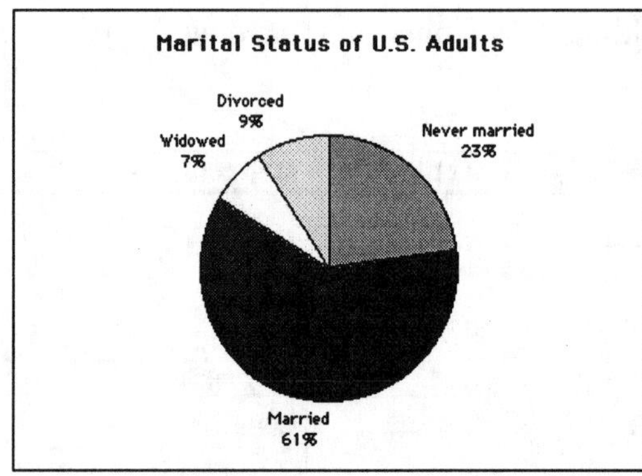

Figure A.9: Pie Chart

A.2 Histograms

The ChartWizard is designed for use with data that are already grouped, for instance, categorical variables or quantitative variables that have been grouped into categories or intervals. For raw numerical data, Excel provides additional commands using the **Analysis ToolPak**.

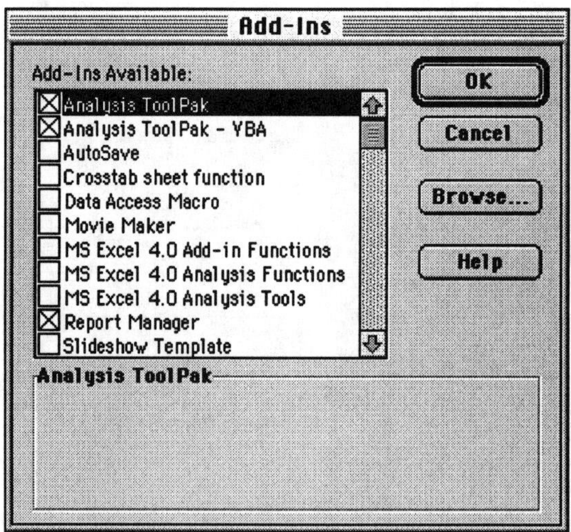

Figure A.10: Add-Ins Dialog Box

To determine whether this toolpak is installed, choose **Tools – Add-Ins** from the Menu Bar for the **Add-Ins** dialog box (Figure A.10). Depending on whether other Add-Ins have been loaded, your box might appear slightly different. If the Analysis ToolPak box is not checked, then select it and click OK. It will now be an option in the pull-down menu when you choose **Tools – Data Analysis**.

	A	B	C	D	E	F	G	H	I	J
1	*Survival Times of Guinea Pigs*									
2	43	45	53	56	56	57	58	66	67	73
3	74	79	80	80	81	81	81	82	83	83
4	84	88	89	91	91	92	92	97	99	99
5	100	100	101	102	102	102	103	104	107	108
6	109	113	114	118	121	123	126	128	137	138
7	139	144	145	147	156	162	174	178	179	184
8	191	198	211	214	243	249	329	380	403	511
9	522	688								

Figure A.11: Survival Times of Guinea Pigs

Histogram from Raw Data

Example A.2. (Exercise 1.80, page 72 in the text.) Make a histogram of survival times of 72 guinea pigs (Figure A.11) after they were injected with tubercle bacilli in a medical experiment.

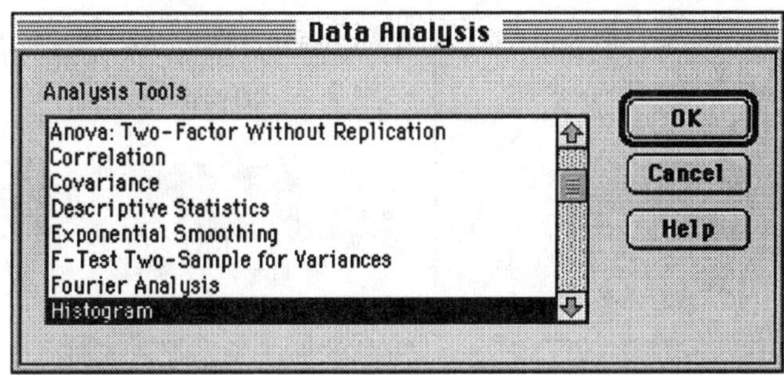

Figure A.12: Analysis ToolPak Dialog Box

Solution. Excel requires a contiguous block of data for the histogram tool.

1. Reenter the data in a block (A2:A73) and type the label "Times" in cell A1.

2. From the Menu Bar choose **Tools – Data Analysis** and scroll to the choice **Histogram** (Figure A.12). Click OK.

3. In the dialog box (Figure A.13) type the range A1:A73 in the **Input Range** area, which is the location on the workbook for the data. As with the Bar Chart you may instead click and drag from cell A1 to A73. The choice depends on whether your preference is for keyboard strokes or mouse clicks. Leave the **Bin Range** blank (because Excel will select the bins), check the **Labels** box because A1 has been included in the Input Range, type C1 for **Output Range**, and check the box **Chart Output**. The option Pareto (sorted histogram) constructs a histogram with the vertical bars sorted from

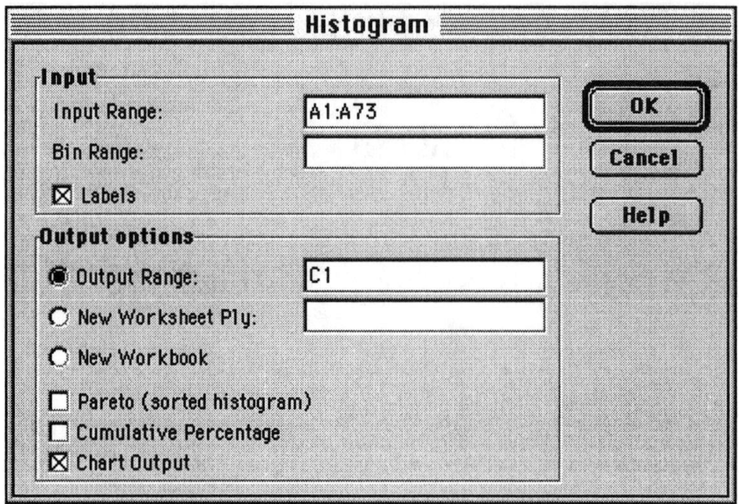

Figure A.13: Histogram Dialog Box

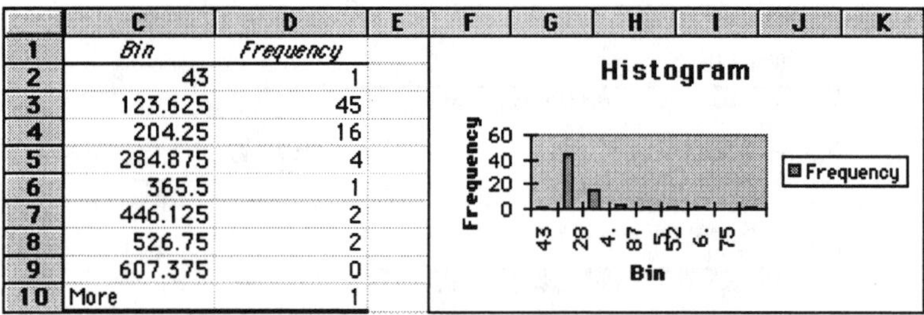

	C	D
1	*Bin*	*Frequency*
2	43	1
3	123.625	45
4	204.25	16
5	284.875	4
6	365.5	1
7	446.125	2
8	526.75	2
9	607.375	0
10	More	1

Figure A.14: Output Table and Default Histogram

left to right in decreasing height. If Cumulative Percentage is checked, the output will include a column of cumulative percentages.

4. The output appears in Figure A.14. The bin interval boundaries (actually, the upper limit for each interval) appear in cells C2:C10, while the corresponding frequencies in cells D2:D10. The histogram appears to the right. We shall shortly modify the histogram by changing the labels and allowing adjacent bars to touch. But first, we explain how to customize the selection of bins.

Changing the Bin Intervals

If the bin intervals are not specified, then Excel creates them automatically so that the number of bins roughly equals the square root of the number of observations beginning and ending at the minimum and maximum, respectively, of the data set.

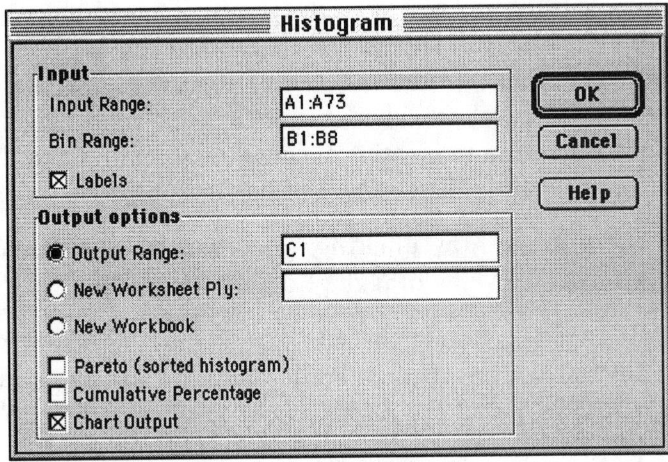

Figure A.15: New Bin Selection

In our example, we will first let Excel choose the default bins and then we will modify the histogram by selecting our own bin intervals.

1. Type "New Bin" (or another appropriate label) in cell B1. Then enter the values 100, 200, 300, 400, 500, 600, and 700 in B2:B8. An easy way to accomplish this is to type 100 and 200 in cells E2 and E3, respectively, then select E2 and E3, click the fill handle in the lower right corner of B3, then drag the fill handle down to cell B8 and release the mouse button.

	C	D
1	*New Bin*	*Frequency*
2	100	32
3	200	30
4	300	4
5	400	2
6	500	1
7	600	2
8	700	1
9	More	0

Figure A.16: Output Table and New Bin Histogram

2. Repeat the earlier procedure for creating a histogram, only this time type B1:B8 in the text area for **Bin Range** (Figure A.15). Before the output appears, you will be prompted with a warning that you are overwriting existing data. Continue and the new output (Figure A.16) will replace Figure A.14 with your selected bin intervals.

Enhancing the Histogram

While the default histogram captures the overall features of the data set, it is inadequate for presentation. Excel provides a set of tools for enhancing the histogram. These tools are too numerous to mention them all here, but a few will be discussed with reference to the example. The other options may be invoked analogously.

Resize. Select the histogram by **clicking once** within its boundary and resize using the handles. Move by dragging to a new location.

Bar Width. Adjacent bars do not touch in the default. To adjust the bar width, **double-click** the chart so that the border becomes a thick grey cross-hatched line (Figure A.17). Select the X axis by clicking it once (Figure A.18). From

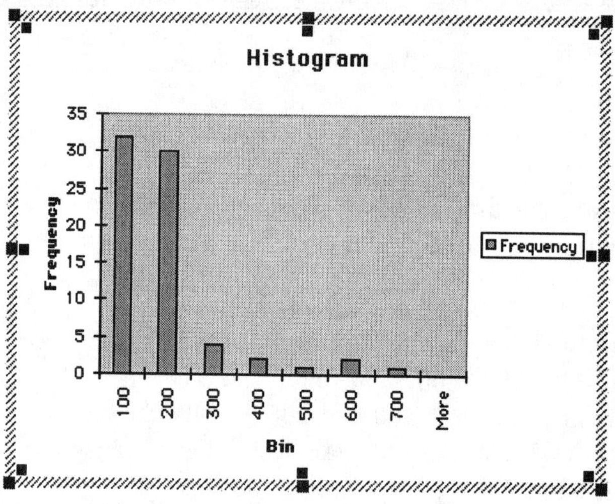

Figure A.17: Editing the Histogram

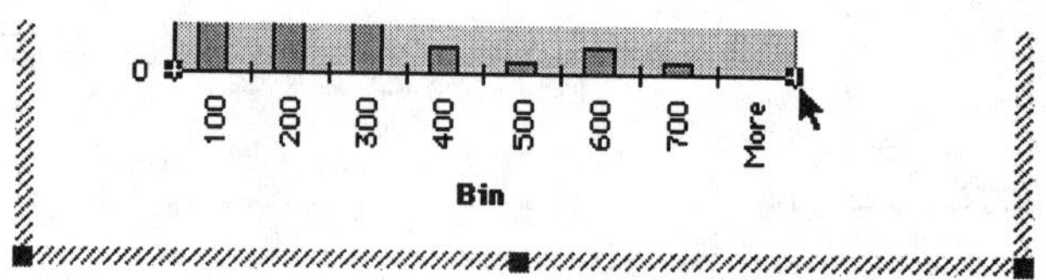

Figure A.18: Selecting the X Axis

the Menu Bar choose **Format – Column Group** and click the **Options** tab. Change the **GapWidth** to "0" and watch the histogram display change so that adjacent bars touch (Figure A.19).

Chart Title. Click on the title word "Histogram." A rectangular grey border with handles will surround the word, indicating that it is selected for editing.

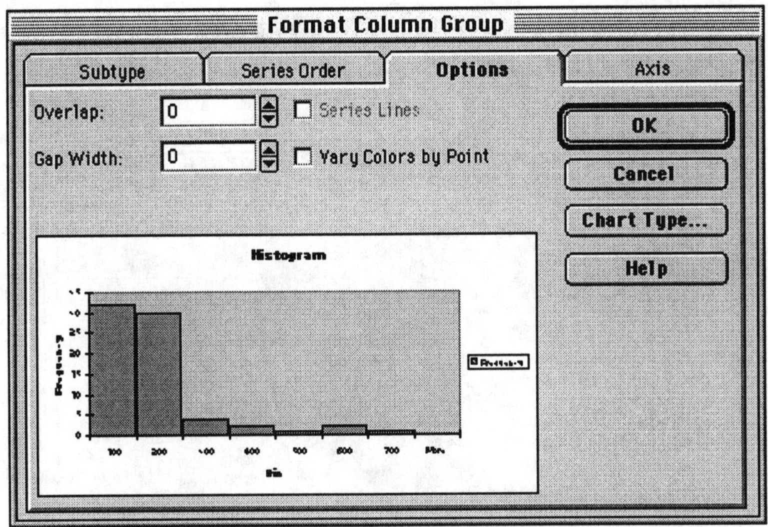

Figure A.19: Formatting Histogram Columns

Begin typing "Survival Times (Days) of Guinea Pigs," hold down the **Alt (Windows)** or the **Command (Macintosh)** key, and press enter. You may now type a second line of text in the **Formula Bar** entry area (Figure A.20). Continue typing "in a Medical Experiment," then click the Bold and the Italic buttons in the **Formatting Toolbar**.

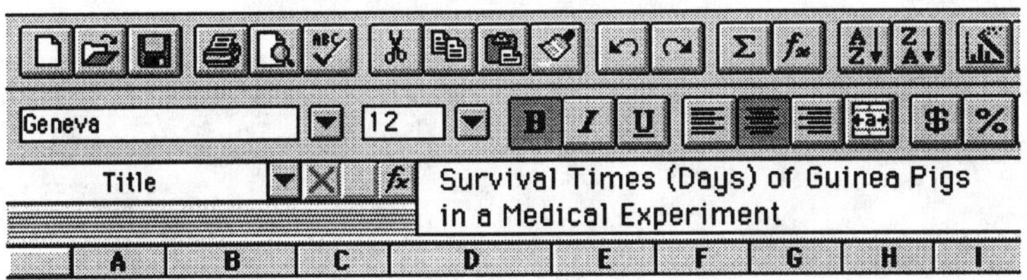

Figure A.20: Editing Chart Title

X axis Title. Click on the word Bin at the bottom of the chart (Figure A.17), type "Survival Time (Days)," and click Bold and Italic in the **Formatting Toolbar**.

Y axis Title. Click on the word "Frequency" on the left side and then click Bold and Italic.

X axis Format. Select the X axis by clicking it once (as illustrated already in Figure A.18). From the Menu Bar choose **Format – Selected Axis....** In the ensuing **Format Axis** dialog box (Figure A.21), you can click on various tabs to change the appearance of the X axis. Select the **Alignment** tab and change the orientation from Automatic to **horizontal** Text. Click OK.

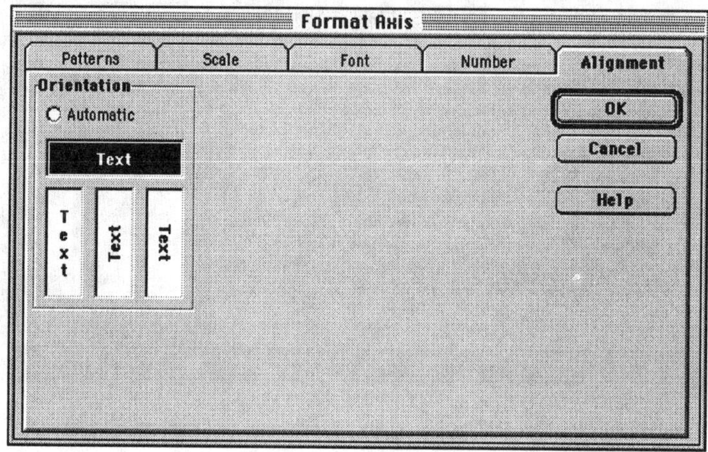

Figure A.21: Formatting the X Axis

Y axis Format. Select the Y axis by clicking it once. From the Menu Bar choose **Format – Selected Axis....** Click on the tab **Scale** and change the Maximum to 50 (Figure A.22). Click OK.

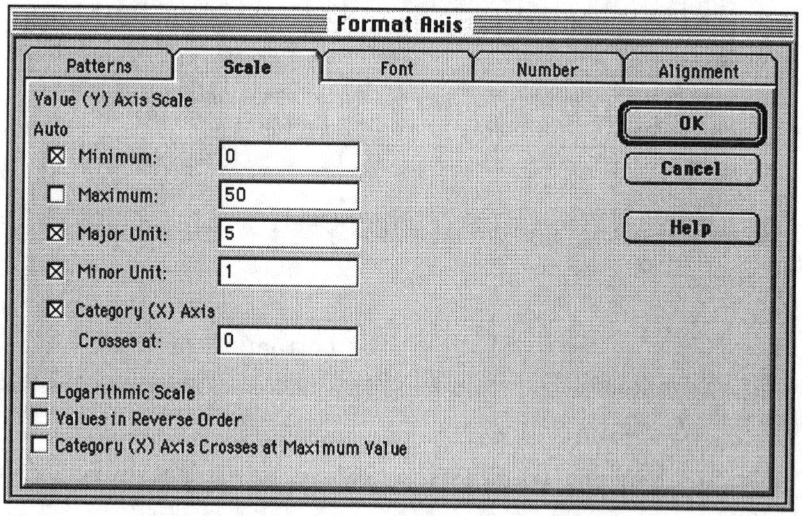

Figure A.22: Formatting the Y Axis

Legend. Remove the legend (which is not needed here) by clicking on it (the word "frequency" on the right in Figure A.17) and then pressing the delete key or choosing **Edit – Clear – All** from the Menu Bar.

More Interval. (Optional) Click any histogram bar, choose **Format – Selected Data Series...**, and in the **Format Data Series** box change the X values from \$D\$9 to \$D\$8, thereby removing the "More" interval. Click OK.

At the conclusion of the formatting, the histogram will appear as in Figure A.23.

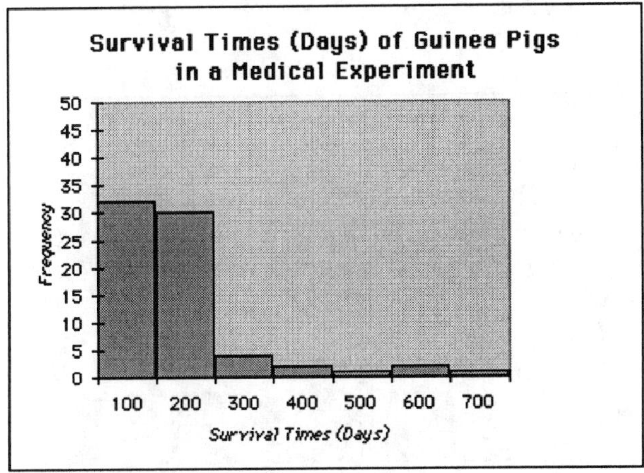

Figure A.23: Final Histogram after Editing

Histogram from Grouped Data

The **Histogram** tool requires the raw data as input. When numerical data has already been grouped into a frequency table, it is the **ChartWizard** which is the appropriate tool. First use it to obtain a bar chart and then modify it exactly as you would enhance a histogram.

> **Example A.3.** (Figure 1.13, page 46 in the text.) Figure A.24 gives the frequencies of vocabulary scores of all 947 seventh graders in Gary, Indiana, on the vocabulary part of the Iowa Test of Basic Skills. Column A is the bin interval and column B is the label for the histogram. To construct a histogram, select B1:C12, click on the **ChartWizard**, then follow the steps presented earlier with Figures A.3–A.7. The final histogram is shown in Figure A.25.

	A	B	C
1	Class	Bin	Number of Students
2	2.0 - 2.9	3	9
3	3.0 - 3.9	4	28
4	4.0 - 4.9	5	59
5	5.0 - 5.9	6	165
6	6.0 - 6.9	7	244
7	7.0 - 7.9	8	206
8	8.0 - 8.9	9	146
9	9.0 - 9.9	10	60
10	10.0 - 10.9	11	24
11	11.0 - 11.9	12	5
12	12.0 - 12.9	13	1
13	Total		947

Figure A.24: Vocabulary Scores

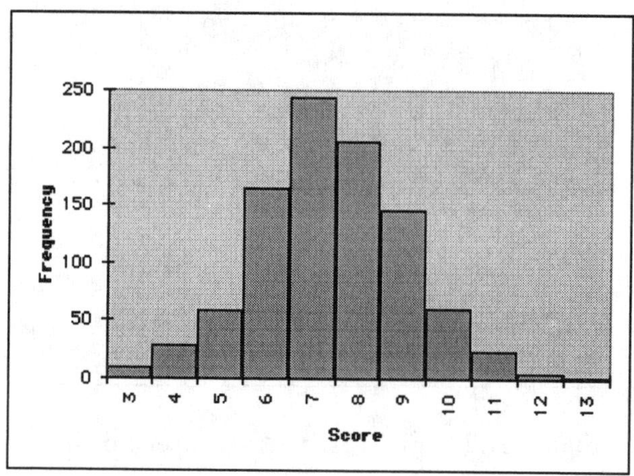

Figure A.25: Histogram from Grouped Data

A.3 The Function Wizard

The **Function Wizard** assists in entering formulas and functions—particularly complex ones—included in Excel. These functions can perform decision-making, action-taking, or value-returning operations. All formulas are entered in the **Formula Bar**, and the Function Wizard simplifies this process by guiding you step by step. There are three ways to invoke the Function Wizard: click on the button in the **Standard Toolbar**; click on the button in the **Formula Bar** after you have begun to enter a formula; choose **Insert – Function** from the Menu Bar. With all three methods a dialog box **Function Wizard – Step 1 of 2** appears. Here you select a function from **Function Category** and **Function Name** lists. The definition and syntax of a function's arguments appear when a function is highlighted. There is a help button that describes the function in greater detail. After you select a function, click Next for Step 2 of 2 in which you enter the arguments.

	A	B	C	D
1	Beef Hot Dogs		*Five Number Summary*	
2				
3	Calories			
4	186	Min	111	=MIN(A4:A23)
5	181	Q1	140.5	=QUARTILE(A4:A23,1)
6	176	Med	152.5	=MEDIAN(A4:A23)
7	149	Q3	177.25	=QUARTILE(A4:A23,3)
8	184	Max	190	=MAX(A4:A23)
9	190			
10	158			
11	139			
12	175			
13	148			
14	152			
15	111			
16	141			
17	153			
18	190			
19	157			
20	131			
21	149			
22	135			
23	132			

Figure A.26: Five-Number Summary

These can be typed directly or *referenced* by using the mouse to point to data by clicking and dragging over cells.

We illustrate use of the **Function Wizard** by deriving the five-number summary of the calories data set shown in Figure 1.24 in Section 1.3.

The Five-Number Summary

Example A.4. (See Exercise 1.39, page 42 in the text.) Figure 1.24 shows the calories and sodium levels measured in three types of hot dogs: beef, meat (mainly pork and beef), and poultry. Find the five-number summary {minimum, first quartile, median, third quartile, maximum} for the calorie distribution of the hot dogs.

Solution. For illustration purposes we consider only the beef calories data.

1. Referring to Figure A.26 where we have entered the beef calorie data in cells A4:A23 of a workbook, enter the labels "Min," "Q1," "Med," "Q3," and "Max" in cells B4:B8. Select cell C5 to enter the quartile function.

2. Click on the **Function Wizard** button in the **Standard Toolbar** and select **Statistical** under Function Category and `QUARTILE` under Function Name in the first dialog box. Click Next.

3. The second dialog box brings up the **QUARTILE** box. Enter A4:A23 for the data **array** and "1" for the **quart** to indicate the first quartile (Figure A.27). The upper portion of Figure A.27 shows the completed formula,

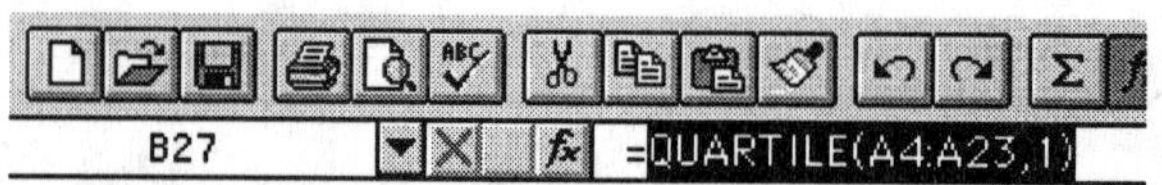

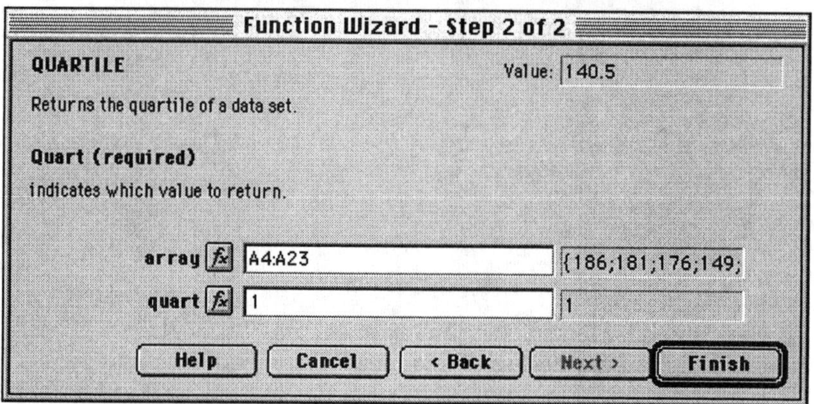

Figure A.27: Function Wizard Dialog Box—Quartile Function

which appears automatically in the **Formula Toolbar**. Click Finish and the Function Wizard constructs the function and prints the value 140.5 in cell C5.

4. Continue with the rest of the five-number summary, either using the **Function Wizard** or entering the formulas by hand. Cells C4:C8 present the syntax while the values are in B4:B8.

The five-number summary is {111, 140.5, 152.5, 177.25, 190}. Note that Excel uses a slightly different definition of quartiles for a finite data set than the text.

A.4 Scatterplot

Statistical studies are often carried out to learn whether or how much one measurement (an explanatory variable x) can be used to predict the value of another measurement (a response variable y). Once data are collected, through either a controlled experiment or an observational study, it is useful to examine graphically whether any relationship is justified. We might plot in Cartesian coordinates all pairs (x_i, y_i) of observed values. The resulting graph is called a **Scatterplot**.

> **Example A.5.** (Exercise 2.9, page 121 in the text.) In 1974 the Franklin National Bank failed. Franklin was one of the 20 largest banks in the nation and the largest to fail. Could Franklin's weakened condition have been detected in advance by simple data analysis? Table A.1 gives the total assets (in billions of dollars) and net income (in millions

Table A.1: Assets and Income for Banks

Bank	1	2	3	4	5	6	7	8	9	10
Assets	49.0	42.3	36.3	16.4	14.9	14.2	13.5	13.4	13.2	11.8
Income	218.8	265.6	170.9	85.9	88.1	63.6	96.9	60.9	144.2	53.6

Bank	11	12	13	14	15	16	17	18	19	20
Assets	11.6	9.5	9.4	7.5	7.2	6.7	6.0	4.6	3.8	3.4
Income	42.9	32.4	68.3	48.6	32.3	42.7	28.9	40.7	13.8	22.2

of dollars) for the 20 largest banks in 1973, the year before Franklin failed. Franklin is bank number 19. Make a scatterplot of these data that displays the relation between assets and income.

Creating a Scatterplot

The steps involved in creating a scatterplot are similar to those for producing a **Histogram** using the **ChartWizard**, except that we use the **Scatterplot** chart type in the ChartWizard.

1. Enter the data from Table A.1 into cells A2:A21 and B2:B21 of a workbook with the labels "Assets" and "Income," referring to Figure A.31 later in this section.

2. Select cells A2:B21, click on the **ChartWizard**, and then click and drag the cross hair from one empty cell to another to select the rectangle in which the scatterplot will be displayed (block C1:H26 here).

3. In Step 1 dialog boxes appear (as in Figures A.3–A.7 in this chapter) beginning by asking you to confirm the data range selected. Click Next.

4. In Step 2 select **XY (Scatter)** as the chart type from the selection offered. Click Next.

5. In Step 3 specify Format **3** (Figure A.28).

6. In Step 4 the dialog box (Figure A.29) shows a preview of the chart with the options selected. Confirm that the radio button for Data Series in **Columns** is selected, that the spin box for Column(s) for X Data contains "1," which tells Excel to use Column B for the x (horizontally plotted) variable, and that "0" appears in the box Row(s) for Legend Text. A legend is not appropriate because each point represents two values (x_i, y_i) (Figure A.29). Click Next.

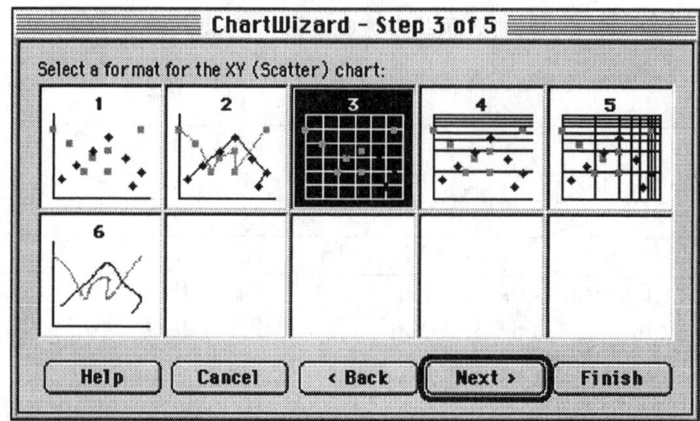

Figure A.28: Selecting Scatterplot in ChartWizard—Step 3

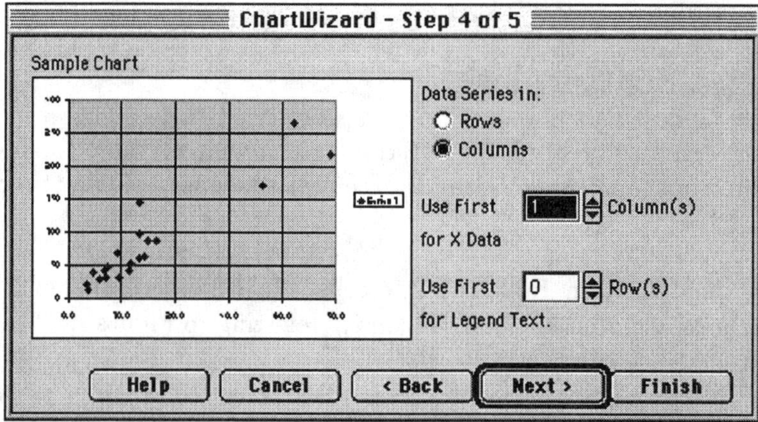

Figure A.29: Preview of Scatterplot—Step 4

7. In Step 5 select radio button **No** for Add a legend? and note that you can also change titles in this step (Figure A.30).

8. The scatterplot appears embedded on your workbook (Figure A.31).

Note: Excel uses a range from 0 to 100% as the default, and sometimes the scatterplot will show unwanted blank space. That is not the case here, but to change the horizontal scale select the X axis by clicking it once, from the Menu Bar choose **Format – Selected Axis...**, and complete the dialog box. Similarly select the Y axis for editing. Refer to the discussion for enhancing a histogram in Section A.2 and Section 2.1.

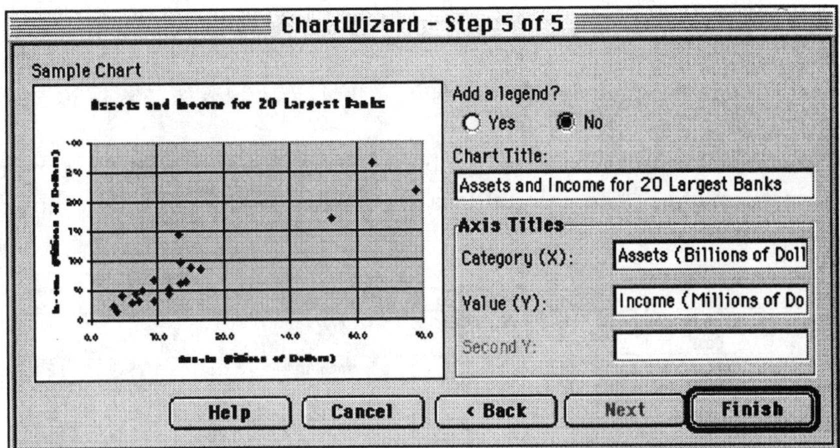

Figure A.30: ChartWizard—Step 5

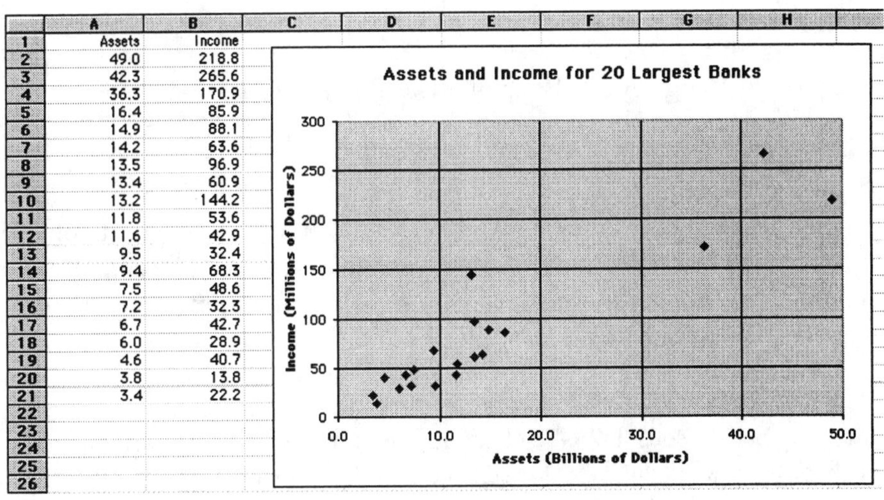

	A	B
1	Assets	Income
2	49.0	218.8
3	42.3	265.6
4	36.3	170.9
5	16.4	85.9
6	14.9	88.1
7	14.2	63.6
8	13.5	96.9
9	13.4	60.9
10	13.2	144.2
11	11.8	53.6
12	11.6	42.9
13	9.5	32.4
14	9.4	68.3
15	7.5	48.6
16	7.2	32.3
17	6.7	42.7
18	6.0	28.9
19	4.6	40.7
20	3.8	13.8
21	3.4	22.2

Figure A.31: Bank Data and Scatterplot Embedded on Sheet

A.5 Boxplots

Excel does not provide a boxplot. However the Microsoft Personal Support Center has a web page "How to Create a BoxPlot – Box and Whisker Chart" located at

http : //support.microsoft.com/support/kb/articles/q155/1/30.asp

with instructions for creating a reasonable boxplot. We illustrate by constructing side-by-side boxplots of the calorie data for beef, meat, and poultry hot dogs in Example A.3.

	A	B	C	D	E	F	G
1			*Boxplots of Hot Dog Calories*				
2	Beef	Meat	Poultry		Beef	Meat	Poultry
3	186	173	129	median	152.5	153	129
4	181	191	132	Q1	140.5	139	102
5	176	182	102	min	111	107	86
6	149	190	106	max	190	195	170
7	184	172	94	Q3	177.25	179	143
8	190	147	102				
9	158	146	87				
10	139	139	99	*Formulas for Beef Column*			
11	175	175	170	median	=MEDIAN(A3:A22)		
12	148	136	113	Q1	=QUARTILE(A3:A22,1)		
13	152	179	135	min	=MIN(A3:A22)		
14	111	153	142	max	=MAX(A3:A22)		
15	141	107	86	Q3	=QUARTILE(A3:A22,3)		
16	153	195	143				
17	190	135	152				
18	157	140	146				
19	131	138	144				
20	149						
21	135						
22	132						

Figure A.32: Boxplot—Data and Preparation

Step 1. Enter the calorie data into three columns of a workbook (Figure A.32), then find and enter the five-number summary into another three columns **in the order** median, first quartile, minimum, maximum, third quartile. We have entered this information, including labels in block D3:G7 in Figure A.32.

Step 2. Select cells D2:G7, click on the **ChartWizard** button, and then click on the worksheet to locate the cell for the upper left corner of your boxplot. If you want to place the boxplot on a new sheet, then after you have selected D2:G7, choose **Insert – Chart – As New Sheet** from the Menu Bar. We have located the boxplot to begin in cell H1. Click Next (Figure A.33).

Step 3. Select the **Combination** chart type and click Next.

Step 4. Select chart style number **6** and click Next. An Alert Box appears with the following warning: **A volume-open-high-low-close stock chart must contain five series.** Click OK.

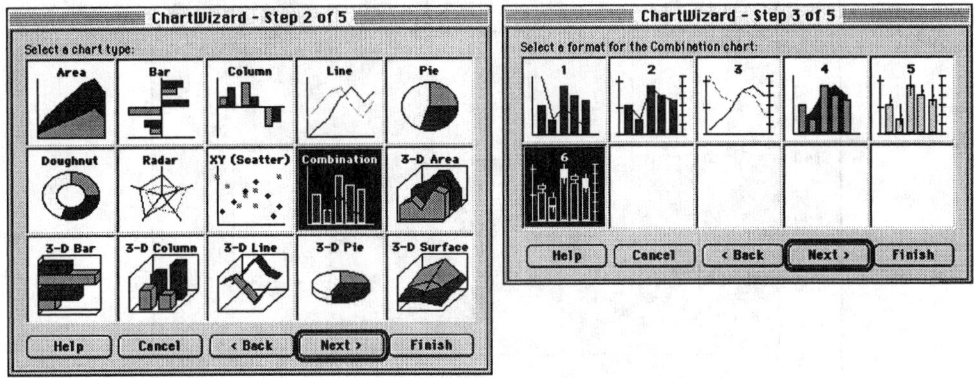

Figure A.33: Boxplot—ChartWizard Steps 2 and 3

Step 5. In the next dialog box check the radio button **Rows** for Data Series in: and click Next (Figure A.34).

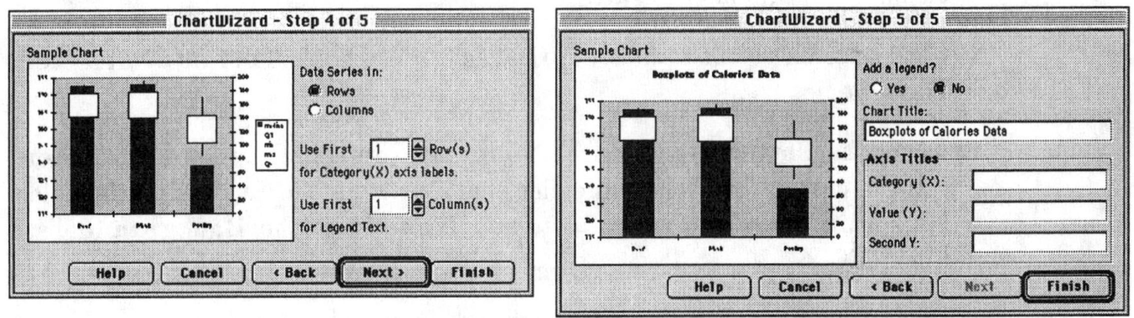

Figure A.34: Boxplot—ChartWizard Steps 4 and 5

Step 6. In the final dialog box check the radio button **No** for Add a legend? and add a Chart Title "Boxplots of Calories Data."

Next, we edit this chart.

1. Double-click the chart to activate it. From the Menu Bar choose **Insert – Axes** and clear the check box next to **Value (Y) Axis** under **Secondary Axis**. Click OK.

2. Click once on any one of the colored columns to select the series. Do not click one of the white columns. From the Menu Bar choose **Format – Chart**

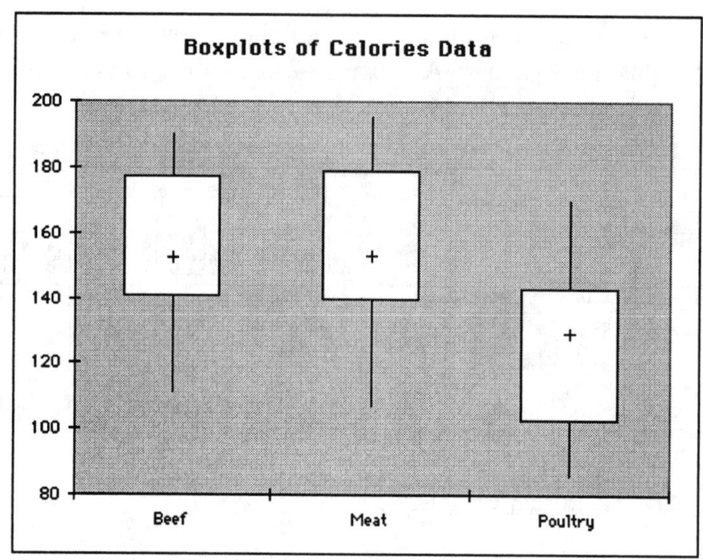

Figure A.35: Boxplot of Calories Data

Type..., click **Line**, and then click OK. A line that connects the three white columns appears in the chart.

3. Click once on the line and from the Menu Bar choose **Format – Selected Data Series...**.

4. Under the **Patterns tab**, select **None** for **Line** and **Custom** for **Marker.** For the custom marker choose the plus sign from the **Style** list, the color black from the **Foreground** list, and **None** from the **Background** list. Click OK.

5. Double-click the **Y axis** and under the **Scale** tab set the Minimum to 80 and click OK. The final boxplot appears on your sheet (Figure A.35).

Index

ASCUS

SYRIA

MODERN JERUSAL
See pp

IR

THE MUSLIM QUARTER
See pp60–75

JORDAN

JERUSALEM

MODERN JERUSALEM

THE MUSLIM QUARTER

THE CHRISTIAN AND ARMENIAN QUARTERS

THE JEWISH QUARTER

THE MOUNT OF OLIVES AND MOUNT ZION

0 metres 800

0 yards 800

THE CHRISTIAN AND ARMENIAN QUARTERS
See pp88–107

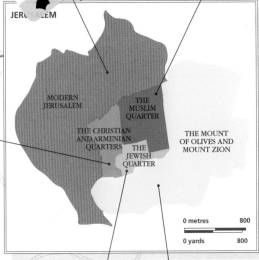

SAUDI ARABIA

FURTHER AFIELD
See pp128–139

THE JEWISH QUARTER
See pp76–85

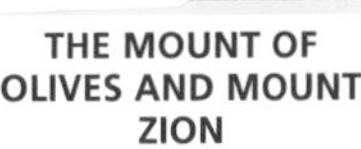

THE MOUNT OF OLIVES AND MOUNT ZION
See pp108–117

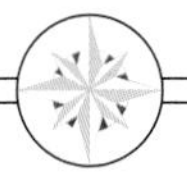

EYEWITNESS TRAVEL

JERUSALEM,

ISRAEL, PETRA & SINAI

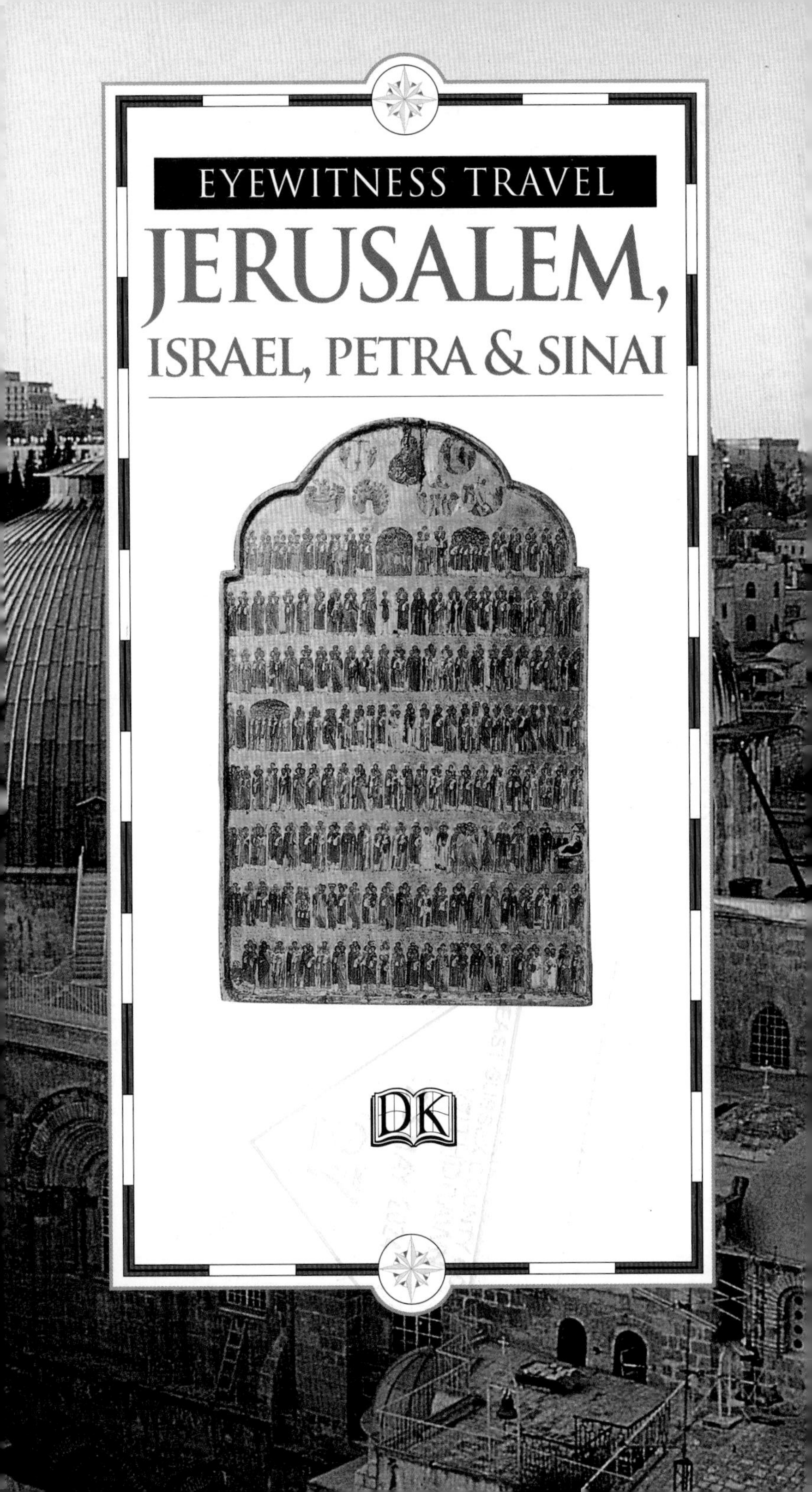
EYEWITNESS TRAVEL
JERUSALEM,
ISRAEL, PETRA & SINAI
DK

LONDON, NEW YORK,
MELBOURNE, MUNICH AND DELHI
www.dk.com

PROJECT EDITORS Nick Inman, Ferdie McDonald
ART EDITORS Jo Doran, Paul Jackson
COMMISSIONING EDITOR Giovanni Francesio
at Fabio Ratti Editoria S.r.l.
EDITORS Elizabeth Atherton, Cathy Day, Simon Hall,
Freddy Hamilton, Andrew Humphreys
DESIGNERS Chris Lee Jones, Anthony Limerick,
Sue Metcalfe-Megginson, Rebecca Milner, Johnny Pau
PICTURE RESEARCH Monica Allende, Katherine Mesquita
MAP CO-ORDINATOR Dave Pugh
DTP DESIGNER Maite Lantaron
RESEARCHER Karen Ben-Zoor

MAIN CONTRIBUTORS
Fabrizio Ardito, Cristina Gambaro, Massimo Acanfora Torrefranca

PHOTOGRAPHY
Eddie Gerald, Hanan Isachar, Richard Nowitz,
Magnus Rew, Visions of the Land

ILLUSTRATORS
Isidoro Gonzáles-Adalid Cabezas (Acanto Arquitectura y Urbanismo S.L.), Stephen Conlin, Gary Cross, Chris Forsey, Andrew MacDonald, Maltings Partnership, Jill Munford, Chris Orr & Associates, Pat Thorne, John Woodcock

Reproduced by Colourscan, Singapore
Printed and bound by South China Printing Co. Ltd, China

First published in Great Britain in 2000
by Dorling Kindersley Limited, 80 Strand, London WC2R 0RL

Reprinted with revisions 2002, 2007, 2010

A CIP CATALOGUE RECORD IS AVAILABLE FROM THE BRITISH LIBRARY.

ISBN 978-1-4053-5856-9

FLOORS ARE REFERRED TO THROUGHOUT INACCORDANCE WITH EUROPEAN USAGE; IE THE "FIRST FLOOR" IS THE FLOOR ABOVE GROUND LEVEL.

NOTE: THE TERM "HOLY LAND" HAS BEEN USED TO DESCRIBE THE AREAS COVERED BY THIS GUIDE.

Front cover main image: Dome of the Rock, Temple Mount, Jerusalem

◁ View over the rooftops of Jerusalem's Christian Quarter

Mount of Olives, Jerusalem

CONTENTS

Old Jaffa's attractive waterfront

JERUSALEM AREA BY AREA

Window detail, Dome of the Rock

ISRAEL, PETRA & SINAI REGION BY REGION

Bedouin camel, Western Jordan

TRAVELLERS' NEEDS

Middle Eastern handicrafts

SURVIVAL GUIDE

Pomegranates

The remote St Catherine's Monastery in Sinai

HOW TO USE THIS GUIDE

This guide helps you to get the most from your visit to Jerusalem and the Holy Land, by providing detailed practical information. *Introducing Jerusalem, Israel, Petra & Sinai* maps the region and sets it in its historical and cultural context. The Jerusalem section and the four regional chapters describe important sights, using maps, photographs and illustrations. Features cover topics from food to wildlife. Recommended hotels and restaurants are listed in *Travellers' Needs*, while the *Survival Guide* has tips on travel, money and other practical matters.

JERUSALEM AREA BY AREA

The city is divided into five areas, each with its own chapter. A last chapter, *Further Afield*, covers peripheral sights. All sights are numbered and plotted on the chapter's area map. The detailed descriptions of the sights are easy to locate, as they follow the numerical order on the map.

A locator map shows where you are in relation to other areas of the city centre.

Each area of Jerusalem has its own colour-coded thumb tab, as shown inside the front cover.

Sights at a Glance lists the chapter's sights by category: Holy Places, Historic Districts, Museums and Archaeological Sites.

1 Area Map
For easy reference, sights are numbered and located on a map. The central sights are also marked on the Street Finder maps *on pages 156–59.*

2 Street-by-Street Map
This gives a bird's-eye view of the key area in each chapter.

Stars indicate the sights that no visitor should miss.

Walking routes, shown in red, suggest where to visit on foot.

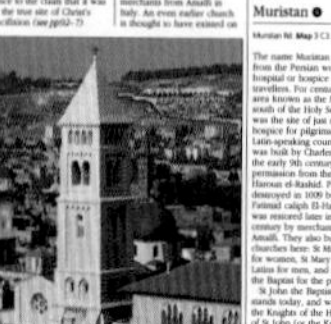

3 Detailed information
The main sights in the city are described individually. Addresses, telephone numbers and opening hours are given, as well as information on admission charges, guided tours, photography, wheelchair access and public transport.

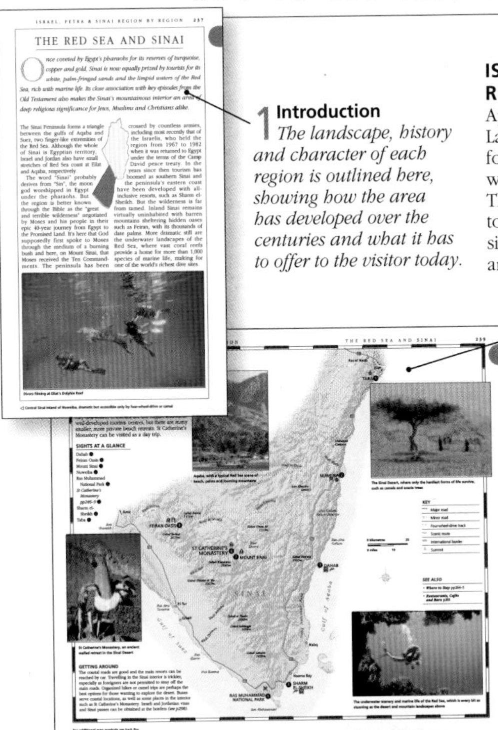

ISRAEL, PETRA & SINAI REGION BY REGION

Apart from Jerusalem, the Holy Land has been divided into four other regions, each of which has a separate chapter. The most interesting cities, towns, historical and religious sites, and other places of interest, are located on a *Regional Map*.

1 Introduction
The landscape, history and character of each region is outlined here, showing how the area has developed over the centuries and what it has to offer to the visitor today.

2 Regional Map
This shows the road network and gives an illustrated overview of the whole region. Interesting places to visit are numbered and there are also useful tips on getting to and around the region by car and public transport.

Each region of the Holy Land can be quickly identified by its colour-coded thumb tabs (see inside front cover).

3 Detailed information
All the important towns and other places to visit are described individually. They are listed in order, following the numbering on the Regional Map. *Within each town or city, there is detailed information on important buildings and other sights.*

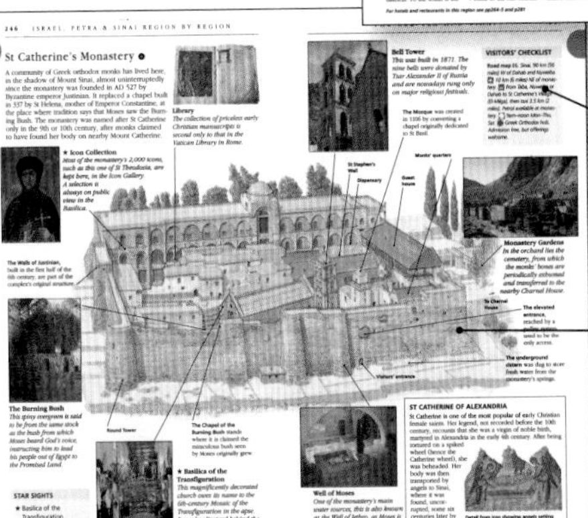

For all major sights, a Visitors' Checklist provides the practical information you will need to plan your visit.

4 The Top Sights
These are given two or more full pages. Historic buildings are dissected to reveal their interiors. Other interesting sights and areas are mapped or shown in bird's-eye view, with the most important features described.

INTRODUCING JERUSALEM, ISRAEL, PETRA & SINAI

DISCOVERING THE HOLY LAND

The "Holy Land" encompasses Israel and large regions of Jordan and Egypt. Rich in associations with three of the world's major faiths – Christianity, Judaism and Islam – it is a fascinating and diverse destination for pilgrims and holidaymakers alike. Religious highlights include the biblical sites of Jerusalem, Galilee and Mount Sinai, and an array of churches, monasteries and mosques. This is also an area of great natural beauty, from the desert landscapes of Jordan and Sinai to the lush greenery of northern Israel and the white sands of the Mediterranean and Red Sea coasts. These two pages are designed to help visitors pinpoint the highlights of this exciting region.

Mosaic in the Jewish Quarter

Jerusalem's Old City walls, built by Suleyman the Magnificent

JERUSALEM

- **Biblical sites**
- **The Western Wall and Dome of the Rock**
- **Museum of the Holocaust**

It's hard to overstate the historical significance of Jerusalem. Any trip begins with an exploration of the tightly walled Old City, home to the cornerstones of three faiths. It has the **Western Wall** *(see p85)* of Judaism; the Christian sites of the **Via Dolorosa** *(see pp30–31)* and **Church of the Holy Sepulchre** *(see pp92–5)*; and the third holiest site of Islam, the **Dome of the Rock** *(see pp72–3)*. Beyond these are many more attractions of similar significance, including the Mount of Olives, with its marvellous views over the city, not to mention more churches, synagogues and mosques, Roman and Byzantine remains, medieval walls and gates, and colourful markets and bazaars.

Visits to the **Mea Shearim** *(see p125)* quarter of the new city, the Holocaust museum of **Yad Vashem** *(see p138)*, and an evening in the 19th-century neighbourhood of **Nakhalat Shiva** *(see p123)* bring the Jewish Jerusalem experience up-to-date.

THE COAST AND GALILEE

- **Beach life in Tel Aviv**
- **The Crusader port of Akko**
- **The Sea of Galilee**

Tel Aviv *(see pp168–73)* is worlds apart from Jerusalem. Jerusalem is a millennia-old hill-top city, weighted with religious significance. Tel Aviv is a secular beachfront city that basks beneath a Mediterranean sun and is barely a century old. Visit Tel Aviv for the superb **Beit Hatfutsot (Museum of the Jewish People)** *(see p168)* and the similarly impressive **Tel Aviv Museum of Art** *(see p170)*, and for its unrivalled heritage of white-washed **Bauhaus architecture** *(see p171)*. Also visit for the shopping, dining and nightlife, in which the city excels. Don't miss the neighbouring ancient port of **Jaffa** *(see pp174–5)* with its attractive harbour-side buildings, several of which house good seafood restaurants.

North along the coast, **Akko** *(see pp178–9)* is another old Arab port, although heavily shaped by the Crusaders, for whom this was one of their principal strongholds. It remains perhaps the most attractive old town in the entire Holy Land. Away from the coast, the **Sea of Galilee** *(see pp182–3)* is Israel's largest freshwater body. It has significant biblical links (it is where Jesus is said to have walked on the water), as well as a beautiful setting ringed by green hills.

The Mediterranean Sea laps at the beaches of central Tel Aviv

THE DEAD SEA AND THE NEGEV DESERT

- **Float on the Dead Sea**
- **Waterfalls and wildlife at Ein Gedi**
- **The legendary fortress of Masada**

Floating on the highly saline waters of the **Dead Sea** *(see p197)*, reading a book, is the oddest of sensations, and one every visitor should experience for themselves. Most people choose to go to Ein Gedi, where there is a wide beach popular with bathers, and showers to remove the water's filmy residue. **Ein Gedi** is also home to a nature reserve *(see p196)* with lush vegetation, twin gorges, waterfalls and abundant wildlife. Further south is **Masada** *(see pp200–201)*, a mountain-top fortress constructed by King Herod but famous for the Jewish defenders who killed themselves rather than be captured by the Romans.

The ancient mountain-top citadel of Masada in the Judaean desert

PETRA AND WESTERN JORDAN

- **Roman ruins at Jerash**
- **The rock-cut, secret city of Petra**
- **Wadi Rum's desert landscapes**

Jordan's capital, **Amman** *(see pp212–14)*, boasts some Roman ruins of its own, but it also makes a good base for a day trip to the even more impressive ruins at **Jerash** *(see pp210–11)*. This is one of the best-preserved Roman cities in the Middle East, with an almost complete theatre that is still used during the annual Jordan Festival.

South of Amman, the town of **Madaba** *(see pp216–17)* is worth visiting for its unique Byzantine-era mosaic map. However, the real reason that most people visit Jordan lies farther south still: **Petra** *(see pp220–31)*. The legendary "Rose City" is one of the most spectacular of archaeological sites, and ranks alongside India's Taj Mahal and the Pyramids of Egypt as one of the world's must-see sights. It is possible to see the highlights in one day but there is so much to see that Petra rewards repeated visits. Be sure to allow time for **Wadi Rum** *(see pp232–4)*, with its wide landscapes of red sands and towering mountains of wind-eroded sandstone.

Bedouin guides lead their camels through Jordan's Wadi Rum

THE RED SEA AND SINAI

- **Dive among magnificent coral reefs**
- **Visit one of the world's oldest monasteries**
- **Watch the sun rise over the Sinai desert**

The appeal for most visitors to the Sinai lies not on the land but in the dramatic underwater landscapes of the **Red Sea** *(see pp240–1)*. Here, vast coral reefs provide a home to a magical array of multi-hued marine life. This is one of the world's top diving locations, but a simple snorkel and flippers can be enough to experience this aquatic wonderland. Several resort towns provide beach-front accommodation and water-sport opportunities.

Another of Sinai's attractions is **St Catherine's Monastery** *(see pp246–8)*, where a community of Orthodox monks has lived in a walled compound since the sixth century. Visitors are allowed inside to visit parts of the holy retreat.

Behind St Catherine's rises **Mount Sinai** *(see p249)*, where, according to tradition, Moses encountered the "burning bush" and received the Ten Commandments. Modern-day pilgrims ascend the 3,700 steps to the summit to witness the sun rise over the peaks of the peninsula.

Scuba divers wading out from the beach on the Sinai coast

Putting the Holy Land on the Map

The crossroads of three continents – Africa to the south, Asia to the east and Europe to the west – the Holy Land encompasses the whole of Israel and the Palestinian Autonomous Territories, and parts of Jordan and Egypt. Its boundaries could be said to stretch from the Mediterranean in the west, inland to the Jordanian deserts, and from Galilee in the north to the southern tip of the Sinai peninsula. At the core of the Holy Land is Jerusalem, an ancient walled city which stands on the Judaean hills, just to the west of the Dead Sea, the lowest point on earth.

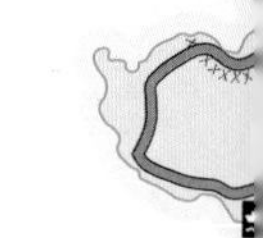

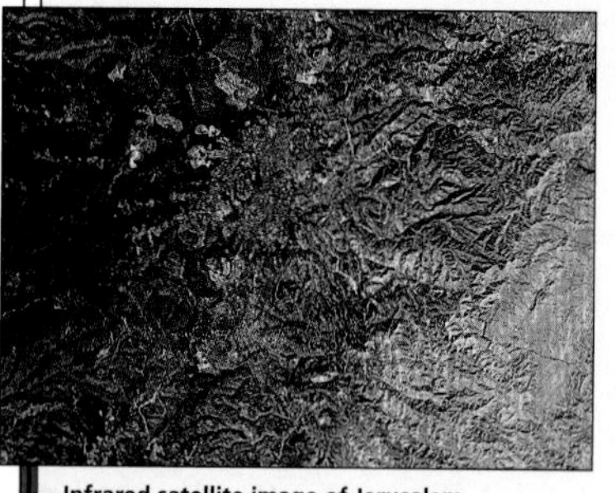

Infrared satellite image of Jerusalem

TU
Anam
MEDITERRANEAN SEA
Marsa Matruh
Baltim
Alexandria
Dumyat
Port Said
Damanhur
Nile Delta
Tanta
El-Aris
Suez Canal
Great Bitter Lake
Suez
Nakhl
EGYPT
SINAI
Gulf of Suez
Mount Sinai
El-Tur
Sharm el-Sheikh
Hurghada
Port Safaga
Nile
0 kilometres 100
0 miles 100

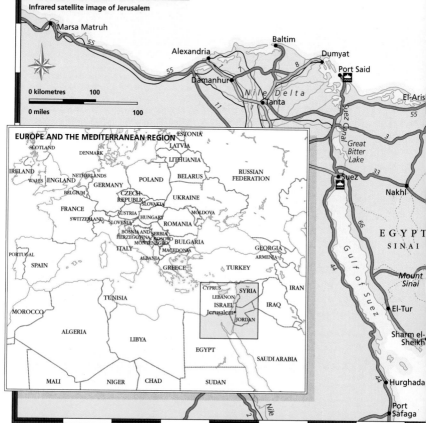

EUROPE AND THE MEDITERRANEAN REGION
SCOTLAND
DENMARK
ESTONIA
LATVIA
LITHUANIA
IRELAND
WALES
ENGLAND
NETHERLANDS
GERMANY
POLAND
BELARUS
RUSSIAN FEDERATION
BELGIUM
CZECH REPUBLIC
SLOVAKIA
UKRAINE
FRANCE
AUSTRIA
HUNGARY
MOLDOVA
SWITZERLAND
SLOVENIA
ROMANIA
BOSNIA AND HERZEGOVINA
SERBIA
KOSOVO
MONTENEGRO
BULGARIA
ITALY
MACEDONIA
GEORGIA
ARMENIA
PORTUGAL
SPAIN
ALBANIA
GREECE
TURKEY
CYPRUS
SYRIA
IRAN
TUNISIA
LEBANON
ISRAEL
Jerusalem
JORDAN
IRAQ
MOROCCO
ALGERIA
LIBYA
EGYPT
SAUDI ARABIA
MALI
NIGER
CHAD
SUDAN

◁ The Monastery at Petra *(see p230)* in a 19th-century engraving by David Roberts

KEY
International airport
Domestic airport
Major port
Motorway
Major road
Minor road
Rail line
Wadi
International boundary
Boundary of disputed area
Mersin
Adana
Ceyhan
Gaziantep
Birecik
Silifke
Goksu
Iskenderun
Azaz
Euphrates
ALEPPO
Lake Assad
Antakya
Idlib
Latakia
Orontes
NICOSIA
Famagusta
YPRUS
Hama
Tartus
Homs
Tripoli
Palmyra
LEBANON
BEIRUT
SYRIA
Sidon
Litani
Tyre
DAMASCUS
SRAEL
Haifa
Sea of Galilee
Tiberias
Suweida
El-Rutbah
Khadera
Deraa
Irbid
Jordan
Netanya
TEL AVIV
Jaffa
WEST BANK
JERUSALEM
AMMAN
IRAQ
Gaza
Dead Sea
JORDAN
Turayf
Beersheva
Kerak
El-Nabk
El-Jalamid
SAUDI ARABIA
Maan
Eilat
Aqaba
El-Humaudah
Gulf of Aqaba
Tabuk
Sheikh Humeid
Duba
RED SEA
TEL AVIV AND JERUSALEM
Herzliya
Kalkilya
TEL AVIV
Rosh ha-Ayin
Jaffa
Ben Gurion
Rishon le-Zion
Lod
Mazraa el-Sharqiya
Ramla
Ramallah
Rekhovot
Yavne
JERUSALEM
Gedera
Qiryat Malakhi
Bethlehem

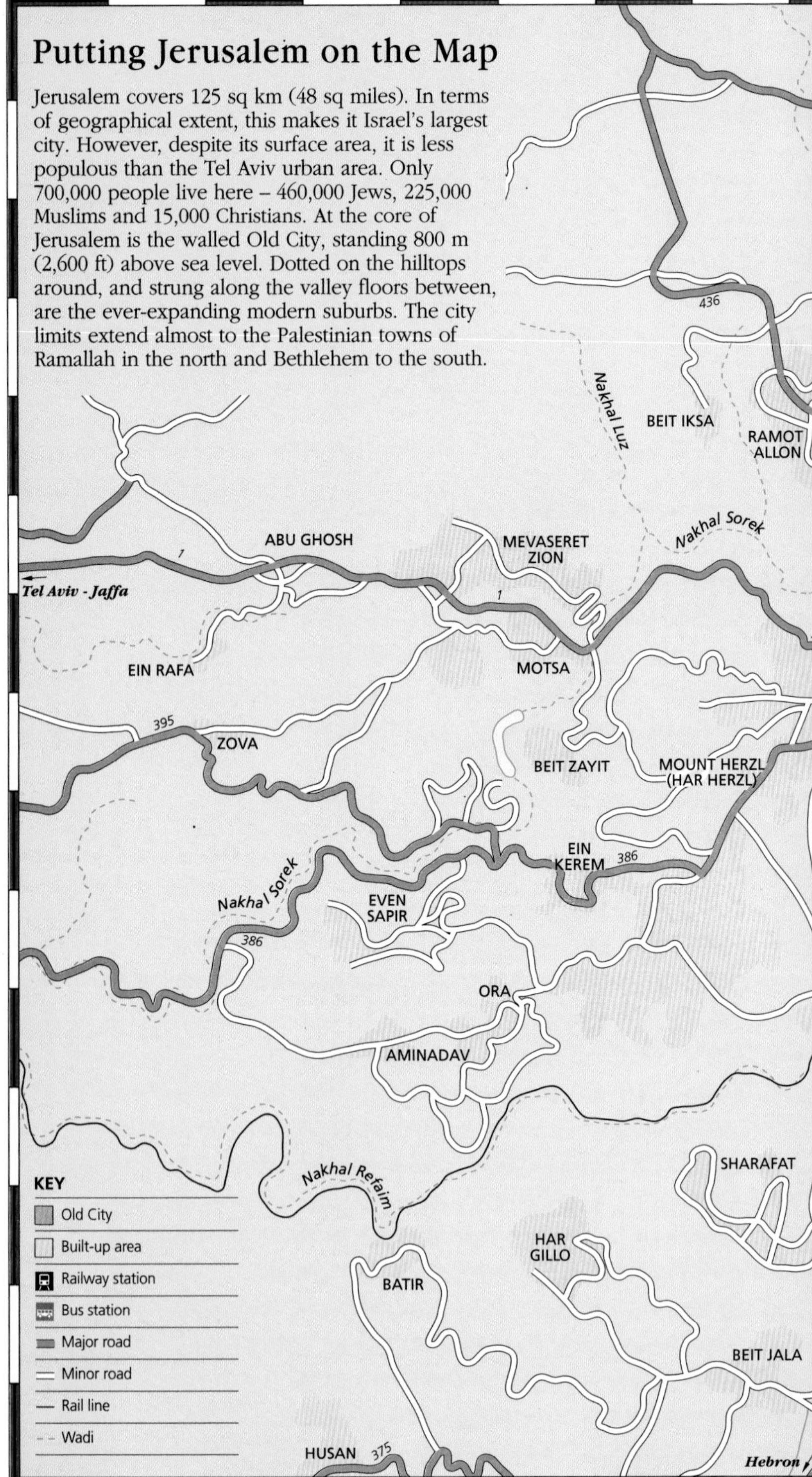
Putting Jerusalem on the Map
Jerusalem covers 125 sq km (48 sq miles). In terms of geographical extent, this makes it Israel's largest city. However, despite its surface area, it is less populous than the Tel Aviv urban area. Only 700,000 people live here – 460,000 Jews, 225,000 Muslims and 15,000 Christians. At the core of Jerusalem is the walled Old City, standing 800 m (2,600 ft) above sea level. Dotted on the hilltops around, and strung along the valley floors between, are the ever-expanding modern suburbs. The city limits extend almost to the Palestinian towns of Ramallah in the north and Bethlehem to the south.
436
Nakhal Luz
BEIT IKSA
RAMOT ALLON
Nakhal Sorek
ABU GHOSH
MEVASERET ZION
1
Tel Aviv - Jaffa
EIN RAFA
MOTSA
395
ZOVA
BEIT ZAYIT
MOUNT HERZL (HAR HERZL)
EIN KEREM
386
Nakhal Sorek
EVEN SAPIR
386
ORA
AMINADAV
Nakhal Refaim
SHARAFAT
HAR GILLO
BATIR
BEIT JALA
HUSAN
375
Hebron
KEY
Old City
Built-up area
Railway station
Bus station
Major road
Minor road
Rail line
Wadi

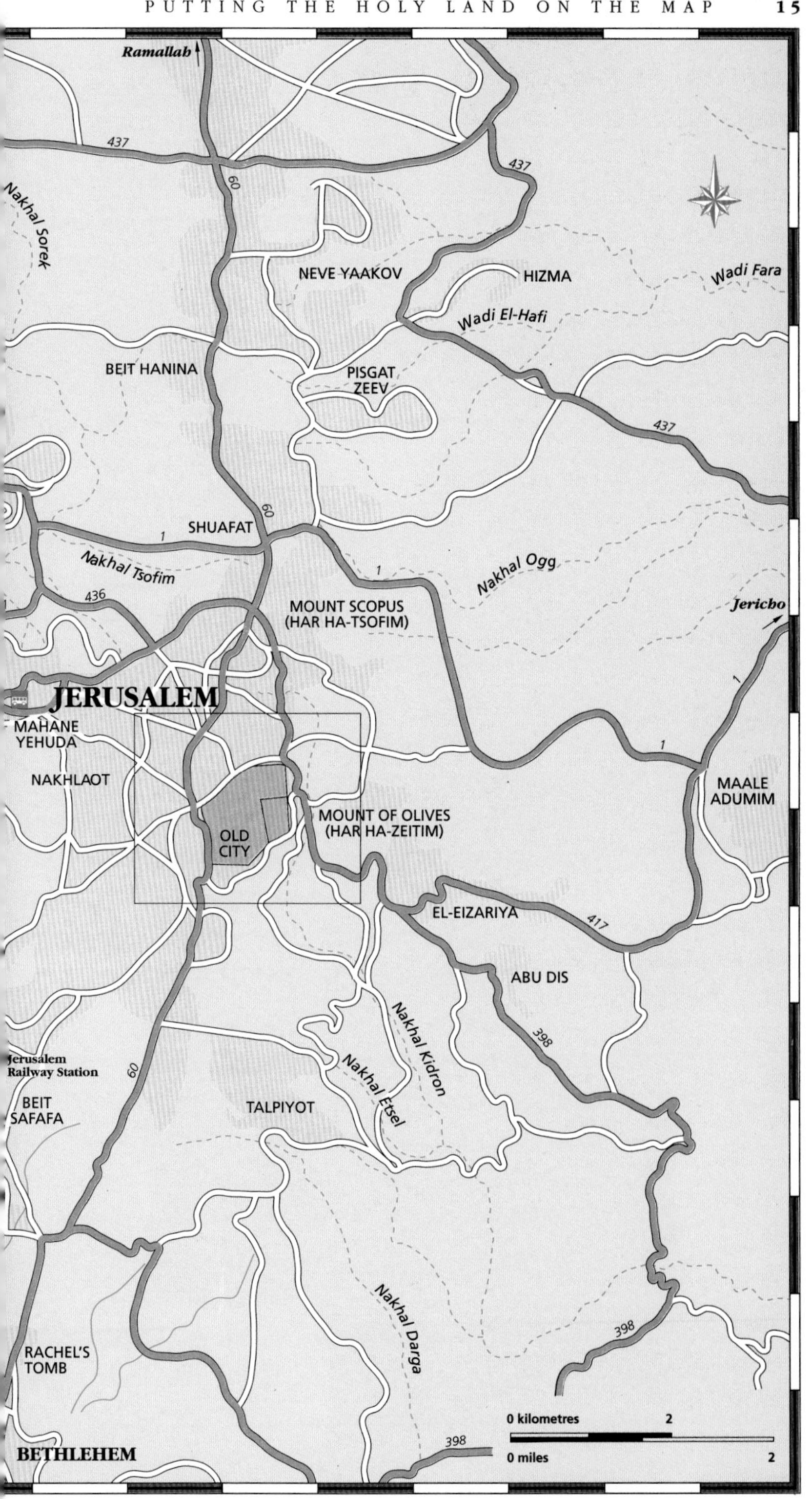

Ramallah
437
60
Nakhal Sorek
NEVE YAAKOV
HIZMA
Wadi Fara
Wadi El-Hafi
BEIT HANINA
PISGAT ZEEV
437
SHUAFAT
1
Nakhal Tsofim
Nakhal Ogg
436
MOUNT SCOPUS (HAR HA-TSOFIM)
Jericho
JERUSALEM
MAHANE YEHUDA
NAKHLAOT
MAALE ADUMIM
MOUNT OF OLIVES (HAR HA-ZEITIM)
OLD CITY
EL-EIZARIYA
417
ABU DIS
Nakhal Kidron
398
Jerusalem Railway Station
Nakhal Etsel
BEIT SAFAFA
TALPIYOT
Nakhal Darga
RACHEL'S TOMB
0 kilometres 2
0 miles 2
BETHLEHEM

A PORTRAIT OF THE HOLY LAND

A Jew growing up in New York, a Christian in Lisbon and a Muslim in Jakarta will have childhoods as different as can be imagined, but one thing they will share is a common set of reference points, which will include names such as Abraham and Moses, and, above all, Jerusalem and the Holy Land.

For around 2,000 years this narrow corridor of land on the eastern shore of the Mediterranean has exercised an influence on world culture far out of proportion to its modest size. Events that are said to have taken place here in antiquity gave rise to the three great monotheistic religions. As these religions extended their influence throughout the world, so the Holy Land in general, and Jerusalem in particular, became overburdened with spiritual significance. Tradition has it that Jerusalem is where Solomon built his great temple, Christ was crucified, and the Prophet Muhammad visited on his Night Journey. It comes as a mild shock to some to discover that this spiritual world centre is no bigger than an average city neighbourhood. Those who come to Jerusalem expecting architectural grandeur to match the stature of these spiritual highlights will be disappointed. The city's churches don't begin to compare with the soaring Gothic cathedrals of Europe. The glorious Dome of the Rock aside, the buildings are quite humble. But the effect this has is to bestow on the city an altogether appropriate air of humility and authenticity, pleasingly at odds with the hyperbole and oversell of the new millennium.

Mural at a Palestinian school in Jerusalem

Bedouin encampment in the desert scenery of Wadi Rum, southern Jordan

◁ Greek Orthodox priest at the Church of the Holy Sepulchre, Jerusalem

The Old City of Jerusalem, viewed from the Jewish cemetery on the Mount of Olives

While Jerusalem is a city rooted in ancient history, at the same time it lies at the heart of a region which possesses a distinctly youthful nature. Both Israel and Jordan, the two countries which, along with Egypt's Sinai peninsula, make up what we know as the Holy Land, are barely more than half a century old. It is a greatly over-used travel cliché, but here it is difficult to avoid commenting on the striking mix of the ancient and modern. In Jerusalem, ultra-Orthodox Jews wearing clothes that were fashionable in Eastern Europe 300 years ago mingle with Christian pilgrims armed with state-of-the-art digital cameras. In the wilderness of the Negev Desert, Bedouin tribesmen speak nonchalantly on mobile phones, while in Galilee Palestinian farmers lead oxen to fields that lie in the shadow of huge biotechnology plants.

Young boy playing football at the Dome of the Rock

Equally striking is the mix of peoples. The modern state of Israel has drawn its citizens from virtually every continent, embracing a worldwide roll call of Jewry, from Minnesota to Murmansk, Adelaide to Addis Ababa. Side by side with the Jews – and Arabs – are such minority peoples as the Druze, a mysterious offshoot sect of Islam, and the Samaritans, who speak Arabic but pray in Hebrew and number around 600.

In this land of diversity, even the one common element shared by the majority of Israelis, the Jewish faith, is not the uniting factor it might be. The notion of what it is to be Jewish and, more pertinently, what form a Jewish state should take, are subjects of great contention. There are large, and increasingly influential, sections of society that believe Israel should adhere strictly to the laws prescribed in the Torah. The greater part of society, however, views the notion of a religious state with horror. The gulf between the two standpoints is best

Souk stall-holder displaying fresh vegetables

illustrated by the phenomenon of Dana International, the flamboyant transsexual singer who won the 1998 Eurovision Song Contest. It was a victory greeted with pride by a part of the nation, while to the religious sector it served only to confirm "the secular sickness of Israel".

An even more contentious issue is ownership of the land. Israel bases its right to exist on an ancient covenant with God, related in the Old Testament, in which this land was promised to the descendants of the Jewish patriarch Abraham, as well as a 3,000-year connection to the land and the political sovereignty granted to them by the United Nations in 1947. The Palestinian Arabs have their own claims on the territory, based on centuries of occupancy. During the 20th century four major wars were fought between the Arabs and the Jews. The problem is still far from being resolved.

Divisive Dana International

Since the Hebrew tribes first emerged from the desert around the 12th century BC, this has been one of the world's most turbulent neighbourhoods. Every major Near Eastern empire fought here. This has resulted in a fantastic legacy of historical remains, including Roman cities, Byzantine churches and early Islamic palaces. Archaeologists are constantly at work to uncover what other riches this troubled land might yield. Often, their aims go far beyond the academic: some expeditions search for evidence to support territorial claims; others seek fabled artifacts such as the Holy Grail or the Ark of the Covenant, which they believe may hold the key to human existence.

Amidst all this hullabaloo, one should not forget that the Holy Land is a marvellous region for the visitor. It is not necessary to have an advanced grasp of history to appreciate the magnificence of the region's ancient cities, isolated monasteries and hilltop fortresses, while the desert scenery of Wadi Rum is a setting in which to live out fantasies, and the diving in the Red Sea is reckoned by some to be unsurpassed anywhere in the world. Added to this, there is plenty of fine dining and comfortable accommodation. It is quite possible to visit the Holy Land and find that the only issue of concern is getting a decent spot on the beach.

Beach life at Tel Aviv, the vibrant cultural and commercial capital of Israel

Old Testament Sites in the Holy Land

Many of the stories recounted in the Old Testament are located within Egypt, Sinai and the "Land of Canaan", which corresponds roughly to present-day Israel. The Bible gives plenty of precise geographical references. Some places, such as Jerusalem and Jericho, still exist and have yielded archaeological evidence confirming some, but by no means all, of the references to them in the Old Testament. Other sites were only attached to their biblical episodes much later. Touring these sites, the visitor cannot but be aware of the contrast between the importance of the events and the often insignificant and all-too-human scale of the places in which they are said to have occurred.

The Destruction of Sodom ①
When Sodom was destroyed by God (see p202) *only Lot and his family were spared, but his wife looked back and was turned into a pillar of salt.*

The Sacrifice of Isaac ②
God asked Abraham to sacrifice his son, Isaac. The patriarch was about to obey when an angel stayed his hand and instructed him to slaughter a ram instead (Genesis 22). Tradition identifies the place of sacrifice as Mount Moriah, later a part of Jerusalem, and the site on which Solomon's Temple is said to have been subsequently built (see p41).

Gaza

The Tombs of the Patriarchs ③
Acquired as a burial place for his wife Sarah, the Machpelah cave was the first plot in the Land of Canaan purchased by Abraham (Genesis 23). A mosque/synagogue now occupies the traditional site of the tomb, located in the present-day town of Hebron (see p196).

0 kilometres 100

0 miles 50

Moses Receives the Ten Commandments ④
Since the 4th century, Mount Sinai (see pp246–7) *has been associated with the story of Moses and the Ten Commandments (Exodus 20). The Bible places Mount Sinai in a region called Horeb, but the location of Horeb has never been identified.*

④ Mount Sinai

GULF OF AQAB

The Death of Moses ⑤
Moses is said to have seen the Promised Land from the summit of Mount Nebo and died in the same place. Christian tradition identifies Mount Nebo (see p215) *as being just southwest of modern-day Amman. As the Bible states, the whereabouts of Moses' tomb is unknown (Deuteronomy 34: 1–7).*

Joshua Conquers Jericho ⑥
The Old Testament story tells how the walls of Jericho (see p190) fell to the blast of horns (Joshua 6). This ancient oasis was the first city conquered by the Israelites, led by Joshua, after they emerged from their 40 years in the wilderness.

The Ark of the Covenant ⑦
At Shiloh the Jews built the first temple and placed in it the Ark of the Covenant, the sacred container of the tablets of the Ten Commandments. The Ark is shown here in a 13th-century illumination being carried by two angels.

Samson and Delilah ⑧
The climax of this story, in which Samson pulls down the Philistines' temple, killing himself and his enemies, is described as taking place in Gaza (Judges 14–16).

⑩ Mount Carmel
SEA OF GALILEE
Megiddo •
Shiloh ⑦
Jordan River
GILEAD
Jerusalem ②
⑥ Jericho
⑤ Mount Nebo
⑨ Ha-Ela Valley
Hebron ③
DEAD SEA
• Beersheva
MOAB
Sodom ①

David Defeats Goliath ⑨
As the champion of the Israelites during the reign of King Saul, David defeated Goliath and routed the Philistines (I Samuel 17). The site of the battle is given as the Ha-Ela Valley, northwest of Hebron.

Elijah and the Prophets of Baal ⑩
Elijah challenged the prophets of the Canaanite god Baal (left). *An altar was set up and sacrifices prepared. Only Elijah's offering burst into flames, showing it had been acknowledged and proving who the true God was (I Kings 18). The traditional site of this event is Mount Carmel, at Haifa* (see p177).

THE OLD TESTAMENT AS HISTORY

Unlike Mesopotamia or Egypt, where ancient texts have allowed the development of a detailed historical framework, the Holy Land has yielded few written archives. The only such resource is the Bible. The later books, which describe events not too far removed from the time they were written, may be relatively accurate. For example, events recounted in Kings I and II can be corroborated by contemporary Assyrian inscriptions. However, the historical basis of stories such as those relating to Abraham, Moses or Solomon, must be viewed with caution. The Old Testament as we know it was compiled from a variety of sources, no earlier than the 6th century BC. These narratives might well contain kernels of historical reality, but by the time they came to be set down they were essentially no more than folk tales.

Assyrian obelisk (825 BC) showing Israelite King Jehu (I Kings 19)

Judaism

Jewishness is not just a matter of religion but of belonging to a people. Jews believe themselves to be descended from Abraham, to whom God promised a land "unto thee, and to thy seed after thee". Judaism traditionally passes through the female line or by conversion, different Jewish movements (Orthodox, Conservative, Reform) having different requirements. Practising Jews conduct their life by the *Torah*, which can be translated as "instruction" or "guidance". Its core is the Five Books of Moses, but the Torah also includes all the teachings and laws within the Hebrew Bible (Old Testament) and subsequent interpretations by rabbinic scholars. The creation of the State of Israel has presented the Jewish people with new political and religious challenges.

The menorah, *a seven-branched candlestick, derives from the candlestick that originally stood in Solomon's Temple.*

THE WESTERN WALL

This is all that remains of the Jews' great Temple *(see pp44–5)*, built to hold the Ark of the Covenant *(see p21)*. It is the holiest of all Jewish sites and a major centre of pilgrimage *(see p85)*.

THE SCROLLS OF THE TORAH

The Torah is traditionally inscribed on scrolls. During a synagogue service the scrolls are ceremonially raised to the congregation before being read. It is an honour to read them. A boy of 13 years of age or a girl of 12 is *bar* or *bat mitzvah*, a "child of the commandment". During a *bar/bat mitzvah* service the boys and girls (Reform Jews only) read from the scrolls.

The Scrolls, *when not in use, are placed in the ark. They may be kept in an ornamental box* (right) *or else tied with a binder inside a decorated cover, adorned with a breast-plate, yad, bells or crown.*

The yad *("hand") is a pointer used to avoid touching the sacred text. It is also meant to direct the reader's attention to the precise word and to encourage clear and correct pronunciation.*

Traditional Jewish life *is measured by the regular weekly day of rest,* Shabbat *(from sundown Friday to sundown Saturday), and a great many festivals* (see pp36–9). *The blowing of the* shofar *(a ram's horn trumpet) marks* Rosh ha-Shanah, *the Jewish New Year.*

DIVISIONS IN JUDAISM

As a result of their history of dispersion and exile, there are Jewish communities in most countries of the world. Over the centuries, different customs have developed in the various communities. The two main strands, with their own distinctive customs, are the Sephardim, descendants of Spanish Jews expelled from Spain in 1492, and the Ashkenazim, descendants of Eastern European Jews. In Western Europe and the US, some Jews adapted their faith to the conditions of modern life, by such steps as altering the roles of women. This divided the faith into Reform (modernizers) and Orthodox (traditionalists), with Conservative Jews somewhere in between. Israeli Jews are frequently secular or maintain only some ritual practices. The ultra-Orthodox, or *haredim*, adhere to an uncompromising form of Judaism, living in separate communities.

Yemenite Jewess in wedding dress

Ultra-Orthodox Jews in Jerusalem's Mea Shearim district in distinctive black garb

THE SYNAGOGUE

Synagogue architecture generally reflects the architecture of the host community, but with many standard elements. There must be an ark, symbolizing the Ark of the Covenant, which is always placed against the wall facing Jerusalem. In front of the ark hangs an eternal light *(ner tamid)*. The liturgy is read from the lectern at the *bimah*, the platform in front of the ark. The congregation sits around the hall, although in some synagogues women are segregated. Traditionally, a full service cannot take place without a *minyan*: a group of 10 men.

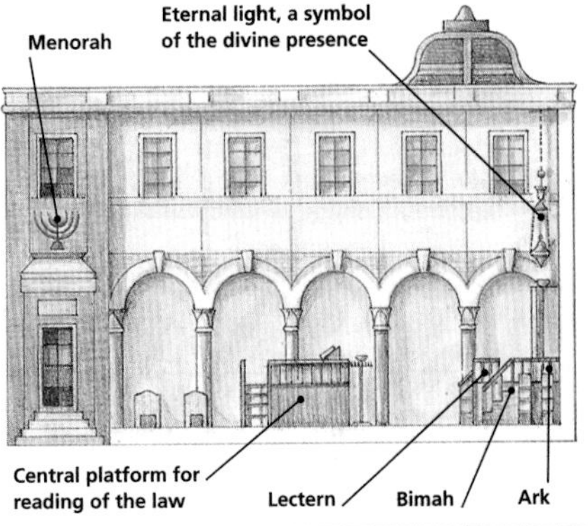

Christianity

To his followers, Jesus of Nazareth was more than just a prophet, he was the Son of God and bringer of a new covenant replacing the one given by God to Abraham *(see p22)*. His Crucifixion in Jerusalem came to be seen as self-sacrifice for the salvation of humankind and inspired a new religious movement based on his teachings. At first this existed as a subsect of Judaism; Jesus came to be known as Christ (*Christos*, the anointed one, in Greek), as he was held to be the Messiah of Jewish prophecies. However, the new religion spread far beyond Judaea. It saw persecution, then recognition by the Roman Empire, eventually becoming its dominant religion in the 4th century AD.

The cross *is a symbol of the Crucifixion of Christ. An empty cross shows that he has risen from the dead.*

THE EUCHARIST (MASS)

Greek Orthodox priests celebrate the Eucharist, the taking of bread and wine, representing the body and blood of Christ. One of the central sacraments of Christianity, it was instituted by Jesus himself at the Last Supper *(see p117)*.

cxix.

The Goſpell off
Sancte Jhon.
The fyrſt Chapter.

IN the begynnynge was that worde/ād that worde was with god:and god was thatt worde. The ſame was in the begynnynge wyth god. All thyngſ were made by it/and with out it/was made noo thīge/that made was. In it was lyfe/And lyfe was the light of mē/And the light ſhyneth ī darckness/ād darckness cōprehēded it not. There was a mā ſent from god/whoſe name was Jhon. The ſame cā as a witnes/to beare witnes of the light/that all men through hī myght beleve. He was nott that light: but to beare witnes of the light. That was a true light/whych lighteneth all men that come ī to the worlde. He was in the worlde/ād the worlde by hī was made:and the worlde knewe hym not.

He cā ī to his awne/ād his receaved hī not. vnto as meny as receaved hī/gave he power to be the ſōnes of god:ī that they beleved ō his name: which were borne not of bloude nor of the will of the fleſſhe/nor yet of the will of men:but of god. And that worde was made fleſſhe/and dwelt amonge vs/and we ſawe the glory off yt/as the glory off the only begotten ſonne off the father/

The Christian Bible *is in two parts: the Old Testament consists of Jewish sacred texts; the New Testament relates the life and teaching of Jesus and his Apostles. The latter was written from the mid-1st century. Most early texts were in Greek; a definitive Latin version by St Jerome* (see p195) *appeared in about AD 404. The Protestant Reformation inspired translations into many other languages, such as this English version, from the 16th century.*

Icons *play a major role in the Greek and Russian Orthodox churches. This example from St Catherine's Monastery* (see pp246–9) *shows* Christ in Majesty. *Usually painted on wood, they are used as aids to devotion, bringing the worshipper into the presence of the subject.*

The Virgin and Child *is a favourite Christian image. Depictions of the baby Jesus emphasize the human side of his nature, while the cult of his mother, the Virgin Mary, allows the faithful to identify with the joys and suffering of motherhood.*

A Palm Sunday procession *recreates Christ's entry into Jerusalem. This is a prelude to Holy Week, the most important Christian festival, commemorating the Crucifixion on Good Friday and Christ's Resurrection on Easter Sunday.*

CHRISTIAN DENOMINATIONS

Almost all the major Christian churches are represented in Jerusalem. The Greek Orthodox *(see p100)* and Syrian churches were the first to be established in the city. Other ancient Christian communities include the Armenians *(see p107)*, Copts and Ethiopians. The Roman Catholic Church established its own Patriarchate here in the wake of the Crusades, and the most recent arrivals were the Protestants. The Greek Orthodox, Greek Catholic and Roman Catholic churches have large congregations, mostly of Palestinian Arabs, while priests and officials tend to be Greek and Italian.

Syrian Orthodox Christmas in Bethlehem

Procession of Ethiopian priests in Jerusalem

Armenian priests in their black hooded copes

CHURCHES IN THE HOLY LAND

The first churches did not appear in the Holy Land until around AD 200 – the earliest Christians gathered together in each other's homes. Roman suspicion of unauthorized sects kept these churches underground. However, the conversion to Christianity of the Roman emperor Constantine signalled a rash of building on the sites connected with the life of Christ. The usual type of Byzantine church was the basilica, a longitudinal structure with a nave (central aisle) lit by windows in the walls of the side aisles. The apse area, containing the altar, was frequently concealed by an iconostasis, a three-panelled screen adorned with icons.

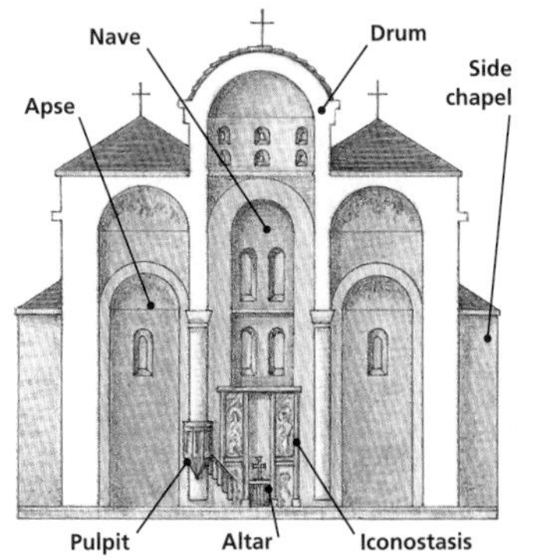

Islam

Islam was founded by Muhammad, a former merchant from Mecca in Arabia. Born around AD 570, at the age of 40 he began to receive revelations of the word of Allah. These continued for the rest of his life and were transcribed as the Quran. Muhammad's preachings were not well received in Mecca and in 622 he and his followers were forced to flee for Medina. This flight, or *hejira*, constitutes year zero in the Islamic calendar. Before Muhammad died in 632, he had returned to conquer Mecca. Within a further four years, the armies of Islam had swept out of the Arabian desert and conquered the Holy Land.

The crescent moon, *the symbol of Islam, has resonances of the lunar calendar, which orders Muslim religious life.*

DOME OF THE ROCK
One of the oldest and most beautiful of all mosques, the richly decorated Dome *(see pp70–73)* is the third most holy site of Islam after the Prophet's cities of Mecca and Medina.

The Quran, *the holy book of Islam, is regarded as the exact word of Allah. Muslims believe that it can never be truly understood unless read in Arabic: translations into other languages can only ever paraphrase. The Quran is divided into 114 chapters, or* suras, *covering many topics, including matters relating to family, marriage, and legal and ethical concerns.*

THE FIVE PILLARS OF FAITH

Islam rests on what are known as the "five pillars of faith". The first of these, known as the *Shahada*, is a simple declaration that "There is no god but Allah and Muhammad is his Prophet". The second pillar is the set daily prayers, performed in the direction of Mecca five times a day (though many Muslims don't completely observe this). The third pillar is the fasting during daylight hours that takes place for the whole of the holy month of Ramadan, and the fourth is the giving of alms. The fifth pillar is *Haj*: at least once in their lifetime all Muslims must, if they are able, make the pilgrimage to Mecca, birthplace of Muhammad.

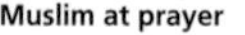

Muslim at prayer

House decorated with pilgrimage scenes, indicating the owner has made the *Haj*

Muslim festivals *are relatively infrequent, with just four major dates in the calendar* (see p38)*. The most important of these are* Eid el-Adha *(which commemorates Abraham's covenant with God), marking the time of the pilgrimage, or* Haj, *and* Eid el-Fitr, *which marks the end of Ramadan. Celebrations tend to be communal.*

The imam *is an Islamic teacher, usually attached to a particular mosque. He delivers the khutba, or sermon, at the midday prayers on Friday. These prayers are always the best attended of the week.*

The Night Journey *was one of the defining episodes in the life of the Prophet Muhammad. He was carried during the night from Mecca to Jerusalem and from there made the* Miraj, *the ascent through the heavens to God's presence, returning to Mecca in the morning.*

THE MOSQUE

Mosques come in many shapes and sizes but they all share some common characteristics. Chief of these is the mihrab, the niche that indicates the direction of Mecca. Most mosques also have a *minbar*, from which the imam delivers his Friday sermon. A dome usually covers the prayer hall. The minaret serves as a platform for the delivery of the call to prayer, once made by a *muezzin*, but these days more often a prerecorded cassette broadcast through a loudspeaker.

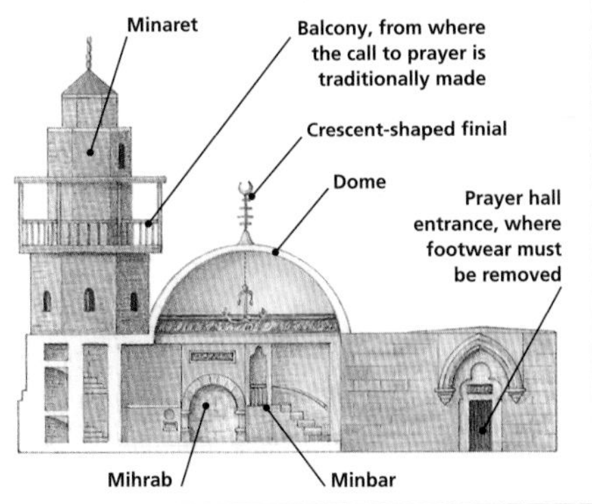

Sites of the New Testament

The life of Jesus Christ, as narrated in the gospels, was played out in a relatively small geographical arena. He was born in Bethlehem; he grew up in Nazareth; his baptism took place at the Jordan River near Jericho; most of his public activity was carried out around the shores of the Sea of Galilee, where he preached, narrated parables and worked miracles; and his crucifixion, resurrection and ascension all occurred in Jerusalem. Unlike the sites of the Old Testament, those of the New Testament saw the rise of sanctuaries, churches and chapels built within two or three centuries of the death of Jesus. For this reason, a number of these sites have some claim to authenticity, although, as with so much in the Holy Land, nothing is beyond dispute.

The Annunciation ①
At Nazareth Mary was visited by the angel Gabriel and told of her forthcoming child (Luke 1: 26–38). The episode is commemorated by the Basilica of the Annunciation (see p180).

The Birth of Jesus ②
In Bethlehem Jesus was born in a grotto and an angel appeared to shepherds in nearby fields, telling them of the birth (Luke 2: 1-20). A church was first built on the site in the 4th century (see pp194–5) *and a star marks the alleged site of the Nativity.*

The Wedding at Cana ③
Jesus performed his first miracle at this small village near Nazareth, at a wedding where he turned water into wine (John 2: 1–11).

Joppa (Jaffa)

The Baptism of Christ ④
John the Baptist, a cousin of Jesus, baptized and preached the coming of the Messiah on the shores of the Jordan River. John recognized Jesus as the "Lamb of God" (Matthew 3). The site traditionally identified with the baptism, known as Qasr el-Yehud, is east of Jericho on the Jordanian border. It lies in a military zone and is accessible to pilgrims on certain days of the week.

0 kilometres 50

0 miles 30

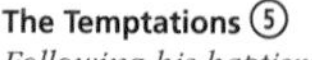

The Temptations ⑤
Following his baptism, Jesus went into the desert, where the Devil tried to tempt him from his 40-day fast (Matthew 4: 1–11). The Greek Orthodox Monastery of the Temptation on Mount Quarntal, just north of Jericho, marks the site of the supposed encounter (see p190).

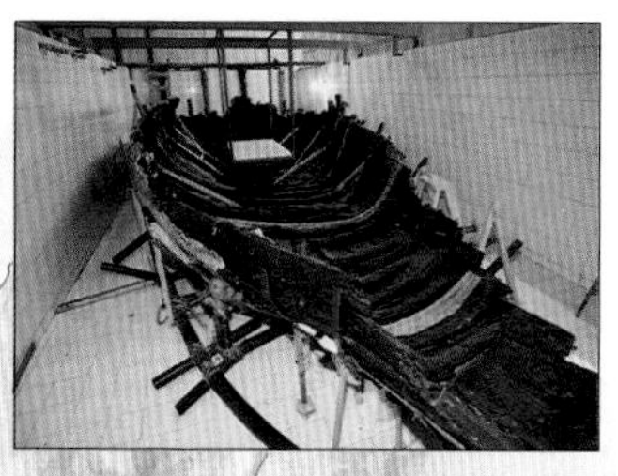

The First Disciples ⑥
Christ's first Disciples were fishermen he encountered on the banks of the Sea of Galilee. He persuaded them to leave their nets to become "fishers of men" (Matthew 5: 18–22). In the mid-1980s a fishing boat was discovered in the mud of the lake. It dates back to the 1st century AD, roughly the time of Christ, and is on display at Kibbutz Ginosar (see pp182–3).

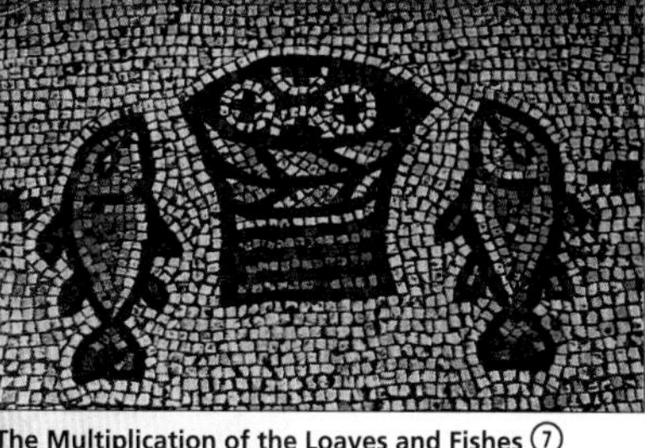

The Multiplication of the Loaves and Fishes ⑦
The gospels locate this famous miracle, more colourfully known as the "feeding of the 5,000" (Matthew 15: 32–39), on the shores of the Sea of Galilee. The episode is commemorated in a church at Tabgha on the lake shore (see p184), *which has a mosaic in front of the altar showing a basket of bread flanked by fish.*

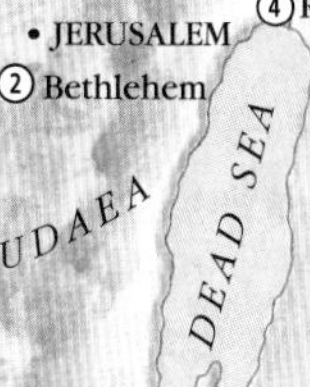

The Sermon on the Mount ⑧
The longest and one of the key sermons in the teachings of Jesus, the Sermon on the Mount, begins with the Beatitudes: "Blessed are the meek for they shall inherit the earth…" (Matthew 5–7). Tradition has it delivered on a small rise at Tabgha. It is celebrated by the nearby, octagonal Church of the Beatitudes (see p184).

JESUS IN JERUSALEM

In what was to be the last week of his life, Jesus made a triumphal entrance into Jerusalem shortly before the Jewish feast of Passover. He proceeded to the Temple where he drove out the money changers (Matthew 21: 12–13). He gathered his Disciples to eat a Passover meal; this was to be the Last Supper. After the meal they went to the Garden of Gethsemane *(see p114)* where Jesus was arrested (Matthew 26: 36–56). Condemned by the Jewish authorities, he was put on trial before Pontius Pilate, possibly in the Antonia Fortress or the Citadel *(see p65)*. After being paraded through the city *(see pp30–31)*, he was crucified and buried at Golgotha, traditionally identified with the site of the Holy Sepulchre church. Following his Resurrection, Jesus departed earth with his Ascension from the Mount of Olives *(see p112)*.

The Last Supper (Matthew 26: 18–30), traditionally associated with a room on Mount Zion *(see p117)*

Via Dolorosa

Via Dolorosa street sign

The Via Dolorosa in Jerusalem traditionally traces the last steps of Jesus Christ *(see pp64–5)*, from where he was tried to Calvary, where he was crucified, and the tomb in the Church of the Holy Sepulchre, where he is said to have been buried. There is no historical basis for the route, which has changed over the centuries. However, the tradition is so strong that countless pilgrims walk the route, identifying with Jesus's suffering as they stop at the 14 Stations of the Cross. The walk is not done the week after Easter or Christmas.

LOCATOR MAP

Via Dolorosa

Jerusalem City Walls

Fourteenth Station
The last Station of the Cross is the Holy Sepulchre itself. The tomb belonged to Joseph of Arimathea, who asked Pilate for Jesus's body.

Sixth Station
Veronica wipes away Jesus's blood and sweat, and her handkerchief reveals an impression of his face. The Chapel of St Veronica commemorates the story, which is not recorded in the gospels.

Seventh Station Jesus falls for the second time. A large Roman column in a Franciscan chapel indicates this station.

Eighth Station
Jesus consoles the women of Jerusalem (Luke 23: 28). The spot is marked by a Latin cross on the wall of a Greek Orthodox Monastery.

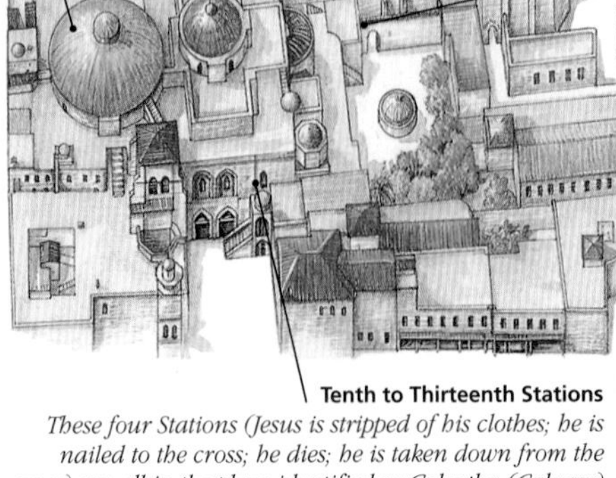

Ninth Station
Jesus falls for the third time. The place is marked by part of the shaft of a Roman column at the entrance to the Ethiopian Monastery *(see pp93–5)*.

Steps to Ninth Station

Tenth to Thirteenth Stations
These four Stations (Jesus is stripped of his clothes; he is nailed to the cross; he dies; he is taken down from the cross) are all in the place identified as Golgotha (Calvary) within the Church of the Holy Sepulchre (see pp92–5).

First Station
Jesus is condemned to death. The traditional site of the Roman fortress where this took place lies inside a Muslim college, the Madrasa el-Omariyya (see p68). *Franciscan friars begin their walk along the Via Dolorosa here every Friday.*

Second Station Jesus takes up the cross, after being flogged, and crowned with thorns. This station is in front of the Franciscan Monastery of the Flagellation *(see p64).*

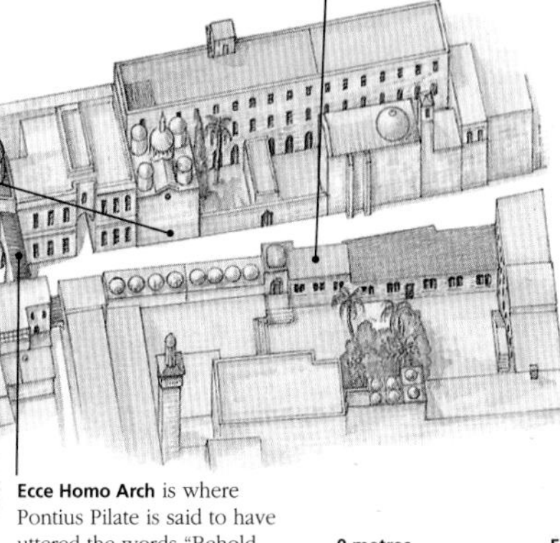

Ecce Homo Arch is where Pontius Pilate is said to have uttered the words "Behold the Man" *(see p64).*

0 metres 50
0 yards 50

Fourth Station
Jesus meets his mother Mary. This point is in front of the Armenian Church of Our Lady of the Spasm, which is built over an earlier Crusader church. This sculpture above the door shows the grief of Mary as she sees her son walking to his death.

Third Station
Jesus falls beneath the weight of the cross for the first time. This is commemorated by a small chapel with a marble relief above the door.

Fifth Station
Simon of Cyrene is ordered by the Roman soldiers to help Jesus carry the cross (Mark 15: 21). A Franciscan oratory marks this point on the Via Dolorosa, which is the start of the ascent to Calvary. This painting also shows St Veronica (see Sixth Station).

Celebrated Visitors

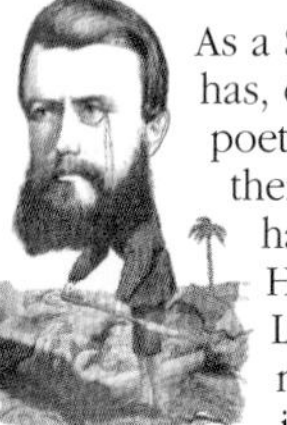

Archaeologist Charles Warren

As a Spiritual or Utopian concept, Jerusalem has, over the centuries, been celebrated by poets and artists who have never been there, and who would perhaps hardly have known where it was on the map. However, the Holy City and the Holy Land have also been the subject of a no less impressive number of accounts, journals and paintings by a great many well-known travellers, writers and artists who did visit. From the early 19th century, the region also became a magnet for a steady flow of archaeologists and biblical scholars.

EARLY PILGRIMS AND TRAVELLERS

The establishment of Christianity as the religion of the Roman Empire in the 4th century AD triggered a wave of visitors, drawn by the region's biblical associations. One of the first pilgrims we know of is a nun named Egeria, who was perhaps Spanish, and visited the Holy Land from AD 380 to 415. An 11th-century manuscript found in Italy in 1884 contained a copy of her travel diary, which makes frequent mention of places such as Sinai and Jerusalem. Present-day writer William Dalrymple used a similar historical account (the journal of John Moschos, a 6th-century monk who wandered the Byzantine world) as the basis for his own Holy Land travels recounted in *From the Holy Mountain* (1996).

Early travellers also visited the Holy Land for trade. The most famous of the merchants was Marco Polo who, in the course of his extensive travels, was entertained by the Crusaders in their halls at Akko.

The works of early Muslim travellers include some lively descriptions of the Holy City. The 10th-century historian El-Muqaddasi described Jerusalem as "a golden basin filled with scorpions". The Moroccan scholar Ibn Batuta who, in the 14th century, travelled over 120,000 km (75,000 miles), also visited Palestine. His journals describe the Tombs of the Prophets in Hebron *(see p196)*, and Jerusalem's Dome of the Rock *(see pp72–3)*, of which he wrote, "It glows like a mass of light and flashes with the gleam of lightning."

Pilgrims in Jerusalem from the *Book of Marvels* on Marco Polo's travels

REDISCOVERING THE HOLY LAND

In the wake of Napoleon's invasion of Egypt (1798) and subsequent expedition into Palestine, and the interest it generated in the Orient, Europeans began to visit the Holy Land. First to arrive were the explorers and adventuring archaeologists, typified by Johann Ludwig Burckhardt *(see p222)*, who was one of the first Westerners ever to visit Jerash, and who discovered Petra in 1812. Lady Hester Stanhope was an eccentric British aristocrat who escaped from her high-society existence to live in Palestine. Although she did conduct some haphazard excavations in Ashkelon (north of Gaza) in 1814, she is more famous for wearing men's clothing in order to avoid wearing the veil.

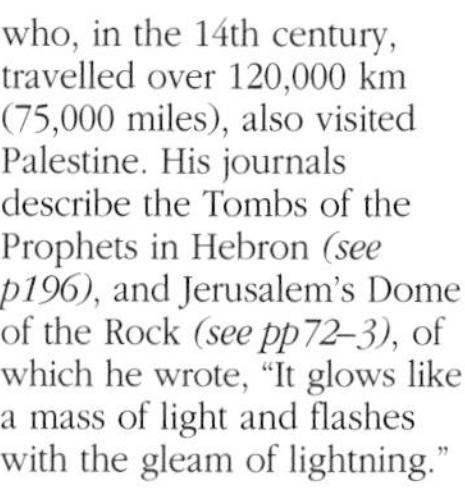

Lady Hester Stanhope

In 1838, Edward Robinson, an American Protestant clergyman with an interest in biblical geography, was the first to make a proper critical study of supposed holy sites; his name is commemorated in Robinson's Arch south of the Western Wall *(see p91)*. In 1867–70, excavations south of the Haram esh-Sharif were carried out by Lieutenant Charles Warren of the Royal Engineers, a man who, some 20 years later, would lead the investigations into the infamous Jack the Ripper serial murders in

Jerusalem from the Mount of Olives **(1859) by Edward Lear**

London. He is remembered in Jerusalem today through "Warren's Shaft", the popular name for the Jebusite well at the City of David archaeological site *(see p115)*.

THE WRITERS

As the ground was broken by the early explorers, a steady stream of adventurous travellers followed in their wake, recording their experiences for eager audiences back in the West. François René de Chateaubriand's brief sojourn in Jaffa, Jerusalem, Bethlehem, Jericho and the Dead Sea area as related in his *Journey from Paris to Jerusalem* (1811) initiated the fashion for travel journals and descriptions of the Holy Land among 19th-century literati. The French poet Alphonse de Lamartine followed in his tracks in 1832, recording his experiences in *Remembrances of a Journey to the East*. In 1850 the creator of Madame Bovary, Gustave Flaubert, visited Palestine and Egypt, but found Jerusalem oppressive, writing in his diary, "It seems as if the Lord's curse hovers over the city." American authors Herman Melville and Mark Twain, both visiting in the mid-19th century were hardly any more enamoured. Melville, author of *Moby Dick*, thought the Holy Sepulchre church "a sickening cheat". Twain was even more caustic, commenting in his 1895 book *The Innocents Abroad*, "There will be no Second Coming. Jesus has been to Jerusalem once and he will not come again." The tradition of scathing comment continued in the 20th century with George Bernard Shaw advising Zionists in the 1930s to erect notices at popular holy sites stating, "Do not bother to stop here, it isn't genuine." More recent writers have been kinder: Nobel laureate Saul Bellow produced a warm-hearted account of the city in *To Jerusalem and Back* (1976).

Mark Twain

THE ARTISTS

With the writers came the artists, the best-known and most prolific of whom was David Roberts, a Scot who visited the Holy Land in 1839. He produced an enormous volume of very precise lithographs, collected and published in 1842, which ensured him fame in his own lifetime. His work remains ubiquitous today, adorning almost every book published on the Holy Land *(see pp8–9)*. Better known for his whimsical verse, artist, writer and traveller Edward Lear (1812–63) spent time in the Holy Land, painting a fine series of watercolours.

The English evangelical painter William Holman Hunt, who belonged to the Pre-Raphaelite movement, settled on Ha-Neviim Street in Jerusalem in 1854, where he painted several of his most famous works. This century, Russian-born Jewish artist Marc Chagall (1887–1985) has become closely identified with Jerusalem. His naïve-styled work, with its strong Jewish themes can be seen at the Israel Museum *(see pp132–7)*, in tapestry form at the Knesset *(see p131)*, and in stained-glass windows at the synagogue of the Hadassah Hospital *(see p139)*.

The Finding of the Saviour in the Temple **(1854–60) by William Holman Hunt**

The Landscape and Wildlife of the Holy Land

Asian buttercup

From the life-giving Jordan River in the north to the scattered oases of the Negev and Sinai deserts in the south, water is precious in the Holy Land. In Israel it is rare to see water that is not used for irrigating land or creating fishponds. Away from the cultivated areas of Galilee and the coast, visitors will encounter a great variety of environments: mountains in the Golan Heights, green hills in Galilee, stony desert in the Negev and sandy desert in southern Jordan. Then there are the strange lifeless waters of the Dead Sea *(see p197)* and the astonishing abundance of life on the reefs of the Red Sea *(see pp240–41)*.

The Jordan River, which flows from the Golan Heights to the Dead Sea

THE DESERT

Much of the Holy Land is desert. South of the Dead Sea, the landscape changes from scrubby steppe to rocky desert with spectacular craters such as Makhtesh Ramon *(see p203)*. The one common tree is the hardy acacia. Animals such as gazelles, ibexes and hyraxes are found at wadis and oases, but the predators that hunted them, the striped hyena and the wolf, are now extremely rare. A more common sight is that of a wheeling vulture or eagle.

Acacia trees growing in the Negev Desert

The fleet-footed Dorcas gazelle *is found in the southern part of Israel and the Sinai peninsula, but in dwindling numbers.*

Oases *are rare in the deserts of this region. Those with plentiful water, like this one planted with date palms near the Dead Sea, are exploited to the full. Others act as magnets for the wildlife of the region.*

A rock hyrax *basks in the hot sun. Hyraxes are hard to spot as they remain hidden among the rocks if it is overcast or cold.*

Ice plants *are succulents that thrive in desert conditions, surviving drought by storing water in their fleshy leaves.*

Wadis *are riverbeds, dry for much of the year. After spring rains, they can fill rapidly with torrents of water, causing a brief explosion of flowers and grasses. Trees that manage to survive in these unpredictable conditions include the acacia and terebinth.*

MOUNTAINS, HILLS AND CLIFFS

The highest mountains in the region are those on the Sinai peninsula and Mount Hermon in the Golan Heights. Trees on the lower slopes in the Golan include Aleppo pine and Syrian juniper. Vegetation in Sinai is very sparse as it is in the spectacular, rocky cliffs and gorges in the Judaean Hills and around the Dead Sea.

The Golan Heights

Egyptian vultures *are found in many of the wilder areas, such as the Negev and the mountains of northern Israel and northwestern Jordan.*

Ibexes *live high in the mountains, descending, in the cool of the morning and late afternoon, to wadis and oases to graze and drink.*

The Madonna lily's *beautiful white flowers symbolize purity. A number of Holy Land plants have names inspired by the Bible.*

Prickly pears *thrive in the hot dry climate. Introduced originally from the Americas, they are much appreciated for their sweet refreshing fruit.*

Oranges *are one of many fruits grown in the fertile areas; they constitute a major export for Israel.*

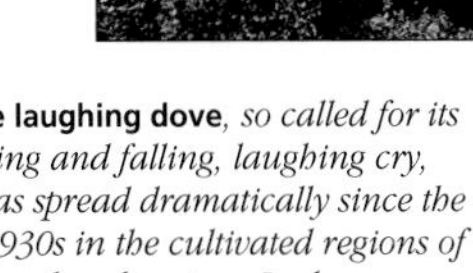

The laughing dove, *so called for its rising and falling, laughing cry, has spread dramatically since the 1930s in the cultivated regions of Israel and western Jordan.*

CULTIVATED AREAS

Israel makes maximum use of the land available for agriculture, even using irrigation to create artificial oases in the desert. There are extensive plantations of oranges and other citrus fruits, avocados, bananas and dates. Jordan is less fortunate, its only fertile area being along the eastern side of the Jordan Valley. In Sinai there are only rare oases such as Feiran *(see p249)*.

Neatly cultivated fields at Migdal on the western shore of the Sea of Galilee

White pelicans taking off from a field near the Hula Reserve

BIRDWATCHING IN THE HOLY LAND

Israel lies on one of the most important routes for migratory birds that winter in Africa then return to Europe and Asia to nest in the spring. Larger species include both black and white storks and many birds of prey. In terms of the number of species that can be seen, the area around Eilat *(see p205)* on the Gulf of Aqaba is reckoned the best place for watching migrating birds in the world. Another popular destination for birdwatchers is the Hula Reserve, an area of protected wetlands north of the Sea of Galilee.

Migrating stork

THE HOLY LAND THROUGH THE YEAR

Shared as it is by Jews, Christians and Muslims, Jerusalem has an overabundance of religious holidays. Add to these secular holidays, commemorations and cultural festivals, and rarely a week passes in which some significant event is not taking place. While visitors may want to time their visit to coincide with some of these events, they may equally want to avoid others. During religious holidays such as Passover (and Ramadan in Israel's Arab areas and in Jordan) many shops, restaurants and museums close for the duration or open only for limited hours, and lodging is hard to find and pricey. The dates of religious and other holidays vary each year so you should check these when planning holidays. The Holy Land has year-round warm weather, but the heat in July and August can be extreme.

Kaparot ritual, eve of Yom Kippur

SPRING

Spring in Jerusalem usually arrives in the latter part of March. This coincides with the Christian Easter and Jewish Passover celebrations, when the city is filled to bursting with pilgrims. The religious festivities are accompanied by cultural events, which increase in frequency as summer approaches. The weather is mild, and this is the best time for trips to Israel's many parks, even though around the Dead Sea the thermometer is already regularly above 30° C (86° F).

MARCH

International Book Fair, Jerusalem. This annual event attracts visitors from more than 40 countries. The Jerusalem Prize is awarded.

Easter falls from late March to April for Catholics and Protestants; the Orthodox and Armenian churches celebrate a week later. Jerusalem's Easter week begins with a Palm Sunday procession from the Mount of Olives to St Anne's *(see p67)*. The most striking ceremony is the Holy Fire *(see p93)*, held on the Saturday of the Orthodox Easter.

Palm Sunday procession in Jerusalem moving along the Via Dolorosa

APRIL

Passover, or Pesach, falls sometime from late March to late April. It celebrates the liberation from slavery under the pharaohs. During the week of the festival, shops and restaurants close, and public transport is limited.

Boombamela Festival *(1st week)*, Ashkelon, Israel. An alternative arts festival held on the beach.

Armenian Holocaust Day *(24 Apr)*, Jerusalem. Marked with a procession, then a service at St James's Cathedral in memory of the Turkish massacres *(see pp106–7)*.

Mimouna is celebrated the day after Passover ends by North African Jews, with festivities throughout Israel.

Music Festival *(Passover)*, Jaffa *(see pp174–5)*. This classical music festival takes place from May to July.

Holocaust Day. Periodically throughout the day sirens signal for two minutes' silence in remembrance of the victims of the Holocaust.

Remembrance Day. In the same fashion as Holocaust Day, this day honours the Israeli dead from past wars.

Spring in Israel, the perfect time for exploring the countryside

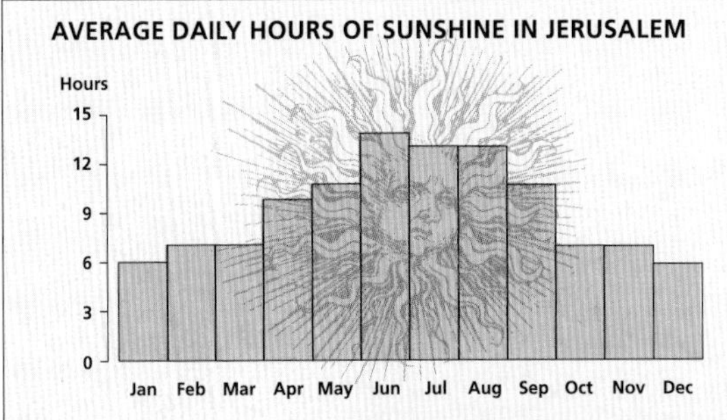

Sunshine Chart
Even during the winter, most days have some sunshine. The summer sun can be very fierce and adequate precautions against sunburn and sunstroke should be taken. Sun screen, a hat and sunglasses are recommended. Drinking plenty of water reduces the risk of dehydration.

Independence Day. Israeli statehood is commemorated with parades and concerts.
South Sinai Camel Festival *(Apr/May)*, Sharm el-Sheikh, Egypt. The Bedouin tribes of Sinai bring their camels to this huge desert race meeting.

Crowds watch an Independence Day air display on Tel Aviv's sea front

MAY

Festival of Israel *(May/Jun)*. The most important cultural event in Israel: three weeks of music, dance and theatre in Tel Aviv, Jerusalem, Haifa and the Roman theatres at Caesarea *(see p176)* and Beth Shean *(see p185)*.

SUMMER

With fewer religious festivals, the attention over summer shifts to the coast, where the soaring temperatures are tempered by sea breezes, and to the towns of Galilee, where the altitude partially counteracts the heat.

JUNE

Ascension falls 40 days after Easter. It celebrates Christ's ascent to Heaven and in Jerusalem it is marked by prayers on the Mount of Olives *(see pp110–11)*.
Beach Festival *(all summer)*, Tel Aviv *(see pp168–73)*. The city-centre beaches are the venue for rock concerts and free open-air cinema.

JULY

Film Festival *(early Jul)*, Jerusalem. Held at the Cinematheque *(see p122)*, this features the work of Israeli and foreign directors.
Jaffa Nights *(1st week)*, Tel Aviv. Two weeks of open-air concerts and shows in the setting of old Jaffa.
Jazz Festival *(Jul–Aug)*, Eilat. Held on the shores of the Red Sea, this festival draws international musicians.
Jordan Festival *(late Jul and Aug)*, Jerash. Jordan's most important festival is held in the spectacular setting of the Roman ruins *(see pp210–11)*. It includes folk dance, ballet, opera, poetry competitions, theatre, classical music and displays of local handicrafts.

Performance by visiting Shakespearean company at the Jerash Festival

AUGUST

Puppet Festival, Jerusalem. This is a festival aimed at the young, with shows in various venues, notably the Train Theatre in the Liberty Bell Gardens.
Klezmer Festival, Safed *(see p181)*. A festival devoted to traditional Eastern European Jewish music.

JEWISH HOLIDAYS

The Jewish calendar is lunar, meaning that each month begins and ends at the new moon. Jewish holidays therefore fall on a different date each year compared to the Western calendar; however, they do remain roughly fixed about a certain time of the year.

Jewish girl dressed for Mimouna

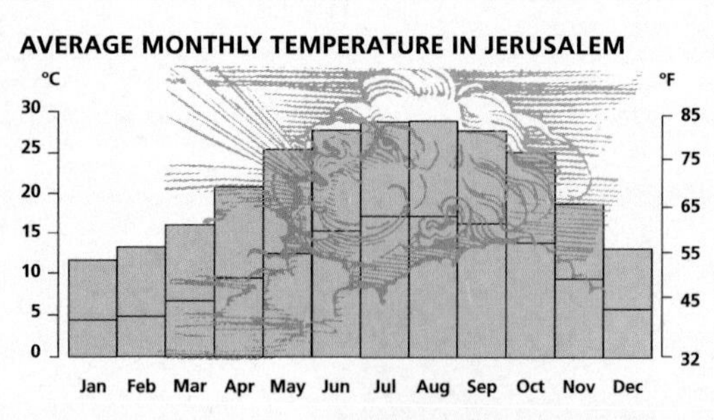

Temperature
Summers in Jerusalem are hot, temperatures frequently climbing to over 30° C (86° F). In winter, the thermometer can drop to near freezing, with even the occasional snowfall. The chart (left) *shows average daily maximum and average daily minimum temperatures for each month.*

AUTUMN

In terms of the weather, autumn is the ideal time to visit Jerusalem. However, several major Jewish holidays occur in September and October, seriously disrupting public transport and reducing opening hours for shops and restaurants. It is also necessary to make hotel reservations well in advance.

Sukkoth booths, in which meals are taken for the feast's duration

SEPTEMBER

Rosh ha-Shanah. The Jewish New Year. It marks the start of ten days of prayer that end with Yom Kippur. On the penultimate day Jews used to perform Kaparot, a ceremony in which a live fowl is waved over the head to absorb sins, although this practice is no longer allowed. The *shofar*, ram's horn, is sounded at services.
Yom Kippur. The Day of Atonement, the holiest day of the year, which Jews observe by fasting for 25 hours and spending most of the day in intensive prayer at their synagogue. The whole country comes to a virtual standstill.
Sukkoth. Commemoration of the Israelites' 40 years in the wilderness after leaving Egypt. Makeshift "booths" are built outside where meals are eaten for seven days. Orthodox Jews even sleep in them.
Haifa International Film Festival, Haifa, Israel. Held annually during the holiday of Sukkoth *(see above)*, the biggest and most important film event in Israel hosts more than 200 screenings over eight days.

OCTOBER

Fringe Theatre Festival, Akko *(see pp178–9)*. This festival in the ancient city of Akko involves local and international avant-garde groups performing in various venues.

NOVEMBER

Jerusalem Marathon *(late Oct/early Nov)*. One of the major sports events in Israel with hundreds of Israelis and foreigners participating.

MUSLIM FESTIVALS

Eid el-Fitr and Eid el-Adha are the major feasts, both lasting two or three days, and celebrated by the slaughter of sheep. Eid el-Fitr marks the end of Ramadan, the month of fasting, observed by all devout Muslims. Eid el-Adha (Festival of Sacrifice) commemorates Abraham's willingness to sacrifice his son for Allah. Other significant days include the Prophet's Birthday (Moulid en-Nabi) and Islamic New Year (Ras el-Sana). The Islamic year is lunar and 11 days shorter than the Western year. This means that in terms of the Western calendar Islamic festivals fall 11 days earlier each year.

Muslim at prayer

WINTER

Christmas is obviously a good time to visit Bethlehem and Nazareth, especially if you can attend one of the special church services. It does occasionally snow in Jerusalem, and snow on the Golan Heights sees the ski-lifts operating.

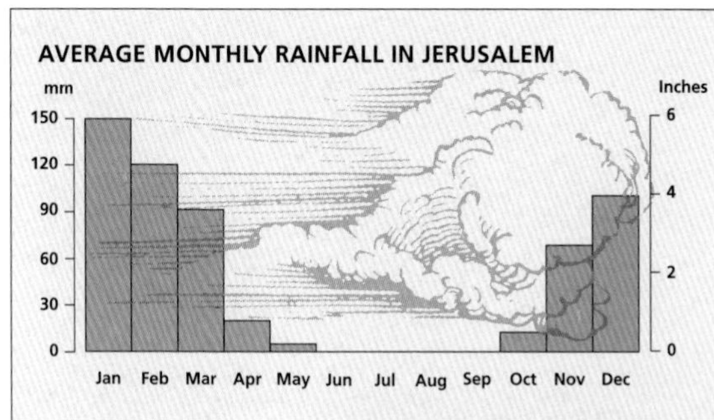

Rainfall
There is virtually no rainfall in Jerusalem from April to October. Showers begin to occur in autumn and winter, and during January and February skies are often filled with threatening grey clouds. Visitors at this time would be wise to go armed with an umbrella.

DECEMBER

Hanukkah. The Jewish Festival of Lights, this commemorates the reconsecration of the Temple in 164 BC *(see p42)*. It lasts eight days, and is celebrated by the lighting of candles in a special eight-branched menorah.

Christmas *(24–25 Dec)*. A Christmas Eve procession from Jerusalem arrives in Bethlehem for midnight mass at the Church of the Nativity *(see pp194–5)*. To attend this service you must book in advance at the Christian Information Centre in Jerusalem *(see p101)*. The mass is also projected on a huge screen in Manger Square. The service at Abu Ghosh *(see p139)* is also impressive. In Nazareth a procession is held on the afternoon of Christmas Eve, which ends with services held in the town's six churches.

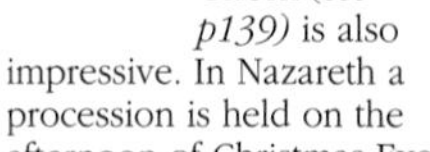
Hanukkah candles

Skiing on Mount Hermon, possible during January and February

International Choir Festival *(26 Dec)*, Nazareth. In the days following the choir festival, the town plays host to sacred music concerts.

Tiberias Marathon *(Dec– Feb)*. Less well-known than the Jerusalem Marathon, this attracts many runners because of the scenery along the route *(see pp182–3)*.

Jewish children dressed up as part of Purim festivities

JANUARY

Orthodox Christmas *(7 Jan)*, Jerusalem. This is celebrated on Christmas Eve with a service at the Holy Trinity Church in the Russian Compound *(see p124)*.

Armenian Christmas *(19 Jan)*, Jerusalem. This is celebrated with a Christmas Eve mass at St James's Cathedral in Jerusalem's Old City *(see pp106–7)*.

FEBRUARY

Purim. This festival celebrates the salvation of the Jews in Persia from threatened genocide (related in the Old Testament Book of Esther). The Scroll of Esther is read publicly in the morning and on the evening of Purim. Children wear fancy dress costumes, while adults participate with the giving of gifts to the poor and to friends, feasting and drinking.

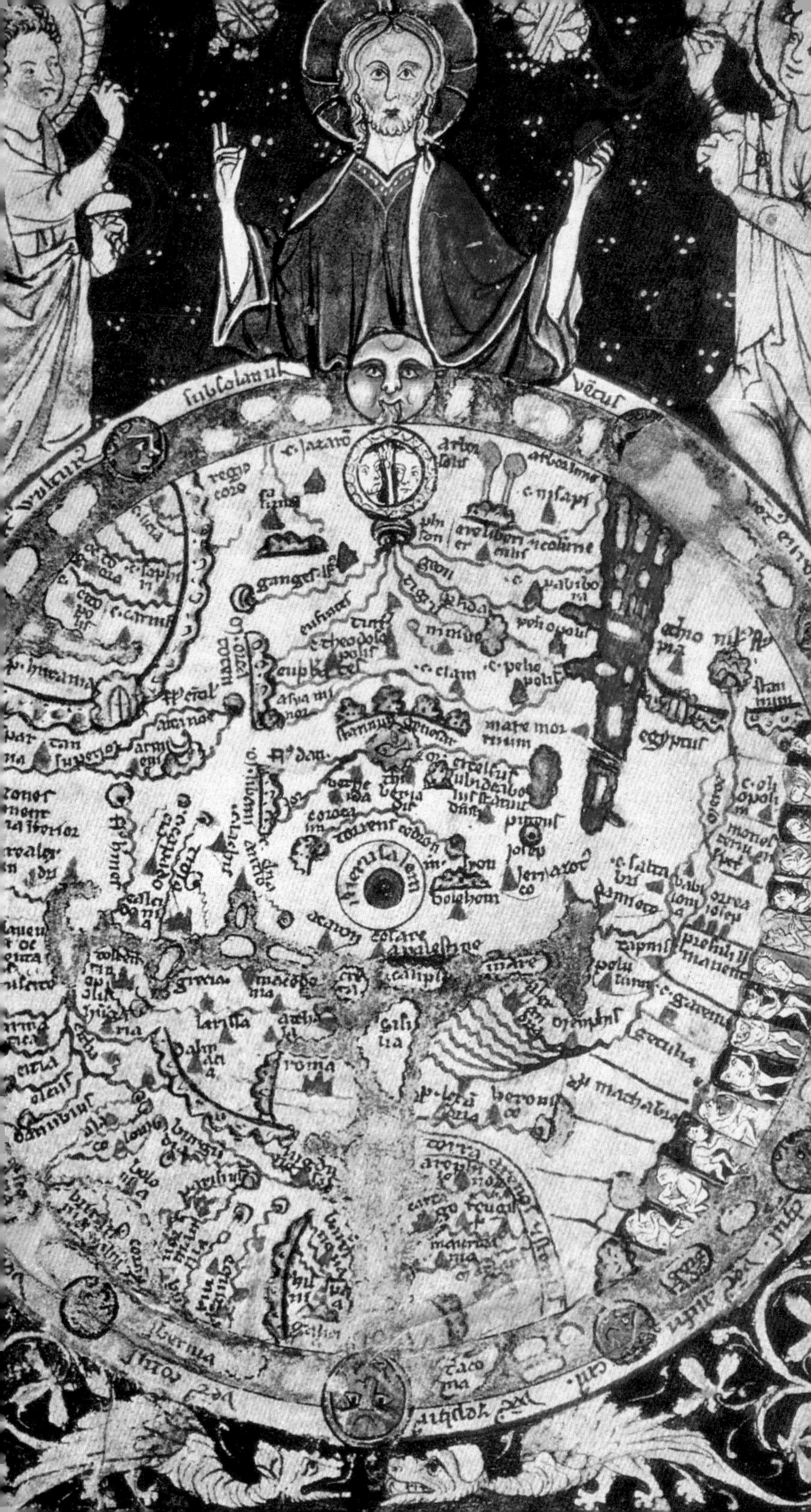
roma
grecia

THE HISTORY OF THE HOLY LAND

Since prehistoric times the fertile plains and scattered oases between the Nile and the rivers of Mesopotamia have been colonized by countless different peoples. The ebb and flow of nations continues to this day; as independent countries, both Israel and Jordan are barely half a century old, with the Jewish state composed of a great many nationalities, all united by their shared faith.

Much of our knowledge of the early prehistory of the Holy Land comes from the site of Jericho, just north of the Dead Sea. Excavations have uncovered a series of settlements dating back to about 10,000 BC, when Stone Age hunters first abandoned their nomadic way of life. In settling, these people took the all-important step which led to cultivating crops and domesticating animals – a process known as the "Neolithic revolution". During the following 3,000 years small farming villages sprang up all over the region.

Philistine sarcophagus lid, 12th century BC

In the 3rd millennium BC the coastal plains witnessed the rise of a fairly uniform culture, known as the Canaanite civilization. There may never have been a single Canaanite nation; rather the Canaanites were probably organized in a series of city-states. A Canaanite army was defeated at Megiddo by the pharaoh Thutmose (1468 BC) and all the city-states were then subject to Egypt. The Canaanites nevertheless survived for two millennia – during which time they developed the world's first alphabet – until their culture was brought to an end by the rise of two new peoples. The first were invaders who came from the sea around 1200 BC; these were the Philistines, after whom the area was called Palestine ("land of the Philistines"). The second were the Hebrew tribes, who, between about 1200 and 1000 BC, coalesced into a political entity known as Israel.

There are several theories as to how the Hebrews came to control Palestine: through hard-won battles, or possibly by peaceful infiltration. There are no historical sources to verify events, but the Old Testament tells how these tribes formed a confederation that eventually led to the birth of a united kingdom whose first sovereign was Saul. His successors, David (whose rule is traditionally given as from around 1010 to 970 BC) and Solomon (c.970–930 BC), laid the foundations for the Jewish nation. It was David, according to the Bible, who captured Jerusalem and made it the Israelite capital, and Solomon who built the Jews' First Temple there.

TIMELINE

10,000–8000 BC First permanent settlements in the region

7000 BC Walled settlement exists at Jericho

Copper crown from Ein Gedi, c.4000 BC

c.1200 BC Arrival of the Philistines and Hebrew tribes

9000 BC | 7000 BC | 5000 BC | 3000 BC | 1000 BC

Skull with cowrie shell eyes from Jericho, c.7000 BC

3200 BC Emergence of Canaanite civilization

c.1010–970 BC Reign of David

c.7000–4000 BC Growth of agricultural communities

c.970–930 BC Reign of Solomon

◁ **Medieval European map, showing the holy city of Jerusalem as the centre of the world**

BABYLONIAN CAPTIVITY

According to the Bible, after Solomon died, conflicts led to the division of the Jewish nation into two separate parts: the Kingdom of Israel in the north and the Kingdom of Judaea in the south. Two centuries later, the Assyrians conquered the north, and many of the Jews of Israel were deported. When Judaea withheld tribute, it too was invaded and defeated at the battle of Lachish. The Assyrians, in turn, were defeated by the Babylonians who, in 587 BC, captured Jerusalem and destroyed Solomon's Temple, forcing the Jews of Judaea into exile. During the brief period of Babylonian captivity the Jews maintained and even strengthened their cultural and religious identity. Defeated by the Persians under Cyrus the Great in 538 BC, the Babylonians disappeared from history and the Jews were allowed to return to their land.

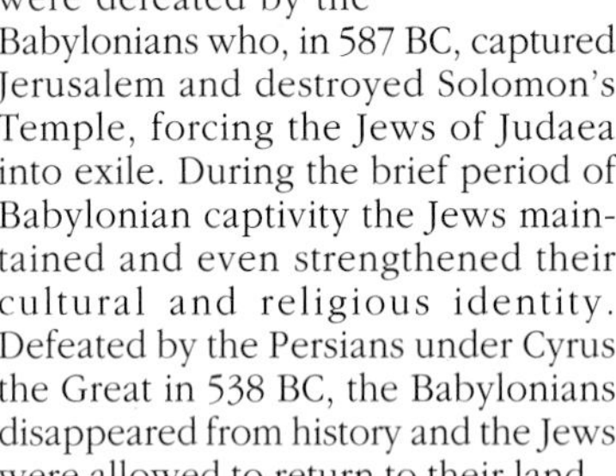

Israelite prisoners leaving Lachish after its fall to the Assyrians in 701 BC

THE SECOND TEMPLE

Returning to Jerusalem, in the 6th century BC, the Jews built a new temple on the same site as the first. This event in the history of Jerusalem marks the beginning of what is referred to as the "Second Temple" period.

The Persians remained dominant in the region until their empire was torn apart by the armies of Alexander the Great. Judaea was swallowed up in the wake of the Macedonian's triumphant progress into Egypt. On the death of Alexander, his empire was split between three generals; the dynasties they founded proceeded to fight over the spoils, with Palestine eventually going to the Syria-based Seleucids. The culture of the Greeks spread throughout the region. This era saw the rise of the Decapolis ("ten cities" in Greek), a loose grouping of Hellenistic city-states in an otherwise Semitic landscape, which included Philadelphia (Amman), Gerasa (Jerash) and Scythopolis (Beth Shean). But Jerusalem resisted. The response of the Seleucid king Antiochus IV Epiphanes (175–164 BC) was to rededicate the Jews' temple in Jerusalem to Zeus and make observance of Hebrew law punishable by death. Led by Judas Maccabeus, a priest of the Hasmonean family, the Jews rebelled in 164 BC. They defeated the Seleucids, took complete control of Jerusalem and reconsecrated their Temple.

Rule of Judaea was assumed by the Hasmoneans. However, independence for the Jews did not ensure peace. There was bitter conflict between the Hasmoneans and the Pharisees, a rival priestly sect who propounded strict observance of Hebrew religious tradition. In the struggle for influence,

The recapture of the Temple by Judas Maccabeus in his successful revolt against the Seleucids, 164 BC

TIMELINE

722 BC Assyria conquers the Kingdom of Israel and sends the Israelites into exile

587 BC The Babylonians conquer Jerusalem and destroy the First Temple

515 BC The founding of the Second Temple

Alexander the Great, whose successors Hellenized Palestine

800 BC | 700 BC | 600 BC | 500 BC | 400 BC

The seal of Jeroboam, a 9th-century Jewish king

538 BC Cyrus the Great frees the Jews in exile in Babylon

332 BC Alexander the Great conquers Palestine

both factions asked for help from the new political and military power of the period – Rome.

THE ROMANS AND JEWISH UPRISINGS

The Romans lost no time in taking advantage of this opportunity: in 63 BC their legions took Jerusalem. The Hasmoneans were superseded by a series of Roman governors, known as procurators. Anxious not to offend local religious sensibilities, the Romans had the Jewish Herod (the Great) rule as a client king in Palestine (37–4 BC). Allowed a relatively free hand in domestic affairs, the ambitious Herod expanded his frontiers and promoted architectural projects such as the Masada and Herodion fortress complexes, the port-city of Caesarea and the grand reconstruction of the Jews' Second Temple in Jerusalem.

On Herod's death his kingdom was ruled for a brief period by his three sons before being governed directly by the Romans. A heavy tax burden, insensitive administration and the imposition of Roman culture were responsible for growing discontent among the Jews. Large numbers of Messianic claimants, revolutionary prophets and apocalyptic preachers only served to inflame the situation further. This was the political climate into which Jesus Christ was born, as described in the biblical New Testament.

Hadrian, builder of Aelia Capitolina

Jewish clashes with Rome broke out repeatedly, culminating in a full-scale revolt in AD 66. It took the Romans four years to gain victory in this First Jewish War. When in AD 70 they finally captured Jerusalem, they destroyed the city and demolished the Temple *(see pp44–5)*. The final subjugation of the Jews occurred three years later at Masada. Judaea once again became a Roman province, but the Jews refused to be subdued and before long a second major revolt broke out.

Jerash, a former Decapolis city which flourished under the Romans

THE EXILE OF THE JEWS

After the Second Jewish War (AD 132–5), Hadrian rebuilt Jerusalem as Aelia Capitolina, a Roman city, which Jews were forbidden to enter. Their communities were broken up and great numbers were sold into slavery and sent to Rome. Others fled, south into Egypt and across North Africa, or east to join the existing Jewish community in Babylon who had settled there after the destruction of the First Temple. This great scattering of the Jews is known as the Diaspora.

164 BC The Maccabean Revolt results in Jewish independence

37–4 BC Herod the Great reigns in Judaea

AD 66–70 First Jewish War and the destruction of the Second Temple

132–5 Second Jewish War led by Simon Bar-Kokhba

0 BC | **200 BC** | **100 BC** | **AD 1** | **AD 100** | **AD 200**

3rd century BC Growth of the Decapolis

1st century BC Petra-based Nabataean empire at its height

63 BC Roman legions under Pompey conquer Jerusalem

AD 73 Fall of Masada

Coin minted by the Jewish rebels at Masada

The Destruction of the Second Temple

Titus

During the Jewish Revolt of AD 66, the Romans suffered early defeats until the emperor Vespasian sent his son Titus to Jerusalem with four legions. The siege of the city was bitterly fought. Eventually, after five months, on 29 August AD 70, the city's defenders were forced to surrender. In *The Jewish War*, historian Flavius Josephus describes how the Temple was set ablaze in the heat of battle. "When the flames rose up," he writes, "the Jews let out a terrific cry and, heedless of mortal danger, ran to put it out." But it was in vain, and the Second Temple was razed to the ground.

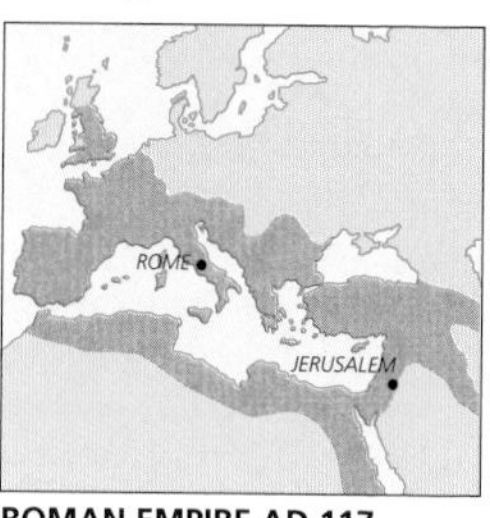

ROMAN EMPIRE AD 117

Maximum extent of the Empire

Arch of Titus
The Romans built the triumphal Arch of Titus in the Forum in Rome, with friezes showing the victorious troops with their booty from the destroyed Temple.

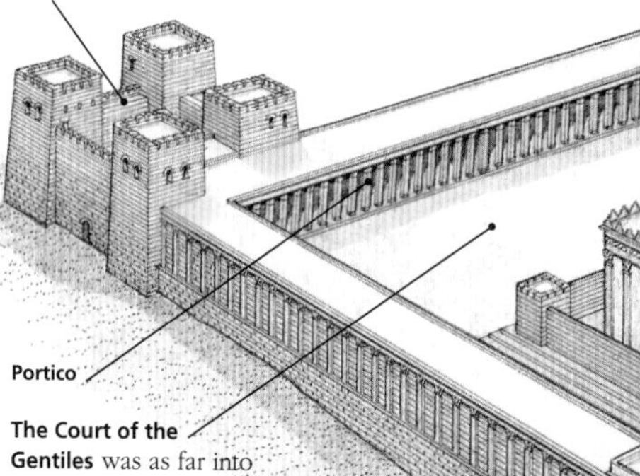

The Antonia Fortress was built by Herod the Great around 37–35 BC to protect the Temple, and named for his patron, Mark Antony. It was the last stronghold of the Jewish rebels in AD 70.

Portico

The Court of the Gentiles was as far into the Temple complex as non-Jews could venture.

The Causeway linked the Temple with the main city gate to the west. Evidence of it remains today in Wilson's Arch *(see p85)*.

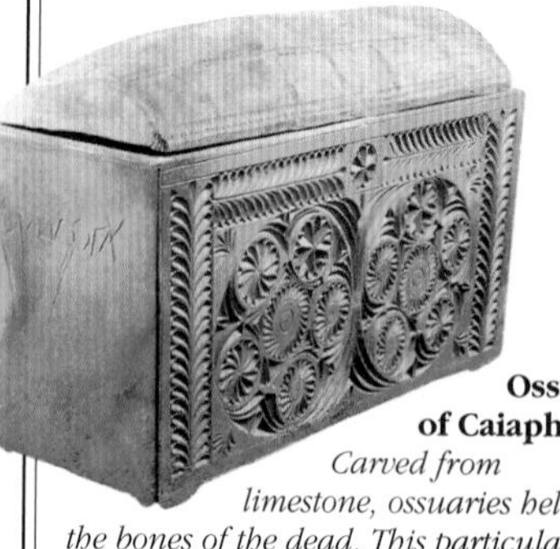

Ossuary of Caiaphas
Carved from limestone, ossuaries held the bones of the dead. This particular ossuary bears the name Caiaphas, which was the name of the Temple High Priest at the time of the crucifixion of Jesus.

The Western Wall
Herod's engineers created the Temple platform by building four walls around a natural hill and filling in. The Western Wall (see p85) *is part of one of those retaining walls.*

Destruction and Sack of the Temple of Jerusalem
Painted by Nicolas Poussin in 1625–6, and now in the collection of the Israel Museum (see pp132–7), *this shows Roman soldiers, directed by Titus on his white horse, emerging from the Inner Temple carrying the Jewish menorah and other treasures.*

The Inner Temple contained the Holy of Holies, an empty chamber meant for the Ark of the Covenant, which was lost when the First Temple was destroyed.

Bronze Helmet
Archaeologists' finds such as this legionary's helmet (c. AD 100) indicate that Rome maintained a strong military presence after the Jewish Revolt.

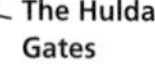

The Royal Stoa was a covered colonnade, 162 columns in length, used for teaching.

The Lower City Steps led to the area known as the City of David. Evidence of them exists in Robinson's Arch *(see p85).*

"Judaea Capta" Coin
A commemorative coin issued after the defeat of the Jewish rebels depicting, on one side, Vespasian and, on the other, Rome standing triumphant over a subdued Judaea.

THE SECOND TEMPLE
Built in the 6th century BC on the same site as the First Temple, which was destroyed by the Babylonians in 587 BC, the Second Temple was greatly expanded by Herod the Great (37–4 BC). He nearly doubled the size of the Inner Temple.

Constantine the Great, the first Christian Roman emperor

PALESTINE UNDER ROMAN RULE

Despite the Jews being banned from Jerusalem, during the 2nd and 3rd centuries their religion and traditions remained very much alive in Palestine, and scholars and religious schools were active throughout Galilee. This was the period in which the academies wrote down Jewish oral law and the commentaries on it, known collectively as the Talmud.

In the early 4th century, the Christians, who had also suffered Roman persecution, were granted freedom of worship by the Emperor Constantine (306–37), himself a convert to the religion. Constantine moved his capital from Rome to Byzantium, which was renamed Constantinople.

This turn of events opened the doors of the Holy Land to pilgrims – first and foremost the devout Helena, mother of Constantine – and Jerusalem regained its former importance. The first Christian churches were built on the sites connected with the life of Christ, and monasticism spread both in the towns and in the deserts of Palestine and Egypt. The first Holy Sepulchre church was dedicated in Jerusalem in 335.

During the rule of Theodosius (379–95) Christianity became the official state religion. Not long after the Roman Empire was divided in 395 between Theodosius's two sons, the Latin-speaking Western Empire fell to Germanic invaders but the Greek-speaking Eastern Empire, thereafter known as the Byzantine Empire, survived.

Byzantine icon of the Madonna and Child, 6th century

THE BYZANTINE ERA

Despite a long series of schisms within the Eastern Church over the nature of Christ *(see p100)*, the Byzantine period was an age of relative stability and prosperity in the Holy Land. The flow of pilgrims continued and monastic life drew ever more adherents. The construction of two important religious buildings, St Catherine's Monastery *(see pp246–8)* in Sinai and the enormous Nea Basilica *(see p82)* in Jerusalem, reflected the confidence of the era. The Holy Land became the land we can see on the early medieval mosaic map at Madaba *(see pp216–17)*. However, upheaval was to arrive in

TIMELINE

AD 300

AD 313 Constantine grants freedom of worship to Christians in the Edict of Milan

Coin of Constantine, AD 320

400

395 The Roman Empire splits into East and West

500

527–65 Reign of Byzantine emperor Justinian

600

638 Battle of Yarmuk River; beginning of Arab dominion in the Holy Land

661 Omayyad dynasty established in Damascus

691 Dome of the Rock completed in Jerusalem

614 in the form of an invading Persian army. Welcomed and supported by the Jews, who hoped for greater religious freedom, the Persians massacred the Christians and desecrated their holy sites before being driven off in 628 by the forces of the Byzantine Empire.

In the same year that the Byzantines reconquered Palestine, in neighbouring Arabia an army led by the Prophet Muhammad conquered Mecca, marking the emergence of a new force in the Near East which, in a little over ten years, would change the face of the Holy Land.

Pilgrimage scroll showing the Haram esh-Sharif

THE ARABS AND ISLAM

In AD 638, only six years after Muhammad's death, the troops of his successor, or *caliph*, Omar defeated the Byzantines at the Yarmuk River, in modern-day Syria. The Muslims became the new rulers of Palestine.

Islam recognizes many of the prophets of the Old Testament, such as Abraham (Ibrahim), and so the Arabs regarded Jerusalem as holy in the same way as the Jews and Christians. The Arabs also believed that the Prophet Muhammad had ascended to Heaven on his Night Journey *(see p27)* from the same rock in Jerusalem on which, according to the Bible, Abraham had been about to sacrifice his son, and over which the Jews had built their temples. Consequently, the rubble in the Temple area was cleared and construction of two mosques began there: the Dome of the Rock (691) and El-Aqsa (705). Access to this "sacred precinct" *(Haram esh-Sharif)*, was forbidden to non-Muslims, but Christians and Jews were permitted to live in the city of Jerusalem on payment of an "infidels" tax.

Groups of Christian pilgrims regularly arrived in the Holy Land from Byzantium and Europe and were given safe passage under the successive Arab dynasties of the Omayyads (661–750), Abbasids (750–974) and, initially, the Fatimids (975–1171). This happy state of affairs ended in 1009 when the third Fatimid caliph El-Hakim initiated the violent persecution of non-Muslims and destroyed the Holy Sepulchre. The situation became critical in 1071 when Jerusalem fell to the Seljuk Turks, who forbade Christians access to the Holy City.

The outraged response of Christian Europe was to take up arms and set off on the first of a series of crusades spread over almost 200 years to recapture the Holy City and biblical sites of Palestine *(see pp48–9)*.

Triumphant group of the feared Muslim cavalry

747 Earthquake drives dwindling populations from Petra and Jerash

Fatimid jewellery

1071 Seljuk Turks capture Jerusalem and bar Christian pilgrims

800	900	1000	1100

1099 The Crusaders take Jerusalem

Dome of the Rock

975 North African Fatimid dynasty rules the Holy Land from Cairo

The Crusades

Crusading emperor Frederick I

"God wills it!" With these words, on 27 November 1095 at the Council of Clermont, Pope Urban II launched an appeal to aid the Byzantines in their wars with the Seljuk Turks and so free the Holy Land. His preachings inspired more than 100,000 men and women from all over Europe to join the armies heading east. They succeeded in creating a Latin kingdom of Jerusalem, but a series of further Crusades meant to reinforce the Western Christian presence in the east were ever less successful. Within 200 years the Crusaders were gone, leaving a legacy of fine ecclesiastical and military architecture.

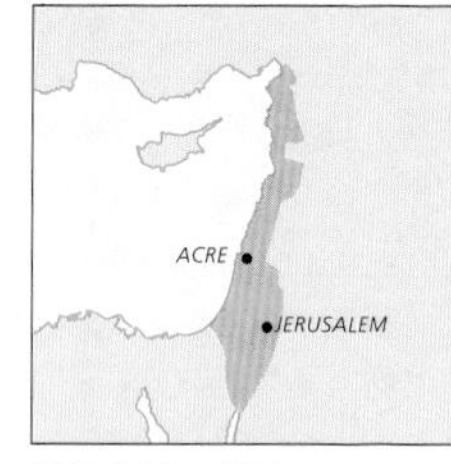

THE HOLY LAND

Crusader domains 1186

The First Crusade

Passing through Constantinople, the Crusaders first engaged the Muslim Seljuks in Anatolia (Turkey). They conquered Nicaea and Antioch before marching down through Syria to Palestine.

The Second Crusade

Most of the Second Crusaders never made it to the Holy Land. Those that did launched a disastrous attack on Damascus and had to withdraw.

THE CAPTURE OF JERUSALEM

On 7 June 1099, the Crusaders laid siege to Jerusalem. The Muslims held out for five weeks until on 15 July the Christian troops breached the walls unleashing a massive slaughter in the streets.

TIMELINE

1099 Crusaders capture Jerusalem; Godfrey of Bouillon becomes "Protector of the Holy Sepulchre"

1100

1119 Founding of the Knights Templar

1120

Templar Knight

1140

1148 Second Crusade defeated while besieging Damascus

1160

Saladin, founder of the Ayyubid dynasty (1169–1250)

1180

1187 Saladin defeats the Crusaders at the Horns of Hattin and takes Jerusalem

1188–92 Third Crusade; after reconquering much of the coas Richard I fails to retake Jerusalem

12

The Third Crusade

The retaking of Jerusalem by Saladin in 1187 prompted the Third Crusade. The Crusade failed to regain the Holy City, but Richard I "the Lionheart" negotiated the right of access for pilgrims.

Richard I and Saladin

The Crucifixion was believed to have taken place on the site occupied by the Holy Sepulchre church.

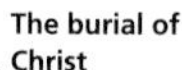

The burial of Christ

The city walls were finally breached by the Crusaders in the north, near Herod's Gate, and also on Mount Zion.

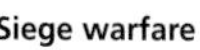

Siege warfare was a major element of the Crusades; siege engines were built on site.

The Fall of Akko

Following a succession of defeats by the Mamelukes, the Crusaders were forced to leave the Holy Land for good in 1291. The last stronghold to fall was Akko, where this coat of arms was discovered.

THE TEMPLARS AND HOSPITALLERS

Much of the defence of Crusader gains in the Holy Land fell to two elite Military Orders of monastic knights, the Hospitallers (see p99) and the Templars, so named because they were headquartered in the former Temple area of Jerusalem. The Orders occupied and refortified Crusader castles in the Holy Land, as well as building new ones of their own.

The Hospitaller castle of Belvoir in the Jordan Valley

1217–21 Fifth Crusade

1244 Jerusalem falls to Muslim mercenaries in the employ of Egypt

1249–50 Louis IX of France leads unsuccessful invasion of Egypt

1260 Mamelukes defeat invading Mongols; Baybars becomes Sultan of Egypt

1270 Last major Crusade, led by Louis IX, ends in his death in Tunis

Louis IX embarking on the last Crusade

1291 Last Latin strongholds in Holy Land, including Akko, fall to Mamelukes

1220 | 1240 | 1260 | 1280 | 1300

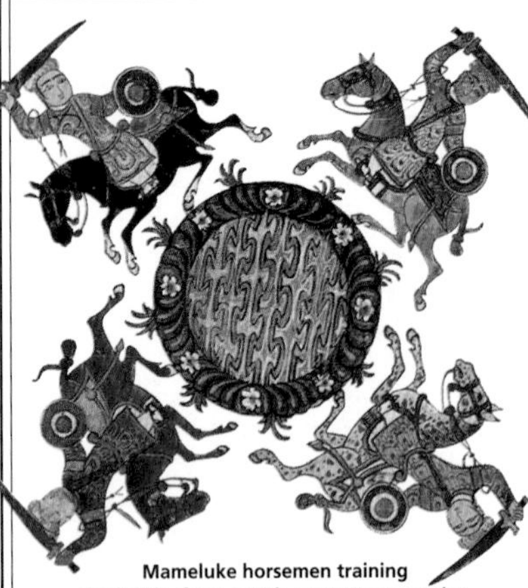

Mameluke horsemen training for battle, from a 15th-century manuscript

PALESTINE UNDER THE MAMELUKES

In the wake of the Crusades, Jerusalem slowly declined to the status of a provincial city. The Mamelukes (former slave guards of Saladin's Ayyubid dynasty) ruled the Holy Land from Egypt, and the Holy City became a place of banishment for officials who fell from court favour in Cairo.

While the Mamelukes had driven the Christian knights from the Holy Land, they did make allowance for Christian pilgrims. In 1333 the Franciscan Friars were permitted a presence in Jerusalem, living in the supposed Hall of the Last Supper. In 1342 Pope Clement VI ratified this mission, which took on the name of the Franciscan Custody of the Holy Land.

The following century saw the beginning of a flow of Jews into Palestine escaping persecution in Europe, a movement that has continued through into the 20th century. In this case, the defeat of the Moors in Spain had given way to the Inquisition and the resultant expulsion of some 100,000 Jews from the country, accused of having too close ties with the vanquished Arabs.

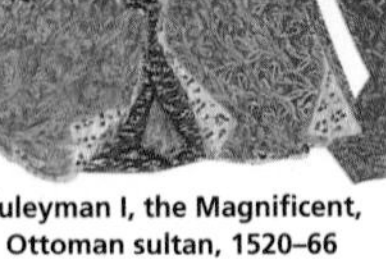

Suleyman I, the Magnificent, Ottoman sultan, 1520–66

THE OTTOMAN EMPIRE

Mameluke control of Palestine ended in 1516 with defeat at the hands of the Ottoman army. Originating in northwest Turkey, the Ottoman Turks had captured Constantinople in 1453, renaming it Istanbul. Under the rule of their greatest sultan, Suleyman the Magnificent (1520–66), vast architectural projects were carried out in Jerusalem, most notably the construction of the city walls and gates.

However, a series of weak sultans meant that by the 18th century the enormous Ottoman empire was no longer so secure, particularly in the provinces where corruption was often a system of administration. This was the case in Palestine, where the people frequently suffered heavy taxes and poor government. But the Jews continued to return, largely because they were safer under Turkish rule than they were in Europe. Many chose to settle in Galilee, around Tiberias and Safed, joining the Sephardic Jewish communities that had fled Spain several centuries earlier. At the same time, Europe was making its first real entry into the region since the Crusades; Napoleon landed in Egypt in 1798 and the following year he had to be repelled from invading at Akko by the Ottoman governor, Ahmed Pasha el-Jazzar.

TIMELINE

14th century Development of the area round the Haram esh-Sharif in Jerusalem

1492 Edict signed by King Ferdinand expelling all professing Jews from Spain

1516 Ottomans defeat the Mamelukes and seize control of Palestine and Egypt

1300 | 1400 | 1500 | 16

1333 Franciscans permitted to settle in Jerusalem

1400 Mamelukes halt westward advance of Mongol ruler Tamerlane

Jaffa Gate, one of seven gates built by Suleyman's engineers

1537 Suleyman the Magnificent orders the construction of the walls of Jerusalem

Akko, rebuilt by successive Ottoman governors

JERUSALEM AND THE COLONIAL POWERS IN THE 19TH CENTURY

With the continuing decline of the Ottoman Empire the European nations, newly empowered by their Industrial Revolution, began to follow in Napoleon's wake – unsuccessful though he had been. When in 1831 the Egyptian ruler Muhammad Ali, the supposed vassal of Istanbul, seized Palestine, it was only with British military help that the Turks regained the territory. A British consul arrived in Jerusalem in 1838, followed closely by diplomatic representatives of France and Prussia. One of the causes of the Crimean War (1854) was a dispute between France and Russia over guardianship of the Holy Places.

All the while, Jewish immigration continued, propelled by virulent anti-Semitism and pogroms in eastern Europe and throughout the Russian Empire. A result of this influx was that in the mid-19th century, Jerusalem overspilled the bounds of its medieval walls with the establishment of a series of small Jewish settlements outside the city gates. The city began to emerge from the lethargy that had characterized it in the preceding centuries.

Over in Europe there had been a growing, but not yet unified, Jewish national movement. In 1839 the British Jew Sir Moses Montefiore had first called for the creation of a Jewish state. This culminated in 1896 with the publication by an Austro-Hungarian Jewish journalist named Theodor Herzl of *Der Judenstat (The Jewish State)*, which proved a rallying cry for Jews worldwide. The following year saw the formation of the World Zionist Organization, with Herzl at its head. Its stated aim was "to create for the Jewish people a home in Palestine". A Jewish National Fund was set up to purchase land for settlement.

However, the Zionist immigrants were laying the foundations for conflict; slogans such as "A land without a people for a people without a land" ignored the large indigenous Arab population of Palestine and the Arab nations' resistance to any form of autonomous Jewish presence there.

The American Colony, one of a great many Western outposts established in 19th-century Jerusalem

Ottoman janissary, soldier of the sultan's guard

1700 **1800** **1900**

1812 Petra rediscovered by Swiss explorer Jean Louis Burckhardt

1831 Egypt's Muhammad Ali takes control of Palestine

1839 British Jew Sir Moses Montefiore first proposes the idea of a Jewish state

1860 Jerusalem's first new Jewish settlements since the Diaspora

1896 Herzl publishes *The Jewish State*

Theodor Herzl

1909 Founding of Tel Aviv and first kibbutz

THE COLLAPSE OF THE OTTOMANS AND THE BRITISH MANDATE

Turkish rule in Palestine ended in 1917, during World War I, when British troops under the command of General Allenby took Jerusalem.

The Arabs, under their leader Faisal, had fought alongside the British and expected Palestine in return. However, with the Balfour Declaration of 1917 the British had let it be known that "His Majesty's government favourably views the creation of a national Jewish home in Palestine". In the event, peace talks in 1920 put Palestine under British authority and this was ratified by the League of Nations on 24 July 1922.

·קרן·קימת·לישראל·
THE JEWISH NATIONAL FUND
REDEEMS THE LAND OF ISRAEL
FOR THE PEOPLE OF ISRAEL

Zionist poster soliciting funds for a homeland in Palestine

The following year, in order to placate Arab discontent, the British recognized Trans-Jordan as an autonomous Arab emirate, ruled by the emir Abdullah, the eldest brother of Faisal, with Amman as its capital. Initially under the supervision of the British in Jerusalem, the territory became totally independent in 1946, with Abdullah confirmed as its king.

ARAB-JEWISH CONFLICT

At the time of World War I, some 500,000 Palestinian Arabs and about 85,000 Jews were living in the Holy Land. In the 20 years between then and the outbreak of World War II about 250,000 more Jews arrived at the ports of Jaffa and Haifa to settle in Palestine. Each new wave of immigrants served to increase the tension between the Palestinian and Jewish communities. In 1929 Palestinian riots culminated in a series of pogroms in Jerusalem, Hebron and Safed. An Arab "revolt" proclaimed in 1936 led to a six-month general strike that brought the country to a standstill.

The *Theodor Herzl* about to dock at Haifa, decks crowded with Jewish immigrants, 1947

TIMELINE

TE Lawrence "of Arabia"

1900	1905	1910	1915	1920	19

1909 Founding of Tel Aviv and first kibbutz in Palestine

1914 War breaks out in Europe; the Ottoman Turks side with Germany

1916 Faisal and the Arabs, encouraged by T E Lawrence, join the British in a desert war against the Turks

1917 General Allenby captures Jerusalem from the Ottoman Turks

24 July 1922 League of Nations ratifies British mandate in Palestine

General Allenby

PROPOSALS FOR PARTITION

By this time, the British were finding rule in Palestine extremely uncomfortable. In 1937, following the deliberations of the Peel Commission, they proposed ending the Mandate and partitioning the country. The Jews accepted but the Arabs refused, claiming that the proposed Jewish homeland occupied the region's most fertile zones.

Allenby Street, in the rapidly expanding Jewish Tel Aviv of the 1930s

Elsewhere, the world was much more concerned with developments in Europe, where war seemed inevitable. In a brazen attempt to improve relations with its potential allies, the Arabs, in 1939, on the eve of war, Britain published a "White Paper" drastically limiting Jewish immigration to Palestine. However, faced with the dangers of Nazism, tens of thousands of Jews continued to arrive, often sneaking in clandestinely by sea. British attempts to check the immigration were, for the most part, in vain.

Ben Gurion witnessing the departure of British troops from Haifa port in 1948

One effect of this new post-war situation was to inspire extremists to attacks on the British. On 22 July 1946 the Jewish military organization Irgun – one of whose leaders was the future prime minister Menachem Begin – bombed British headquarters at the King David Hotel in Jerusalem, killing more than 80 and wounding hundreds more.

Trapped in a no-win situation, the British placed the "Palestine question" before the newly-formed United Nations. On 29 November 1947 the UN voted for the partition of the Holy Land into an Arab state and a Jewish state, with Jerusalem under international administration. Britain announced its intention to pull out of Palestine on 15 May 1948 and leave the Arabs and Jews to fight among themselves.

THE CREATION OF ISRAEL

Skirmishing between the Palestinians and Jews escalated as both sides manoeuvred to control as much territory as possible before the end of the Mandate. Jewish extremists attacked Palestinian villages (most infamously at Deir Yassin, on the road between Tel Aviv and Jerusalem), while armed Palestinians made similar raids against Jewish settlements.

As the British prepared to leave, the Jews were ready to replace them. On 14 May 1948, the eve of departure, David Ben Gurion declared the birth of the State of Israel.

1929 Arab attacks on Jews in Jerusalem, Hebron and Safed

1930

1934 Jews flee central Europe and the threat of Hitler's Germany

1935

1936 Arab Revolt in Palestine

1937 Peel Commission proposes partition of Palestine

1939 Great Britain publishes the "White Paper"

1940

1947 Discovery of the Dead Sea Scrolls at Qumran

14 May 1948 State of Israel declared in Tel Aviv

1950

One of the Dead Sea Scrolls

Refugees crossing the border into Jordan in 1967

THE 1948 WAR

The Arab reaction to the creation of Israel was swift. Lebanon, Syria, Iraq, Jordan and Egypt launched a combined attack with the avowed aim of casting the new-born state into the sea. Fighting continued until an armistice was signed in December 1949. At the cease of hostilities, the Israelis had made great territorial gains at the expense of the Palestinians. Prior to 1948 the Jews owned less than seven per cent of Palestine but at the war's end they occupied about 80 per cent. As a result, some 500,000 to 750,000 Palestinians were made refugees in neighbouring Arab countries and in camps in the Egyptian-controlled Gaza Strip and in the Jordanian-held territories on the west bank of the Jordan River.

Pre-1967 poster, with the West Bank shown as part of Jordan

One of the main objectives of the opposing sides had been the capture of Jerusalem. Neither side had achieved this; the Israelis held the modern quarters of West Jerusalem, the Jordanians held the Old City and East Jerusalem. The city was to remain divided, along what came to be known as the Green Line, for almost 20 years.

THE ARAB-ISRAELI WARS AFTER 1949

After the violent birth of Israel, the infant state sought to strengthen its position by passing the Law of Return. This extended to all Jews throughout the world the right to live in Israel. The first to heed the invitation were communities of Jews from the Arab world, followed by displaced Jews from Europe. Those that followed came from everywhere, from the then-Soviet Union to South America.

Relations with the Arabs remained on a war footing. In 1956, the Israeli army swept into Sinai as part of the French and British plan to seize the Suez Canal, nationalized by Egypt's President Nasser. On this occasion, under pressure from the United States and the United Nations, they were forced to retreat. Eleven years later, in 1967, Israeli tanks rolled into Sinai once again. Alarmed by a build-up of Egyptian forces on the border, Israel launched a pre-emptive attack. Despite then facing the combined forces of all its Arab neighbours, in six days Israel's army had taken the Golan Heights from Syria, the Gaza Strip and Sinai from Egypt, and the West Bank from Jordan. The Israelis also captured the whole of Jerusalem. In what amounted to a face-saving exercise, on 6 October 1973, the Jewish fast of Yom Kippur, Egypt and Syria

TIMELINE

1951 Assassination of King Abdullah of Jordan in Jerusalem by Palestinian extremists

1956 Suez crisis

Golda Meir, Israeli prime minister 1969–74

6 October 1973 Yom Kippur War breaks out

1982 Sinai returned to the Egyptians

1950 | 1955 | 1960 | 1965 | 1970 | 1975 | 1980

14 May 1948 On the declaration of the State of Israel war breaks out with the Arabs

Hussein, crowned king of Jordan in May 1953

5–11 June 1967 Six Day War results in reunification of Jerusalem under the Israelis

1979 Camp David peace treaty signed between Egypt and Israel

The Israeli-built wall, designed to stop Palestinian bombers

launched a surprise attack on Israeli positions. Caught off guard, the Israelis suffered initial losses but they counterattacked and reversed early Arab gains. At the cease of hostilities the action had not altered the territorial state of affairs set six years previously.

The 1973 War did, however, pave the way for the first talks between Egypt and Israel. In 1979 the two countries formally agreed to peace by signing the Camp David agreement. In 1982 Sinai was returned to Egypt.

THE QUEST FOR PEACE

The peace treaty was not welcomed by all parties. The Palestinians saw it as undermining their campaign for self-rule. Groups such as the Palestine Liberation Organisation (PLO) stepped up their anti-Israel guerrilla war. Their tactics won them little sympathy with the international community. That changed in late 1987 with the beginning of the *intifada* ("shaking off"), a grass-roots Palestinian revolt against Israeli occupation in the Gaza Strip and West Bank. Television screens worldwide were filled with images of stone-throwing young Arab boys facing up to well-armed Israeli soldiers. In the wake of 1991's Gulf War, the Americans brokered a meeting between Israeli and Palestinian delegations in Madrid. This seemed to achieve little, but in 1993 it was revealed that the two parties had been meeting in Norway where agreement had been reached. The signing of the "Oslo Accords" was capped that year by a handshake between Israeli prime minister Yitzhak Rabin and PLO president Yasser Arafat on the lawn of the White House. The following year saw Jordan and Israel formally end the state of war that had existed between the two countries since 1948.

Since then, Rabin has been assassinated by a Jewish extremist and Arafat has died. Israel has experienced 60 years of statehood but the Palestinians remain stateless. The Israelis have built a giant wall between themselves and the Palestinians in an attempt to halt the terror bombings that have been a fact of daily life since the 1990s. The cycle of violence continues, but so do the attempts to find a solution that will bring a lasting peace to the region.

Activists on both sides unite for peace

1993 Oslo Accords lead to Rabin and Arafat shaking hands

1995 Israeli prime minister Yitzhak Rabin assassinated

1999 King Hussein of Jordan dies

Yasser Arafat, first president of the Palestinian Authority dies, November 2004

5 | 1990 | 1995 | 2000 | 2005 | 2010 | 2015 | 2020

1987 Eruption of Palestinian *intifada* against Israeli occupation

1994 Palestinians granted limited autonomy

First issue of Palestinian stamps, 1994

2005 Israel withdraws Jewish settlements from the Gaza Strip

2006 Second Lebanon war

2010 Jerusalem Light Rail starts service

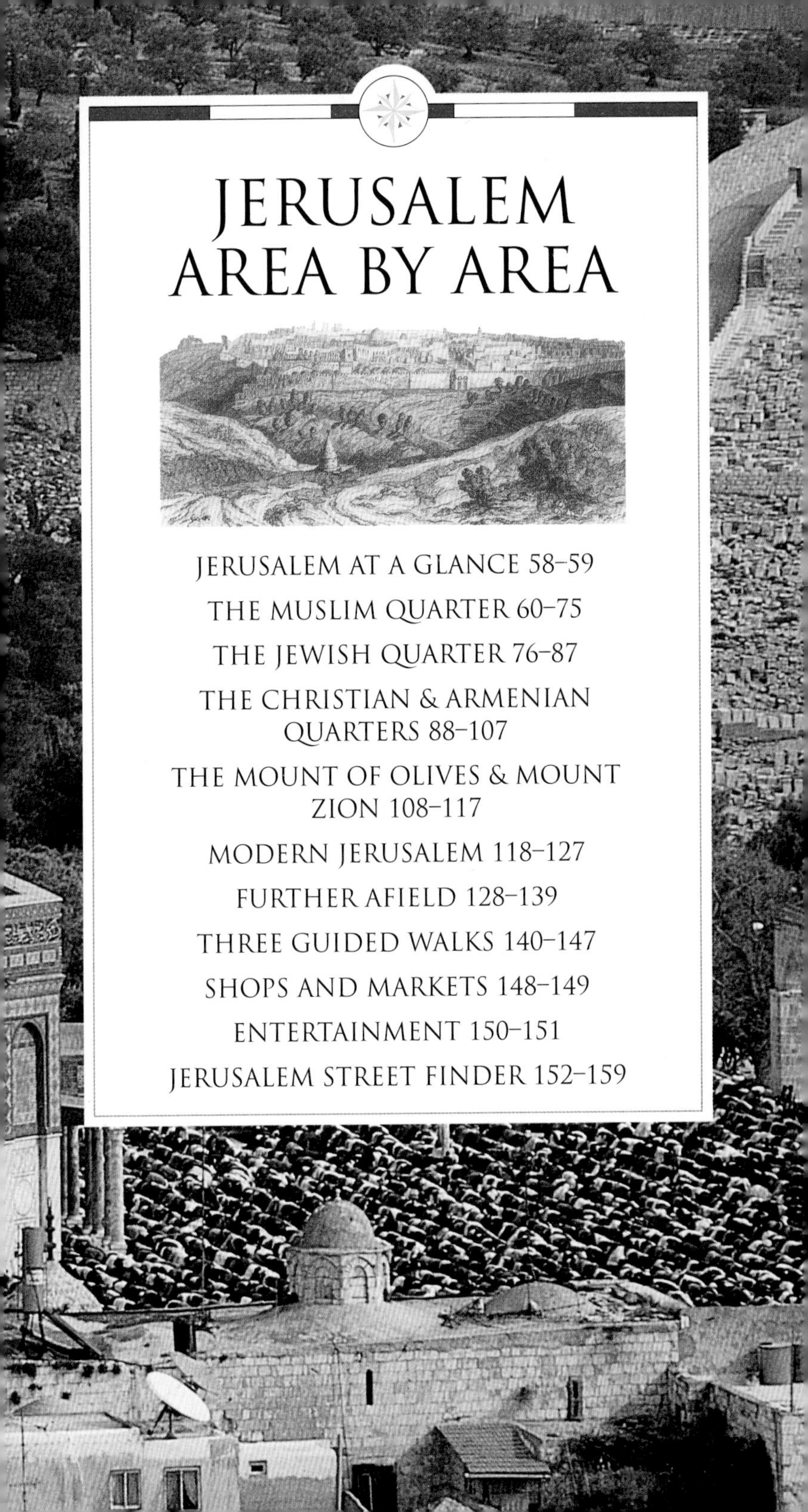

JERUSALEM AREA BY AREA

Jerusalem at a Glance

The old city of Jerusalem has a history that stretches back more than 3,000 years, although the present street plan dates largely from Byzantine times, and the encircling walls are from the 16th century. Within the walls, the Old City divides into four vaguely defined quarters – one each for the Christians, Jews and Muslims, and the fourth occupied by the Armenians. East and south of the Old City are the Mount of Olives and Mount Zion, both places traditionally linked with the last acts of Jesus Christ. To the north and west is modern Jerusalem, liberally endowed with fine examples of late 19th-century architecture.

The Church of the Holy Sepulchre (see pp92–5) *is the most important of the Holy Land's Christian sites. Tradition has it that the church occupies the site of Golgotha, where Jesus Christ was crucified and buried.*

The Citadel (see pp102–4) *is an impressively restored, fortified complex, which has its origins in the 2nd century BC. It now houses an excellent museum devoted to the history of Jerusalem. There are also splendid views of the city from its ramparts.*

MODERN JERUSALEM
(See pp118–127)

THE CHRISTIAN AND ARMENIAN QUARTERS
(See pp88–107)

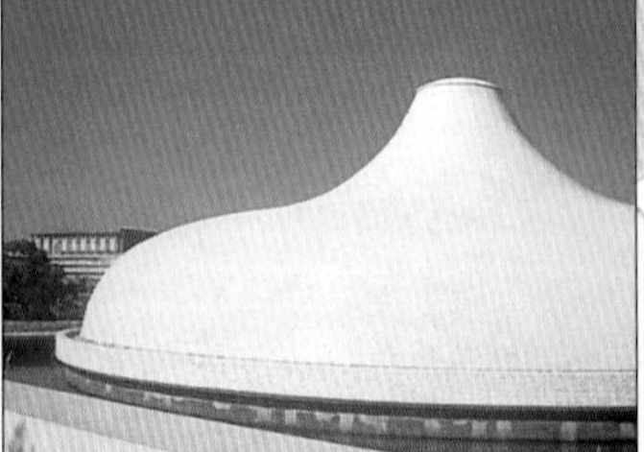

The Israel Museum (see pp132–7) *was purpose-built in the 1960s to house the country's most significant archaeological finds, including the Dead Sea Scrolls, which are displayed in this uniquely shaped hall. The museum was renovated in 2007–10 and is a short distance west of the city centre.*

Yemin Moshe (see pp120–1) *is one of several attractive old quarters in modern Jerusalem, developed in the mid-19th century to escape overcrowding in the Old City. It is distinguished by its windmill and by this communal housing block, known as Mishkenot Shaananim.*

◁ **The Dome of the Rock, with Dominus Flevit Chapel and the Mount of Olives behind**

The Haram esh-Sharif (see pp68–73) *is the focus of the Muslim faith in Jerusalem. A large plateau on the eastern edge of the Old City, it contains some fine Islamic buildings, including the 8th-century El-Aqsa Mosque, and the magnificent Dome of the Rock, with its dazzling interior.*

The Western Wall (see p85) *is Judaism's holiest site. It is believed to be part of the great Temple enclosure built by Herod in the 1st century BC. The plaza in front is busy, day and night, with supplicants at prayer.*

THE MUSLIM QUARTER
(See pp60–75)

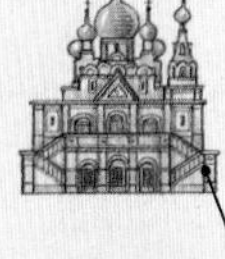

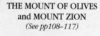

THE JEWISH QUARTER
(See pp76–85)

THE MOUNT OF OLIVES and MOUNT ZION
(See pp108–117)

The Sephardic Synagogues (see p82) *are a group of four synagogues which were at the heart of the 17th-century Sephardic community, once the largest Jewish group in Jerusalem. The Ben Zakkai Synagogue, shown here, was built in 1610.*

The Mount of Olives (see pp108–13) *has several fine churches, including the richly decorated Russian Orthodox Church of St Mary Magdalene.*

0 metres 300

0 yards 300

THE MUSLIM QUARTER

This is the largest and most densely populated quarter of the Old City. It was first developed under Herod the Great and delineated in its present form under the Byzantines. In the 12th century it was taken over by the Crusaders, hence the quarter's wealth of churches and other Christian institutions, such as the Via Dolorosa *(see pp30–31)*. In the 14th and 15th centuries the Mamelukes rebuilt extensively, especially in the areas abutting the Haram esh-Sharif. The quarter has been in decay since the 16th century. Today it contains some of the city's poorest homes. It is also one of the most fascinating and least explored parts of Jerusalem.

Street sign for a Quranic recitation school

SIGHTS AT A GLANCE

Historic Streets, Buildings and Gates

Chain Street ❻
Damascus Gate ❽
Ecce Homo Arch ❷
Herod's Gate ❾
Lady Tunshuq's Palace ❹
Lions' Gate ⓫
Via Dolorosa ❸

Souks and Markets

Central Souk ❼
Cotton Merchants' Market ❺

Holy Places

Haram esh-Sharif pp68–73 ⓬
Monastery of the Flagellation ❶
St Anne's Church ❿

GETTING THERE

The Muslim Quarter is served by Damascus, Herod's and Lions' gates. There are buses from the New City to Damascus Gate *(see p311)*. Alternatively, for visitors with their own transport, there is a car park just outside Lions' Gate.

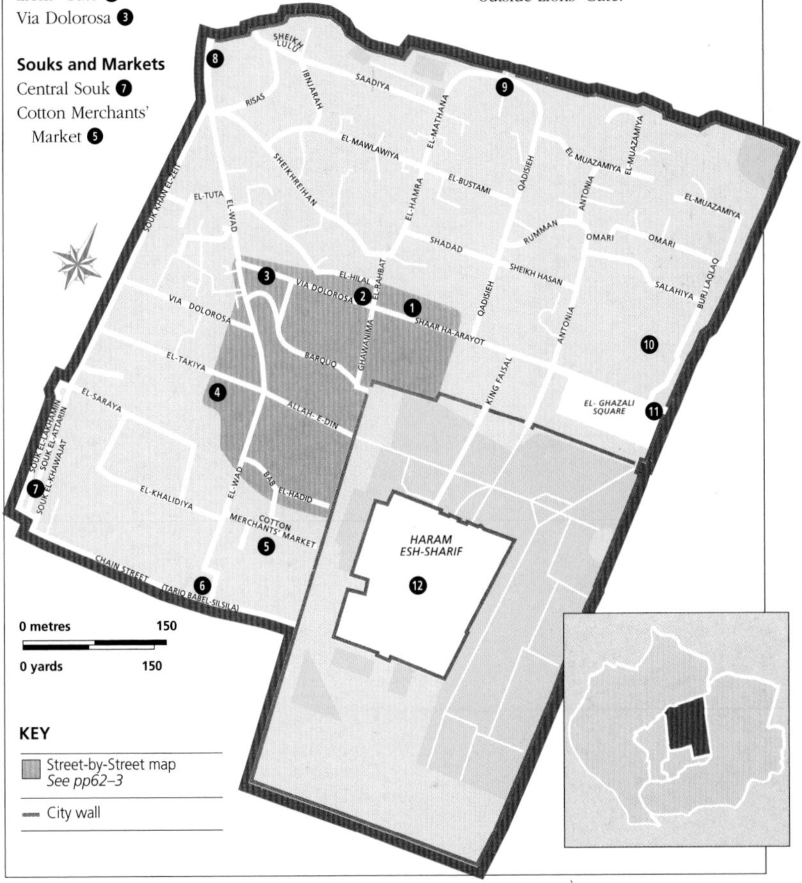

0 metres 150

0 yards 150

KEY

Street-by-Street map *See pp62–3*

City wall

◁ **Muslim women entering the Dome of the Rock, centrepiece of the Haram esh-Sharif**

Street-by-Street: The Muslim Quarter

Studium Museum artifact

The main routes through this busy quarter are along the Via Dolorosa and up and down El-Wad. Both streets are lined with a gaudy array of shops, whose salesmen eagerly press on visitors all manner of ornaments and kitsch, from plastic crucifixes to glass-bowled water pipes. Few people stray from the main thoroughfares, but those who do are richly rewarded. The quiet, winding back alleys contain a wealth of fine medieval Islamic architecture, much of it dating from the Mameluke era (1250–1516). Not all of it is in good condition, but many of these buildings still perform the functions for which they were intended.

The Austrian Hospice was built in 1869 to accommodate Christian pilgrims.

Damascus Gate

Via Dolorosa

Crossing the quarter from east to west, this street is revered by Christian pilgrims as the route taken by Christ as he was led to his crucifixion ❸

VIA DOLOROSA

EL-WAD

Holy Sepulchre church and the Christian Quarter

Abu Shukri restaurant *(see p272)*

El-Takiya Street

A narrow, stepped street at the heart of the quarter, El-Takiya contains some of the city's finest examples of Mameluke architecture.

EL-TAKIYA

Lady Tunshuq's Palace

The banding of different coloured stone and panels of intricate marble inlay typify the decorative style of the Mamelukes ❹

KEY

– – – Suggested route

STAR SIGHTS

★ Monastery of the Flagellation

★ Ecce Homo Arch

For hotels and restaurants in this area see p256 and p272

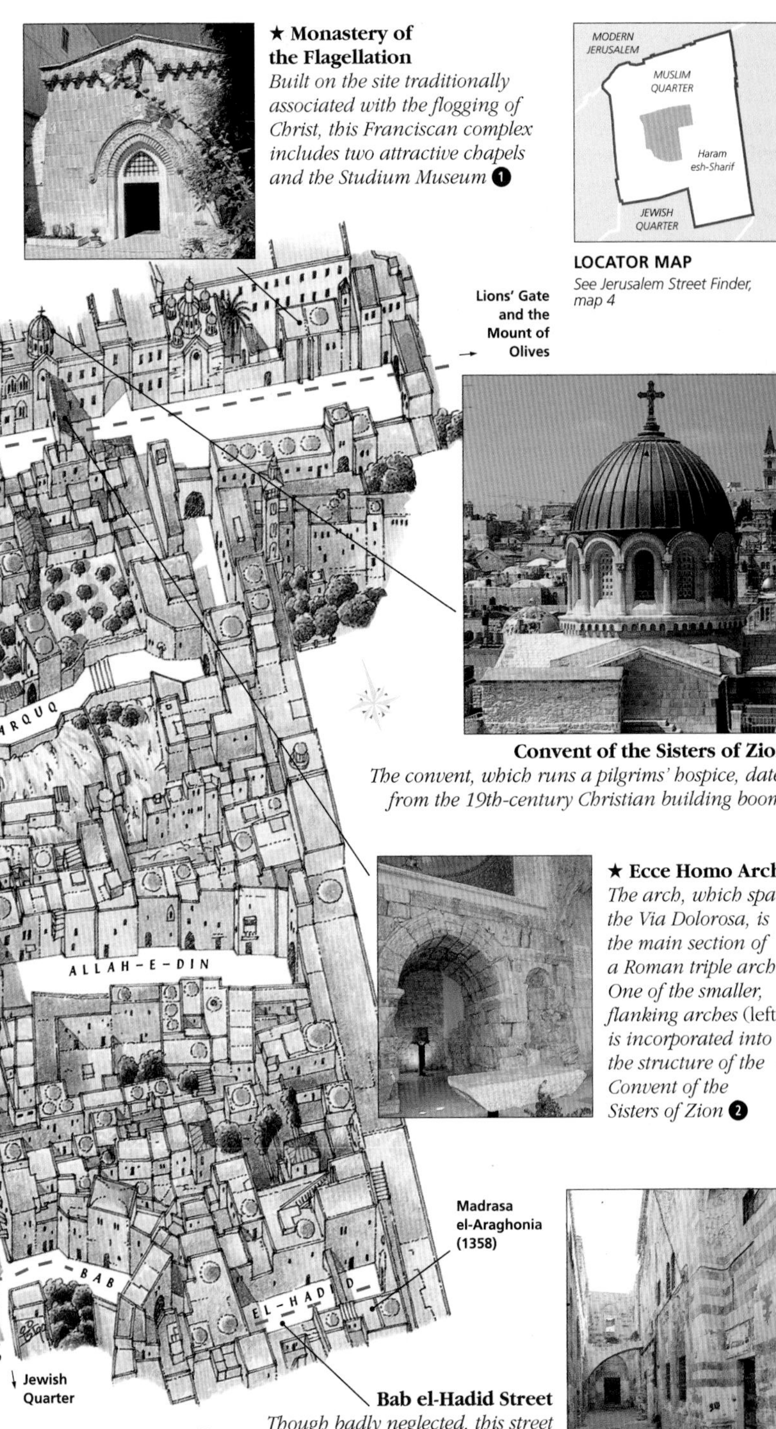

LOCATOR MAP

See Jerusalem Street Finder, map 4

★ Monastery of the Flagellation

Built on the site traditionally associated with the flogging of Christ, this Franciscan complex includes two attractive chapels and the Studium Museum ❶

Convent of the Sisters of Zion

The convent, which runs a pilgrims' hospice, dates from the 19th-century Christian building boom.

★ Ecce Homo Arch

The arch, which spans the Via Dolorosa, is the main section of a Roman triple arch. One of the smaller, flanking arches (left) *is incorporated into the structure of the Convent of the Sisters of Zion* ❷

Bab el-Hadid Street

Though badly neglected, this street has a number of madrasas (see p71) *from the 14th and 15th centuries.*

Monastery of the Flagellation, with the Via Dolorosa behind

Monastery of the Flagellation ❶

Via Dolorosa. **Map** 4 D2. ***Tel*** *(02) 627 0444.* *8am–6pm (winter: 5pm) daily.* **Studium Museum** *9am–11:30am Mon–Sat (phone for appt).*

Owned by the Franciscans, this complex embraces the simple and striking Chapel of the Flagellation, designed in the 1920s by the Italian architect Antonio Barluzzi, who was also responsible for the Dominus Flevit Chapel on the Mount of Olives *(see p113)*. It is located on the site traditionally held to be where Christ was flogged by Roman soldiers prior to his Crucifixion (Matthew 27: 27–30; Mark 15: 16–19).

On the other side of the courtyard is the Chapel of the Condemnation, which also dates from the early 20th century. It is built over the remains of a medieval chapel, on the site popularly identified with the trial of Christ before Pontius Pilate.

The neighbouring monastery buildings house the Studium Biblicum Franciscanum, a prestigious institute of biblical, geographical and archaeological studies. Also part of the complex, the **Studium Museum** contains objects found by the Franciscans in excavations at Capernaum, Nazareth, Bethlehem and various other sites. The most interesting exhibits are Byzantine and Crusader objects, such as fragments of frescoes from the Church of Gethsemane, precursor of the present-day Church of All Nations *(see p114)*, and a 12th-century crozier from the Church of the Nativity in Bethlehem *(see pp194–5)*.

Crusader-era angel's head, Studium Museum

Ecce Homo Arch ❷

Via Dolorosa. **Map** 4 D2. **Convent of the Sisters of Zion** ***Tel*** *(02) 643 0887.* *9am–noon & 2–5:30pm (winter: 5pm) Mon–Fri, 9am–5pm Sat.*

This arch that spans the Via Dolorosa was built by the Romans in AD 70 to support a ramp being laid against the Antonia Fortress, in which Jewish rebels were barricaded *(see p44)*. When the Romans rebuilt Jerusalem in AD 135 in the wake of the Second Jewish War *(see p43)*, the arch was reconstructed as a monument to victory, with two smaller arches flanking a large central bay. It is the central bay that you see spanning the street.

The Ecce Homo Arch bridging the Via Dolorosa

One of the side arches is also still visible, incorporated into the interior of the neighbouring **Convent of the Sisters of Zion**. Built in the 1860s, the convent also contains the remains of the vast Pool of the Sparrow (Struthion), an ancient reservoir which collected rainwater directed from the rooftops. The pool was originally covered with a stone pavement *(lithostrothon)* and it was on this flagstone plaza, Christian tradition has it, that Pilate presented Christ to the crowds and uttered the words "Ecce homo" ("Behold the man"). However, archaeology refutes this, dating the pavement to the 2nd century AD, long after the time of Christ. Within a railed section you can see marks scratched into the stone. Historians speculate that they may have been carved by bored Roman guards as a kind of street game.

Via Dolorosa ❸

Map 3 C3 & 4 D2.

The identification of the Via Dolorosa *(see pp30–31)* with the ancient "Way of Sorrows" walked by Christ on the way to his Crucifixion has more to do with religious tradition than historical fact. It nevertheless continues to draw huge numbers of pilgrims every day. The streets through which they walk are much like any others in the Muslim Quarter, lined with small shops and stalls, but the route is marked out by 14 "Stations of the Cross", linked with events that occurred on Christ's last, fateful walk. Some of the Stations are commemorated only by wall plaques, which can be difficult to spot among the religious souvenir stalls. Others are located inside buildings. The last five Stations are all within the Holy Sepulchre church *(see pp92–5)*.

Friday is the main day for pilgrims, when, at 3pm in winter and 4pm in summer, the Franciscans lead a procession along the route.

In fact, the more likely route for the original Via Dolorosa begins at what is now the Citadel *(see pp102–103)* but was at the time the royal palace. This is where Pontius Pilate resided when in Jerusalem, making it a more likely location for the trial of Christ. From here, the condemned would probably have been led down what is now David Street, through the present-day Central Souk *(see p66)*, out of the then city gate and to the hill of Golgotha, the presumed site of which is now occupied by the Holy Sepulchre church.

An unusually quiet Via Dolorosa, leading down from Ecce Homo Arch to El-Wad Road

Stalactite stone carvings above a window on Lady Tunshuq's Palace

Lady Tunshuq's Palace ❹

El-Takiya St. **Map** 4 D3.
■ *to public.*

Lady Tunshuq, of Mongolian or Turkish origin, was the wife, or mistress, of a Kurdish nobleman. She arrived in Jerusalem some time in the 14th century and had this edifice built for herself. It is one of the loveliest examples of Mameluke architecture in Jerusalem. Unfortunately the narrowness of the street prevents you from standing back and appreciating the building as a whole, but you can admire the three great doorways with their beautiful inlaid-marble decoration. The upper portion of a window recess also displays some fine carved-stone, stalactite-like decoration, a form known as *muqarnas*. The former palace now serves as an orphanage and is not open to the public.

When Lady Tunshuq died, she was buried in a small tomb across from the palace. The fine decoration on the tomb includes panels of different coloured marble, intricately shaped and slotted together like a jigsaw – a typical Mameluke feature known as "joggling".

If you head east and across El-Wad Road, you will enter a narrow alley called Ala ed-Din, which contains more fine Mameluke architecture. Most of the façades are composed of bands of different hues of stone, a strikingly beautiful Mameluke decorative technique known as *ablaq*.

Cotton Merchants' Market ❺

Off El-Wad Rd. **Map** 4 D3.

Known in Arabic as the Souk el-Qattanin, this is a covered market with next to no natural light but lots of small softly-lit shops. It is possibly the most atmospheric street in all the Old City. Its construction was begun by the Crusaders. They intended the market as a free-standing structure but later, in the first half of the 14th century, the Mamelukes connected it to the Haram esh-Sharif *(see pp68–73)* via a splendidly ornate gate facing the Dome of the Rock. (But note, non-Muslims are not allowed to enter the Haram esh-Sharif by this gate, although you can depart this way.)

As well as some 50 shop units, the market also has two bathhouses, the Hammam el-Ain and the Hammam el-Shifa. One of these has been undergoing restoration with a view to its being eventually opened to the public. Between the two bathhouses is a former merchants' hostel called Khan Tankiz, also being restored.

Less than 50 m (160 ft) south of the Cotton Merchants' Market on El-Wad Road is a small public drinking fountain, or *sabil*, one of several such erected during the reign of Suleyman the Magnificent.

The tunnel-like interior of the Cotton Merchants' Market

Chain Street ❻

Map 4 D4.

Some of the many and varied spices on sale at the Central Souk

The Arabic name for this street is Tariq Bab el-Silsila, which means "Street of the Gate of the Chain". The name refers to the magnificent entrance gate to the Haram esh-Sharif *(see pp68–73)* situated at its eastern end. The street is a continuation of David Street, and together the two streets run the width of the Old City from Jaffa Gate to the Haram esh-Sharif.

Chain Street has several noteworthy buildings commissioned by Mameluke emirs in the 14th century. Heading eastwards from David Street, the first is the Khan el-Sultan caravanserai, a restored travellers' inn. Further along on the right is Tashtamuriyya Madrasa, with its elegant balcony. It houses the tomb of the emir Tashtamur, and is one of many final resting places built here in the 14th and 15th centuries in order to be close to the Haram esh-Sharif. On the same side of the street is the tomb of the brutal Tartar emir Barka Khan, father-in-law of the Mameluke ruler Baybars, who drove the Crusaders out of the Holy Land *(see pp48–9)*. This building, with its intriguing façade decoration, now houses the Khalidi Library.

Window on Khalidi Library

Opposite the Khalidi Library are two small mausoleums. Of the two, that of emir Kilan stands out for its austere, well-proportioned façade. Further along on the same side is the tomb of Tartar pilgrim Turkan Khatun, easily recognizable by the splendid arabesques on its façade. Opposite the Gate of the Chain is the impressive entrance to the 14th-century Tankiziyya Madrasa. In the inscription, three symbols in the shape of a cup show that emir Tankiz, who built the college, held the important office of cupbearer. Nearby is a drinking fountain, or *sabil*, from the reign of Suleyman the Magnificent, which combines Roman and Crusader motifs.

Central Souk ❼

David St/Chain St. **Map** 3 C4. *8am–7pm Sat–Thu.*

The Central Souk consists of three parallel covered streets at the intersection of David Street and Chain Street. They once formed part of the Roman Cardo *(see p80)*. Today's markets sell mostly clothes and souvenirs, although the section called the Butchers' Market (Souk el-Lakhamin in Arabic), restored in the 1970s, still offers all the excitement of an eastern bazaar. It is not for the faint-hearted, however, as the pungent aromas of spices and freshly slaughtered meat can be overwhelming.

Damascus Gate ❽

Map 3 C1. *1, 2.* **Roman Square Excavations** *9am–5pm Sat–Thu, 9am–3pm Fri.*

Spotting this gate is easy, not only because it is the most monumental in the Old City, but also because of the perpetual bustle of activity in the area outside the gate.

Arabs call it Bab el-Amud, the Gate of the Column. This could refer to a large column topped with a statue of the emperor Hadrian *(see p43)* which, in Roman times, stood just inside the gate. For Jews it is Shaar Shkhem, the gate which leads to the biblical city of Shechem, better known by its Arabic name – Nablus.

The present-day gate was built over the remains of the original Roman gate and parts of the Roman city. Outside the gate and to the west of the raised walkway, steps lead down to the excavation area. In the first section are remains of a Crusader chapel with frescoes, part of a medieval roadway and an ancient sign marking the presence of the Roman 10th Legion. Further in, metal steps lead down to the single surviving arch of

Crowds of visitors and market traders outside Damascus Gate

the Roman gate, which gives access to the **Roman Square Excavations**. Here, the fascinating remains of the original Roman plaza, the starting point of the Roman Cardo, include a gaming board engraved in the paving stones. A hologram depicts Hadrian's column in the main plaza. It is possible to explore the upper levels of the gate as part of the ramparts walk *(see pp142–3)*.

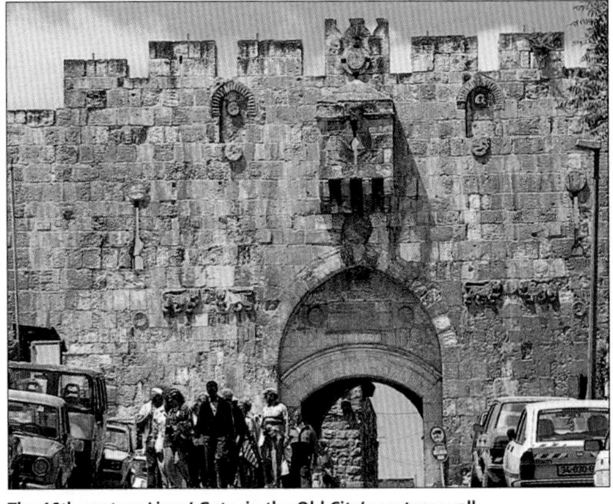

The 16th-century Lions' Gate, in the Old City's eastern wall

Herod's Gate ❾

Map 4 D1.

The Arabic and Hebrew names for this gate, Bab el-Zahra and Shaar ha-Prakhim respectively, both mean "Gate of Flowers", referring to the rosette above the arch. It came to be known as Herod's Gate in the 1500s, when Christian pilgrims wrongly thought that the house inside the gate was the palace of Herod the Great's son. It was via the original, now closed, entrance further east that the Crusaders entered the city and conquered it on 15 July 1099 *(see pp48–9)*.

St Anne's Church ❿

2 Shaar ha-Arayot St. **Map** 4 E2. ***Tel*** *(02) 628 3285.* *8am–noon & 2–6pm (winter: 5pm) daily.*

This beautiful Crusader church is a superb example of Romanesque architecture. It was constructed between 1131 and 1138 to replace a previous Byzantine church, and exists today in more or less its original form. It is traditionally believed that the church stands on the spot where Anne and Joachim, the parents of the Virgin Mary, lived. The supposed remains of their house are in the crypt, which is also noted for its remarkable acoustics.

Shortly after the church was built, it was made larger by moving the façade forwards by several metres. The connection with the original church can still be seen in the first row of columns. In 1192, Saladin *(see pp48–9)* turned the church into a Muslim theological school. There is an inscription to this effect above the church's entrance. Later abandoned, the church fell into ruins, until the Ottomans donated it to France in 1856 and it was restored.

Next to the church are two cisterns that once lay outside the city walls. They were built in the 8th and 3rd centuries BC to collect rainwater. Some time later, under Herod the Great they were turned into curative baths. Ruins of a Roman temple, thought to have been to the god of medicine, can be seen here, as can those of a later Byzantine church built over the temple. It is also widely believed that this is the site of the Pool of Bethesda, described in St John's account of Christ curing a paralysed man (John 5: 1–15).

Archaeological site in front of St Anne's Church

Lions' Gate ⓫

Map 4 F2.

Suleyman the Magnificent built this gate in 1538. Its Arabic name, Bab Sitti Maryam (Gate of the Virgin Mary), refers to the Tomb of the Virgin in the nearby Valley of Jehoshaphat *(see p115)*. The Hebrew name, Shaar ha-Arayot, or Lions' Gate, refers to the two emblematic lions on either side of the gateway, although one school of thought insists that they are panthers. There are many different stories to explain the significance of the lions. One is that Suleyman the Magnificent had them carved in honour of the Mameluke emir Baybars and his successful campaign to rid the Holy Land of Crusaders. Also known as St Stephen's Gate, this name was adopted in the Middle Ages by Christians who believed that the first Christian martyr, St Stephen, was executed here. Prior to that, it was thought that St Stephen had been stoned to death outside Damascus Gate.

Lion detail from Lions' Gate

The gate is also significant because of its more recent history, for it was through it that the Arab Legion penetrated the Old City in 1948 *(see p54)* and where Israeli paratroopers entered in 1967 *(see p54)*. It is an excellent starting point for the walk along the Via Dolorosa *(see p64)*.

Haram esh-Sharif ⓬

Dome of the Prophet

Haram esh-Sharif, the "Noble Sanctuary" or Temple Mount, is a vast rectangular esplanade in the south-eastern part of the Old City. Traditionally the site of Solomon's Temple, it later housed the Second Temple, enlarged by Herod the Great and destroyed by the Romans *(see pp44–5)*. Left in ruins for more than half a century, the site became an Islamic shrine in AD 691 with the building of the Dome of the Rock. Over the centuries other buildings have been added to this, the third most important Islamic religious sanctuary.

★ Dome of the Rock
This is the crowning glory not just of the Haram esh-Sharif but of all Jerusalem (see pp72–3).

Madrasa el-Omariyya is one of several Mameluke-era schools on the Haram.

Madrasa el-Isardiyya

Dome of the Prophet

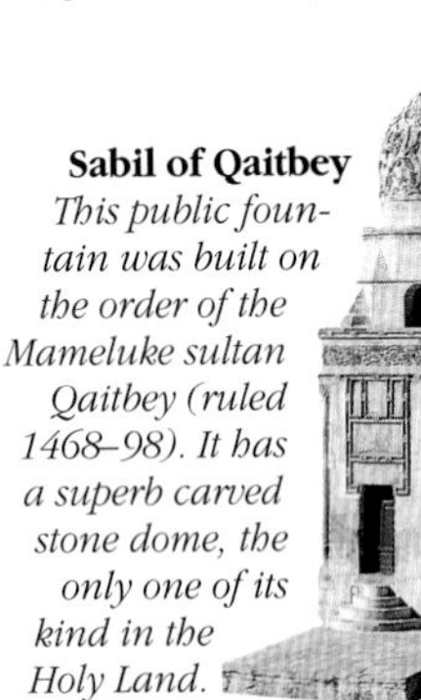
Sabil of Qaitbey
This public fountain was built on the order of the Mameluke sultan Qaitbey (ruled 1468–98). It has a superb carved stone dome, the only one of its kind in the Holy Land.

Cotton Merchants' Gate is a strikingly decorated Mameluke portal giving access to the market of the same name *(see p65)*.

Chain Gate (Ha-Shalshelet)

Western Wall *(see p85)*

Moors' Gate (Bab el-Magharibа) is one of only two gates that non-Muslims may use to enter the Haram.

Grammar College
Also known as "The Dome of Learning", this still serves as a Quranic teaching school. The doorway on the north side is flanked by some unusual candy-twist columns dating from the Ayyubid era (1169–1250).

Museum of Islamic Art
This engraved Mameluke vessel is part of a collection of artifacts, largely from the Middle Ages, that includes Qurans, textiles, ceramics and weaponry (see p70).

For hotels and restaurants in this area see p256 and p272

★ Dome of the Chain
This small dome (see p71) *stands at the approximate centre of the Haram esh-Sharif, which, according to one theory, equated to the centre of the world. The 13th-century tiling on the interior surpasses even that of the Dome of the Rock.*

VISITORS' CHECKLIST

Entrance via Chain Gate and Mugrabi Gate only. **Map** 4 E3. *summer: 7:30–11:30am, 1:30–2:30pm Sun–Thu; winter: 7:30–10am, 12:30–1:30pm Sun–Thu.* *Fri, Sat, Muslim hols.* *Advanced booking required: (02) 622 6250.* *inside Dome of the Rock and El-Aqsa Mosque. Check security with tourist office before visiting. Currently mosques are not open to visitors.*

Qanatir
Each of the eight flights of steps up to the platform of the Dome of the Rock is topped by a qanatir, *or free-standing arcade* (see p70). *Some of the column capitals were recycled from Roman-era buildings.*

Asbat Minaret

Gate of the Tribes (Bab el-Asbat) leads to the Via Dolorosa.

Golden Gate is one of the original city gates *(see p71)* but was sealed up by the Muslims in the 7th century. The area is out of bounds.

★ El-Aqsa Mosque
Originally built in the early years of the 8th century (see p70), *El-Aqsa remains the main place of Islamic worship in Jerusalem and draws huge crowds of devout Muslims each Friday for noon prayer.*

Crusader-built tower

Women's mosque

El-Kas Fountain
Carved from a single block of stone and dating from 1320, this is the largest of the Haram's many old but still functioning ablutions fountains.

STAR FEATURES

- ★ Dome of the Rock
- ★ Dome of the Chain
- ★ El-Aqsa Mosque

Exploring the Haram esh-Sharif

Stone window, El-Aqsa

Although the undoubted main attraction is the Dome of the Rock, the Haram esh-Sharif has a great many other features that are worthy of attention. The esplanade acts as a virtual museum of Islamic architecture, beginning with the Dome, which dates back to the Omayyad era and is the earliest structure, and running through the Ayyubid (Grammar College), Mameluke (numerous *madrasas*) and Ottoman periods. Visitors should be aware that certain parts of the Haram esh-Sharif are out of bounds, notably the area south of the Gate of the Tribes and east of El-Aqsa.

The much reconstructed interior of the El-Aqsa Mosque

EL-AQSA MOSQUE

Construction of El-Aqsa was begun less than 20 years after the completion of the Dome of the Rock. However, unlike the Dome, whose structure and interior have remained intact over the centuries, El-Aqsa has undergone great changes. In the first 60 years of its existence the mosque was twice razed to the ground by earthquakes. Its present form dates from the early 11th century. When the Crusaders captured Jerusalem in 1099, El-Aqsa became the headquarters of the Templars *(see p49)*; their legacy remains in the three central bays of the main façade. As it appears today, the façade has seven bays; in the mid-14th century the Mamelukes added an extra two on either side of the original Crusader porch.

The interior is dominated by mid-20th century additions, notably ranks of marble columns, donated by Benito Mussolini, and an elaborately painted ceiling paid for by King Farouk of Egypt. Older elements include the mihrab, decorated in 1187 under the patronage of Saladin, and the mosaics above the central aisle arch and around the drum of the dome, dating from 1035. Until 1969, the mosque had a fine carved pulpit *(minbar)*, also dating from the time of Saladin, but this was lost in a fire started by a deranged visitor.

Antiquity-strewn area in front of the Museum of Islamic Art

MUSEUM OF ISLAMIC ART

Housed in the Crusader-era refectory of the Knights Templar, this sparsely-filled museum contains objects donated to the Haram esh-Sharif over the centuries, as well as architectural remnants from many of the Haram's buildings. Worthy of mention are the precious large Qurans, with pages adorned by fine Islamic calligraphy; part of a carved cypress-wood ceiling from El-Aqsa, dating from the 7th century and removed in 1948; and fine 15th-century copper doors from the Dome of the Rock. Admission to the museum is included in the fee for the Dome of the Rock and El-Aqsa Mosque.

Visitors with an interest in Islamic art should also visit the LA Mayer Museum in the new city *(see p130)*.

THE QANATIRS

Eight short flights of steps lead up to the platform on which the Dome of the Rock sits. All these stairways are of different sizes and lengths, and they all date from different periods. The flight opposite the Sabil of Qaitbey, leading up to the main entrance of the Dome, is unique in that it is carved out of the stone of the platform. Each flight is crowned by a slender arcade known as a *qanatir*. An alternative name for the arches is *mawazin*, or scales, because according to a

A *qanatir*, topping a flight of steps up to the Dome

widely-accepted Muslim belief, on the day of the Last Judgment, the scales used by God to weigh the souls of humankind will be hung from these arches on the Haram.

JERUSALEM AND ISLAM

The Dome of the Rock and neighbouring El-Aqsa Mosque represent the first great religious complex in the history of Islam. Although Muslims venerate many of the same prophets as the Jews and Christians, notably Abraham (Ibrahim to the Muslims), Jerusalem itself is never mentioned in the Quran. The choice of this site was more likely a political issue. In locating his mosque on the site of the Temple, the caliph Abd el-Malik meant to reinforce the idea that the new religion of Islam, and its worldly empire, was the successor and continuation of those of the Jews and the Christians. It was only later that Jerusalem came to be tied into Islamic tradition through the story of the Night Journey *(see p27)*. In this, Muhammad visits *el-masjid el-aqsa*, which means literally "the most distant mosque", and this name was retroactively applied to the whole Haram esh-Sharif before later being restricted to the mosque only.

Angel with Muhammad's robe on the Night Journey

DOME OF THE CHAIN

Beside the dome of the Rock, the Haram has many other, smaller domes. The most impressive is the Dome of the Chain, immediately to the east of the Dome of the Rock. It is a simple structure of a domed roof supported on 17 columns. It originally had 20 columns but was remodelled to its current form by the Mameluke emir Baybars in the 13th century. The interior tiling is splendid *(see p69)*. Some mystery exists over the purpose of the dome, but it is likely that it was a treasury. Its name derives from the legend that a chain once hung from the roof, and whoever told a lie while holding it would be struck dead by lightning.

THE MADRASAS

Most of the buildings fringing the Haram are *madrasas* – Islamic colleges. Of these, the **Ashrafiyya** on the western side of the Haram, built in 1482 by Sultan Qaitbey, is a masterpiece of Islamic architecture. It has an especially ornate doorway exhibiting all the best elements of Mameluke design, including bands of different coloured stone, stalactite carvings above the doorway and, on the benches on either side, intricate, interlocking stones known as "joggling".

Adjoining the Ashrafiyya to the north, close to the Sabil of Qaitbey, is another *madrasa*, the **Uthmaniyya**. Its upper section has beautiful wheel-shaped decorations formed by inlays of yellow and red stone. Along the northern edge of the Haram are two more, the triple-domed **Isardiyya** and adjacent **Malekiyya**. Both date from the 14th century. West of these two, in the corner, is the **Omariyya** college, which is held to contain the First Station of the Cross, but can only be entered from the Via Dolorosa *(see pp30–31)*.

GOLDEN GATE

Also known as the Gate of Mercy (Bab el-Rahma), the Golden Gate was one of the original Herodian city gates. According to Jewish tradition, the Messiah will enter Jerusalem through this gate, which is said to be the reason why the Muslims walled it up in the 7th century. The existing structure dates to the Omayyad period and is best viewed from outside the city walls.

The domed fountain, the Sabil of Qaitbey, with part of the Ashrafiyya Madrasa in the background

Dome of the Rock

Tile above the south entrance

One of the first and greatest achievements of Islamic architecture, the Dome of the Rock was built in AD 688–91 by the Omayyad caliph Abd el-Malik. Intended to proclaim the superiority of Islam and provide an Islamic focal point in the Holy City, the majestic structure now dominates Jerusalem and has become a symbol of the city. More a shrine than a mosque, the mathematically harmonious building echoes elements of Classical and Byzantine architecture, including the rotunda of the Holy Sepulchre *(see pp92–5)*.

View of the Dome of the Rock with the Muslim Quarter in the background

The drum is decorated with tiles and verses from the Quran which tell of Muhammad's Night Journey.

★ Tilework

The multicoloured tiles that adorn the exterior are faithful copies of Persian tiles that Suleyman the Magnificent added in 1545 to replace the badly damaged original mosaics.

Quranic verses

The octagonal arcade is adorned with original mosaics (AD 692) and an inscription inviting Christians to recognize the truth of Islam.

Marble panel

Inner Ambulatory

The space between the inner and outer arcades forms an ambulatory around the Rock. The shrine's two ambulatories recall the ritual circular movement of pilgrims around the Qaaba in Mecca.

STAR FEATURES

★ Interior of Dome

★ Tilework

For hotels and restaurants in this area see p256 and p272

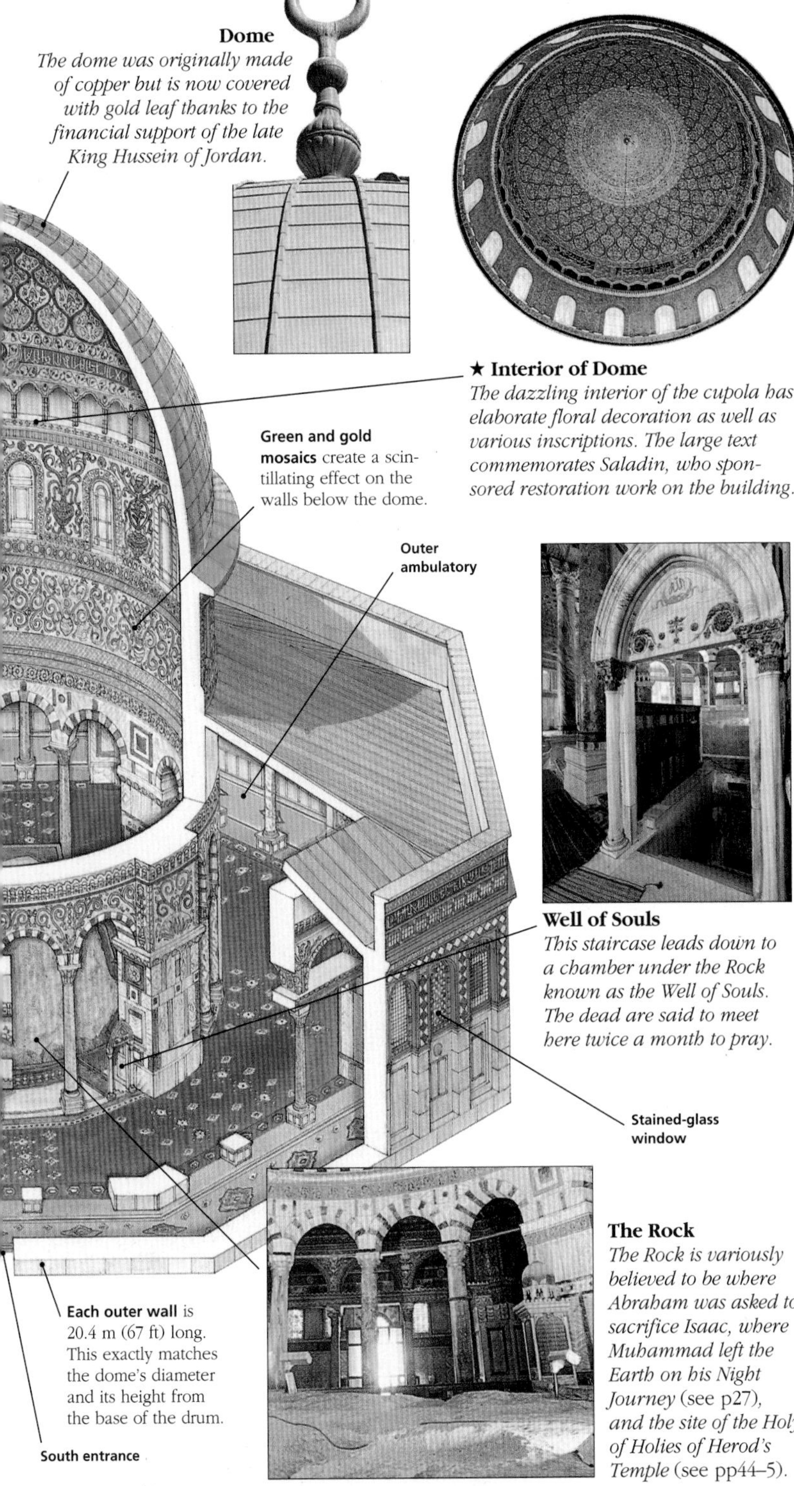

Dome
The dome was originally made of copper but is now covered with gold leaf thanks to the financial support of the late King Hussein of Jordan.
★ Interior of Dome
The dazzling interior of the cupola has elaborate floral decoration as well as various inscriptions. The large text commemorates Saladin, who sponsored restoration work on the building.
Green and gold mosaics create a scintillating effect on the walls below the dome.
Outer ambulatory
Well of Souls
This staircase leads down to a chamber under the Rock known as the Well of Souls. The dead are said to meet here twice a month to pray.
Stained-glass window
The Rock
The Rock is variously believed to be where Abraham was asked to sacrifice Isaac, where Muhammad left the Earth on his Night Journey (see p27), and the site of the Holy of Holies of Herod's Temple (see pp44–5).
Each outer wall is 20.4 m (67 ft) long. This exactly matches the dome's diameter and its height from the base of the drum.
South entrance

Geometric tiling and verses from the Quran on the exterior of the Dome of the Rock ▷

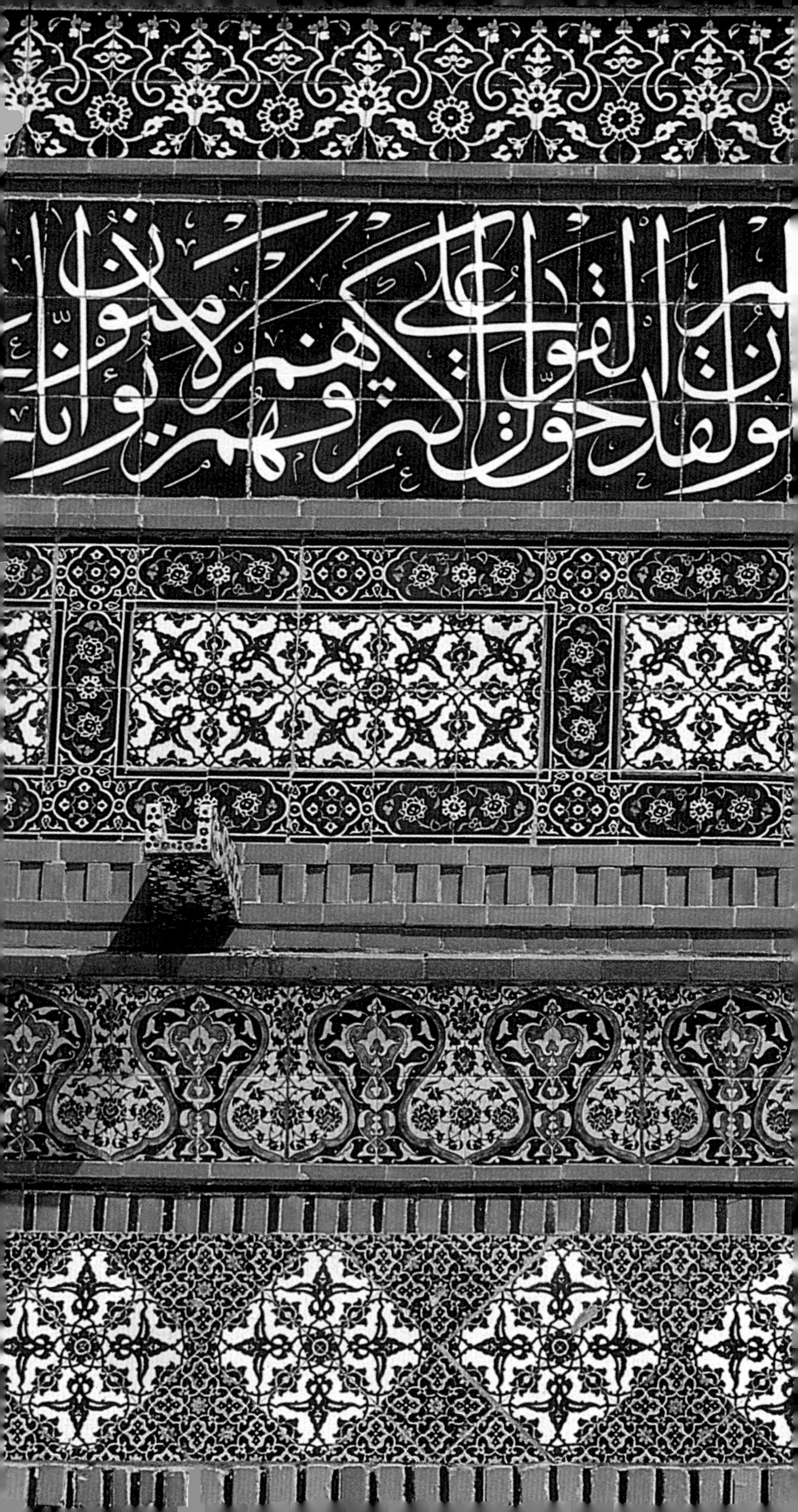

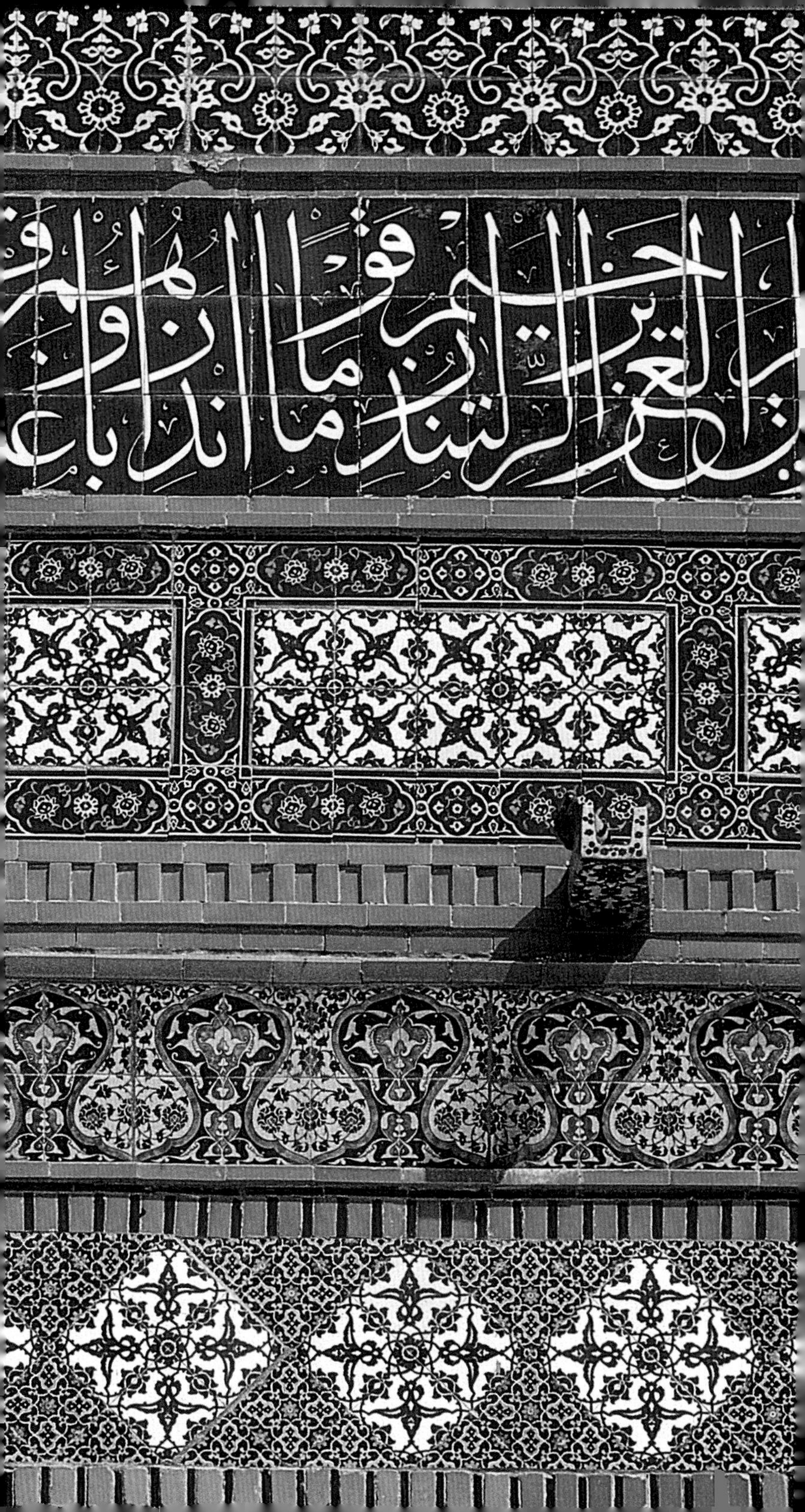

THE JEWISH QUARTER

In Herodian times this area abutted the Temple enclosure *(see pp44–5)* and was occupied by the priestly elite. In the late Roman period, Jews were forbidden from living in Jerusalem, but under the more tolerant Arab rule a small community was re-established here. The district became prevalently Jewish during Ottoman rule, when it acquired its present name. By the 16th century, pilgrimage to the Western Wall – the only surviving remnant of the Temple – had become a strong tradition. After the destruction wrought in the 1948 War and the subsequent years of Jordanian occupation, the Jewish Quarter was liberated by Israeli troops in 1967, and reconstruction work began soon afterwards. A great many ruins from ancient periods were uncovered below more recent buildings. These remains were made accessible to the public, so that the Jewish Quarter of today stands as a fascinating, living mix of more than 3,000 years of Jerusalem Jewry.

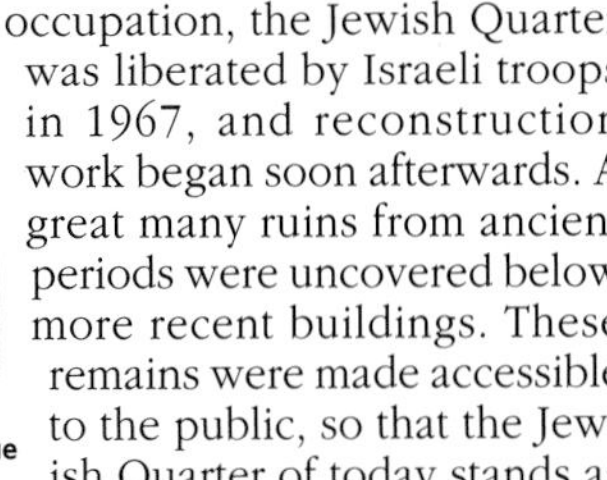

Ark in a Jewish Quarter synagogue

SIGHTS AT A GLANCE

Archaeological Sites
- The Broad Wall 2
- The Cardo 1
- Israelite Tower 10
- Jerusalem Archaeological Park 15
- St Mary of the Germans 13

Museums
- Ariel Centre for Jerusalem in the First Temple Period 11
- The Burnt House 12
- Old Yishuv Court Museum 9
- Wohl Archaeological Museum 6

Holy Places
- Ramban Synagogue 4
- The Sephardic Synagogues 8
- The Western Wall 16

Streets and Squares
- Batei Makhase Square 7
- Dung Gate 14
- Hurva Square 3
- Tiferet Yisrael Street 5

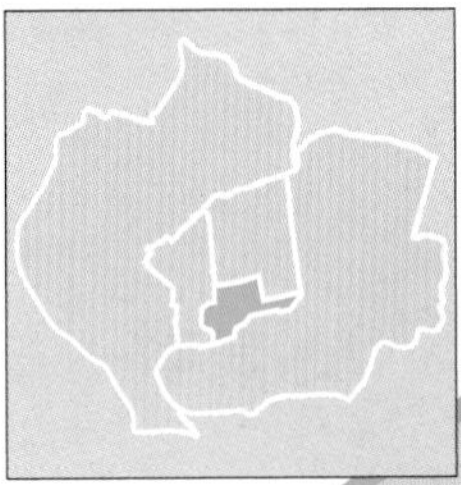

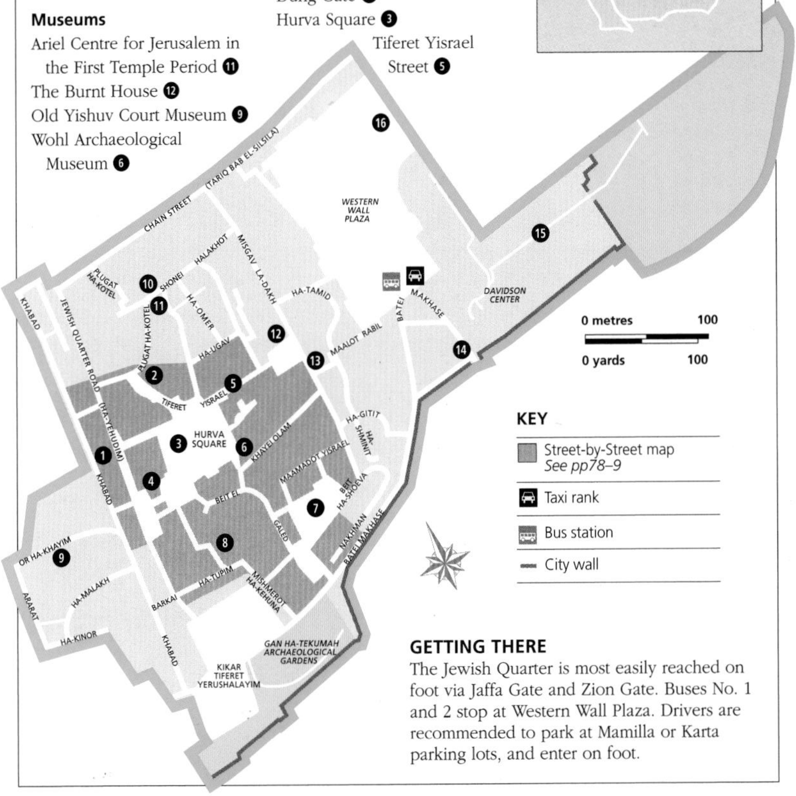

0 metres 100
0 yards 100

KEY
- Street-by-Street map *See pp78–9*
- Taxi rank
- Bus station
- City wall

GETTING THERE

The Jewish Quarter is most easily reached on foot via Jaffa Gate and Zion Gate. Buses No. 1 and 2 stop at Western Wall Plaza. Drivers are recommended to park at Mamilla or Karta parking lots, and enter on foot.

◁ **Orthodox Jews praying at the Western Wall**

Street-by-Street: Around Hurva Square

Jewish Quarter sign

Extensively reconstructed since 1967 and largely residential, the Jewish Quarter is noticeably more orderly than the rest of the Old City, though it is also frequented by large groups of tourists. The focal point for the local community is Hurva Square. This has a few small shops and cafés with outdoor seating. Most of the interesting sights in the quarter are just a few minutes' walk from here. Another hub of the district is the Cardo and Jewish Quarter Road area, which is filled with souvenir shops and more places to eat.

Looking towards Hurva Square from Jewish Quarter Road

Cardo shopping arcade

The Sidna Omar minaret is all that remains of a 14th-century mosque.

★ The Cardo

This is an excavated and partially reconstructed section of the main street of Byzantine-era Jerusalem ❶

The Sephardic Synagogues

Two of these four synagogues date back to the early 17th century. They all contain much ornate decoration ❽

Rothschild House

Batei Makhase Square

A small secluded square, this is favoured by local children as a play area. Its most notable feature is the elegant 19th-century Rothschild House, with its arcaded façade ❼

Shelter Houses *(see p82)*

Remains of Nea Basilica *(see p82)*

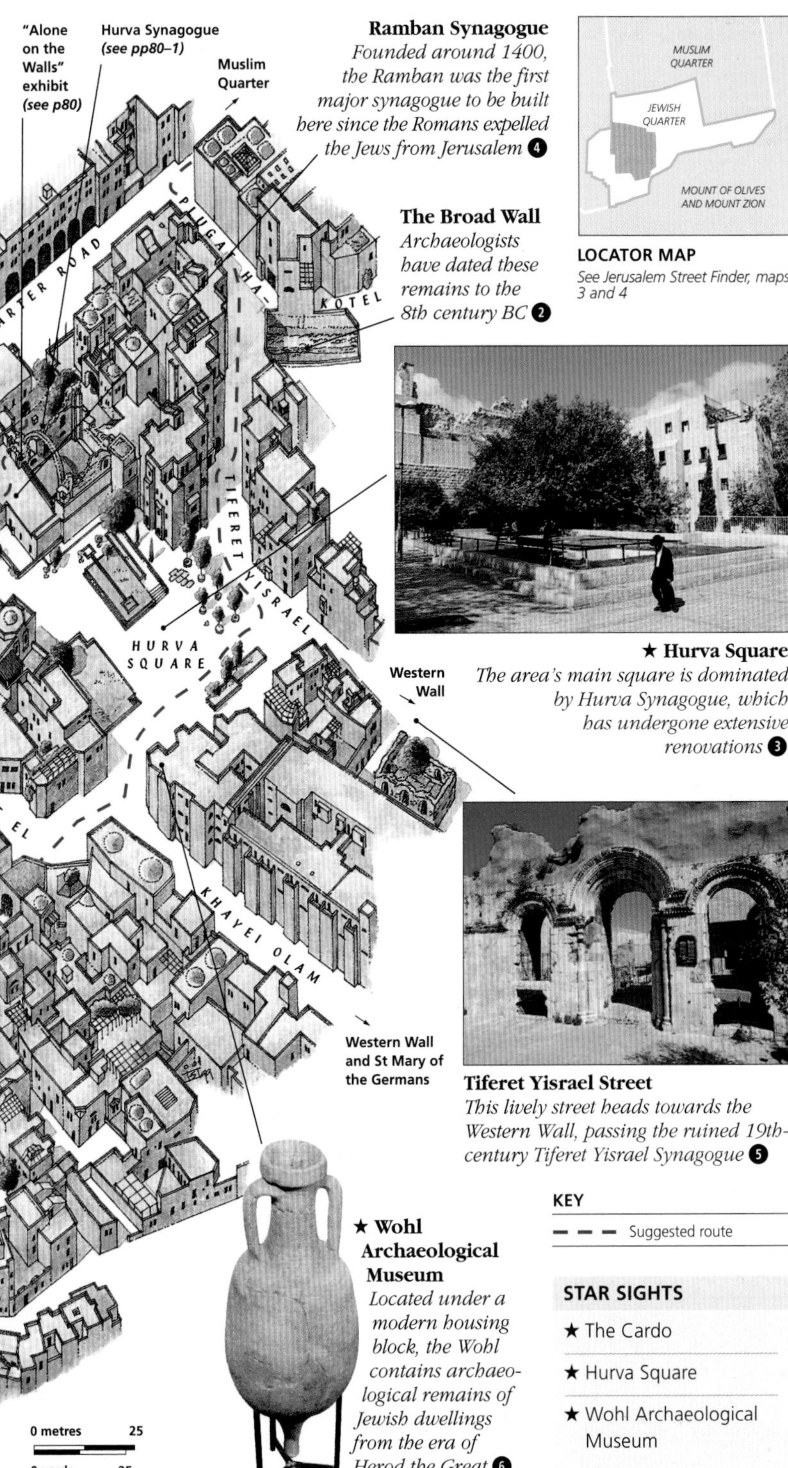

Ramban Synagogue
Founded around 1400, the Ramban was the first major synagogue to be built here since the Romans expelled the Jews from Jerusalem ❹

The Broad Wall
Archaeologists have dated these remains to the 8th century BC ❷

LOCATOR MAP
See Jerusalem Street Finder, maps 3 and 4

★ Hurva Square
The area's main square is dominated by Hurva Synagogue, which has undergone extensive renovations ❸

Tiferet Yisrael Street
This lively street heads towards the Western Wall, passing the ruined 19th-century Tiferet Yisrael Synagogue ❺

★ Wohl Archaeological Museum
Located under a modern housing block, the Wohl contains archaeological remains of Jewish dwellings from the era of Herod the Great ❻

KEY

– – – Suggested route

STAR SIGHTS

★ The Cardo

★ Hurva Square

★ Wohl Archaeological Museum

The Cardo ❶

Map 3 C4.

Now in part an exclusive shopping arcade, the Cardo was Jerusalem's main thoroughfare in the Byzantine era. It was originally laid by the Romans, then extended in the 4th century as Christian pilgrims began to flock to Jerusalem and the city expanded accordingly. The Byzantine extension, which remains in evidence today, linked the two major places of worship of the time, the Church of the Holy Sepulchre *(see pp92–7)* in the north and the long-since-vanished Nea Basilica *(see p82)* in the south.

The central roadway of the Byzantine Cardo was 12.5 m (41 ft) wide. This was flanked by broad porticoed pavements and lined with shops. You can visit a reconstructed section, which runs for almost 200 m (650 ft) along Jewish Quarter (Ha-Yehudim) Road.

The Cardo's continued importance during the reign of Justinian in the 6th century is attested to by its prominent appearance on the famous Madaba map *(see pp216–17)*. Some 500 years later, in the Crusader era, the Cardo was converted into a covered market; the northern section is now preserved as an arcade of smart galleries and boutiques.

An exhibition on Jewish Quarter Road entitled "Alone on the Walls" displays photographs that document the fall of the Jewish Quarter to a regiment of the Arab Legion in 1947–8, in which 68 residents lost their lives.

The Broad Wall, part of the city's 8th-century BC fortifications

The Broad Wall ❷

Plugat ha-Kotel Street. **Map** 3 C4.

The Jewish quarter was largely destroyed during the 1948 War and allowed to deteriorate further under Jordanian occupation. Following the 1967 Israeli victory, a vast reconstruction programme resulted in many significant archaeological finds. One of these was the unearthing of the foundations of a wall 7 m (22 ft) thick and 65 m (215 ft) long. This was possibly part of fortifications built by King Hezekiah in the 8th century BC to enclose a new quarter outside the previous city wall. The need for expansion was probably brought about by a flood of refugees after the Assyrian invasion of 722 BC.

On the building next to the exposed wall, a clearly visible line indicates what archaeologists think was the original height of the wall. Also visible are the remains of housing from the same period, demolished to make way for the wall, as described in the Book of Isaiah (22: 10), "And ye have numbered the houses of Jerusalem, and the houses have ye broken down to fortify the wall".

Hurva Square ❸

Map 3 C4.

This is the heart and social centre of the present-day Jewish Quarter. In the maze of narrow, winding streets which, though modern, follow the topography of the quarter before its destruction, Hurva Square is one of the few open spaces in the area. It has cafés, souvenir shops and a few snack bars that have small tables outside when the weather is good. Also here is the **Jewish Students' Information Centre**, which provides help with accommodation and invitations to Shabbat (Sabbath) dinners for visiting young Jews.

Sidna Omar minaret

On the west side of the square is the minaret of the long-since vanished 14th-century Mameluke **Mosque of Sidna Omar**, along with the historic Hurva and Ramban synagogue complexes. Hurva means "ruins" and the history of the **Hurva Synagogue** more than justifies its name. In the 18th century a group of a few hundred Ashkenazi Jews from Poland came to Jerusalem and founded a synagogue on this site. However, it was burnt down by creditors angered by the community's unpaid debts. The synagogue was rebuilt in 1864 in a Neo-Byzantine style. However, during the fighting that took place in 1948 between the Arab and Jewish armies, the synagogue was destroyed. After the Israelis recaptured

The Cardo, the main street of Byzantine-era Jerusalem

Hurva Square, the social and commercial hub of the Jewish Quarter

the Old City in 1967, a single arch of the synagogue's main façade was reconstructed. The structure underwent further renovation and has now been reconstructed in the same style as the 1864 Neo-Byzantine building.

Ramban Synagogue ❹

Hurva Square. **Map** 3 C4. *for morning and evening prayers.*

When the Spanish rabbi and scholar Moses Ben Nahman (Nahmanides) arrived in Jerusalem in 1267, he was shocked to find only a handful of Jews in the city. He dedicated himself to nurturing a Jewish community and bought land near King David's Tomb on Mount Zion in order to build a synagogue. Some time around 1400, the synagogue was moved to its present site. It was perhaps the first time there had been a Jewish presence in this quarter of the Old City since the exile of the Jews in AD 135. The synagogue had to be rebuilt in 1523 after it collapsed. It is believed that, at this time, it was probably the only Jewish place of worship in what was then Ottoman-controlled Jerusalem. In 1599 the authorities banned the Jews from worship in the synagogue and the building became a workshop. It was not until the Israelis took control of the Old City in 1967 that it was restored as a place of worship.

Tiferet Yisrael Street ❺

Map 4 D4.

This is one of the busiest streets in the Jewish Quarter. It connects Hurva Square with the stairs that descend towards the Western Wall. Partway along is the shell of the ruined **Tiferet Yisrael Synagogue**, destroyed in the 1948 War and left gutted as a memorial. Sectarian feelings run high around here, and local souvenir shops stock contentious items such as Israeli Army T-shirts and postcards of the Haram esh-Sharif with its mosques replaced by the "future Third Temple". The street ends in an attractive tree-shaded square which has several snack bars and cafés, including the popular Quarter Café, which serves kosher food and offers great views of the Haram esh-Sharif and Dome of the Rock from its terrace.

Tiferet Yisrael Street, one of the liveliest thoroughfares in the Jewish Quarter

Wohl Archaeological Museum ❻

1 Ha-Karaim Street. **Map** 4 D4. ***Tel*** *(02) 626 5922.* *9am–5pm Sun–Thu, 9am–1pm Fri.*

In the era of Herod the Great (37–4 BC), the area of the present-day Jewish Quarter was part of a wealthy "Upper City", occupied for the most part by the families of important Jewish priests. During post-1967 redevelopment, the remains of several large houses were unearthed here. This rediscovered Herodian quarter now lies from 3 to 7 m (10 to 22 ft) below street level, underneath a modern building, and is preserved as the Wohl Archaeological Museum.

The museum is remarkable for its vivid evocation of everyday life 2,000 years ago. All the houses had an inner courtyard, ritual baths, and cisterns to collect rain, which was the only source of water at the time. The first part of the museum, called the Western House, has a mosaic in the vestibule and a well-preserved ritual bath *(mikveh)*. Beyond this is the Middle Complex, the remains of two separate houses where archaeologists found a maze-pattern mosaic floor covered in burnt wood; this, they surmised, was fire damage from the Roman siege of Jerusalem in AD 70. The most complete of all the Herodian buildings is the Palatial Mansion, with more splendid mosaic floors and ritual baths.

The entrance fee to the Wohl Museum also covers admission to the Burnt House *(see p84)*.

Batei Makhase Square ❼

Map 4 D5.

This quiet square is named after the so-called Shelter Houses (Batei Makhase), which lie just south of it. They were built in 1862 by Jews from Germany and Holland for destitute immigrants from central Europe. Severe damage in the 1948 and 1967 wars made restoration necessary.

The work brought to light the first remains of the Nea (New) Basilica, whose existence had previously been known only from the Madaba map *(see pp216–17)* and literary sources. Built by Byzantine emperor Justinian in AD 543, it was at the time the largest basilica in the Holy Land. The remains of one of the apses can be seen near the square's southwest corner. Archaeologists have now traced the basilica's full extent – an enormous 116 m (380 ft) by 52 m (171 ft). More impressive remains can be found in the cellar of a house to the north of the square.

The handsome, arcaded building on the western side of the square was built for the Rothschild family in 1871. In front of it are parts of Roman columns, whose original provenance is unknown.

The 17th-century Ben Zakkai Synagogue

The Sephardic Synagogues ❽

Ha-Tupim Street. **Map** 3 C5. ***Tel*** *(02) 628 0592. 9:30am–4pm Sun–Thu, 9am–1pm Fri.*

The four synagogues in this group became the spiritual centre of the area's Sephardic community in the 17th century.

The Sephardim were descended from the Jews expelled from Spain in 1492 and Portugal in 1497. They had first settled in the Ottoman Empire and then moved to Palestine when the latter was conquered by the Turks in 1516. When the first two synagogues were built, the Sephardim formed the largest Jewish community in Jerusalem. The synagogue floors were laid well below street level to allow sufficient height for the buildings, as Ottoman law stated that synagogues should not rise above the surrounding houses.

Bimah **from the Istambuli Synagogue**

The **Ben Zakkai Synagogue** was built in 1610. Its courtyard, with a matroneum, or gallery for women worshippers, was converted into the **Central Synagogue**, whose present form dates from the 1830s. The **Prophet Elijah Synagogue**, created from a study hall built in 1625, was consecrated in 1702. Legend has it that during prayers to mark Yom Kippur, Elijah appeared as the 10th adult male worshipper needed for synagogue prayer – hence the building's name. The **Istambuli Synagogue** was built in 1857 and, like the other three, contains furnishings salvaged from Italian synagogues damaged in World War II.

Old Yishuv Court Museum ❾

6 Or ha-Khayim Street. **Map** 3 C5. ***Tel*** *(02) 627 6319. 10am–5pm Sun–Thu, 10am–1pm Fri.*

This small museum, devoted to the history of the city's Jewish community from the mid-19th century to the end of Ottoman rule in 1917, occupies one of the oldest complexes of rooms in the Jewish Quarter. Of Turkish construction, thought to date from the 15th or 16th centuries, it was once part of a private home. The exhibits, consisting largely of reconstructed interiors, memorabilia and photographs, also include the Ari Synagogue on the ground floor. This was used by a Sephardic congregation during most of the Ottoman period. Badly damaged in the fighting of 1936, it fell into disuse until 1967, when it was restored. On the top floor is the

Rothschild House and a Roman column base and capital in Batei Makhase Square

Household objects on display at the Old Yishuv Court Museum

18th-century Or ha-Khayim Synagogue, used by Ashkenazi Jews in the 19th century. Closed between 1948 and 1967, it is now a functioning synagogue once more.

Israelite Tower ⑩

Shonei Halakhot Street. **Map** 4 D4. ***Tel*** *(02) 628 8141.* *call ahead for opening hours.*

Steps at the corner of Shonei Halakhot and Plugat ha-Kotel streets lead underneath a modern apartment block to the remains of a tower of the 7th century BC. The tower, the walls of which are over 4 m (13 ft) thick and survive to a height of 8 m (26 ft), is believed to have been part of a gateway in the Israelite city wall. At its foot were found the heads of Israelite and Babylonian arrows, as well as evidence of burning. These finds are thought to date from the Babylonian conquest of Jerusalem in 586 BC *(see p42)* and may identify the gate as the one through which Babylonian troops entered the city (Jeremiah 39: 3). The other visible remains belong to the 2nd-century BC Hasmonean city wall, another section of which can be seen at the Citadel *(see pp102–5)*.

The apartment block above was built on stilts, as were other modern buildings in the Jewish Quarter, to allow access by archaeologists. However, the need to rebuild rapidly after the 1948 War meant that there was insufficient time to uncover many of the remains and draw a complete plan of the area's fortifications.

Ariel Centre for Jerusalem in the First Temple Period ⑪

Bonei Hahomah Street. **Map** 4 D4. ***Tel*** *(02) 628 6288.* *9am–4pm Sun–Thu (Aug 9am–6pm Sun–Thu).* **www**.ybz.org.il

The principal exhibit here is a model of all the archaeological remains of First Temple Period Jerusalem (around the 8th century BC). It illustrates the relationship between remains, which can be difficult to interpret when they are seen on the ground surrounded by other buildings. It also shows the original topography of the area before valleys were filled in and occupation layers built up. An audiovisual show describes the city's history from 1000 to 586 BC.

There is also a display of finds from a secret dig carried out in 1909–11 by English archaeologist Captain Montague Parker. His team of excavators penetrated underneath the Haram esh-Sharif in search of a chamber that reputedly contained King Solomon's treasure. When news of the dig got out, violent demonstrations by Jews and Muslims, united in their opposition to the desecration of their holy site, forced Parker to flee the city.

JEWISH QUARTER ARCHITECTURE

Heavily damaged during the 1948 War, the Jewish Quarter has been almost totally reconstructed in recent times. While there is no distinct "Jewish style", the quarter's modern architecture belongs to a well-defined Jerusalem tradition. First and foremost, everything is constructed of the pale local stone. Use of this stone has been mandatory in Jerusalem since a law to this effect was passed by the British military governor, Ronald Storrs, in 1917. Buildings and street patterns are deliberately asymmetrical to evoke haphazard historical development. Streets are also narrow and cobbled, with many small courtyards and external staircases to upper levels. Buildings make great use of traditional Middle Eastern elements such as arches, domes and oriels (the high bay windows supported on brackets, much favoured by Mameluke builders). A jumble of different heights means that the roof of one building is often the terrace of another. The result is a very contemporary look, which is at the same time firmly rooted in the past.

Modern additions harmonise with traditional styles

The Burnt House ⓬

Tiferet Yisrael Street. **Map** 4 D4.
Tel *(02) 626 5902.* *9am–5pm Sun–Thu, 9am–1pm Fri.*

In AD 70 the Romans took Jerusalem and destroyed the Temple and Lower City to the south. A month later they rampaged through the wealthy Upper City, setting fire to the houses. The charred walls and a coin dated to AD 69 discovered during excavations show that this was one of those houses.

A stone weight found among the debris bears the inscription "son of Kathros", indicating that the house belonged to a wealthy family of high priests. They are known from a subsequent reference to them in the Babylonian Talmud, written between the 3rd and 6th century AD.

The rooms on view, introduced by a moving sound and light show with commentary, comprise a kitchen, four rooms that may have been bedrooms, and a bathroom with a ritual bath. It is believed that these formed part of a much larger residence, but further excavations cannot be undertaken as the remains lie beneath present-day neighbouring houses.

The entrance fee also discounts the Wohl Archaeological Museum *(see p81)*.

The outline of rooms and some of the artifacts unearthed at the Burnt House

Surviving walls of the Crusader-built St Mary of the Germans

St Mary of the Germans ⓭

Misgav la-Dakh Street. **Map** 4 D4.
daily.

Immediately below the terrace of Tiferet Yisrael's Quarter Café are the original walls of St Mary of the Germans. This early 12th-century Crusader church was part of a complex that included a pilgrims' hospice (no longer in existence) and a hospital. It was built by the Knights Hospitallers *(see p49)* and run by their German members. This was in response to the influx of German-speaking pilgrims unfamiliar with French, the lingua franca, or Latin, the official language, of the new Latin Kingdom of Jerusalem. Activity ceased when Jerusalem fell to the Muslims in 1187, but the church and the hospital were again used during the brief period when Jerusalem was once more under Christian rule (1229–44).

Today the church is roofless. However, the walls survive to a considerable height, showing clearly the three apses of the typical basilica plan so widely used in the Holy Land from early Byzantine times.

Beside the church is a flight of steps down to the Western Wall Plaza. These provide wonderful views of the Western Wall, the Dome of the Rock and the Mount of Olives behind.

Dung Gate ⓮

Map 4 D5.

In old photographs the Dung Gate is shown to be hardly any larger than a doorway in the average domestic house. Its name in Hebrew is Shaar ha-Ashpot, and it is mentioned in the Book of Nehemiah (2: 13) in the Old Testament. It is probably named after the ash that was taken from the Temple to be deposited outside the city walls. The Arab name is Bab Silwan, because this is the gate that leads to the Arab village of Silwan.

Dung Gate, leading to the Western Wall

The gate was enlarged by the Jordanians in 1948 to allow vehicles to pass through. It is now the main entrance and exit for the Jewish Quarter, but it still remains the smallest of all the Old City gates. It retains its old Ottoman carved arch with a stone flower above.

Jerusalem Archaeological Park ⓯

See pp86–7.

The Western Wall ⑯

Western Wall Plaza. **Map** 4 D4. *1, 2, 38.* *on Sabbath.* **Chain of the Generations Centre** ***Tel*** *(02) 627 1333.* *8am–evening Sun–Thu, 8am–noon Fri. Visits must be booked in advance.* *Jewish hols.* *compulsory.* **Western Wall Tunnel** ***Tel*** *(02) 627 1333.* *7am–evening Sun–Thu; 7am–noon Fri. Visits must be booked in advance.* *Jewish hols.* *compulsory.* **www**.thekotel.org

A massive, blank wall built of huge stone blocks, the Western Wall (Ha-Kotel in Hebrew) is Judaism's holiest site, and the plaza in front of it is a permanent place of worship. The wall is part of the retaining wall of the Temple Mount and was built by Herod the Great during his expansion of the Temple enclosure *(see pp44–5)*. The huge, lower stones are Herodian, while those higher up date from early Islamic times.

During the Ottoman period, the wall became where Jews came to lament the destruction of the Second Temple. For this reason it was for centuries known as the Wailing Wall.

Houses covered most of what's now the Western Wall Plaza until relatively recently. When the Israelis gained control of the Old City after the 1967 war, they levelled the neighbouring Arab district.

WORSHIP AT THE WESTERN WALL

The Western Wall Plaza functions as a large, open-air synagogue where groups gather to recite the daily, Shabbat (Sabbath) and festival services of the Jewish faith. Special events are also celebrated here, such as the religious coming of age of a boy or girl (*bar* or *bat mitzvah*). Some worshippers visit the wall daily to recite the entire Book of Psalms; others, who believe that petitions to God made at the wall are specially effective, insert written prayers into the stones. On Tisha B'Av, the ninth day of the month of Av, which falls in either July or August, a fast is held commemorating the destruction of both Temples *(see pp42–5)*. People sit on the ground reciting the Book of Lamentations and liturgical dirges called *kinot*. Since the plaza is essentially a public space, conflicts arise over such issues as the relative size of the men's and women's sections and the wish of non-Orthodox groups to hold services in which men and women participate together.

Prayers inserted into gaps between the stones of the Western Wall

Orthodox Jew at prayer beside the Western Wall

Non-Jews can approach the wall, provided they dress appropriately and cover their heads *(see pp298–9)*.

At the left-hand corner of the men's prayer section is Wilson's Arch (named after a 19th-century archaeologist). Now contained within a building that functions as a synagogue, it originally carried the Causeway to the Temple. From the arch, archaeologists have dug the **Western Wall Tunnel** to explore the wall's foundations. It follows the base of the outside face of the Temple wall along a Herodian street, below today's street level, and emerges on the Via Dolorosa. The **Chain of the Generations Centre** tells the story of the Jewish people. Access to this and the Tunnel is by tour only; book well in advance.

The Western Wall Plaza, with the men's prayer section to the left and women's to the right

Jerusalem Archaeological Park 15

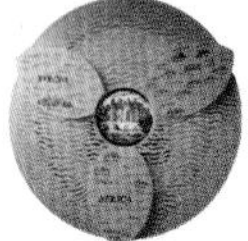
Exhibit at the Davidson Center

The area south of the Western Wall and Haram esh-Sharif is one of the most important archaeological sites in all Jerusalem. Excavations, ongoing here since 1968, have uncovered remains dating back to the First and Second Temple periods *(see pp41–2)*, and through Byzantine times to the Omayyad era. In this one small, L-shaped site, the entire sweep of the history of the ancient city is revealed. The new Davidson Center provides a multi-media introduction to the site and contextualizes the archaeologists' findings.

The Western Wall Plaza *(see p85)*

Robinson's Arch
A row of stones projecting from the wall is the remains of an arch that once supported a flight of stairs, as shown in this model at the Tower of David Museum (see pp102–5).

Ritual Bath (Mikveh)
The baths are where worshippers purified themselves before approaching the Temple. The divider, running down the centre of the stairs, ensured the separation of the clean and the unclean.

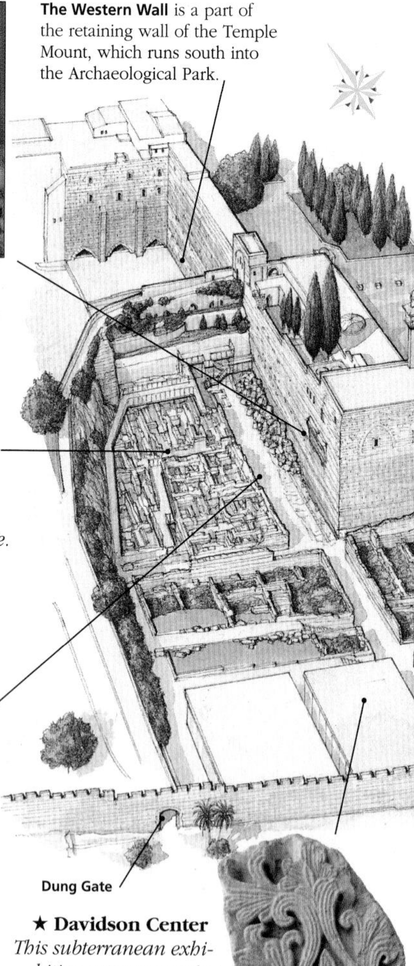

The Western Wall is a part of the retaining wall of the Temple Mount, which runs south into the Archaeological Park.

Dung Gate

Herodian Street
At the base of the Temple Mount is a flagged street dating from the time of the Second Temple. It would have been lined with shops – four small doorways have been reconstructed.

★ Davidson Center
This subterranean exhibition centre contains artifacts from the site and screens two informative films, plus a computer-animated recreation of the Second Temple.

EARLY EXCAVATORS

Before the archaeologists, the Temple Mount area drew the attentions of 19th-century biblical scholars. The American Edward Robinson (1794–1863) was the first to identify the huge arch that is now named after him. The first serious excavations were made by the British officer Captain Charles Warren, who discovered a series of underground tunnels, as well as the nearby water shaft that carries his name *(see p115)*.

Charles Warren, 1840–1927

VISITORS' CHECKLIST

Batei Makhase Street, Jewish Quarter. ***Tel*** *(02) 627 7550.* **www**.archpark.org.il *8am–5pm Sun–Thu, 8am– 2pm Fri.* *Sat & Jewish holidays.* *Guided tours last 1 hr and must be booked in advance. The computer-animated reconstruction of the Second Temple screened in the Davidson Center may only be viewed as part of a guided tour.*

Temple Mount
The great retaining wall of the Temple Mount dates from the reign of Herod (37–4 BC). To see what the complex would have looked like at this time, turn to pages 44–5.

★ Hulda Gates
At the top of a monumental flight of steps, a Double Gate and Triple Gate (together known as the Hulda Gates) provided access to the precincts of the Second Temple. They were later walled up by the Romans.

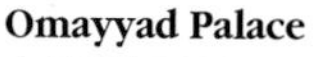

Omayyad Palace
A canopy covers what was the central courtyard of an Omayyad-era palace. The building would have filled the area between the Temple Mount and the city walls.

STAR SIGHTS

★ Davidson Center

★ Hulda Gates

THE CHRISTIAN AND ARMENIAN QUARTERS

Under Byzantine rule the Christian community of Jerusalem expanded rapidly. Settlement was concentrated in the northwest corner of the city, in the shadow of the great basilica of the Holy Sepulchre. Bounded by Souk Khan el-Zeit and David Street, the modern quarter remains filled with the churches, patriarchates and hospices of the city's many Christian denominations. To the south is the area traditionally inhabited by the Armenians, who have a long history in Jerusalem. It is one of the quietest parts of the Old City.

Old City sign made of Armenian tiles

SIGHTS AT A GLANCE

Museums
The Citadel pp102–4 ❾
Mardigian Museum ⓮
Museum of the Greek Orthodox Patriarchate ❼

Churches
Alexander Hospice ❷
Church of the Holy Sepulchre pp92–5 ❶
Church of St John the Baptist ❺
Lutheran Church of the Redeemer ❸
St James's Cathedral ⓭
St Mark's Church ⓬

Historic Areas, Streets and Gates
Christian Quarter Road ❻
Jaffa Gate ❽
Muristan ❹
Omar ibn el-Khattab Square ❿
Zion Gate ⓯

Walks
A Walk on the Roofs ⓫

GETTING THERE

These two quarters are served mainly by Jaffa Gate; a great many buses from the New City halt just outside. The area can also be entered from Zion and New gates. Drivers are recommended to park at Mamilla or Karta parking lots.

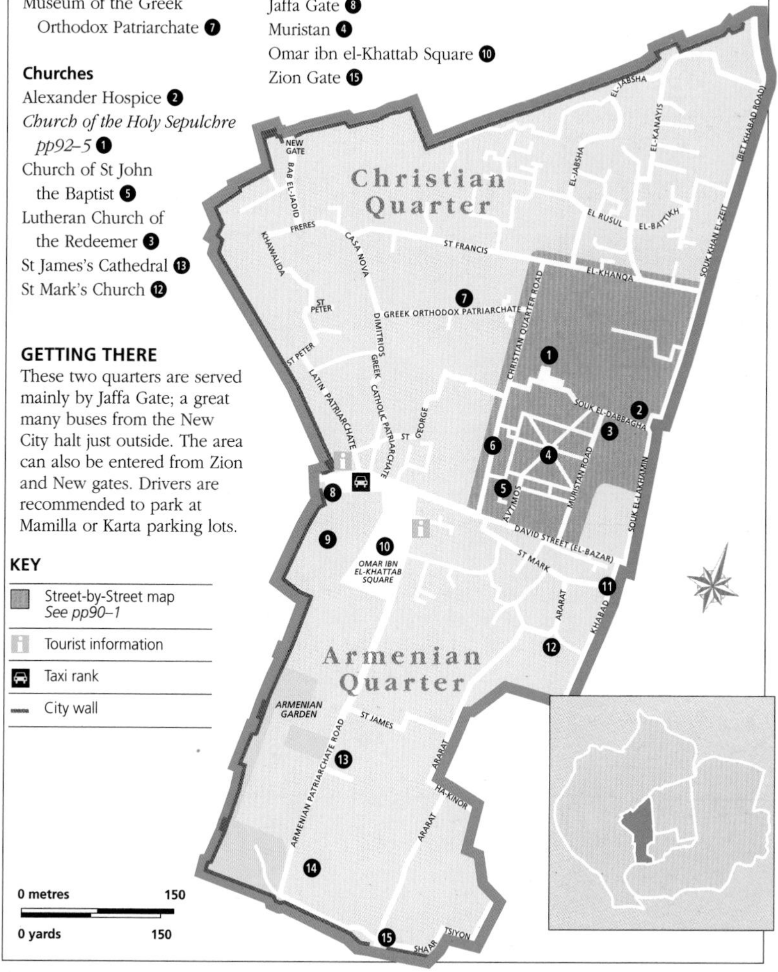

KEY

Street-by-Street map *See pp90–1*

Tourist information

Taxi rank

City wall

0 metres 150
0 yards 150

◁ **Pilgrims crowding outside the main doorway of the Church of the Holy Sepulchre**

Street-by-Street: The Christian Quarter

Capital from the Church of the Redeemer

The most visited part of the Old City, the Christian Quarter is a head-on collision between commerce and spirituality. At its heart is the Church of the Holy Sepulchre, the most sacred of all Christian sites. It is surrounded by such a clutter of churches and hospices that all one can see of its exterior are the domes and entrance façade. The nearby streets are filled with shops and stalls that thrive on the pilgrim trade. Respite from the crowds can be found in the cafés of Muristan Road.

The Christian Quarter, centred on the Holy Sepulchre

Church of St John the Baptist
The founding of the Crusader Knights Hospitallers is connected with this small church. A carved stone cross echoes the order's historic emblem 5

Christian Quarter Road
Along with David Street, this is the quarter's main shopping thoroughfare. It specializes in religious items and quality handicrafts 6

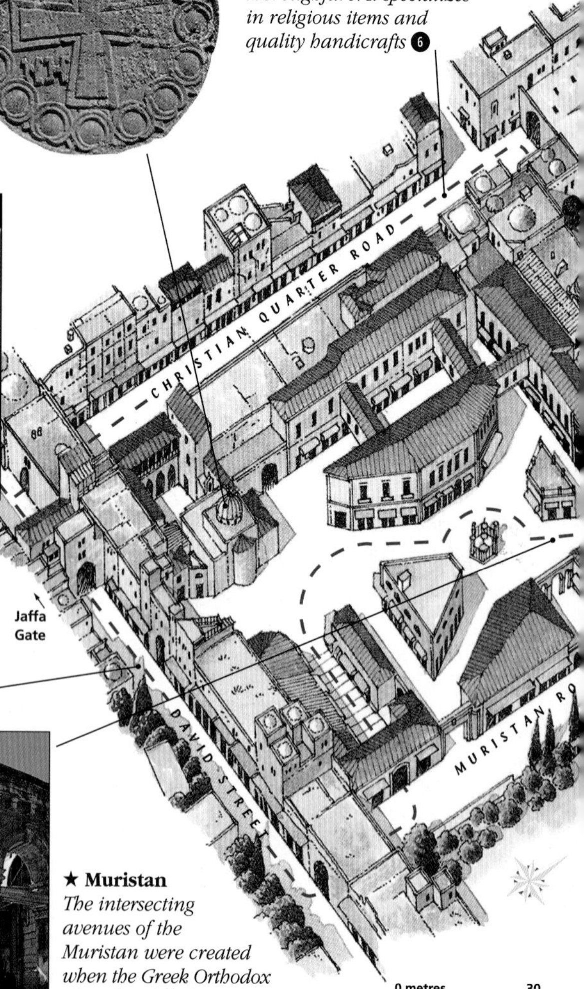

CHRISTIAN QUARTER ROAD

David Street
From the Jaffa Gate area, David Street is the main route down through the Old City. This cramped, stepped alley doubles as a busy tourist bazaar.

Jaffa Gate

DAVID STREET

MURISTAN RO

★ Muristan
The intersecting avenues of the Muristan were created when the Greek Orthodox Church redeveloped the area in 1903 4

0 metres 30

0 yards 30

For hotels in this area see p256

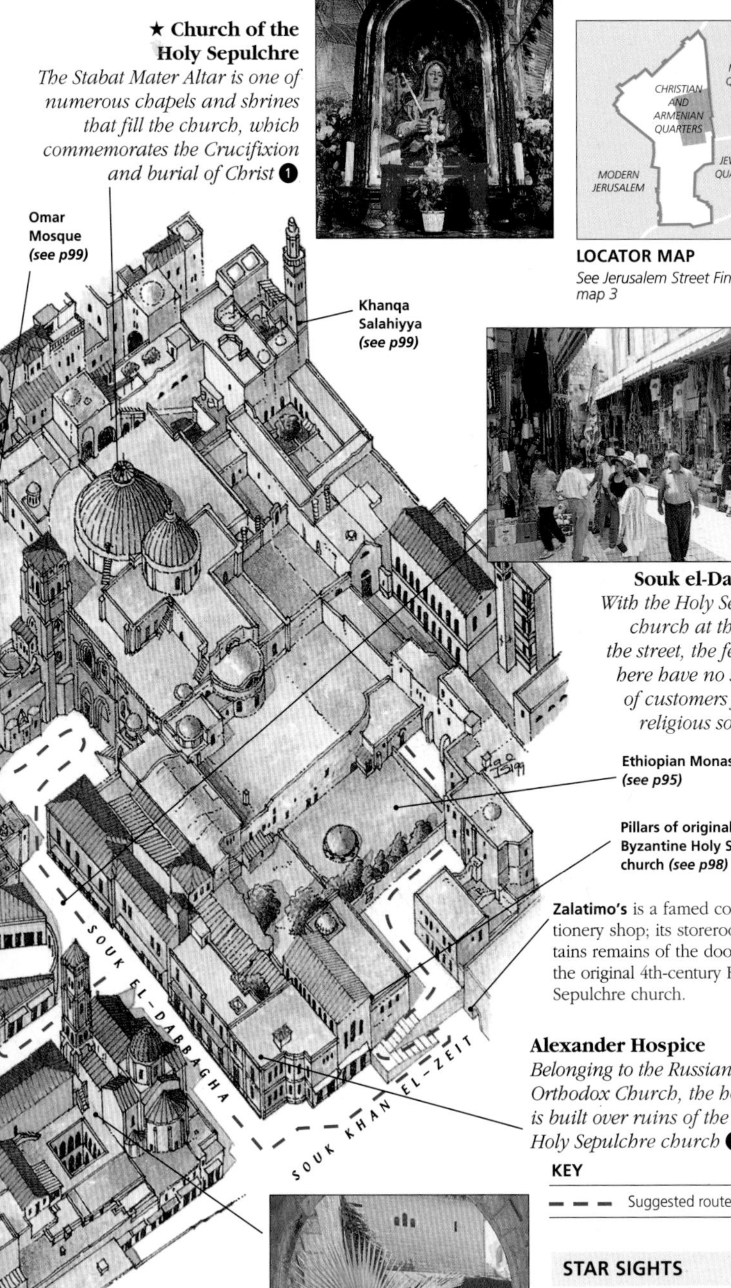

★ Church of the Holy Sepulchre
The Stabat Mater Altar is one of numerous chapels and shrines that fill the church, which commemorates the Crucifixion and burial of Christ ❶

LOCATOR MAP
See Jerusalem Street Finder, map 3

Omar Mosque ***(see p99)***

Khanqa Salahiyya ***(see p99)***

Souk el-Dabbagha
With the Holy Sepulchre church at the end of the street, the few shops here have no shortage of customers for their religious souvenirs.

Ethiopian Monastery ***(see p95)***

Pillars of original Byzantine Holy Sepulchre church ***(see p98)***

Zalatimo's is a famed confectionery shop; its storeroom contains remains of the doorway of the original 4th-century Holy Sepulchre church.

Alexander Hospice
Belonging to the Russian Orthodox Church, the hospice is built over ruins of the early Holy Sepulchre church ❷

★ Lutheran Church of the Redeemer
This church has an attractive medieval cloister, but most people visit for the views from the bell tower ❸

KEY

– – – Suggested route

STAR SIGHTS

- ★ Church of the Holy Sepulchre
- ★ Lutheran Church of the Redeemer
- ★ Muristan

Church of the Holy Sepulchre ❶

Built around what is believed to be the site of Christ's Crucifixion, burial and Resurrection, this complex church is the most important in Christendom. The first basilica here was built by Roman emperor Constantine between AD 326 and 335 at the suggestion of his mother, St Helena. It was rebuilt on a smaller scale by Byzantine emperor Constantine Monomachus in the 1040s following its destruction by Fatimid sultan Hakim in 1009, but was much enlarged again by the Crusaders between 1114 and 1170. A disastrous fire in 1808 and an earthquake in 1927 necessitated extensive repairs.

The mosaic of roofs and domes of the Church of the Holy Sepulchre

The Rotunda, heavily rebuilt after the 1808 fire, is the most majestic part of the church.

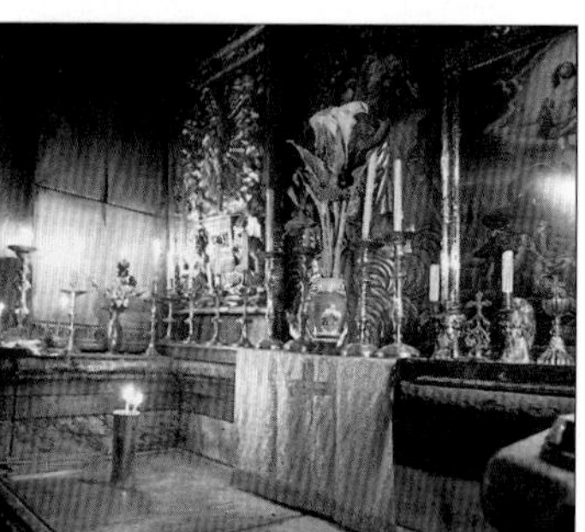

★ Christ's Tomb

For Christians, this is the most sacred site of all. Inside the 1810 monument, a marble slab covers the rock on which Christ's body is believed to have been laid.

The Crusader bell tower was reduced by two storeys in 1719.

Stone of Unction

This is where the anointing and wrapping of Christ's body after his death has been commemorated since medieval times. The present stone dates from 1810.

Chapel of the Franks

The main entrance is early 12th century. The right-hand door was blocked up late in the same century.

Courtyard

The main entrance courtyard is flanked by chapels. The disused steps opposite the bell tower once led to the Chapel of the Franks, the Crusaders' ceremonial entrance to Golgotha.

For hotels in this area see p256

THE HOLY FIRE

On the Saturday of Orthodox Easter, all the church's lamps are put out and the faithful stand in the dark, a symbol of the darkness at the Crucifixion. A candle is lit at Christ's Tomb, then another and another, until the entire basilica and courtyard are ablaze with light to symbolize the Resurrection. Legend says the fire comes from heaven.

The Easter ceremony of the Holy Fire

VISITORS' CHECKLIST

Entrance from Souk el-Dabbagha. **Map** 3 C3. ***Tel*** *(02) 626 7011.* *summer: 5am–9pm daily; winter: 4am–7pm daily.*

Catholikon Dome
Rebuilt after the 1927 earthquake and decorated with an image of Christ, this dome covers the central nave of the Crusader church. This part of the building is now used for Greek Orthodox services.

The Seven Arches of the Virgin are the remains of an 11th-century colonnaded courtyard.

The Centre of the World, according to ancient map-makers *(see p40)*, is marked here by a stone basin.

★ Golgotha
Through the glass around the Greek Orthodox altar can be seen the outcrop of rock venerated as the site of the Crucifixion.

Chapel of Adam ***(see p94)***

Rock of Golgotha ***(see p94)***

The Chapel of St Helena is now dedicated to St Gregory the Illuminator, patron of the Armenians.

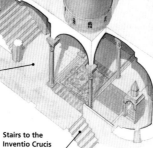

Ethiopian Monastery
A cluster of small buildings on the roof of the Chapel of St Helena is inhabited by a community of Ethiopian monks.

Stairs to the Inventio Crucis Chapel ***(see p95)***

STAR FEATURES

- ★ Christ's Tomb
- ★ Golgotha

Exploring the Church of the Holy Sepulchre

Chapel door, main courtyard

The reconstructions and additions that have shaped this church over the centuries make it a complex building to explore. Its division into chapels and spaces allotted to six different denominations adds a further sense of confusion. The interior is dimly lit, and queues often form at Christ's Tomb, so that the time each person can spend inside the shrine may be limited to just a few minutes. Nonetheless, the experience of standing on Christianity's most hallowed ground inspires many visitors with a deep sense of awe.

The Roman Catholic altar on Golgotha

GOLGOTHA

Just inside the church's main entrance, on the right, two staircases lead up to Golgotha, which in Hebrew means "Place of the Skull" and was translated into Latin as Calvary. The space here is divided into two chapels. On the left is the Greek Orthodox chapel, with its altar placed directly over the rocky outcrop on which the cross of Christ's Crucifixion is believed to have stood. The softer surrounding rock was quarried away when the church was built and the remaining, fissured, so-called Rock of Golgotha can now be seen through the protective glass around the altar. It can be touched through a hole in the floor under the altar. The 12th Station of the Cross *(see p30)* is commemorated here.

To the right is the Roman Catholic chapel, containing the 10th and 11th Stations of the Cross. The silver and bronze altar was given by Ferdinand de Medici in 1588. The 1937 mosaics encircle a Crusader-era medallion of the Ascension on the ceiling. The window looks into the Chapel of the Franks *(see p92)*.

Between these altars is the Altar of the Stabat Mater, commemorating Mary's sorrow as she stood at the foot of the cross. It marks the 13th Station of the Cross. The wooden bust of the Virgin is 18th century.

Archaeological evidence that the church rests on a possible site of the Crucifixion is scant, but positive. Excavations show that the site lay outside the city walls until new ones encompassed it in AD 43; that in the early 1st century it was a disused quarry in which an area of cracked rock had been left untouched; and that rock-hewn tombs were in use here in the 1st centuries BC and AD. This all tallies with Gospel accounts of the Crucifixion.

CHAPEL OF ADAM

Immediately beneath the Greek Orthodox chapel on Golgotha, this chapel is built against the Rock of Golgotha. It is the medieval replacement of a previous Chapel of Adam that was part of Constantine's 4th-century basilica. It was so called because tradition told that Christ was crucified over the burial place of Adam's skull – a tradition first recorded by the Alexandrian theologian Origen (c.AD 185–245).

The crack in the Rock of Golgotha, clearly visible in the apse, is held by believers to have been caused by the earthquake that followed Christ's death (Matthew 27: 51).

11th-century apse, Chapel of Adam, built against the Rock of Golgotha

THE STATUS QUO

Fierce disputes, lasting centuries, between Christian creeds *(see p100)* over ownership of the church were largely resolved by an Ottoman decree issued in 1852. Still in force and known as the Status Quo, it divides custody among Armenians, Greeks, Copts, Roman Catholics, Ethiopians and Syrians. Some areas are administered communally. Every day, the church is unlocked by a Muslim keyholder acting as a "neutral" intermediary. This ceremonial task has been performed by a member of the same family for several generations.

Coptic priest in ceremonial vestments

CHRIST'S TOMB

The present-day shrine around the tomb of Christ was built in 1809–10, after the severe fire of 1808. It replaced one dating from 1555, commissioned by the Franciscan friar Bonifacio da Ragusa. Before that, there had been a succession of shrines replacing the original 4th-century one destroyed by the sultan Hakim in 1009. Constantine's builders had dug away the hillside to leave the presumed rock-hewn tomb of Christ isolated and with enough room to build a church around it. They had also had to clear the remains of an AD 135 Hadrianic temple from the site, as well as the material with which an old quarry had been filled to provide the temple's foundations. In so doing, the Rock of Golgotha was also found.

Today the shrine, owned by the Greek Orthodox, Roman Catholic and Armenian communities, contains two chapels. The outer Chapel of the Angel has a low pilaster incorporating a piece of the stone said to have been rolled from the mouth of Christ's Tomb by angels. It serves as a Greek Orthodox altar. A low door leads to the tiny inner Chapel of the Holy Sepulchre with the 14th Station of the Cross. A marble slab covers the place where Christ's body was supposedly laid. The slab was installed in the 1555 reconstruction and purposely cracked to deter Ottoman looters.

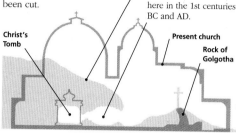

SITE OF CHRIST'S TOMB

In the first century AD, this site consisted of a small, rocky rise just outside the city walls and a disused stone quarry into whose rock face tombs had been cut.

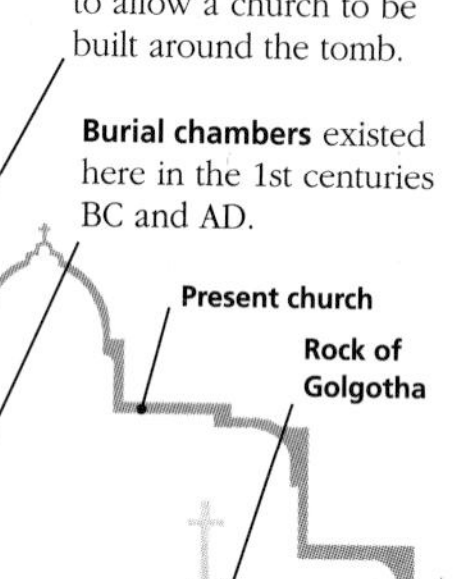
The hillside was dug away in the 4th century to allow a church to be built around the tomb.
Burial chambers existed here in the 1st centuries BC and AD.
Christ's Tomb
Present church
Rock of Golgotha

In the Coptic chapel behind the shrine, a piece of polished stone is shown as being part of the tomb itself, but it is granite and not limestone, as the tomb here is known to be.

ROTUNDA AND SYRIAN CHAPEL

The Rotunda is built in Classical Roman style. The outer back wall (now hidden by interior partitions) survives from the 4th-century basilica up to a height of 11 m (36 ft). The 11th-century dome was replaced after the 1808 fire and the two-storey colonnade built. The first two columns on the right, standing with your back to the nave, are replicas of two that survived the fire, but were judged unstable. The originals were made in the 11th century from the two halves of a single, gigantic Roman column – from either the 4th-century basilica or the previous Hadrianic temple. In the Rotunda's back wall is the chapel used by the Syrians. It contains Jewish rock tombs (c.100 BC–AD 100), marking the limit to which the hillside was dug away when the first church was built.

People queuing to enter the shrine containing Christ's Tomb in the church's Rotunda

CHAPELS OF ST HELENA AND THE FINDING OF THE CROSS

From the ambulatory in the Crusader-period apse, now the choir in the Greek Catholikon, steep steps lead down to St Helena's Chapel. The crosses on the walls were carved by pilgrims. Although this crypt was built by the Crusaders, who reused Byzantine columns, the side walls are foundations of the 4th-century basilica. More stairs go down to the Finding of the Cross (Inventio Crucis) Chapel, a former cistern, in which St Helena is said to have found the True Cross. The statue of her is 19th century.

Carvings in St Helena's Chapel

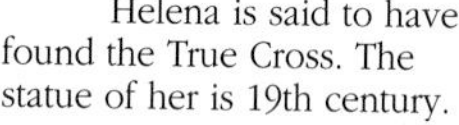

ETHIOPIAN MONASTERY

This simple monastery is approached either through the Ethiopian chapel in the corner of the courtyard, to the right of the main entrance, or from Souk Khan el-Zeit *(see p91)*, up steps beside Zalatimo's, a famous pastry shop.

It occupies a series of small buildings on the roof of St Helena's Chapel, among the ruins of the former Crusader cloister. The Ethiopians were forced here in the 17th century, when, unable to pay Ottoman taxes, they lost ownership of their chapels in the main church to other communities.

View of the Holy Sepulchre church from the roof of St Helena's Chapel ▷

Alexander Hospice ❷

Souk el-Dabbagha. **Map** 3 C3.
Tel *(02) 627 4952.*
Excavations *9am–6pm daily.*

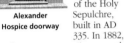

Home to St Alexander's Church, the central place of worship for Jerusalem's Russian Orthodox community, the Alexander Hospice also houses some important excavations. When the hospice was founded in 1859, the site was already known to contain ruins of the original church of the Holy Sepulchre, built in AD 335. In 1882, however, excavations revealed remains of a Herodian city wall. This finally proved that the site of the Holy Sepulchre church was outside the ancient city walls, which added credence to the claim that it was on the true site of Christ's crucifixion *(see pp92–7).*

Alexander Hospice doorway

Also preserved here are remnants of a colonnaded street and, in the church, part of a triumphal arch from Hadrian's forum, begun in AD 135. The excavations are open to the public, but only parts of the church can be visited.

Lutheran Church of the Redeemer ❸

24 Muristan Rd. **Map** 3 C3.
Tel *(02) 627 6111. 9am–1pm & 1:30–5pm (winter: 4pm) Mon–Sat.*
for bell tower only.

This Neo-Romanesque church was built for the German Kaiser Wilhelm II, and completed in 1898. Renewed interest in the Holy Land by Europe during the late 19th century had ushered in a period of restoration and church building, with many nations wanting to establish a religious presence in Jerusalem. The Lutheran Church of the Redeemer was constructed over the remains of the 11th-century church of St Mary of the Latins, built by wealthy merchants from Amalfi in Italy. An even earlier church is thought to have existed on the site from the 5th century. Many details from the medieval church have been incorporated into the new building, and the entrance way, decorated with the signs of the zodiac and symbols of the months, is largely original. The attractive cloister, which is inside the adjacent Lutheran hospice, has two tiers of galleries and dates from the 13th–14th centuries. Perhaps the most interesting part of the church though is the bell tower. After climbing the 177 steps, visitors are rewarded with some great views over the Old City.

The dominating tower of the Lutheran Church of the Redeemer

One of the many souvenir shops in the Muristan

Muristan ❹

Muristan Rd. **Map** 3 C3.

The name Muristan derives from the Persian word for a hospital or hospice for travellers. For centuries the area known as the Muristan, south of the Holy Sepulchre, was the site of just such a hospice for pilgrims from Latin-speaking countries. It was built by Charlemagne in the early 9th century, with permission from the caliph Haroun el-Rashid. Partly destroyed in 1009 by the Fatimid caliph El-Hakim, it was restored later in the 11th century by merchants from Amalfi. They also built three churches here: St Mary Minor for women, St Mary of the Latins for men, and St John the Baptist for the poor.

St John the Baptist still stands today, and was where the Knights of the Hospital of St John (or the Knights Hospitallers) were founded. They were to take over much of the Muristan area as their

For hotels in this area see p256

The fountain square, at the heart of the Muristan

headquarters, later building their own huge hospital to the north of the church. During the Crusades it was reported that there could often be up to 2,000 people under their care here at any one time.

By the 16th century the Muristan had fallen into ruins and Suleyman the Magnificent had its stones used to rebuild Jerusalem's city walls.

Today the Muristan is very different from how it once looked, most traces of the original buildings having long since disappeared. It is now characterized by its quiet lanes and attractive pink-stone buildings. The lanes converge at the ornate fountain in the main square – site of the original hospice. The surrounding streets are packed with small shops selling souvenirs, handicrafts and antiques. Along the nearby Muristan Road you will also find a number of outdoor cafés where you can sit and absorb the atmosphere.

The distinctive dome of the Church of St John the Baptist

Church of St John the Baptist ❺

Christian Quarter Rd. **Map** 3 C4.
to the public.

The silvery dome of the Church of St John the Baptist is clearly visible above the rooftops of the Muristan, but the entrance is harder to spot among the hordes of people along busy Christian Quarter Road. A small doorway leads into a courtyard, which in turn gives access to the neighbouring Greek Orthodox monastery and the church proper.

Founded in the 5th century, the Church of St John the Baptist is one of the most ancient churches in Jerusalem. After falling into ruin, it was extensively rebuilt in the 11th century, and aside from the two bell towers which are a later addition, the modern church is little changed.

In 1099 many Christian knights who were wounded during the siege of Jerusalem were taken care of in this church. After their recovery they decided to dedicate themselves to helping the sick and protecting the pilgrims visiting Jerusalem. Founding the Knights of the Hospital of St John, they later developed into the military order of the Hospitallers and played a key role in the defence of the Holy Land *(see pp48–9)*.

Christian Quarter Road ❻

Map 3 B3.

Together with David Street, which runs from Jaffa Gate towards the Muristan, Christian Quarter Road is one of the main streets in the Christian Quarter. Marking off the Muristan zone, it passes by the western side of the Holy Sepulchre, and parallel to Souk Khan el-Zeit. This busy road is lined with shops selling antiques, Palestinian handicrafts (embroidery, leather goods and Hebron glass), and religious articles (icons, carved olive-wood crucifixes and rosaries).

Midway up the road on the right, down an alley signposted for the Holy Sepulchre, a short stairway descends to the modest **Omar Mosque**, with its distinctive square minaret. Its name commemorates the caliph Omar, the person generally credited with saving the Holy Sepulchre from falling into Muslim control after Jerusalem passed under Muslim dominion in February 638. Asked to go and pray inside the church, which would almost certainly have meant its being converted into a mosque, he instead prayed on the steps outside, thus allowing the church to remain a Christian site. The Omar mosque was built later, in 1193, by Saladin's son Aphdal Ali, beside the old Hospital of the Knights of St John.

Glassware on sale on Christian Quarter Road

The unassuming **Khanqa Salahiyya** is at the top of Christian Quarter Road. Built by Saladin between 1187 and 1189 as a monastery for Sufi mystics, it is on the site of the old Crusader Patriarchate of Jerusalem. Its ornate entrance way may be as close as you are allowed, however, as it is not open to non-Muslims. Along the north side of the mosque is El-Khanqa Street. This attractive, old, stepped street is lined with interesting shops, and runs up one of the Old City's many hills.

Museum of the Greek Orthodox Patriarchate ❼

Greek Orthodox Patriarchate Road. **Map** 3B3. ***Tel*** *(02) 627 4941.* *8am–3pm Mon–Sat.*

Tucked away in the back alleys of the Christian Quarter, this museum houses a collection of ecclesiastical items that includes icons, embroidered vestments, mitres, chalices and filigree objects. It also has a fine array of archaeological finds.

Of most interest are two white-stone sarcophagi found at the end of the 19th century in a tomb near the present-day King David Hotel *(see p122)*. They are considered to belong to the family of Herod the Great *(see p120)*, and are covered in wonderfully elaborate floral decoration, which represents some of the finest Herodian-era funerary art ever found. The museum also displays Crusader objects, including a 12th-century carved capital from Nazareth, and artifacts found in the tomb of Baldwin I (king of Jerusalem, 1100–18) in the Church of the Holy Sepulchre. Other treasures include a 12th-century mitre carved from rock crystal, with bands of copper around the base and set with gems, which may once have contained relics of the Holy Cross.

Among a collection of historical firmans (imperial edicts), is one that purports to have been issued by the caliph Omar in AD 638, granting the Greek Orthodox Church custody Jerusalem's holy places.

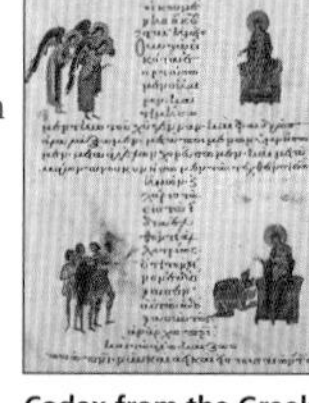

Codex from the Greek Patriarchate Museum

Jaffa Gate ❽

Map 3 B4. *1, 13, 20.*

This is the busiest of the seven Old City gates. It is the main gate for traffic and pedestrians coming from modern West Jerusalem via Mamilla. Despite the gate's great size, the entrance tunnel is narrow; it is also L-shaped – both measures meant to slow attackers. It was constructed during the reign of Suleyman the Magnificent – an exact date of 1538 is given in a dedication within the arch on the outside of the gate. The breach in the wall through which cars now pass was made in 1898, in order to allow the visiting Kaiser Wilhelm II of Germany to enter the city in his carriage.

Jaffa Gate, the main way into the Old City from West Jerusalem

Immediately inside the gate, set into the wall behind some railings on the left, are two graves. Tour guides like to tell how these belong to Suleyman's architects, executed because they failed to incorporate Mount Zion within the city walls. An alternative legend has it that they were killed to prevent them ever building such grand walls for anyone else. In fact, they are the graves of a prominent citizen and his wife.

Jaffa Gate is one of the places where visitors can access the ramparts to walk along the city walls *(see pp142–3)*. To the Arabs this gate is known as Bab el-Khalil, from the Arabic name for Hebron (El-Khalil). The old road to the town started here.

EASTERN CHRISTIANITY AND THE PATRIARCHATES

Jerusalem's Greek Orthodox Patriarch

There are no fewer than 17 churches represented in Jerusalem, a result of a great many historical schisms. As Christianity spread in the 2nd and 3rd centuries, patriarchates were established in Alexandria, Antioch, Constantinople, Jerusalem and Rome. Their heads, the patriarchs, claimed lineage from the Apostles, which gave them the authority to pronounce on correct doctrine. The first major schism came when the Council of Chalcedon (AD 451) proclaimed the dual "divine and human" nature of Christ, and in so doing estranged the Armenian, Ethiopian, Coptic and Syrian churches from the Roman Catholic and mainstream Orthodoxy. Eastern and Western Christianity split in 1054, when the Eastern churches refused to acknowledge the primacy of the Pope and the Roman church. Today there are four patriarchs (a position akin to that of an archbishop) resident in Jerusalem: those of the Greek Orthodox, Armenian, Greek Catholic and Latin (Roman Catholic) churches. The Ethiopians and Copts have a building called a patriarchate, but without the figure of the patriarch.

Syrian Orthodox priest

Armenian priest

Omar ibn el-Khattab Square, just inside Jaffa Gate

The Citadel 9

See pp102–5.

Omar ibn el-Khattab Square 10

Map 3 B4.

Not so much a square as a widening of the road as it passes around the Citadel, this area just inside Jaffa Gate is a focal point of Old City life. Arab boys selling street food solicit black-garbed Orthodox Jews heading for the Western Wall, and priests in cassocks pose for the cameras of the tourist groups, who pick up their tour guides here.

The square takes its name from the caliph Omar, who captured Jerusalem for Islam in AD 638. The Muslim name is misleading, as most of the property around the square is owned by the Greek Orthodox Patriarchate. In the late 19th century, the patriarchate built the hotels and shops on the north side, including the Neo-Classical **Imperial Hotel**. These days the hotel suffers badly from neglect and has appeal only for those who value atmosphere over comfort.

At a street junction behind the hotel is a **Roman column**, erected around AD 200 in honour of the prefect of Judaea and commander of the 10th Legion. This was one of the legions that participated in the recapture of Jerusalem in AD 70 *(see p43)*, and was subsequently quartered in the city. The column now supports a street light.

Several cafés with pavement tables fringe the east side of the square. Next to the cafés is the Christian Information Centre, and, opposite the entrance to the Citadel, the Anglican Christ Church compound. Its Neo-Gothic church (1849) was the first Protestant building in the Holy Land.

A Walk on the Roofs 11

Map 3 C4.

At the corner of St Mark's Road and Khabad Street, in an area where the Jewish, Christian and Muslim Quarters all overlap, an iron staircase leads up to the Old City rooftops. From here it is possible to walk above the central souk area, peering down through ventilation grilles to the bustling street below. It is possible to walk for some distance, between satellite dishes and dividing walls. There is even a ramshackle children's playground up here. Locals use the rooftops as a short cut; for visitors the appeal is in the unusual views the terrace affords of the Church of the Holy Sepulchre and Dome of the Rock. It is also worth coming up here in the evening to see the rooftop skyline thrown into silhouette by moonlight. A second set of stairs leads down past a *yeshiva* (Jewish religious school) onto El-Saraya Street in the Muslim Quarter.

St Mark's Church 12

5 Ararat Street. **Map** 3 C4. ***Tel*** *(02) 628 3304.* *9am–5pm (winter: 4pm) Mon–Sat, 11am–4pm Sun.*

This small church is the centre of the Syrian Orthodox community in Jerusalem. It is a place rich in biblical associations, albeit of suspect authenticity. According to tradition the church was built on the site of the house of Mary, mother of St Mark the Evangelist. A stone font in the church is supposedly that in which the Virgin Mary was baptized, and the church also has a painting on parchment of the Virgin and Child that is often attributed to St Luke. Of course, historians identify it as dating from a much later period. Some scholars do believe, however, that a small cellar room here was the true site of the Last Supper, not Mount Zion *(see p117)*.

Orthodox Jews cross the rooftops of the Old City

The Citadel ❾

Ruined arch in the courtyard

Now occupied by the Tower of David Museum of the History of Jerusalem *(see p104)*, the Citadel is an imposing bastion just inside the city wall. The present-day structure dates principally from the 14th century and includes additions made in 1532 by Suleyman the Magnificent. However, excavations have revealed remains dating back to the 2nd century BC, and indicate that there was a fortress here from Herodian times. This supports the view that this is the most likely site of Christ's trial and condemnation.

View of the Citadel and the Dome of the Rock behind, from the New City

Base of an early Islamic tower

The mosque was built by the Mamelukes above a Crusader hall.

The Hasmonean city wall (2nd century BC) is one of the oldest finds. Part of the same wall can be seen in the Jewish Quarter *(see p83)*.

Southeast Tower

East Tower

The entrance was built with an L-shaped hallway to impede the progress of attackers.

Open-air mosque

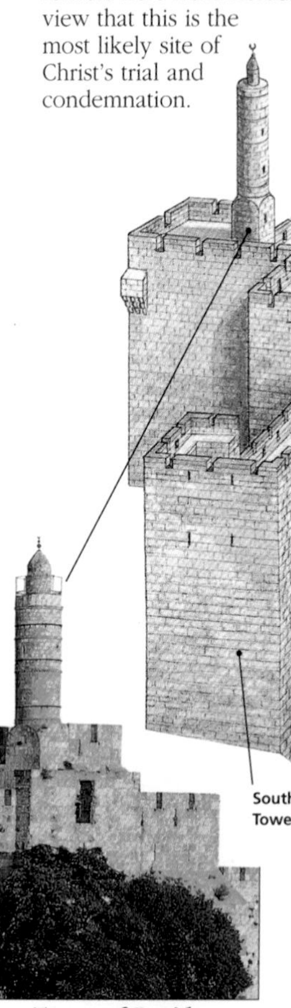

Tower of David

The Citadel is also called the Tower of David. The misnomer dates back to Byzantine confusion over the geographical layout of the city. Today it is also applied to this minaret, added in 1655.

Triple-arched Gateway

This ornamental gate was built in the 16th century. It was on the steps in front that General Allenby accepted the city's surrender in 1917 (see p52).

★ Ramparts
The crenellated walls have the same outline as in Crusader times, but date largely from the 14th century. It is possible to walk almost the whole circuit, taking in views of the city in all directions.

VISITORS' CHECKLIST

Jaffa Gate. **Map** 3 B4. ***Tel*** *(02) 626 5333.* *10am–4pm Sun–Thu, 10am–2pm Sat (also Fri in Jul & Aug); call in advance during hols.*
www.towerofdavid.org.il

An 1873 model of Jerusalem is on display in an underground cistern.

The courtyard within the Citadel has archaeological remains from almost every era from the 2nd century BC to the 12th century AD.

Entrance to café

★ Phasael's Tower
Herod the Great built a huge defensive tower here, naming it after his brother Phasael. It was demolished by Hadrian in AD 135 and partly rebuilt in the 14th century. The top offers spectacular views of the Old City.

Traces of the Byzantine city wall can be seen at the base of this section of wall.

The massive blocks at the base of Phasael's Tower are part of the original Herodian structure. This section is solid all the way through.

Moat

Mameluke Cupola
This small cupola and the hexagonal room beneath are part of the Mameluke rebuilding that took place around 1310. The tour of the museum starts on this rooftop.

STAR FEATURES

★ Phasael's Tower

★ Ramparts

Exploring the Citadel

Statue of a Crusader

There is a lot to see in the Citadel's Tower of David Museum. To help the visitor, there are three well-signposted routes: the Observation Route runs along the ramparts for the best panoramic views of the city, both Old and New; the Excavation Route concentrates on the archaeological remains in the courtyard; and the Exhibition Route takes visitors through a series of rooms tracing the history of the city. This takes the form of displays, dioramas and models, rather than a collection of historical artifacts. Visitors can join a free English tour of the route departing at 11am Sunday to Friday, and lasting one and a half hours.

Three-dimensional representation of the Second Temple

PHASAEL'S TOWER

The Exhibition Route begins in Phasael's Tower with a short, animated film. From here, exit to the roof of the octagonal entrance chamber where there is the first in a series of models placed throughout the museum that depict Jerusalem at various stages during its history. This one shows the topography of the site before the founding of the city. If you then ascend Phasael's Tower, you can see the pattern of hills and valleys for yourself.

THE CANAANITES AND THE FIRST TEMPLE

Heading clockwise from Phasael's Tower, the first two sections deal with the origins of Jerusalem, covering the period from 3150 to 587 BC, the year the First Temple was destroyed. The Canaanite era is explained in three display boards outside the East Tower, while the First Temple-era exhibits are inside the tower. These include a replica of a 19th-century BC Egyptian statuette bearing the first written reference to Jerusalem. There is also a model of the 10th-century City of David, prior to the building of the Temple, a hologram of the Temple itself, and an informative animation showing how the ancient city's water system worked. The latter is very useful for anyone who intends visiting Hezekiah's Tunnel and the Pool of Siloam *(see p115)*.

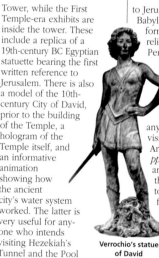

Verrochio's statue of David

RETURN TO ZION AND THE SECOND TEMPLE

The next series of rooms, in a lower level of the East Tower, traces the return of the Jews to Jerusalem from exile in Babylon – illustrated in the form of interpretative reliefs in Babylonian and Persian style. One room features a three-dimensional portrait of the Second Temple, which is worth studying closely by anyone who intends visiting the Ophel Archaeological Park *(see pp86–7)*. There is also an illustration of the three original Herodian towers – one of these forms the base of Phasael's Tower, visited at the start of this route, which still has some of the stone used to build the lower part of the 2nd-century BC structure. The destruction of the Temple is represented by a reproduction of a frieze from the Arch of Titus, erected in Rome in AD 81 to celebrate the triumph over the Jews 11 years earlier. It shows Roman soldiers carrying off Jewish treasures including a menorah and trumpets.

Between here and the next exhibition room is a bronze copy of Verrochio's *David*, a Renaissance sculpture of the young king. David, in fact, had nothing to do with the Citadel or the tower that bears his name *(see p102)*, as the fortress dates from the time of Herod, a thousand years after the time of David. The statue was a gift to Jerusalem from the city of Florence in Italy.

Phasael's Tower (left) seen across the Citadel's courtyard

For hotels in this area see p256

The sabil of Suleyman in a finely detailed model in the Ottoman room

LATE ROMAN AND BYZANTINE PERIODS

A small room in the Southeast Tower deals with the creation of Aelia Capitolina, the Roman city, built on the ruins of Second Temple-era Jerusalem. The room has floors based on mosaics from Hadrian's Villa in Rome and the St Martyrius Monastery near Jerusalem. There is also a splendid model, 1.5 m (5 ft) long, of the Church of the Holy Sepulchre as it is thought to have looked when it was first built in the 4th century, on the orders of Helena, mother of the Emperor Constantine.

The prayer niche and pulpit in the Citadel's former mosque

EARLY ISLAM AND THE CRUSADES

Appropriately enough, the early Islamic exhibits are housed in the Citadel's former mosque. This is the most striking room in the whole Citadel complex, with a still intact *mihrab* (niche indicating the direction of Mecca) and *minbar* (pulpit). At the centre of the room is a large, detailed, sectioned model of the Dome of the Rock. The model apparently took two years to construct. An aluminium model at the centre of the room shows that by this time the Old City had taken on the form in which it appears today. There is also a diorama of the Crusader Church of St Anne's and life-size statues of the Western knights, as well as a brightly coloured diorama depicting the famed conqueror of the Crusaders, Saladin (Salah ed-Din in Arabic), in his tent outside the city walls.

Members of Saladin's retinue

THE MAMELUKES AND OTTOMANS

The final exhibition rooms are housed in the large, northwest tower. The Mamelukes (1260-1516), a dynasty of former slaves who ruled from Egypt, endowed Jerusalem with some of its most distinctive and beautiful buildings. Their contribution is represented by drawings and a scale reconstruction of a street of distinctive striped-stone *(ablaq)* architecture. You can see similar examples today at Lady Tunshuq's Palace in the Old City *(see p65)*. Illustrating Ottoman Jerusalem is a large-scale model of a fountain *(sabil)* erected by Suleyman the Magnificent – the real thing survives today on Chain Street in the Muslim Quarter.

END OF THE OTTOMANS AND THE BRITISH MANDATE

This last room is a brief race through the city's more recent history. The story it tells is of the mass influx of Christian pilgrims and Jewish immigrants who began to settle for the first time outside the security of the walls of the Old City and, in doing so, established what is now the modern city of Jerusalem. A video wall with nine screens depicts 30 years of British mandate from 1917 until 1948. There is also some rare 1896 Lumière Brothers footage of the Jerusalem–Jaffa railway.

In a separate hall is a vast and superb model of late 19th-century Jerusalem, made by a Hungarian artist in 1873. It was exhibited throughout Europe before going into storage and being forgotten for a century until its rediscovery and removal here in the early 1980s.

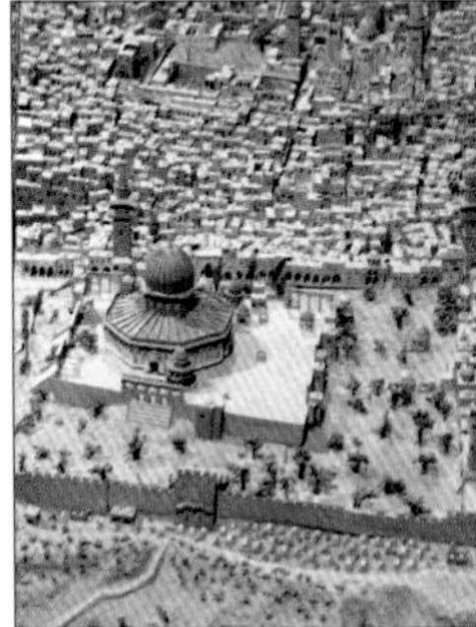

Detail of the enormous model of Jerusalem constructed in 1873

Wrought-iron gate framing the ornate main entrance to St James's Cathedral

St James's Cathedral ⓭

Armenian Patriarchate Rd. **Map** 3 B5. ***Tel*** *(02) 628 2331. 6–7:30am & 3–3:30pm Sun–Fri, 6:30–9:30 am Sat.*

The Armenian Cathedral is one of the most beautiful of all Jerusalem's sacred buildings. It was originally constructed in the 11th and 12th centuries over the reputed tomb of St James the Great, the Apostle, killed by Herod Agrippa I (AD 37–44). Many alterations and additions have since been made, most notably in the 18th century, when much of the existing decoration was added.

Entrance to the cathedral is via a small courtyard with a 19th-century fountain. On the western wall of the courtyard are inscriptions in Armenian, one of which dates from 1151. Hanging in the vaulted porch are wooden bars. Each afternoon a priest strikes these with a wooden mallet known as a *nakus*, to signal the start of the service.

The cathedral interior is enchanting. It is only dimly illuminated by a forest of oil lamps hung from the ceiling. There are no seats; instead the floors are thickly laid with Oriental rugs. Four great square piers divide the main space into three aisles. These piers, along with the walls, are covered in blue-and-white tiles with floral and abstract patterns. In the apses at the end of each of the three aisles are altars, separated from the rest of the church by the iconostasis screen. Two thrones stand in the choir; the one nearest the pier is said to be that of St James the Less, traditionally held to have been a step-brother of Christ and the first bishop of Jerusalem. It is used only once a year, in early January, on the occasion of his feast day. The other throne is the one normally used by the patriarch.

The cathedral contains many small shrines and chapels. The third on the left as you enter is the most important: it supposedly holds the head of St James the Great. Off to the right, the Etchmiadzin Chapel has some beautiful tiling.

Mardigian Museum ⓮

Armenian Patriarchate Rd. **Map** 3 B5. ***Tel*** *(02) 628 2331. 10am–4:30pm Mon–Sat.*

Dating from 1863, this was originally the seminary of the nearby Armenian patriarchate. It is now a museum dedicated to the history and culture of the Armenian people. The building is attractive, with a long central courtyard flanked by porticoes. The oldest finds in the collection are fragments of 1st-century frescoes from the courtyard of the so-called House of Caiaphas on Mount Zion, and remains from Byzantine-period Armenian churches unearthed near Damascus Gate. The pride of the museum is its collection of early manuscripts. In addition, there are also a great many liturgical objects, many of which were donated to St James's Cathedral by Armenian pilgrims. There are also examples of the pottery for which the Armenians have always been famous.

Other interesting objects are examples of the first books printed in the first print shop in Jerusalem, which has been active since 1833 inside the Armenian monastery.

17th-century jug, Mardigian Museum

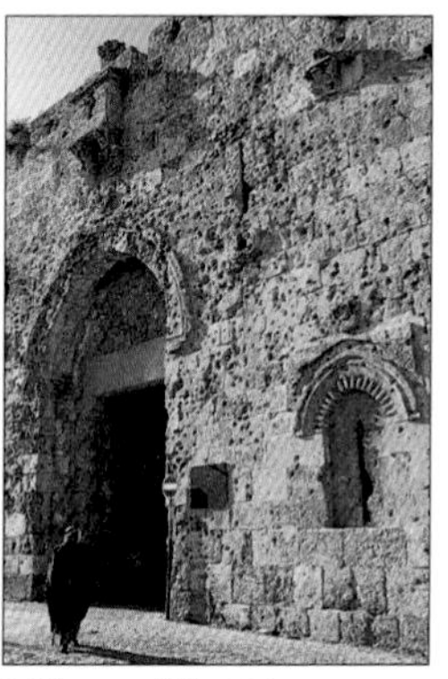

Battle-scarred Zion Gate

Zion Gate ⓯

Map 3 C5.

Zion Gate was constructed by Suleyman the Magnificent's engineers *(see p105)* in 1540. It allowed direct access from the city to the holy sites on Mount Zion. Fighting was particularly fierce here in 1948, when Israeli soldiers were desperate to breach the walls to relieve the Jewish Quarter inside, under siege by the Jordanians. The outside of the gate is terribly pockmarked by bulletholes. A short distance to the west of the gate there is conspicuous damage to the base of the wall where soldiers tried to blast their way through with explosives.

In Arabic, the gate is known as Bab el-Nabi Daud (Gate of the Prophet David), because of its proximity to the place traditionally known as King David's Tomb *(see p117)*.

The Armenians in Jerusalem

The kingdom of Armenia was the first country to make Christianity the state religion, when in AD 301 its king was converted. Armenian pilgrims began to visit the Holy City soon after. In the 12th century they purchased St James's Cathedral from the Georgians, and this became the focal point of their community in Jerusalem. The Armenian Quarter grew to its current size in the 17th and 18th centuries, during the rule of the Turks. In the early 20th century Armenian numbers were swollen by refugees who had fled from the 1915 persecution in Turkey, a terrible genocide in which some one and a half million Armenians were exterminated. But from a peak of around 16,000 in 1948, the Armenian population of Jerusalem has since dwindled to less than 2,000, largely due to emigration. After the 1967 war, the Jews also started to encroach into the area, and the fear now is that other than in name, the Armenian Quarter may one day disappear altogether.

Detail from an Armenian carpet

Tiling *adorns the interior of St James's Cathedral. The tiles were made in the early 18th century in Kütahya, a town around 125 km (75 miles) southeast of Constantinople, and renowned as the foremost Armenian ceramic centre in the Ottoman Empire.*

The Armenian Church *is one of the three major guardians of the Christian places in the Holy Land. Among the sites they have at least partial jurisdiction over are the Church of the Holy Sepulchre, the Tomb of the Virgin Mary at the foot of the Mount of Olives, the Church of the Nativity in Bethlehem and, of course, St James's Cathedral* (above).

Mosaics represent *the finest legacy of ancient Armenian art. This 5th- or 6th-century example was unearthed just outside Damascus Gate.*

Armenian-language manuscripts, *such as this 13th-century example, are held in huge numbers at the Gulbenkian Library, next to St James's Cathedral.*

Giant pots *for wine or oil, dating from around 1700, are displayed at the Mardigian Museum.*

TCHEQUE
Otče náš,
jenž jsi na nebesích,
posvěť se jméno tvé.
Přijď království tvé.
Buď vůle tvá, jako
v nebi tak i na zemi.
Chléb náš vezdejší
dej nám dnes. A odpusť
nám naše viny, jako
i my odpouštíme našim
viníkům. A neuveď nás
v pokušení, ale zbav
nás od zlého.
Amen.

THE MOUNT OF OLIVES AND MOUNT ZION

The Mount of Olives is the hill that rises to the east of the Old City. Its slopes have been used as a place of burial since the 3rd millennium BC. The hill is also dotted with sites connected with the last days of Jesus Christ, but the highlight for many visitors is the superb view of the Old City from the summit. Between the city walls and the hill is the Valley of Jehoshaphat, with several tombs from the 1st and 2nd centuries BC. At the southern end of the valley is the site of the 3,000-year-old settlement that was to become Jerusalem (the City of David). The land rises again to the west to Mount Zion, an area of the city traditionally linked with the Last Supper.

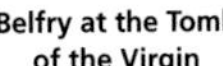

Belfry at the Tomb of the Virgin

SIGHTS AT A GLANCE

Holy Places
Basilica of the Agony 7
Church of the Dormition 13
Church of the Paternoster 3
Church of St Mary Magdalene 6
Dominus Flevit Chapel 5
Hall of the Last Supper 14
Mosque of the Ascension 2
Russian Church of the Ascension 1
St Peter in Gallicantu 11
Tomb of the Virgin 8

Archaeological Sites
City of David 10

Historic Areas
Mount Zion 12

Tombs
King David's Tomb 15
Schindler's Tomb 16
Tombs of the Prophets 4
Valley of Jehoshaphat 9

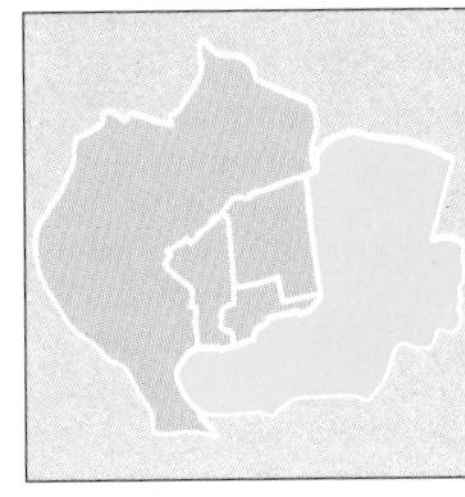

KEY

Mount of Olives
See pp110–11

GETTING THERE

The best way to see the Mount of Olives is to take a bus (No. 5A from the City Hall complex area) or a taxi to the summit and walk down. Walking from the Old City involves a strenuous uphill climb. Mount Zion is most easily reached via Zion Gate in the Old City.

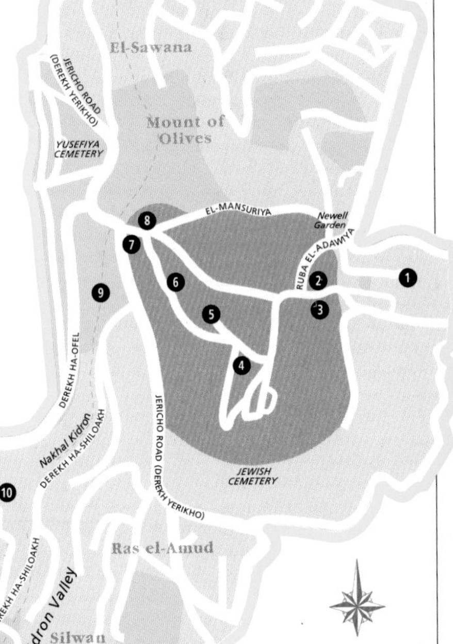

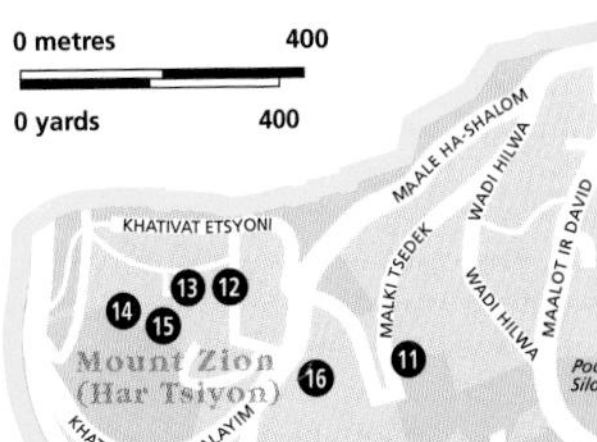

◁ The cloister of the Church of the Paternoster, which displays the Lord's Prayer in over 60 languages

The Mount of Olives

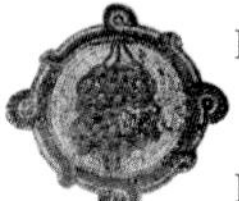
Mosaic, Dominus Flevit Chapel

Rising on the eastern side of Jerusalem, the Mount of Olives offers magnificent views of the Dome of the Rock and the Old City. Now best known as the scene of Christ's Agony and betrayal in the Garden of Gethsemane and his Ascension into Heaven, this prominent hill has always been a holy place to the inhabitants of the city. The Jebusites dug tombs here as early as 2400 BC, as later did Jews, Christians and Muslims. To take in all the sights it is wisest to start at the top, near the Mosque of the Ascension, and walk downhill to the Tomb of the Virgin. The Old City views are best in the morning.

Dominus Flevit Chapel
The chapel's west window frames a breathtaking view of the Old City 5

The Cave of Gethsemane is the traditional site of Christ's betrayal by Judas.

Garden of Gethsemane

★ Tomb of the Virgin
An impressive flight of Crusader steps leads into the cruciform underground church. Tradition says this is where the Virgin Mary was laid to rest 8

★ Basilica of the Agony
Mosaics, predominantly in blues and greens, decorate the 12 domes of this church, built in 1924 with donations from many countries 7

Jericho Road

STAR SIGHTS

- ★ Tomb of the Virgin
- ★ Basilica of the Agony
- ★ Church of the Paternoster

Church of St Mary Magdalene
This Russian Orthodox Church, with typically Muscovite gilded onion domes, was built by Tsar Alexander III in memory of his mother, whose patron saint was Mary Magdalene 6

For hotels in this area see p256

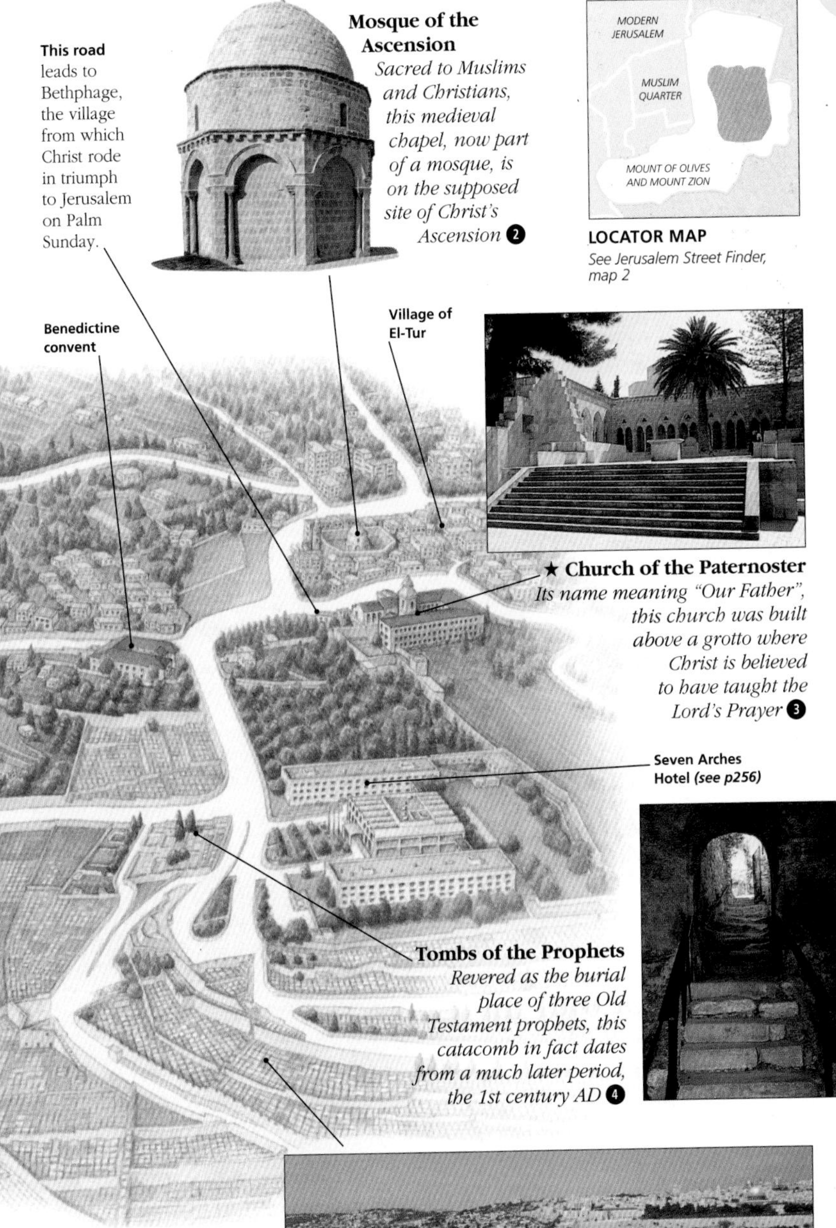

LOCATOR MAP
See Jerusalem Street Finder, map 2

This road leads to Bethphage, the village from which Christ rode in triumph to Jerusalem on Palm Sunday.

Mosque of the Ascension
Sacred to Muslims and Christians, this medieval chapel, now part of a mosque, is on the supposed site of Christ's Ascension ❷

Benedictine convent

Village of El-Tur

★ Church of the Paternoster
Its name meaning "Our Father", this church was built above a grotto where Christ is believed to have taught the Lord's Prayer ❸

Seven Arches Hotel *(see p256)*

Tombs of the Prophets
Revered as the burial place of three Old Testament prophets, this catacomb in fact dates from a much later period, the 1st century AD ❹

Jewish Cemeteries
Many Jews wish to be buried on the Mount of Olives so as to be close to the Valley of Jehoshaphat, where it is said mankind will be resurrected on the Day of Judgment.

Church of the Ascension's bell tower in the quiet convent gardens

Russian Church of the Ascension ❶

Off Ruba el-Adawiya St, Mount of Olives. **Map** 2 F3. ***Tel*** *(02) 628 4373.* *9am–noon Tue & Thu.*

This is the church of a still active Russian Orthodox convent built between 1870 and 1887. The bell tower, a prominent landmark on the Mount of Olives, was built tall enough to allow pilgrims too infirm to walk to the River Jordan to see it from afar. The 8-tonne bell was hauled from Jaffa by Russian pilgrims.

Two Armenian mosaics were found during construction. A small museum was built over the most beautiful, which is fragmentary and dates from the 5th century AD; the other, complete and of slightly later date, is in the Chapel of the Head of John the Baptist, inside the church. An iron cage on the floor shows where John's head was supposedly found.

Mosque of the Ascension ❷

Off Ruba el-Adawiya St, Mount of Olives. **Map** 2 F3. *daily (if closed, ring bell).*

Poemenia, a Christian noblewoman, built the first chapel here around AD 380 to commemorate Christ's Ascension. It had three concentric porticoes around an uncovered space, where the dust miraculously formed the image of Christ's footprints. The Crusaders rebuilt the chapel as an octagon and the column bases of a surrounding Crusader portico are still visible outside. By this time, the footprints, now set in stone, were venerated here and the right imprint remains to this day. The capitals were carved in the 1140s and the two depicting animals and leaves are particularly beautiful.

The chapel became a Muslim shrine after Saladin's conquest in 1187. In 1200 it was roofed with a dome, the arches were walled in, a mihrab added and a surrounding wall built. The outer wall today is largely rebuilt. The adjacent minaret and mosque are 17th century.

The underground tomb near the entrance is venerated by Jews as belonging to the Old Testament prophetess Huldah, by Christians as St Pelagia's and by Muslims as that of the holy woman Rabia el-Adawiya.

Church of the Paternoster ❸

Mount of Olives. **Map** 2 F4. ***Tel*** *(02) 626 4904.* *8:30am–noon & 2:30–5pm Mon–Sat.*

This church stands next to the partly restored ruins of one commissioned by the Emperor Constantine, who sent his mother, St Helena, to supervise construction in AD 326. Called Eleona (*elaion* in Greek meaning "of olives"), it was sited above a grotto where the Ascension was commemorated. By Crusader times, the church had been rebuilt three times and the grotto was known as the place where Christ had taught the Disciples the Paternoster (meaning "Our Father"), or Lord's Prayer.

Site of Christ's footprint in the Mosque of the Ascension

The present church and a Carmelite monastery were built close by between 1868 and 1872 by the French Princesse de la Tour d'Auvergne. Excavations of the Byzantine church in 1910–11 unearthed a marble plaque engraved in Latin with the Paternoster. In 1920, the grotto was restored, but plans to reconstruct the Byzantine church were never realized through lack of funds.

Today, the 19th-century church and its cloister are famous for the tiled panels inscribed with the Paternoster in more than 60 languages.

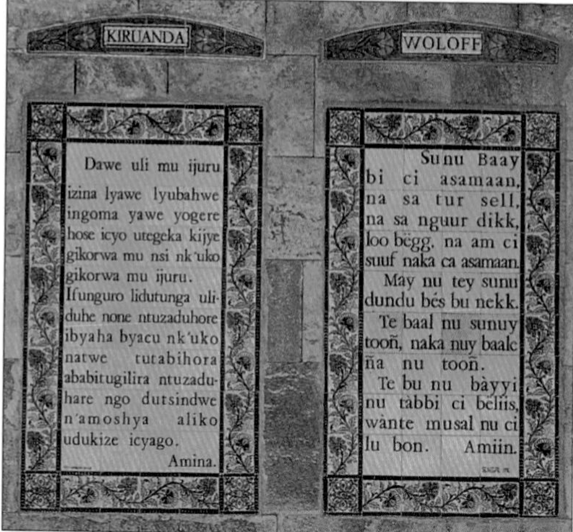

Panels inscribed with the Lord's Prayer, Church of the Paternoster

For hotels in this area see p256

Tombs of the Prophets ❹

Mount of Olives. **Map** 2 F4. *9am–3:30pm Mon–Fri.*

The southwestern slope of the Mount of Olives, facing the Kidron Valley (also known along this stretch as the Valley of Jehoshaphat – *see p115*), is densely occupied by Jewish cemeteries. At the top of the slope, an unusual, fan-shaped catacomb containing *kokhim* (oven-shaped) graves is held by Christian and Jewish tradition to enclose the tombs of the 5th-century BC prophets Haggai, Malachi and Zechariah. The graves actually date from the 1st century AD and were reused in the 4th or 5th.

Dominus Flevit Chapel ❺

Mount of Olives. **Map** 2 F4. ***Tel*** *(02) 626 6450.* 99. *8–11:45am, 2:30–5pm daily.*

Russian Church of St Mary Magdalene, built in Muscovite style

Its name meaning "The Lord Wept", this chapel stands where medieval pilgrims identified a rock as the one on which Jesus sat when he wept over the fate of Jerusalem. The chapel was designed in the shape of a teardrop by Italian architect Antonio Barluzzi and built in 1955 over a 7th-century chapel. Part of the original apse is preserved in the new one. The view of the Dome of the Rock from the altar window is justly famous. A mosaic floor preserved in situ outside is from a 5th-century monastery. The graves on view nearby show the types found in the 1950s in a vast cemetery here, in use periodically from 1600 BC to AD 70. Also on show are some carved stone ossuaries.

Church of St Mary Magdalene ❻

Mount of Olives. **Map** 2 E3. ***Tel*** *(02) 628 4371.* 99. *10am–noon Tue, Thu & Sat.*

In 1885, Tsar Alexander III had this Russian Orthodox church built in memory of his mother, Maria Alexandrovna. It is pleasantly set among trees, and the seven gilded onion domes are among the most striking features of Jerusalem's skyline when viewed from the Old City. The domes and other architectural and decorative features are in 16th–17th-century Muscovite style.

The church was consecrated in 1888 in the presence of Grand Duke Sergei Alexandrovich (Tsar Alexander III's brother) and his wife, Grand Duchess Elizabeth Feodorovna. In 1920, after her murder during the Russian Revolution, her remains were buried here.

THE RUSSIANS IN JERUSALEM

Russia's Christians belong to the Eastern Orthodox church, the centre of which was once Constantinople. In the 19th century, when the European powers were competing to stake their claims on pieces of the crumbling Ottoman Empire, the Russians thus presented themselves as the successors to the Byzantine Empire and the true "defenders of Christianity and the Holy Places". At this time some 200,000 Russian pilgrims were visiting Jerusalem each year. The Russian government purchased land on a grand scale, notably on the Mount of Olives and just west of the Old City, where they built a great cathedral, a consulate, a hospital and several hospices, all enclosed in a walled compound *(see p124)*. In World War I Britain captured Jerusalem and confiscated all Russian property as "enemy institutions". Although some White (Tsarist) Russians did remain after the war.

Russian Orthodox nuns embroidering vestments, Church of the Ascension

Mosaic-decorated, vaulted ceiling in the Basilica of the Agony

Basilica of the Agony ❼

Jericho Rd. **Map** 2 E3. ***Tel*** *(02) 626 6444.* 🚌 *99.* ◯ *8am–noon & 2–5pm (summer: 6pm) daily.*

The Basilica of the Agony, also known as the Church of All Nations, was named for the rock in the Garden of Gethsemane on which it is believed Christ prayed the night before he was arrested.

The 4th-century church built here was destroyed in an earthquake in 747. The Crusaders built a new one, aligned differently to cover three outcrops of rock, recalling Christ's three prayers during the night. It was consecrated in 1170, but fell into disuse after 1345. After excavation of the site in the early 20th century, the present church was designed by Antonio Barluzzi *(see p113)* and built in 1924 with financial contributions from 12 nations – hence the church's other name and its 12 domes decorated with national coats of arms. In the centre of the nave is the rock of the Byzantine church, surrounded by a wrought-iron crown of thorns. The mosaic in the apse represents Christ's agony, while others depicting his arrest and Judas's kiss are at the sides. The plan of the Byzantine church is traced in black marble on the floor, and sections of Byzantine mosaic pavement can also be seen.

Outside, the gilded mosaic scene decorating the pediment also depicts the Agony. Next to the church is the surviving part of the Garden of Gethsemane with its centuries-old olive trees.

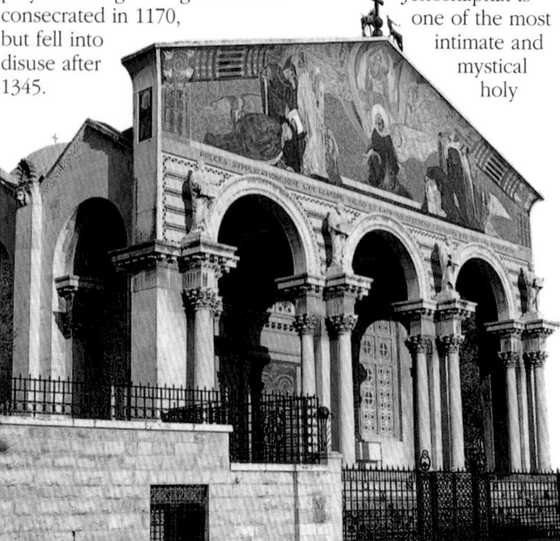

The Basilica of the Agony in the Garden of Gethsemane

Tomb of the Virgin ❽

Jericho Rd. **Map** 2 E3. ***Tel*** *(02) 628 4054.* 🚌 *99.* ◯ *6am–12:30pm & 2–6pm daily.* **Cave of Gethsemane** ◯ *6am–12:30pm & 2–6pm daily.*

Believed to be where the Disciples entombed the Virgin Mary, this underground sanctuary in the Valley of Jehoshaphat is one of the most intimate and mystical holy places in Jerusalem. The façade, the impressive flight of 47 steps and the royal Christian tombs in side niches halfway down all date from the 12th century. The tomb on the right, going down, was originally the burial place of Queen Melisande of Jerusalem, who died in 1161. Her remains were moved into the crypt in the 14th century and the tomb has been venerated since about that time as that of St Anne and St Joachim, Mary's parents.

The first tomb was cut in the hillside here in the 1st century AD. The cruciform crypt as seen today, much of it cut into solid rock, is Byzantine. By the 5th century, an upper chapel had also been built. This was destroyed by the Persians in 614, rebuilt by the Crusaders, but again destroyed by Saladin in 1187. He left the crypt, however, largely intact.

The 12th-century entrance to the atmospheric Tomb of the Virgin

The Tomb of Mary stands in the eastern branch of the crypt, which is decorated with icons and sacred ornaments typical of Orthodox Christian tradition. Today, religious services are held here by Greek, Armenian, Coptic and Syrian Christians.

In the southwestern wall beside the Tomb of Mary is a mihrab installed after Saladin's conquest. The place was sanctified by Muslims because, according to the 15th-century scholar Mujir al-Din, Muhammad saw a light over the tomb of his "sister Mary" during his Night Journey to Jerusalem *(see p27)*. In the opposite wall, a 1st-century tomb is evidence of the site's earliest use for burials.

Outside, to the right of the façade, is the **Cave of Gethsemane**, or Cave of the Betrayal, the traditional place of Judas's betrayal. It was once used for oil pressing, but fragments of 4th–5th-century mosaics bear

The Tomb of Bnei Hezir (left) and the pyramid-roofed Tomb of Zechariah in the Valley of Jehoshaphat

witness to its transformation into a place of worship. The stars on the vaults were painted in Crusader times.

Valley of Jehoshaphat ❾

Map 2 E3.

The Kidron Valley separates the Old City from the Mount of Olives. Near Gethsemane the valley is also known by its Old Testament name, the Valley of Jehoshaphat (meaning "Yahweh judges", Yahweh being the Hebrew name for God), where it was believed the dead would be resurrected on the Day of Judgment (Joel 3: 1–17). For this reason, the valley sides are densely covered with Christian, Jewish and Muslim cemeteries.

At the southern end are several Jewish rock-hewn tombs of the 1st and 2nd centuries BC. Four are particularly fine. Absalom's Tomb, like an inverted funnel, was ascribed in medieval times to King David's rebellious son, Absalom. The so-called Tomb of Jehoshaphat (the 9th-century BC King of Judah) behind it has a carved frieze above the doorway. The pyramid-topped Tomb of Zechariah is actually the above-ground monument of the adjacent Tomb of Bnei Hezir. The latter has a rectangular opening with two Doric columns and was identified by an inscription referring to the "sons of Hezir", a Jewish priestly family.

City of David ❿

Maalot Ir David. **Map** 2 D4. ***Tel*** **6033.* *winter: 8am–5pm Sun–Thu, 8am–1pm Fri & holiday eves; summer: 8am–7pm Sun–Thu, 8am–3pm Fri & holiday eves.* *for Shaft & Tunnel* *phone ahead for times.* **www**.cityofdavid.org.il

South of the Temple Mount (Haram esh-Sharif) a rocky ridge runs beside the Kidron Valley. Its summit was already settled by the Jebusites, a Canaanite *(see p41)* people, in the 20th century BC, making this the oldest part of Jerusalem. It was from them that David supposedly took the city for his capital in about 1000 BC (2 Samuel 5: 6–17).

On the site are remains of buildings up to the city's capture by the Babylonians in 586 BC. They include 13th-century BC walls belonging to the Jebusite acropolis, fragments of a palace attributed to David, and houses burnt in the Babylonian attack. About 100 m (330 ft) from the entrance to the acropolis excavations is **Warren's Shaft**, named after Charles Warren, its 19th-century English discoverer. A sloping tunnel, reached by spiral stairs, leads to the vertical shaft at the bottom of which is a pool fed by the Gihon Spring. The system was built by the Jebusites to ensure a water supply during sieges. Nearby is their 18th-century BC city wall, identified by the large, uncut stone blocks used in its construction. It was sited to bring the entrance to Warren's Shaft within the confines of the city.

The Pool of Siloam, which stored the City of David's water supply

In the 10th century BC a tunnel, later attributed to Solomon, was dug to take water from the Gihon Spring to fields in the Kidron Valley. In the face of Assyrian invasion in about 700 BC, King Hezekiah had a new tunnel built to bring the spring water right into the city, so concealing the source of the supply. **Hezekiah's Tunnel** ran 533 m (1,750 ft) from the spring to a large, new storage pool – the Pool of Siloam – in the south of the city. Not far from the Siloam end an inscription, carved by the engineer, describes the tunnel's construction. The pool is now smaller than it was originally and was rebuilt after the Romans sacked Jerusalem in AD 70 and burnt it "as far as Siloam", as told by contemporary historian Flavius Josephus.

Visitors can wade through the tunnel in thigh-deep water from the Gihon Spring – wear shoes and bring a flashlight.

The beautifully painted interior of St Peter in Gallicantu

St Peter in Gallicantu ⓫

Malki Tsedek Rd. **Map** 2 D5. ***Tel*** *(02) 673 4812.* 🚌 *38.* ◯ *8:30am–5pm Mon–Sat.* ♿

Standing to the east of Mount Zion, on the slopes overlooking the City of David *(see p115)* and the Kidron Valley, this church commemorates the traditional site of St Peter's reported denial of Christ which fulfilled the prophecy, "Before the cock crow twice, thou shalt deny me thrice" (Mark 14: 72). Built in 1931, the church has a modern appearance. In the crypt, however, are ancient caves where, it is said, Christ spent the night before being taken to Pontius Pilate. The remains of some Herodian architecture have been discovered under the church and, in the garden, there still exists part of a Hasmonean stairway, in use in Christ's time, which once connected the city with the Kidron Valley. Mosaics from a previous 5th–6th-century Byzantine church and monastery have also been unearthed.

Mount Zion ⓬

Map 1 C5. 🚌 *1, 2.*

A short walk from Zion Gate is the hill synonymous with biblical Jerusalem and the Promised Land. Believed by many to be the site of King David's tomb and associated with the final days of Christ, Mount Zion is revered by Jews, Muslims and Christians alike.

The hill is bounded to the east by the Kidron Valley, to the south and west by the Hinnom Valley and to the north by the city walls. This makes it seem like an island outside the confines of the Old City. This was not always the case, however, for on the Madaba mosaic map in Jordan *(see pp216–17)* it is shown inside the walls. It appears to have been excluded in 1542 when the walls were rebuilt. Legend has it that Suleyman the Magnificent's architects left it outside by mistake.

Christians began assembling here some time after Christ's death to worship in the Hall of the Last Supper and later at the stone where the Virgin Mary is said to have died. Now the site of the Church of the Dormition, this point marked the ceasefire border from 1949 to 1967 *(see p54)*.

Church of the Dormition ⓭

Mount Zion. **Map** 1 C5. ***Tel*** *(02) 565 5330.* 🚌 *38, 20.* ◯ *8am–5pm Mon–Sat, 10:30am–5pm Sun.* ♿

Crowned by a tall bell tower and a dome with four small corner turrets, the Neo-Romanesque Church of the Dormition dominates the Mount Zion hilltop. The large, airy, white-stone church stands on the site where the Virgin Mary is said to have fallen into an "eternal sleep". After Christ's death, according to Christian tradition, his mother went to live on Mount Zion until she herself died.

The hill soon became a holy site, available information suggesting that there may have been a church here as early as the 4th century AD. It is known with more certainty that around the 6th century a large basilica was built on the site which later fell into ruins. When the Crusaders came, they too erected a church with chapels devoted to the Dormition of the Virgin and the Last Supper.

The present-day church, which includes the Chapel of the Dormition and Dormition Abbey, was built in the early 20th century for Kaiser Wilhelm II and was inspired by the Carolingian cathedral in Aachen, Germany.

During the 1948 and 1967 wars the church was used as a strategic outpost by Israeli soldiers and was damaged in the crossfire of several battles. The main part of the church

The conical dome and bell tower of the Church of the Dormition

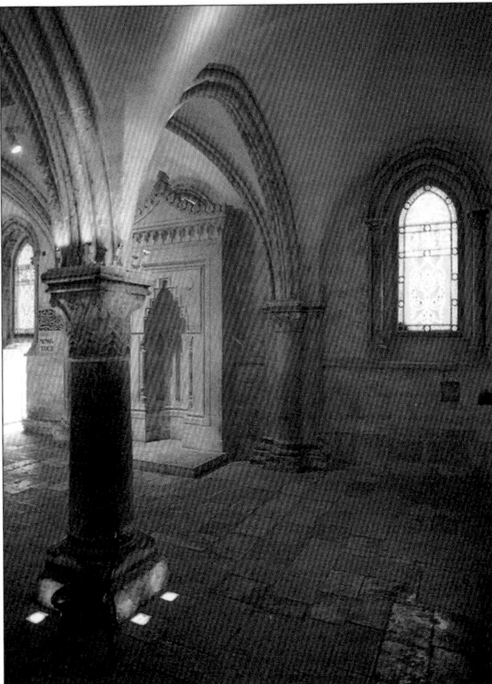

The Crusader-built Hall of the Last Supper, with fine Gothic details

boasts a fine mosaic floor featuring zodiac symbols and the names of saints and prophets. In the crypt is a wood and ivory sculpture of the "sleeping" Virgin, while the walls are adorned with images of women from the Old Testament, including Eve, Judith, Ruth and Esther. In the rooms on the mezzanine are some of the remains from the site's previous churches.

Hall of the Last Supper ⓮

Mount Zion. **Map** 1 C5.
8am–8pm (winter: 6pm) daily.

On the first floor of a Gothic building – all that remains of the large church constructed by the Crusaders to commemorate Mary's Dormition and overshadowed slightly by the more recent Church of the Dormition – is the Hall of the Last Supper, or Coenaculum. Christian tradition maintains that it is on the site of Christ's last meal with his Disciples. The room is unadorned apart from the Gothic arches dividing it.

In the Middle Ages it became part of the adjacent Franciscan monastery, while in the 15th century it was turned into a mosque by the Turks, who added a mihrab and some stained-glass windows.

King David's Tomb ⓯

Mount Zion. **Map** 1 C5. ***Tel*** *(02) 671 9767.* *1, 2, 31, 38.* *summer: 8am–8pm Sat–Thu & hols, 8am–2pm Fri; winter: 8am–sunset Sat–Thu, 8am–1pm Fri.* *Sat.*

Beneath the Hall of the Last Supper, on the lower floor of the Crusader building, are some small chambers venerated as King David's Tomb. The main chamber is bare apart from a cenotaph covered by a drape. The site was first identified as David's tomb in the 11th century AD and in the 15th century was incorporated into a mosque by the Muslims, who consider David one of the true prophets. In spite of doubts about the tomb's authenticity, it is one of the most revered Jewish holy sites. It was particularly so between 1948 and 1967, when the Old City was under Jordanian control. As the Western Wall was inaccessible to Jews, they came here to pray. Today the entrance hall is still used as a synagogue, where there is separate seating for men and women. From the 4th to the 15th centuries, the tomb was associated with Pentecost and the death of the Virgin, and, according to tradition, it was here that Christ washed his Disciples' feet after the Last Supper (John 13: 1–17).

Schindler's Tomb ⓰

Mount Zion. **Map** 1 C5. *1, 2.*

Straight down the hill from Zion Gate, the path forks left past the Chamber of the Holocaust, a small museum commemorating the thousands of Jewish communities wiped out by the Nazis. Across the road at the end of the path is a Christian cemetery. It is here that the grave of German-born Oskar Schindler is located.

Schindler was an industrialist who, during World War II, went out of his way to use Jewish prisoners as labourers in his factory. By doing this, he saved over 1,000 people from the death camps. He became a symbol of the fight against the Holocaust and before he died, in 1974, he asked to be buried in Jerusalem. The story of his courageous stand against the Nazis was told in Steven Spielberg's successful 1993 movie, *Schindler's List*.

Schindler's tomb in the Christian cemetery on Mount Zion

MODERN JERUSALEM

By the 1860s the Old City had become overcrowded, and the need for more space gave rise to a period of unrestricted building activity outside the walls. The earliest developments, such as Yemin Moshe, Nakhalat Shiva and Mea Shearim, were Jewish community projects or, like the Russian Compound, intended to cater for Holy Land pilgrims. The architecture of the new city became increasingly eclectic as colonial builders imported their own national styles. As a result, exotic features such as Muscovite domes and Florentine towers form the backdrop to the equally multi-cultural bustle on the streets of the modern city.

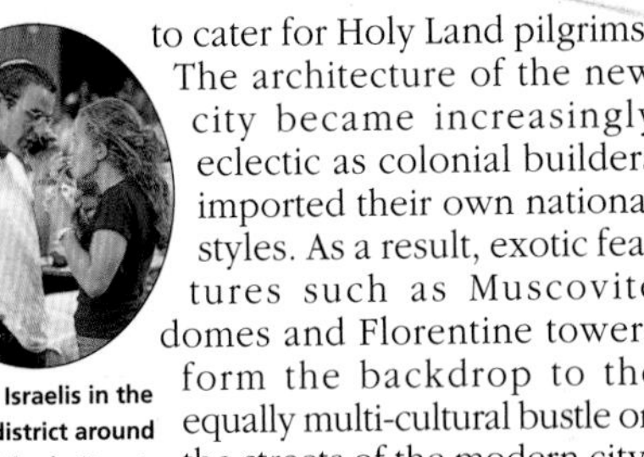

Young Israelis in the lively district around Ben Yehuda Street

SIGHTS AT A GLANCE

Historic Districts
Ben Yehuda and Nakhalat Shiva ❹
Ha-Neviim Street ❾
Mea Shearim ⓫
Russian Compound ❽

Holy Places
Italian Synagogue ❺
St Etienne Monastery ⓮
St George's Cathedral ⓯

Tombs
Garden Tomb ⓭
Kings' Tombs ⓰

Museums and Historic Buildings
American Colony Hotel ⓱
City Hall ❼
Italian Hospital ❿
Jerusalem Time Elevator ❸
King David Hotel ❷
Rockefeller Museum ⓲
Ticho House ❻
YMCA ❶

Archaeological Sites
Solomon's Quarries ⓬

KEY
Street-by-Street map
See pp120–1
Bus station
Light Rail stop
Taxi rank

0 metres 500
0 yards 500

GETTING THERE
There are buses to most of the sights from Ha-Emek Street, just outside Jaffa Gate, and from Nablus Road, just north of Damascus Gate. Taxis can be convenient, but also expensive *(see p310)*.

◁ **The Bloomfield Gardens' Lion Fountain in the Yemin Moshe district**

Street-by-Street: Yemin Moshe

Sir Moses Montefiore, a rich British Jewish philanthropist was so shocked by the living conditions in the squalid Old City that he decided to improve the Jews' lot by building new homes outside the walls. The first project was Mishkenot Shaananim ("Dwellings of Tranquillity"), a communal block of 16 apartments, completed in 1860. Initially, people were afraid to move outside the security of the walls because of bandits, but by the end of the century a small community called Yemin Moshe had been established nearby and was thriving. From this core, the vast spread of modern Jerusalem has grown. Yemin Moshe survives as its beautifully renovated historic heart.

Public Sculptures
Outdoor sculptures, such as these buried cubes, are found all around Yemin Moshe.

Jaffa Road

★ YMCA
Even if a room is beyond your budget, as one of Jerusalem's most elegant and beautiful buildings, both inside and out, the YMCA is well worth looking around ❶

ABA SIKRA

KING DAVID STREET (DAVID HA-MELEKH)

King David Hotel
Still the premier hotel in Jerusalem, and all Israel, the King David has been hosting royalty, politicians and international celebrities since it first opened its doors in the 1930s ❷

BLOOMFIELD

KING DAVID STREET (DAVID HA-MELE

KEY

— — — Suggested route

0 metres 100

0 yards 100

STAR SIGHTS

★ YMCA

★ Yemin Moshe

★ Montefiore's Windmill

Herod's Family Tomb
The splendour of this 1st-century BC tomb, discovered in 1892, suggests that it may be that of Herod's family. The king himself was supposedly buried at the Herodion (see p192).

★ Yemin Moshe
Built on the slope of the valley facing the Old City walls, these early, attractive Oriental-style houses are now some of the most sought-after and exclusive residences in all Jerusalem.

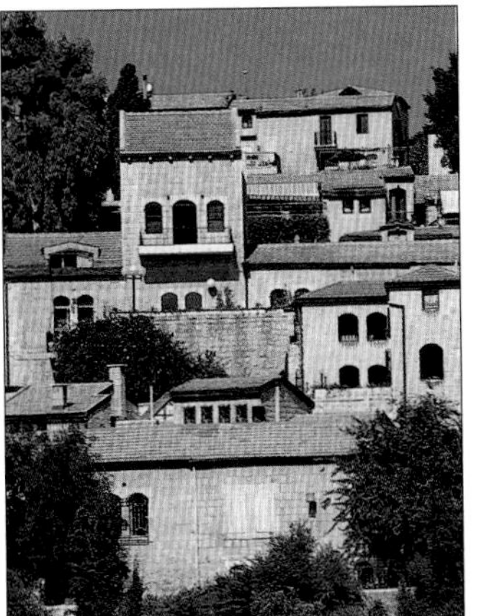

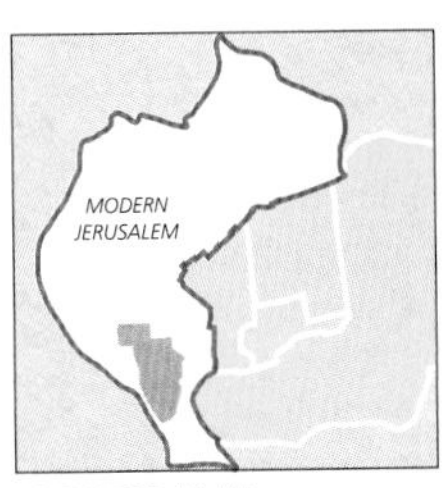

LOCATOR MAP
See Jerusalem Street Finder, map 1

★ Montefiore's Windmill
Montefiore meant Mishkenot Shaananim to be self-sufficient, hence a windmill to grind the settlement's own flour. Unfortunately, there was rarely enough wind to turn the sails.

Mishkenot Shaananim
In the earliest days, lodging in this block had to be offered rent-free in order to attract tenants. Now the place serves as a guesthouse for artists and writers. Saul Bellow, Marc Chagall and Simone de Beauvoir have all been accommodated here.

Bloomfield Gardens
Grassy parks fringe Yemin Moshe. Attractive in their own right, and dotted with ornament, such as the Lion Fountain (right), *the parks also afford great views across the valley to the Old City.*

YMCA ❶

26 King David St. **Map** 1 A4. ***Tel*** *(02) 569 2692.* *7, 8, 30, 38.* **Tower** *8am–8pm Mon–Thu, 8am–12:30pm Fri.*

Built in 1926–33 by Arthur Loomis Harmon, who also created New York's Empire State Building, Jerusalem's YMCA *(see p257)* is one of the city's best-known landmarks. It consists of three sections – the central body, dominated by a bell tower offering extraordinary views of the city, and the two side wings. The stone and wrought-iron decorative elements on the outside of the building, including the 5-m (16.5-ft) bas-relief of one of the six-winged seraphim described in the Old Testament (Isaiah 6: 2–3), reflect a stylized form of Oriental Byzantine design, combined with elements of Romanesque and Islamic art.

Yet the exterior, splendid as it is, does not prepare the visitor for the fabulously elaborate decor on the inside. Here design elements from three different cultures are woven through with symbols from the three main monotheistic religions. In the concert hall, the dome's 12 windows represent the 12 Tribes of Israel, the 12 Disciples of Christ and the 12 Followers of Muhammad, while depicted on the chandelier are the Cross, Crescent and Star of David. The entire decor has a kind of Art Deco gloss, while the ethos of its eclectic design is one of peace and tolerance between faiths and cultures.

The distinctive bell tower of Jerusalem's YMCA

King David Hotel ❷

23 King David St. **Map** 1 B4. ***Tel*** *(02) 620 8888.* *7, 8, 30, 38.*

Eye-catching not least for its pink stone walls and green windows, this impressive 1930s hotel *(see p258)* is a grandiose display of colonial architecture. It was designed by Swiss architect Emile Vogt for the Jewish-Egyptian Mosseri family.

Inside the elegant lobby of the King David Hotel

Inside, the spacious lobbies and public areas, with their discreet period wooden furnishings, reflect a sense of splendour from an altogether different era. The richly ornamental style includes Egyptian, Phoenician, Assyrian and Greek elements, as well as aspects of Islamic art. The hotel boasts an impressive list of former guests, including Winston Churchill and Haile Selassie, and for a long time, part of the British Mandate administration *(see p52)* was housed here. In 1946 it was the target of a bomb attack perpetrated by the Zionist paramilitary terrorist group Irgun, led by Menachem Begin *(see p53)*. It was rebuilt and the two top floors were added later. Nearby is the Alrov Mamilla complex, with exclusive shops and restaurants.

Jerusalem Time Elevator ❸

Beit Agron, 37 Hillel St. **Map** 1 A3. ***Tel*** *(02) 624 8381.* *10am–5pm Sun–Thu, 10am–2pm Fri, noon–6pm Sat.* **www**.time-elevator.co.il

On the southern edge of the neighbourhood of Nakhalat Shiva *(see p123)*, this is a themepark-style ride through 3,000 years of Jerusalem's often-turbulent history. The audience is belted into their seats and given surround-sound headphones for an audiovisual journey enhanced by computer-generated animation and other special effects. It begins in the times of King David and Solomon, and rattles through dramatic highlights of conquest, destruction, earthquake and fire, ending with the Six Day War of 1967 and reunification.

The special "motion" seats jolt and sway through the experience, which culminates in an "aerial" ride over the Jerusalem of today. The ride

The square-set form of the King David Hotel, the choice of many rich and famous visitors to Jerusalem

The Italian Synagogue and Museum of Italian-Jewish Art in a quiet square

lasts about 30 minutes, with shows at 40-minute intervals, and it is a useful introduction to the city's complicated chronology. The Time Elevator ride is not recommended if you do not enjoy rollercoasters.

Ben Yehuda and Nakhalat Shiva ❹

Map 1 A3. 🚌 *20, 23, 27.*

One of the popular streetside cafés and restaurants in Ben Yehuda

At the heart of modern Jerusalem are the pedestrianized precincts of Ben Yehuda Street and Nakhalat Shiva. They constitute one of the liveliest parts of the city, with shops, restaurants, street vendors and musicians coming together to create a rich and varied atmosphere. In the minds of local people, Ben Yehuda Street and Nakhalat Shiva are the embodiment of secular Jerusalem. The contrast with the Orthodox city, just a short distance to the north in Mea Shearim *(see p125)*, is clear.

Ben Yehuda Street was built in the 1920s, and has since been the traditional meeting place for Jewish intellectuals, politicians and journalists. South of Ben Yehuda Street is a series of narrow lanes, with low houses and connecting courtyards. These are collectively known as Nakhalat Shiva, meaning "the Domain of the Seven", which refers to the seven families who built them. Dating back to 1869, this area was the third Jewish residential quarter to appear outside the Old City walls. Despite being threatened with demolition on more than one occasion, the area was finally renovated in the 1980s. Today it is filled with shops, workshops, bars, restaurants and cafés and is invariably busy until the early hours.

Other streets in this locality also have much to interest the visitor. Buildings of varied architectural styles reflect the diverse cultural influences that have shaped the city.

Italian Synagogue ❺

27 Hillel St. **Map** 1 A3. ***Tel*** *(02) 624 1610.* 🚌 *18, 21, 22, 30.* 🕒 *9am–5pm Sun, Tue & Wed, 9am–2pm Mon, 9am–1pm Thu & Fri.* ⬤ *Jewish hols.* 📷 ♿ **www**.jija.org

Originally a German college constructed in the late 19th century, this building now houses an 18th-century synagogue from Conegliano Veneto, near Venice in Italy. In 1952, with no more Jews living there, the synagogue had fallen into disuse. It was decided to dismantle the interior and bring it here. It is arguably the most beautiful synagogue in Israel, and on Saturdays and Jewish holidays the Italian-Jewish community worships here. The building also houses the Museum of Italian-Jewish Art, which has some fascinating items, such as medieval ritual objects. On the lower floor is the Centre of Studies on Italian Judaism and a library on the same subject.

Ticho House ❻

9 Ha-Rav Kook St. **Map** 1 A2. ***Tel*** *(02) 624 5068.* 🚌 *13, 18, 20.* **Museum** 🕒 *10am–5pm Sun, Mon, Wed & Thu, 10am–10pm Tue, 10am–2pm Fri.* ⬤ *Jewish hols.* 📷

Built in the 19th century as the luxurious residence of a wealthy Jerusalem family, this is one of the city's loveliest examples of an Arab mansion. Its large central drawing room is the focal point of both the architecture and the social life of the building. In the early 20th century the house was bought by Dr Abraham Ticho, a famous Jewish ophthalmologist who used to give the poor free treatment, irrespective of their ethnic origin or religion. Dr Ticho's wife, Anna, who grew up and studied in Vienna, was an artist. By day the house was a clinic and by night it was the centre of Jerusalem's social and intellectual life.

Nowadays the house is administered by the Israel Museum *(see pp132–7)*, to which Anna Ticho left more than 2,000 watercolours and drawings. Some of these are exhibited here. The house also has a charming café overlooking a lovely garden.

View over the beautiful garden at the back of Ticho House

City Hall 7

Jaffa Rd. **Map** 1 B3. ***Tel*** *(02) 629 7777 or (02) 629 6666.* *6, 13, 18, 20.* *8:30am–4pm Sun–Thu.* *10am Mon (in English).*

Completed in 1993, the City Hall complex is sited just outside the Old City walls, where Jewish West Jerusalem meets Arab East Jerusalem. Its architecture displays an appropriate spirit of synthesis – the complex includes ten renovated historical buildings, along with two modern blocks that refer subtly to historical models (for example, the banding of different coloured stone echoes the Mameluke buildings of the Old City).

One of the renovated buildings, on Jaffa Road, is the old City Hall. It is still pocked with bullet holes from its days as a frontline Israeli army post when, between 1948 and 1967, the city was divided *(see p53)*.

City Hall, seen through the palms of Safra Square

Russian Compound 8

1 Mishol Hagevura St. **Map** 1 B3. *13, 18, 20.* **Underground Prisoners' Museum 1917–48** ***Tel*** *(02) 623 3166.* *9am–5pm Sun–Thu.* ***Russian Cathedral****: 9am–1pm Tue–Fri, 9am–noon Sat & Sun.*

The Russians were some of the first people to settle outside the Old City in the 19th century *(see p113)*. The process began around 1860 when a few acres of land were acquired a short distance outside the city walls. The Russians built a self-contained compound to provide lodgings for the city's growing number of Russian pilgrims, and erected a cathedral for services. Consecrated in 1864, the Cathedral of the Holy Trinity is fashioned in an unmistakably Muscovite style, with eight drums topped by green domes. Across the plaza, under a pavement grille, is what is known as Herod's Column, a 12-m (40-ft) stone pillar, which historians believe is from the Byzantine period or was intended for the Second Temple before it cracked and was abandoned.

These days the Russians own only the cathedral, as many of the other buildings belonging to the compound were sold off by the Soviet Union in exchange for shipments of Israeli oranges. The building with the crenellated tower – the grandest of the former pilgrims' hostels – is now home to the Agriculture Ministry. The street on which it stands, Heleni ha-Malka, is one of the city's nightlife centres, filled with bars and cafés.

The former women's hostel, behind the cathedral, houses the **Underground Prisoners' Museum 1917–48**, which is dedicated to Jewish underground movements, some members of which were jailed here during the British Mandate *(see pp52–3)*.

The Cathedral of the Holy Trinity, in the Russian Compound

Ha-Neviim Street 9

Map 1 B2. *1.*

One of the oldest streets outside the Old City, Ha-Neviim (Street of the Prophets) marks the dividing line between the religious and secular halves of modern Jerusalem (ultra-Orthodox Mea Shearim lies just to the north; the drinking and dining scene of the Russian Compound is to the south). Once a prestigious address, Ha-Neviim is lined with some grand buildings. At No. 58 is Thabor House, the self-designed home of Conrad Schick, a German who arrived in the Holy Land a Protestant missionary and became the city's most renowned architect of the late 19th century. The house now belongs to the Swedish Theological Institute, but visitors can admire the eccentric fortress-like main gate. Someone will usually answer the bell and admit the curious into the courtyard to admire the building's façade, complete with embedded archaeological finds.

Royal lion above the door, Ethiopian Church

A few steps west at No. 64 is the house once occupied by the Victorian painter

William Holman Hunt *(see p33)*. It is now a private residence and closed to the public. A couple of minutes' walk to the north, along narrow, leafy Etyopya Street, is Ben Yehuda House, named after the man responsible for reviving popular usage of the Hebrew language. This was his residence in the early years of the 20th century.

A little further up the lane is the striking, round form of the Ethiopian Church, which sits in beautifully tended gardens. It was built between 1873 and 1911, and is modelled after churches in Ethiopia, with its sanctuary clearly separated from the main body of the church. Just five minutes' walk away, back on Ha-Neviim Street, the Ethiopians also have their consulate. It is notable for a vivid blue and gold mosaic on the façade depicting the Lion of Judah.

Italian Hospital ⑩

Corner of Ha-Neviim and Shivtei Yisrael Sts. **Map** 1 B2. *1, 50.* *to public.*

The grandest building of all on Ha-Neviim Street is the Italian Hospital. It was built just before World War I to underscore Italian presence in the Holy City, at a time when the colonial powers were using architecture to assert their influence and status. Designed by prolific architect Antonio Barluzzi, the hospital is clearly inspired by the Palazzo Vecchio in Florence. The building now houses the Ministry of Education.

The extravagant Italian Hospital

Mea Shearim, the heartland of Jerusalem's insular ultra-Orthodox community

Mea Shearim ⑪

Map 1 A1. *1.*

Possibly the most unusual district in all Jerusalem, Mea Shearim is a perfectly preserved, living model of 18th-century Jewish Eastern Europe. It is a quarter inhabited exclusively by ultra-Orthodox Jews, where the influence of the outside world is kept to an absolute minimum. Dress is traditional in the extreme; many men wear black stockings and long black coats, and women keep their hair covered beneath a snood. The streets either side of main Mea Shearim Street are narrow alleyways, which squeeze between long, narrow two-storey dwellings, occasionally opening out into washing-strewn communal courtyards. The area is completely self-contained, with its own bakeries, markets, synagogues and, although no longer in use, its own huge cistern.

Mea Shearim was founded in the late 19th century and built in three stages, to a design by Conrad Schick, for Jews from Poland and Lithuania. Until well into this century the quarter was shut off from the rest of the city each night by six gates.

The gates are gone but visitors should bear in mind that this is still a very insular community. Skirts should reach below the knee, and men must not wear shorts or T-shirts. Discretion is advised when taking photographs.

Northwest of Mea Shearim is the Bukharan Quarter, founded in the late 19th century by wealthy Central Asian Jews. Traces of its former grandeur remain in some elegant, if dilapidated, mansions.

ULTRA-ORTHODOX JEWS

The life of the ultra-Orthodox *(haredim)* is grounded in rigorous observance of Judaic law and study of the Torah. Their lifestyle involves an uncompromising rejection of modern life and all its trappings, which means no television, no cars and minimum intrusion by technology. The ultra-Orthodox live and dress strictly according to traditions practised in Eastern Europe several centuries ago. This lifestyle means that they segregate themselves from less observant Jews. More radical factions are opposed to the common use of Hebrew, the "Holy tongue", and instead speak Yiddish; some do not recognize the State of Israel or its laws, even refusing to pay taxes. They claim that there can be no true Jewish state until the coming of the Messiah.

Ultra-Orthodox Jews dressed in everyday attire

Solomon's Quarries ⓬

Sultan Suleyman St. **Map** 4 D1. *1. 9am–4pm Sun–Thu, 9am–2pm Fri.*

This is an enormous empty cave stretching under the Old City, with its entrance at the foot of the wall between Damascus and Herod's gates. Despite the popular name, historians are not convinced that the cave has any connection with Solomon, but it is likely that Herod took stone from here for his many building projects, including his modification of the Second Temple.

The quarry is also known as Zedekiah's cave, after the last king of Judaea who, legend has it, hid here during the Babylonian conquest of Jerusalem in 586 BC.

Tourists visiting the ancient Garden Tomb in its attractive setting

Garden Tomb ⓭

Conrad Schick St. **Map** 3 C1. ***Tel*** *(02) 627 2745. 1, 3. 9am–noon, 2–5:30pm Mon–Sat.* **www**.gardentomb.org

Towards the end of the 19th century the British general, Charles Gordon, of Khartoum fame, was visiting Jerusalem and started a dispute among archaeologists. He argued that this skull-shaped hill was the Golgotha referred to in the New Testament (Mark 15: 22) and that the real burial site of Jesus Christ was here and not at the Holy Sepulchre *(see pp92–5)*. Excavations carried out in 1883 did in fact unearth some ancient tombs, but further study found them to date back to the 9th–7th century BC, with an entirely different configuration from those in use in Christ's time. However, regardless of its authenticity, this place is well worth a visit if only for the lovely garden.

The simple Neo-Romanesque chapel at St Etienne Monastery

St Etienne Monastery ⓮

Nablus Rd. **Map** 1 C2. ***Tel*** *(02) 626 4468. 23. open all day; ring the bell.*

The name of this site relates to the belief that in AD 439 Cyril of Alexandria interred the remains of St Stephen (St Etienne in French), the first Christian martyr, in a basilica built on this spot. The basilica was destroyed by the Persians in AD 614, and a subsequent 7th-century chapel on the same site was also destroyed, this time by the Crusaders holding Jerusalem, who feared Saladin would use it as a base for assaults on the city.

The present monastery was built between 1891 and 1901 by the French Dominicans. Its eclectic design includes an Oriental tower, Romanesque walls and Neo-Gothic flying buttresses. Within are remains of the mosaic floor of the original Byzantine church, as well as the Ecole Biblique, the Holy Land's first school of biblical archaeology.

St George's Cathedral ⓯

30 Nablus Rd. **Map** 1 C1. ***Tel*** *(02) 627 1670. 6, 23. not generally open for visitors so call first.*

This Archetypal Middle England church, with its pretty, cloistered courtyard and connotations of vicars, tweeds and cucumber sandwiches, stands in startling contrast to the chaotic Arab streets of its East Jerusalem neighbourhood.

The cathedral dates from 1910 and is named for the patron saint of England, who was actually a Palestinian conscript in the Roman army, executed in AD 303 for tearing up a copy of the emperor Diocletian's decree forbidding Christianity. He is supposedly buried at Lod (ancient Lydda), now better known as the site of Ben Gurion airport.

In World War I the cathedral was the local headquarters of the Turkish army, and the 1917 truce sanctioning British presence in Palestine was signed in the bishop's quarters.

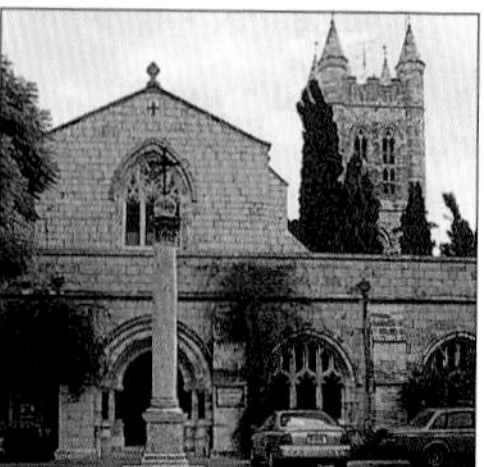

St George's Cathedral, part of Jerusalem's colonial heritage

Kings' Tombs ⑯

Salah ed-Din St. **Map** 1 C1. 23.
8am–5pm Mon–Sat.

Despite the name, this single- but elaborate tomb is thought to have been that of Queen Helena of Adiabene. In the 1st century AD she converted to Judaism and moved to Jerusalem from her kingdom in Mesopotamia. The tomb was named by early explorers who believed that the magnificent tomb housed members of the dynasty of David. A small entrance leads down into a dimly lit maze of chambers with stone doors.

The tomb is one of the places of interest on the East Jerusalem walk *(see pp146–7)*.

Well-worn steps leading to the deceptively named Kings' Tombs

American Colony Hotel ⑰

23 Nablus Rd. **Map** 1 C1.
Tel *(02) 627 9777.* 23.

This elegant hotel *(see p258)* built in 1865–76 has long been a favourite of diplomats and journalists. It started life as the home of a rich Turkish merchant. The name American Colony came about in the late 19th century when Anna and Horatio Spafford of Chicago bought the building and made it the centre of an American religious community dedicated to good works. When the community broke up in the early 20th century, a Baron Ustinov, related to the actor Peter Ustinov, suggested converting the building to accommodate pilgrims to the Holy Land. Soon after, it was turned into a beautiful hotel, which it remains today. If you cannot afford to stay here, it is definitely worth coming for lunch, taken out in the tree-shaded courtyard.

The Rockefeller Museum courtyard

Rockefeller Museum ⑱

27 Sultan Suleyman St. **Map** 2 D2.
Tel *(02) 628 2251.* *1, 2.*
10am–3pm Sun, Mon, Wed & Thu, 10am–2pm Sat.

This museum was made possible by a substantial financial gift made in 1927 by the American oil magnate John D Rockefeller. British architect Austin Harrison designed the building along Neo-Gothic lines. It is vaguely reminiscent of the Alhambra in Spain and runs around a central courtyard. Constructed from the white stone typical of Jerusalem buildings, the Rockefeller has Byzantine- and Islamic-type decorative motifs. It was once one of the most important museums in the Middle East and the first to make a systematic collection of finds from the Holy Land. These days, it is a branch of the Israel Museum *(see pp132–7)*, but still houses a very impressive collection.

Among its many remarkable objects are the stuccowork from Hisham's Palace in Jericho, beams from the Holy Sepulchre church and wooden panels from El-Aqsa mosque. Other exhibits worth seeing include a fascinating portrait modelled on an 8,000-year-old cranium discovered in Jericho; a lovely Bronze Age bull's head; a Canaanite vase in the shape of a human head; sculptures from the time of the Crusades; and Hellenistic and Roman objects found in Judaean desert caves. The museum also holds a number of the Dead Sea Scrolls *(see p137)*.

The delightfully secluded courtyard of the American Colony Hotel

FURTHER AFIELD

Since the creation of the state of Israel in 1948, the boundaries of Jerusalem have greatly expanded in all directions. The city has also been endowed with a great many significant new buildings. Two stand out as being of particular importance: the Israel Museum, a world-class institution that incorporates several collections of priceless treasures, including the famous Dead Sea Scrolls; and the Knesset, the seat of national government.

Another cornerstone in the psyche of Israeli society is Yad Vashem, the moving – and, in parts, harrowing – memorial complex that honours the more than six million Jews who died at the hands of the Nazis during the Holocaust. The site of this memorial is Mount Herzl, named after Theodor Herzl, the founding father of Zionism *(see p51)*. The grassy slopes here are also home to an extensive military cemetery, in which many figures of national importance are buried.

A memorial statue at Yad Vashem

As Jerusalem has expanded, what, not too long ago, were small, isolated villages are now virtually suburbs of the city. They have not, however, lost their character. Places such as Ein Kerem, nestled in the valley below Mount Herzl, and Abu Ghosh, further to the north-west, have a great deal of rural charm, as well as several attractive religious buildings linked with biblical events.

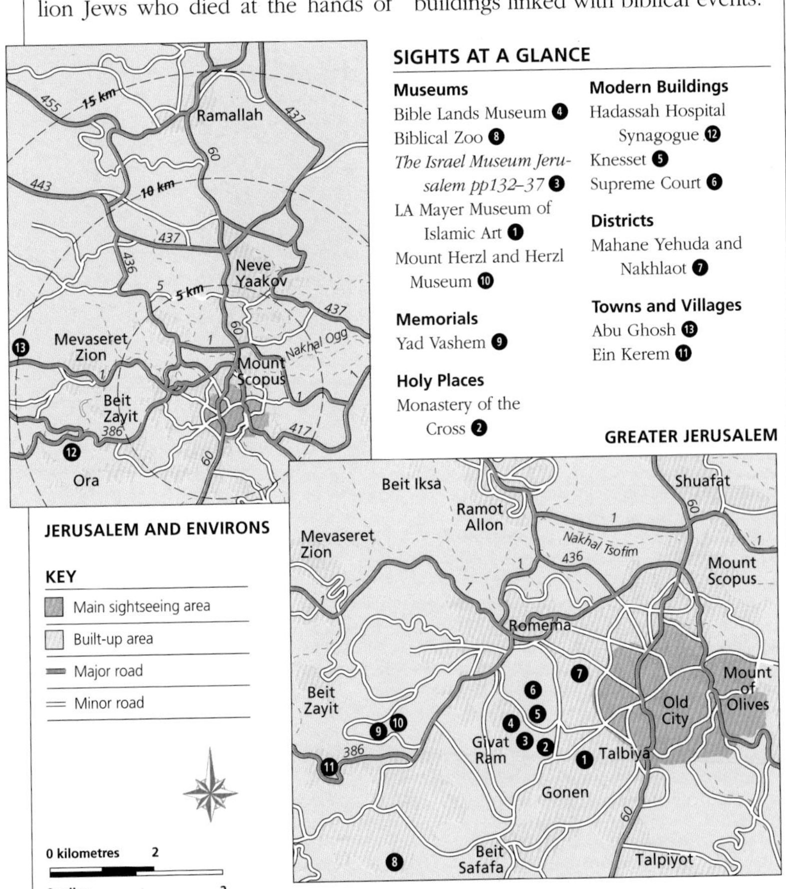

SIGHTS AT A GLANCE

Museums
Bible Lands Museum 4
Biblical Zoo 8
The Israel Museum Jerusalem pp132–37 3
LA Mayer Museum of Islamic Art 1
Mount Herzl and Herzl Museum 10

Memorials
Yad Vashem 9

Holy Places
Monastery of the Cross 2

Modern Buildings
Hadassah Hospital Synagogue 12
Knesset 5
Supreme Court 6

Districts
Mahane Yehuda and Nakhlaot 7

Towns and Villages
Abu Ghosh 13
Ein Kerem 11

◁ Crusader church complex at Abu Ghosh, 13 km (8 miles) west of Jerusalem

The refectory at the Monastery of the Cross

LA Mayer Museum of Islamic Art ❶

2 Ha-Palmakh St, Talbiya. ***Tel*** *(02) 566 1292. 13. 10am–3pm Sun, Mon & Wed, 10am–7pm Tue & Thu, 10am–2pm Fri, 10am–4pm Sat.* **www**.islamicart.co.il

While the cream of Islamic artifacts collected in the Holy Land are to be found in the Rockefeller Museum *(see p127)* and the Museum of Islamic Art on the Haram esh-Sharif *(see p70)*, this modern, purpose-built museum offers a beautifully presented collection of pieces from the greater Islamic world. There are especially attractive examples of Persian tiling and Indian Moghul miniatures, plus an Arabic calligraphy section.

Monastery of the Cross ❷

Shalom St, Neve Granot. ***Tel*** *(052) 221 5144. 32. 10am–4pm Mon–Sat.*

Stranded in the middle of a large area of scrubland, ringed at its outer perimeters by main roads and modern buildings, this solitary Byzantine monastery has the look of a place that time forgot and urban planners ignored. Its high, buttressed walls emphasize still more its seclusion and reflect its once precarious position outside the Old City.

There was a church here in the 5th century, but it was destroyed by the Persians in 614. Part of its mosaic floor can still be seen on one side of the main altar in the present church. The monastery which exists today was built in the 11th century by monks from Mount Athos, with financial backing from King Bagrat of Georgia. According to tradition, it marks the spot where the tree grew that was used to make Christ's cross.

In the 13th century the Georgian poet Shota Rustaveli lived here and commissioned the frescoes in the main church. They were repainted in the 17th century respecting the original style.

By the 14th century the monastery had become the centre of Jerusalem's Georgian community and a major centre of Georgian culture in the region. Gradually, however, their standing declined and by 1685 the monastery had been taken over by the Greek Orthodox Patriarchate.

The church is largely in its original, 11th-century form, while many other parts of the complex have been altered or added to. The courtyard and the late Baroque bell tower display clear signs of 19th-century changes. In the late 1990s large-scale restoration was undertaken. The simple dome is one of the church's most beautiful features. Also remarkable are the frescoes, which show an unusual combination of Christian, pagan and worldly images. Visitors are permitted to wander freely around the complex. Particularly evocative of monastic life are the refectory on the upper floor and the kitchen.

Israel Museum ❸

See pp132–7.

Bible Lands Museum ❹

25 Avraham Granot St, Givat Ram. ***Tel*** *(02) 561 1066. 9, 17, 24, 99. 9:30am–5:30pm Sun–Tue & Thu, 9:30am–9:30pm Wed, 9:30am–2pm Fri & eves of Jewish hols. Sat & Jewish hols. English-speaking guides available.* **www**.blmj.org

Opposite the Israel Museum is this rather unremarkable building which houses an outstanding collection of archaeological finds that reflect the different cultures of the Holy Land region in biblical times. The museum was inaugurated in 1992 with the private collection of Elie Borowski, a passionate scholar of ancient Middle Eastern civilizations. The collection features many finely crafted objects from ancient Egypt, Syria, Anatolia, Mesopotamia and Persia. Among these are a great number of artifacts that shed light on the culture of the Mesopotamian region in

Babylonian tablet, Bible Lands Museum

The Bible Lands Museum, covering the early history of the Middle East

The sculpted menorah near the entrance to the Knesset

the millennia before the Christian era. The many fascinating and unique objects include ancient inscriptions, jewellery, mosaics, seals, ivory carvings and scarabs.

The exhibits are displayed in a way that enables the visitor to build a clear and illuminating picture of the cultural context in which the biblical texts were written. The items are arranged according to both chronology and region. The result is a clear illustration of the way in which different cultures influenced each other and new societies evolved.

Knesset ❺

1 Kaplan, Givat Ram. ***Tel*** *(02) 675 3333. 🚌 9, 24, 99. ◯ 8:30am–2pm Sun & Thu. 🅖 compulsory (ring in advance to book; bring passport).*

The Knesset (Assembly) is the seat of the Israeli Parliament. It takes its name from the Knesset ha-Gedola (Great Assembly) of 120 men that governed the political and civic life of Jews in the Second Temple period *(see p42)*. The building, inaugurated in 1966, was designed by Joseph Klarwin. His design makes use of classical elements and is inspired by the Parthenon in Athens and various reconstructions of the Temple.

Opposite the entrance is a large, seven-branched menorah (candelabrum), symbol of the State of Israel. It is the work of British sculptor Benno Elkan and was a gift from the British parliament. The relief work on its branches depicts crucial moments in Jewish history and is accompanied by biblical quotations. Nearby is a monument with an eternal flame, commemorating the dead of the Holocaust and Israel's wars *(see pp53–5)*.

The reception area inside the Knesset was designed and decorated by the Russian-Jewish artist Marc Chagall *(see p33)*. It is adorned with his mosaics and a triple tapestry which depicts the creation of the world, the exodus of the Israelites from Egypt and the city of Jerusalem. The main chamber ends in a stone wall that is a very clear reference to the Western Wall *(see p85)*.

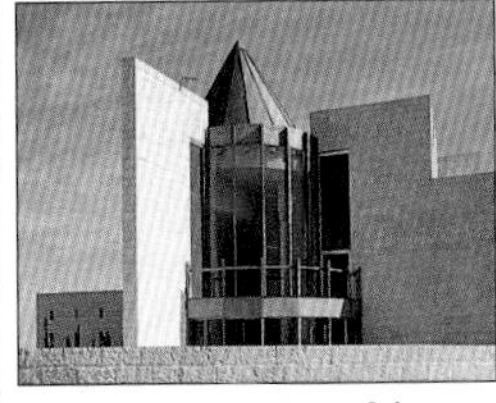

The Supreme Court, one of the city's architectural highlights

Supreme Court ❻

Shaarei Mishpat St, Givat Ram. ***Tel*** *(02) 675 9612. 🚌 9, 24, 99. ◯ 8:30am–2:30pm Sun–Thu. 🅖 noon daily in English (groups must ring in advance to book).*

In the absence of a formal constitution, Israel's Supreme Court plays a pivotal role in the lives of ordinary citizens. Its significance is reflected in the building's design – by Ram Karmi and Ada Karmi-Melamed – which manages to depict the concept of justice in architectural terms. The two copper pyramids on the roof are powerful symbols of the immutable nature of the principles of law. The long sweeping stairway seems to represent the accessibility of the law to ordinary people, and at the top it offers an all-embracing view of Jerusalem.

Motifs from the past, such as the Islamic elements in the inner courtyard and the Byzantine-era mosaic outside the entrance, recall Israel's cultural and historical influences. They are given a modern context to link the past with the present and reflect the universality of justice.

Mahane Yehuda and Nakhlaot ❼

🚌 6, 8, 13, 14, 18, 21.

The district of Mahane Yehuda, which means Field of Judah, was built in 1929 to house Jewish immigrant workers. It is famous for its vibrant and very colourful market, selling mainly foodstuffs. It is also home to a large number of popular local restaurants, which specialize in Middle Eastern salads and kebabs. To the south of Mahane Yehuda is the older district of Nakhlaot. This lively, warren-like jumble of low houses and narrow alleyways is fascinating to explore.

Displays of fruit and vegetables at the market in Mahane Yehuda

The Israel Museum, Jerusalem ❸

Built in 1965 on a ridge overlooking West Jerusalem, the Israel Museum contains some of the country's finest art and archaeology. It was designed by Israeli architects A Mansfeld and D Gad as a modernist reference to traditional Arab hilltop villages. A major renovation was completed in 2009, and the reorganized and expanded collections include synagogue interiors and the world-famous Dead Sea Scrolls.

***Apple Core* (1992), Claes Oldenburg**

★ Shrine of the Book
This innovatively designed underground hall houses the Dead Sea Scrolls. It is the most visited part of the museum (see pp136–7).

★ Beth Shean Mosaic
This 6th-century mosaic floor, from a synagogue at Beth Shean (see p185), *shows the Ark of the Covenant* (see p19) *flanked by two menorahs.*

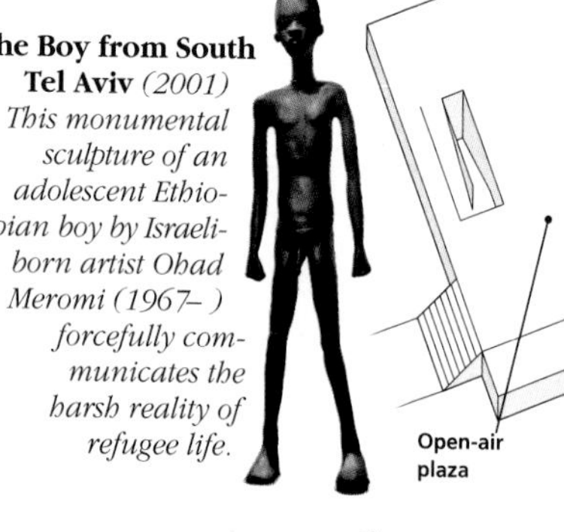

The Boy from South Tel Aviv *(2001)*
This monumental sculpture of an adolescent Ethiopian boy by Israeli-born artist Ohad Meromi (1967–) forcefully communicates the harsh reality of refugee life.

KEY TO FLOORPLAN

- Jewish Art and Life Wing
- Art collections
- Archaeology
- Temporary exhibitions
- Non-exhibition space

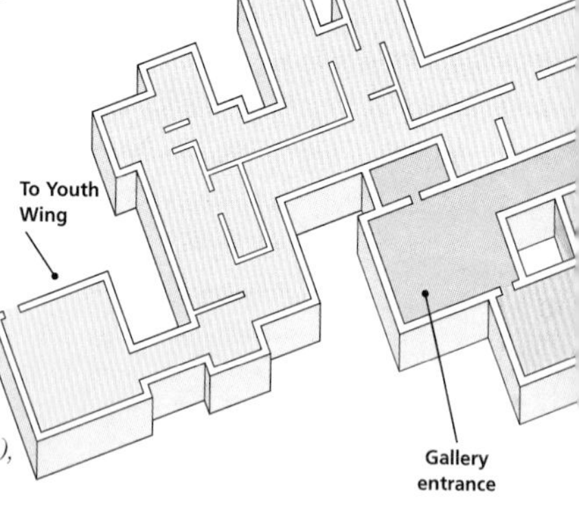

Billy Rose Art Garden
Woman Combing Her Hair *(1914), by Ukrainian-born Alexander Archipenko, is one of the art garden's striking sculptures.*

PLAN OF MUSEUM

KEY

- Entrance pavilion
- Main museum block
- Ruth Youth Wing
- Billy Rose Art Garden
- Shrine of the Book

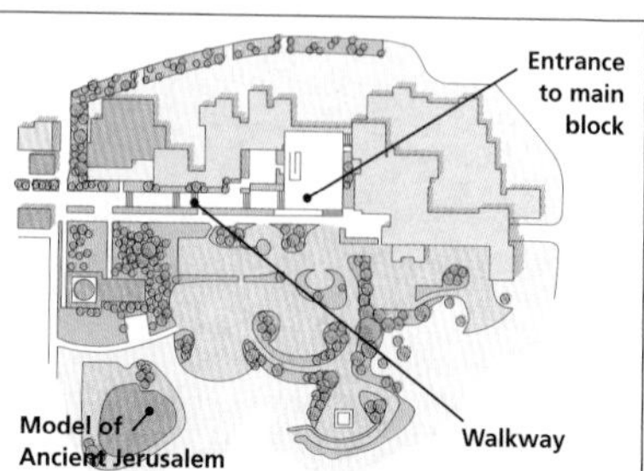

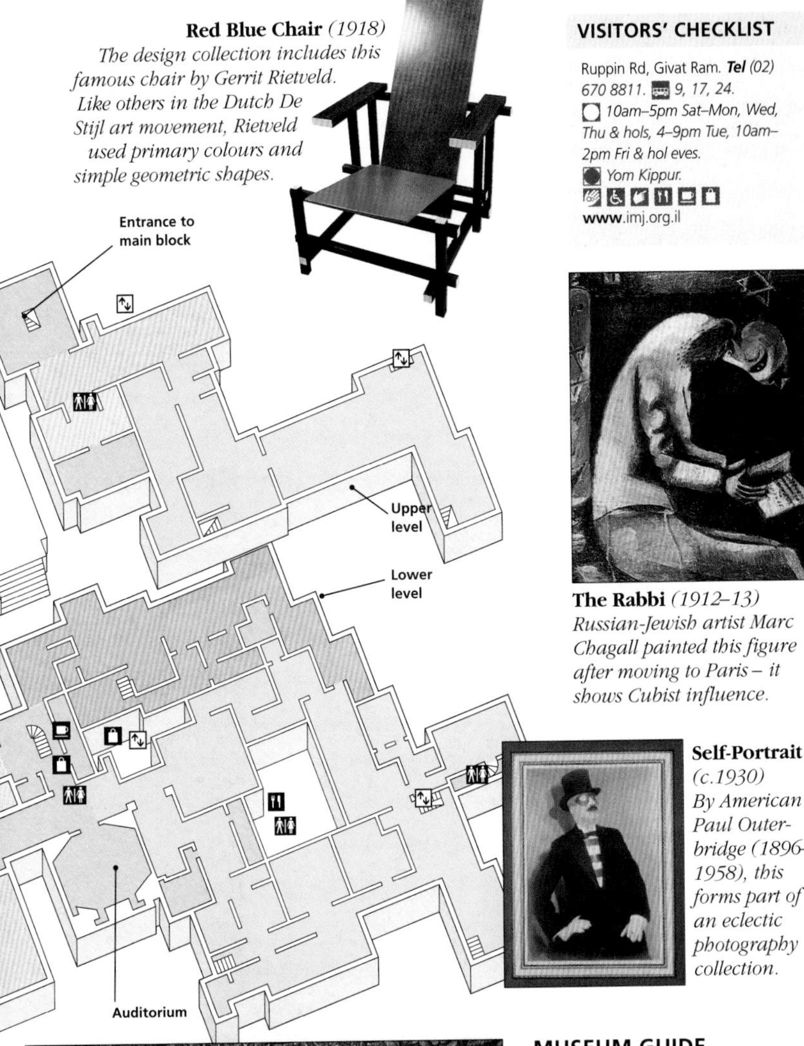

Red Blue Chair *(1918)*
The design collection includes this famous chair by Gerrit Rietveld. Like others in the Dutch De Stijl art movement, Rietveld used primary colours and simple geometric shapes.

The Rabbi *(1912–13)*
Russian-Jewish artist Marc Chagall painted this figure after moving to Paris – it shows Cubist influence.

Self-Portrait *(c.1930)*
By American Paul Outerbridge (1896–1958), this forms part of an eclectic photography collection.

VISITORS' CHECKLIST

Ruppin Rd, Givat Ram. ***Tel*** *(02) 670 8811.* 9, 17, 24.
10am–5pm Sat–Mon, Wed, Thu & hols, 4–9pm Tue, 10am–2pm Fri & hol eves.
Yom Kippur.
www.imj.org.il

MUSEUM GUIDE

The museum's 20-acre campus has extensive gallery space for archaeology, fine arts, and Jewish art and life collections. It also includes a large outdoor sculpture garden, a Youth Wing, which organizes educational programmes and exhibitions, and the Shrine of the Book and Model of the Second Temple of Jerusalem complex.

★ Horb Synagogue
This richly painted synagogue interior from Horb in Germany dates from 1735. The decoration includes flowers, animals and excerpts from traditional prayers.

STAR EXHIBITS

★ Shrine of the Book

★ Beth Shean Mosaic

★ Horb Synagogue

Exploring the Israel Museum

Thanks to its wide variety of sources, the collection is extraordinarily eclectic. Its core was inherited from the Bezalel School and Museum (Israel's first arts academy) and the Israel Department of Antiquities, and this has been supplemented by gifts, loans and acquisitions from around the globe. The biggest draw, though, for most visitors is the Shrine of the Book, which houses the Dead Sea Scrolls *(see pp136–7)*.

Byzantine-era oil lamp

JEWISH ART & LIFE WING

The museum's collection of Judaica and Jewish Ethnography provides a comprehensive picture of the Jewish cultural tradition. The collection spans the period from the Middle Ages to the present, and has exhibits from as far afield as Spain and China. Five main sections integrate the sacred and secular dimensions of Jewish life from different cultures. Among the most precious objects are the medieval illuminated manuscripts. These include a 14th-century German *Haggadah* (the story read at Passover of the Israelites' liberation from Egypt) and the Rothschild Miscellany, a 15th-century collection of biblical, legal and other pieces. Elaborate silverwork includes *hadassim* (spiceboxes used during the ceremony of separation between the Sabbath and the start of the week) and the *rimonim* (pomegranates that decorate Torah scrolls in the synagogue). Another highlight is the large collection of *Hannukkiot* – the oil lamps that are lit for Hanukkah *(see p39)*. There are also four beautiful, complete synagogue interiors, from Italy, Germany, India and Suriname. The daily life of Jewish communities from around the world is also represented in textiles, clothing, jewellery, reconstructions of rooms and ritual articles connected with life events such as birth, circumcision and marriage.

The Rothschild Miscellany

Jeanne Hebuterne, Seated (1918), by Amedeo Modigliani

ART COLLECTIONS

The museum's various art collections cover a wide range of periods and artistic disciplines. Visitors can take in Chinese porcelain, African figurines, Impressionist masterpieces and even an entire 18th-century French salon.

The modern art collection has international works from the 1890s to the 1960s. These include paintings by figures such as Gauguin, Cézanne, Chagall, Matisse and Modigliani. Twentieth-century sculpture is also represented, both here and outdoors in the Billy Rose Art Garden *(see*

The Rothschild Room, an 18th-century Parisian salon donated by Baron Edmond de Rothschild

p136). Other rooms are devoted to design, architecture and contemporary art.

One of the largest collections of Israeli art in the country is also exhibited here on both floors. It begins with paintings and drawings produced in the 19th century, at the beginning of Jewish resettlement *(see p51).* The 1920s and 30s are represented by figurative pieces by artists such as Reuven Rubin and Yitzhak Danziger. The contemporary Israeli art on display mirrors, and sometimes anticipates, tendencies seen elsewhere in the world.

Other rooms are devoted to prints and drawings, Old Master paintings (including a large work by Poussin depicting the sacking of the Second Temple; *see p45*), Islamic and East Asian art and the art of Africa, Oceania and the Americas. The distinguished Levine Photography collection comprises 125 works and builds on the museum's long history of collecting photographs.

Anthropoid sarcophagi, a highlight of the archaeology collection

ARCHAEOLOGY

The archaeology collection constitutes the largest section of the museum. Most pieces are on loan from the Israel Antiquities Authority and come from excavations carried out all over the country, which has the highest concentration of digs in the world. The digs cover a vast period of history – from as far back as 1.5 million BC – and have revealed artifacts from an impressive number of civilizations, from Palaeolithic flint utensils, through Canaanite and Israelite figurines, to Byzantine mosaics and Islamic jewellery. The museum's collection represents most aspects of this cultural spectrum, and visitors will require at least two hours to fully appreciate the range of pieces on display.

The artifacts are arranged chronologically within the renovated gallery, as six "chapters" of an archaeological timeline. Objects to look out for in the first section (Palaeolithic to Chalcolithic periods, 1.5 million–3500 BC) include the jewellery and sculpted figures of the Natufian culture (10th–9th millennium BC), the 6,000-year-old, house-shaped ossuaries at the end of the first gallery and the elegant copperware of the so-called Judaean Desert Treasure (5th millennium BC). Highlights from the Canaanite Period (3500–1200 BC) are the sophisticated gold jewellery and the anthropoid sarcophagi found in a cemetery at Deir el-Balah, in the Gaza Strip.

The Israelite Period (1200–586 BC) starts with the rise of the Israelites in the region and ends with the destruction of Solomon's Temple. Look out for the beautiful Philistine pottery, the ivory pomegranate inscribed with ancient Hebrew (believed to be the only object ever found relating to worship in Solomon's Temple) and the priestly benediction written on a tiny silver amulet – the earliest known fragment of biblical text (7th century BC).

Mosaic from floor of 6th-century AD synagogue at Gaza, showing King David playing the lyre

Finds from the next 300 years are relatively scarce but the Hellenistic, Roman and Byzantine periods (332 BC–AD 636) offer fascinating objects, such as the sarcophagi and ossuaries from various Jewish catacombs, the bronze statue of the emperor Hadrian and the beautiful mosaics from Tsipori (Sepphoris), Kisufim, Gaza and Beth Shean.

In the last room are objects from neighbouring Middle Eastern and Mediterranean civilizations that had some bearing on the history of the Holy Land. The artifacts here include Egyptian cult and game objects, Assyrian and Babylonian reliefs, Greek vases and Roman jewellery.

Throughout the section are interesting models and reconstructions of some of the most important sites in this part of the world. The permanent exhibition is also flanked by temporary displays based on historical themes or particular archaeological sites.

JEWISH ART OF THE DIASPORA

During the many centuries of the Diaspora, Jews around the world directed their artistic talents primarily to ritual objects connected with the life cycle and synagogue liturgy. They produced fine examples of applied art, especially in the fields of gold- and silverware, other metalwork and manuscript decoration. Naturally, the motifs and techniques reflect the place and time in which the objects were produced, but many elements, both functional and iconographic, recur again and again. These recurring themes and local variations can be appreciated among the many exhibits in the museum's Judaica section.

18th-century silver spicebox from Germany

RUTH YOUTH WING

This section of the museum is devoted to interactive art activities. The idea behind it was to introduce children to art and culture. The largest of its kind in the world, the centre has now extended its reach to adults. With ten classrooms, an auditorium, library, recycling workshop and exhibition space, it provides a stimulating environment for children and adults to learn about creative processes. There are regular "hands on" exhibitions, art courses and summer schemes for all ages, as well as tours for groups with special needs.

Children participating in creative activities in the Ruth Youth Wing

BILLY ROSE ART GARDEN

The garden was designed by the American sculptor Isamu Noguchi. It is an extraordinary combination of elements from local history and landscape, motifs from the traditional Zen garden and significant works of modern sculpture. It is laid out as a series of semi-circular terraces echoing those made for centuries by farmers in the Judaean Hills. Indigenous plants such as olive trees, cypresses and rosemary bushes are dotted around the garden.

The garden offers an overview of sculpture through the 20th century. There are stunning early works by Rodin, Maillol, Picasso and Bourdelle. The curvaceous shapes in Henry Moore's pieces contrast with the angular composition of David Smith's *Cubi VI* (1963). Contemporary sculptures include James Turrell's intriguing installation with a large rectangular opening in the top for observing the sky, and Claes Oldenburg's "rotting" apple core, rich in symbolism and existential allusions.

SHRINE OF THE BOOK

Built to house the Dead Sea Scrolls and other important artifacts, the intriguingly shaped Shrine of the Book has become a symbol of the whole museum. The unusual design, by American architects F Kiesler and A Bartos, is inspired by the scrolls themselves. The distinctive dome is intended to imitate the lids of the jars in which the scrolls were found. Near the entrance is a black granite wall. The contrast between the black of the wall and the white of the dome is a reference to the decisive battle between the Children of Darkness and the Children of Light, described in the scroll known as the War Scroll. This final confrontation between good and evil would, the authors believed, herald the coming of the Messiah.

Inside, a long, subtly lit passageway, designed to evoke the catacomb-like environment in which the scrolls were found, has a permanent exhibition on life in Qumran at the time the scrolls were written. It leads into the main chamber under the dome. The imposing showcase directly beneath the dome contains a facsimile of the Great Isaiah Scroll, the only biblical book that survived in its entirety. Its 66 chapters were written on several strips of parchment, which were then sewn together, making it more than 7 m (23 ft) long. One of the surrounding display cases contains part of the real scroll. Also on show are the Psalms Scroll, 28 columns of text consisting of psalms, hymns and a prose passage about the psalms; the War Scroll; the Manual of Discipline; the Temple Scroll; and the 10th-century Aleppo Codex – not one of the Dead Sea Scrolls, but the oldest complete Bible in Hebrew.

Magdalena Abakonowicz's *Negev* (1987), Billy Rose Art Garden

On the Shrine's lower level are 2nd-century AD articles, such as keys and baskets, found in the Cave of Letters, south of Ein Gedi *(see p197)*. Adjacent to the Shrine of the Book is a Second Temple-era model of Jerusalem. Originally constructed on the grounds of the Holyland Hotel on the outskirts of the city, this large-scale model was relocated to the museum in 2006. It offers visitors a three-dimensional view of the landscape of Jerusalem during the 1st century. Mainly built from local limestone, the model covers almost one acre and was constructed at a scale of 1:50, with 2 cm of the model representing one metre of the city.

Symbolic clash of darkness and light at the Shrine of the Book entrance

The Dead Sea Scrolls

In 1947, a Bedouin shepherd, in search of a lost goat near the Dead Sea, entered a cave and discovered jars containing seven ancient scrolls. Over the next two decades fragments of some 800 more were found in 11 caves. At the same time, archaeologists, looking for signs of habitation, uncovered the nearby settlement of Qumran *(see p191)*. The scrolls had been written in the Late Second Temple period, between the 3rd century BC and AD 68. Some contain the oldest existing versions of biblical scriptures. Others are tracts on history, daily life and the messianic predictions of a Hebrew sect generally identified with the separatist and monastic Essenes. Since the discovery of the scrolls, their interpretation, the identity and mission of their authors and the significance of nearby Qumran have been the subject of passionate academic and theological debate.

Jar in which scrolls were found

The Shrine of the Book *is dominated by a dramatic display case, which contains a copy of the Great Isaiah Scroll. It was designed to look like the wooden rods around which the Torah scrolls are rolled for readings at synagogue services.*

Inkwell found at Qumran

The reconstruction *of thousands of scroll fragments is still being carried out by researchers hoping to unravel the mysteries surrounding the scrolls.*

The parchment *on which the scrolls were written was made from sheepskin. Inkwells found near a table at Qumran suggest a scriptorium – a room for copying manuscripts.*

The Great Isaiah Scroll *is the largest and best preserved of the scrolls. Written around 100 BC, it is 1,000 years older than the oldest biblical manuscript known before the finds at Qumran.*

Qumran was excavated *by Roland de Vaux, a French Dominican friar. He believed that the settlement was a communal retreat used by the Essenes.*

Biblical Zoo ❽

Manahat. ***Tel*** *(02) 675 0111. 26, 33, 99. 9am–5pm (7pm Jun–Aug) Sun–Thu, 9am–4:30pm Fri, 10am–5pm (6pm Jun–Aug) Sat; last entry 1 hour before closing.* **www**.jerusalemzoo.org.il

The Jerusalem Biblical Zoo, also known as the Tisch Family Zoological Gardens, is famous for its collection of wildlife featured in the Bible. This group of animals, many of which are no longer naturally present in the Holy Land, includes bears, lions, Arabian oryx and Nile crocodiles. There are also other endangered species from around the world. The zoo occupies an attractive site in the southwestern suburbs of the city. You can gain an overview of the zoo on a train ride around the grounds.

Yad Vashem ❾

Mount Herzl. ***Tel*** *(02) 644 3400. 13, 21, 23, 27. 9am–5pm Sun–Wed, 9am–8pm Thu, 9am–2pm Fri & hols.* **www**.yadvashem.org

Yad Vashem, meaning "a memorial and a name" (from Isaiah 56: 5), is an archive, research institute, museum and, above all, a monument to perpetuate the memory of the more than six million who died in the Nazi Holocaust. More than 20 monuments occupy this hillside site.

Entrance to Yad Vashem is along the Avenue of the Righteous Among Nations, which is lined with plaques bearing the names of Gentiles who helped Jews and, in doing so, put their own lives at risk. Some 23,000 people are recognized, including Oskar Schindler *(see p117)*. The avenue leads to the new Historical Museum, which was designed by Jewish architect Moshe Safdie. The museum is one long corridor, carved into the mountain, with 10 exhibition halls, each dedicated to a different chapter of the Holocaust. Its exhibits include some 2,500 personal items donated by survivors, adding a harrowing first-person dimension to the horrors that began with the rise of the Nazis in 1933 and culminated in the death camps.

The Hall of Remembrance beside the museum is a stark, tomb-like chamber that bears the names of 21 of the main camps on flat, black basalt slabs. At the centre of the vast chamber is a casket of ashes from the cremation ovens; above it is an eternal flame. The Hall of Names inside the Historical Museum records the names of all those Jews who perished, along with as much biographical detail as possible. Yad Vashem also has a museum of Jewish art and a visual centre where films related to the Holocaust may be viewed. Visitors must dress appropriately – no shorts or miniskirts.

Janusz Korczak Memorial, Yad Vashem

Memorial to the Victims in Camps, Yad Vashem

Grave of Rabin, Mount Herzl

Mount Herzl and Herzl Museum ❿

Mount Herzl. ***Tel*** *(02) 632 1515. 13, 17, 18, 20, 21, 23, 24, 26. 8:45am–3:45pm Sun–Thu, 9am–12:30pm Fri (11am winter).*

Mount Herzl (in Hebrew *Har Hertzel*) is a high hill north of central Jerusalem, named after Theodor Herzl, the man considered to have been the founder of Zionism *(see p51)*. The slopes serve as a large cemetary and Herzl's tomb lies at the top of the hill. At the entrance to the site is the Herzl Museum, which opened in 2005. It offers a crash course in Zionist history, with audiovisual presentations and recreations of the founding father's study and library. Mount Herzl is also the burial place of three of Israel's prime ministers and the country's presidents. It is also the site of Israel's main military cemetary.

Ein Kerem ⓫

7 km (4 miles) W of central Jerusalem. *17, 184.*

A picturesque village, Ein Kerem ("the vineyard spring") has strong biblical associations. According to Christian tradition, John the

Baptist was born and lived here. The village boasts several fine churches and monasteries connected with his life. Recognizable by its tall, thin tower, the Franciscan **Church of St John the Baptist** dates from the 17th century, but is built over the ruins of earlier Byzantine and Crusader structures. Steps inside the church lead down into a natural cave, known as the Grotto of the Nativity of St John, which tradition connects with the birth of the Baptist.

The other church of note is the two-tiered **Church of the Visitation**, completed in 1955 to a design by Antonio Barluzzi, architect of the Dominus Flevit Chapel *(see p113)* and the Chapel of the Flagellation *(see p64)*. It commemorates the Virgin Mary's visit to Elizabeth, mother of John the Baptist, who was then pregnant, an episode depicted in mosaic on the church's façade. Within is a natural grotto, in front of which are the remains of Roman-era houses. According to tradition, the grotto is where Elizabeth hid with her infant son to escape from the Massacre of the Innocents (the killing of all first-born sons, ordered by King Herod). The courtyard walls are lined with tiled panels inscribed with the *Magnificat* (Luke 1: 46–55), Mary's hymn of thanks, in 42 languages.

At the bottom of the hill below the church is a small, abandoned mosque. Beside it surfaces the spring (popularly known as the Spring of the Virgin) from which the village takes its name.

One of the other pleasures to savour in Ein Kerem is its tranquil, wooded, valley setting. This can be best appreciated on a beautiful scenic walk that starts beside the sculpture at the beginning of the access road to Yad Vashem, and winds through the trees.

Church of St John the Baptist, Ein Kerem

Hadassah Hospital Synagogue ⓬

Ein Kerem. ***Tel*** *(02) 677 6271.* *19, 27.* *8am–1pm & 2–3:30pm Sun–Thu.*

A splendid cycle of 12 stained-glass windows decorates the synagogue at the otherwise unremarkable Hadassah Hospital. The windows were created in 1960–61 by the Russian-Jewish artist Marc Chagall *(see p33)*, and installed the following year for the inauguration of the building. Each of the windows represents one of the 12 tribes of Israel (Genesis 49). Tradition associates each of the tribes with a symbol, a precious stone and a social role, and these elements are all represented in Chagall's imagery and choice of colour.

Several of the windows were damaged by shrapnel during the 1967 War *(see p54)* and had to be repaired by the artist. However, one of the windows (a green one) bears a small symbolic bullet hole in the lower half, deliberately left there as a testimony to the fighting.

Abu Ghosh ⓭

13 km (8 miles) W of central Jerusalem. *185, 186.*

This Arab village just north of the main Jerusalem-Tel Aviv highway was considered by the Crusaders to be Emmaus, where Christ appeared to two disciples in the days after his Resurrection. The beautiful Romanesque **Crusader Church** was built in the early 12th century by the Knights Hospitallers and stands almost complete in its original form. Its 12th-century frescoes are lovely, but in a poor state of repair. The adjacent early 20th-century monastery belongs to French Olivetan Benedictine monks, who produce pottery. Up on the hill above the village stands the **Church of Notre Dame de l'Arche de l'Alliance**, built in 1924 over the remains of a 5th-century church, whose mosaics are still visible. It is said to occupy the site of the house of Abinadab, where the fabled Ark of the Covenant *(see p21)* rested for 20 years (1 Samuel 7: 1–2) until David took it to Jerusalem.

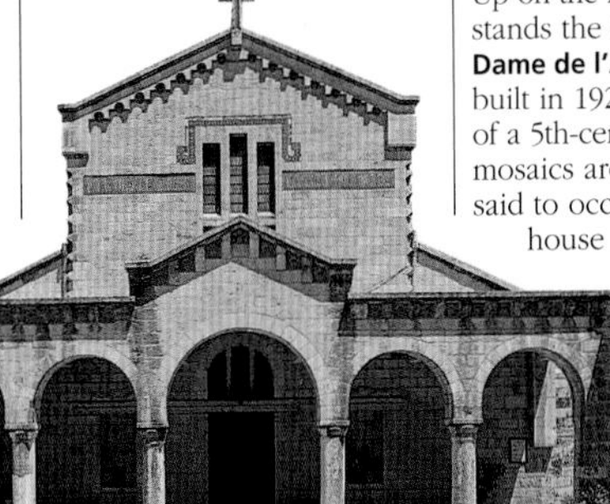

The modern Church of Notre Dame de l'Arche de l'Alliance, Abu Ghosh

THREE GUIDED WALKS

Jerusalem is a perfect city to explore on foot: it is small and compact, and there are plenty of sites to see and places to sit and rest. This is particularly true in the Old City, which, with the exception of just one or two roads, doesn't allow for motor vehicles at all, so dodging traffic is rarely an issue though pavements may be crowded. Most streets are simply too narrow and meandering for motorized traffic, and there are too many steps. It is a place perfectly described by the over-used adjective "labyrinthine"; a place in which getting lost is inevitable. However, this is no bad thing because wandering aimlessly around the Old City is a highly pleasurable activity. For that reason, we have avoided describing any walks within Jerusalem's ancient fortified walls.

Jerusalem shield at New City Hall

Instead, we suggest you get up on the walls themselves, which is something few visitors do, largely because they remain unaware that the opportunity exists.

Similarly, few visitors spend any time exploring the more modern parts of the city and so miss out on some attractive old quarters and some fine architecture. Much of this is non-indigenous, raised at the end of the 19th century, when the great powers of Europe were all vying for political influence in the Holy City. This was expressed through ostentatious examples of their own national architectures. Muscovite churches, English Gothic cathedrals, German hospices and Italian insurance offices all serve as reminders of the central role Jerusalem has always played in the Western consciousness.

CHOOSING A WALK

Three Walks

The routes of the three walks are marked on this map, which shows the main areas of Jerusalem.

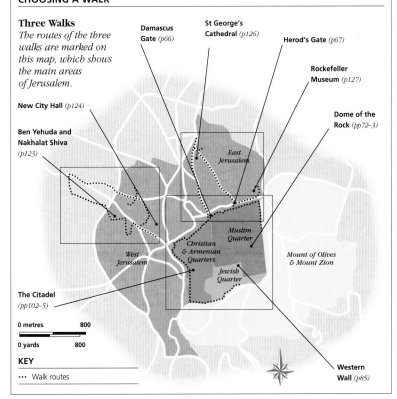

Damascus Gate *(p66)*

St George's Cathedral *(p126)*

Herod's Gate *(p67)*

Rockefeller Museum *(p127)*

New City Hall *(p124)*

Dome of the Rock *(pp72–3)*

Ben Yehuda and Nakhalat Shiva *(p123)*

East Jerusalem

Muslim Quarter

Christian & Armenian Quarters

West Jerusalem

Mount of Olives & Mount Zion

Jewish Quarter

The Citadel *(pp102–5)*

0 metres 800

0 yards 800

KEY

··· Walk routes

Western Wall *(p85)*

◁ **Snow covers graves on the Mount of Olives, which faces the Old City across the Kidron Valley**

A 90-Minute Walk around the Old City Walls

The Old City of Jerusalem may occupy a relatively small area geographically, but its compactness and uneven topography make it a frequently confusing place to explore. One good way to gain an overview is to take to the ramparts and view the crush of alleys, domes and towers from the top of the walls that enclose them. Visitors can walk along two sections of wall: from Jaffa Gate clockwise to Lions' Gate, and from Jaffa Gate anti-clockwise to the Dung Gate. The section between Lions' Gate and the Dung Gate is closed to the public. Many steep flights of steps mean that this is not a walk for the elderly or infirm.

The city walls wrap around the Tower of David

Clockwise from Jaffa Gate

Jerusalem's walls were built in the first half of the 16th century (in part on the line of earlier walls) on the order of the Ottoman sultan Suleyman the Magnificent. They are pierced by eight gates, of which seven remain in use. Until as recently as 1870, the gates were all closed from sunset to sunrise.

A section of the ramparts just east of New Gate ②

TIPS FOR WALKERS

Starting point: *Jaffa Gate.*
Length: *Jaffa Gate to Lions' Gate 1.5 miles (2.4 km); Jaffa Gate to Dung Gate 0.75 miles (1.2 km).*
Open: *summer: 9am–5pm daily; winter: 9am–4pm Sat–Thu, 9am–2pm Fri (south side open Sat only). Admission fee.*
Stopping-off points: *There are several small cafés on Omar ibn el-Khattab Square, just inside the Jaffa Gate. Otherwise, when you descend at Lions' Gate, walk west along the Via Dolorosa and then left onto El-Wad Road for Abu Shukri, which serves the best houmous in town.*

Start the walk by climbing the steps that are immediately inside the **Jaffa Gate** ① *(see p100)*, to your left as you enter the Old City. After paying admission, you pass through a gate and ascend a steep flight of steps leading to the top of the gatehouse. Heading north brings you to the first of some 35 watchtowers that punctuate the circuit of the walls. This one has a raised platform which allows walkers to step up for a view of the large new shopping and office development that is currently taking shape outside the city walls. Looking into the Old City, you will see the backs of buildings belonging to the Latin Patriarchate, the centre of Roman Catholicism in Jerusalem.

A short distance on and you'll notice that the third watchtower along has been reinforced with side walls; this was done by the Jordanians when they were in occupation of the Old City between 1948 and 1967, and Jerusalem was divided between Arabs and Jews. After skirting around three sides of a crescent-topped dome, the ramparts pass over **New Gate** ②. This was added in 1889 to allow pilgrims in the compounds outside the walls direct access to the Christian Quarter.

From here the ramparts drop, following the slope of the land. Notice the profusion of aerials and satellite dishes inside the walls, evidence of the large number of people who continue to live in the Old City. At a certain point the level of the rooftops falls below that of the ramparts, affording a fine view of the golden Dome of the Rock.

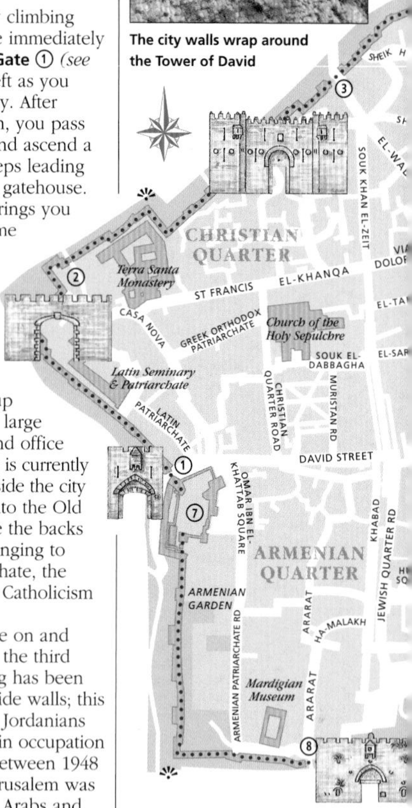

KEY

••• Walk route

Viewpoint

Damascus Gate to Lions' Gate

The ramparts now climb over **Damascus Gate** ③ *(see p66)*, the grandest of all the Old City gates. From up here you can survey the vaulted roof over the gate's defensive dogleg entrance tunnel and the crowds on El-Wad Road. Continuing east, you will encounter a rapid succession of

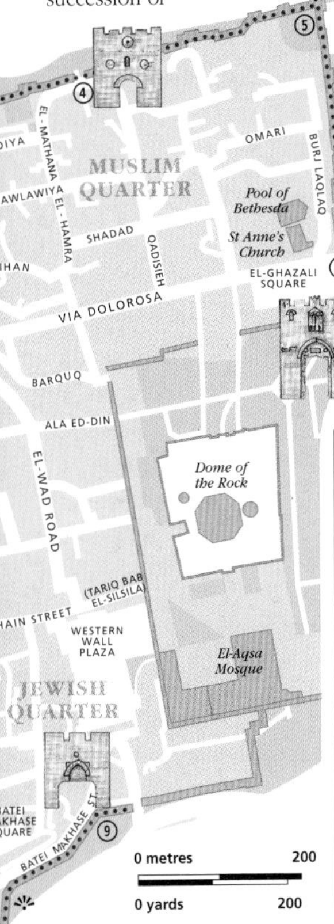

towers, because attacks on Jerusalem have traditionally always come from the north, where the approach is flattest (the approaches to the east, south and west are protected by deep valleys).

It was the north wall, just east of the next gate, **Herod's Gate** ④ *(see p67)*, that the Crusader army breached on 15 July 1099 to capture Jerusalem from the Muslims. Look outwards from the gate and you are facing down Salah ed-Din Street, the main street of Arab East Jerusalem.

The view from the ramparts between New Gate and Damascus Gate

At **Storks' Tower** ⑤, with its views to the northeast of the Hebrew University's Mount Scopus campus, the wall swings through 90° to run due south. From the ramparts here, you overlook the tombs that fill the Kidron Valley below and the slopes of the Mount of Olives *(pp110–111)*. As you approach the final gate, to your right, just inside the walls, are the remains of the complex of the biblical Pool of Bethseda and, beside them, the Crusader-built St Anne's Church *(see p67)*.

The walk ends at **Lions' Gate** ⑥ *(see p67)*, built by Suleyman the Magnificent, where you descend to street level. The beginning of the Via Dolorosa *(see pp30–31)* is just ahead, which, if followed, leads back towards the Jaffa Gate area. Energy permitting, you can then embark on another short ramparts walk.

Crenellations on Damascus Gate ③

The modern amphitheatre outside Damascus Gate ③

Anti-clockwise from Jaffa Gate

The access to this section of the ramparts is from outside the city walls, just south of the **Citadel** ⑦ *(see pp102–5)*. The initial stretch southwards is like a trench, with a high stone wall on either side of the walkway. This arrangement was fashioned by the Jordanian army between 1948 and 1967. Occasional vantage points allow you to look out across the Hinnom Valley below to the red rooftops of the early Jewish settlement of Mishkenot Shaananim *(see p121)* and the cliff-like bulk of the King David Hotel *(see p122)*. At the southwestern corner you have a good view of Sultan's Pool, an ancient reservoir, now dry and used as an outdoor concert venue.

As the ramparts run east, they pass close by the Church of the Dormition *(see p116)* before passing over the **Zion Gate** ⑧ *(see p106)*. The gate is riddled with bulletholes from the fighting in 1948, although, of course, you can't see this from above.

The final stretch affords wonderful views of the Arab village of Silwan, before the rampart walk ends on Batei Makhase Street, which you can follow down to the **Dung Gate** ⑨ *(see p84)*. This is the smallest of the city gates, despite being widened for cars by the Jordanians. The name indicates that what is now the main access to the Western Wall was probably once the site of a refuse tip.

A 90-Minute Walk around West Jerusalem

The heart of West Jerusalem, centred on Jaffa Road, was largely developed during the years of the British Mandate (1917–48). So, while it is nowhere near as ancient as the Old City, it does carry a weight of recent history related to the founding of the Jewish state of Israel. Aside from the scattering of historic buildings and monuments, this is also the heart of the modern city, with pedestrianised streets of cafés, restaurants and shops, cultural centres and busy markets. It is a highly rewarding area to explore.

Passing time on pedestrianised Ben Yehuda Street ⑤

Water sculpture on Safra Square at the City Hall complex ①

Jaffa Road

Until Tel Aviv got its own port in the 1930s, Jews arriving in Palestine would disembark at Jaffa, entering Jerusalem on the Jaffa Road. It ran right up to the Old City and the correspondingly named Jaffa Gate. The road now ends just short of the city walls, which is where this walk begins, at the rounded façade of the **Former Barclays Bank** ① (look for the "BB" in the iron window grilles). The building was on the line that divided Arabs and Jews between 1948 and 1967 and still bears the scars left by bullets.

Walk west, past two British Mandate-era post boxes, and almost immediately you come to palm-filled Safra Square, forecourt to the **City Hall complex** ② *(see p124)*, also home to the main tourist information office. Cross to the lefthand side of the road at the next junction to pass **Feingold House** ③, built in 1895, with its series of arched shop fronts and one arched entrance to a passageway containing the fine bar-restaurant Barood *(see p150)*. Look back to spot the winged lion on top of the Generali Building, trademark of the Italian insurance company that once had its offices here.

Continue along Jaffa Road, taking the next left into Rivlin Street and **Nakhalat Shiva** ④ *(see p123)*. This is one of the oldest parts of the modern city (founded 1869) but also one of the liveliest. Its attractive two-storey buildings are home to trendy eateries and late-night bars. At the bottom of Rivlin turn right, then head up Salomon to Zion Square, the traditional gathering point for protests and demonstrations. Running west from here, **Ben Yehuda Street** ⑤ *(see p123)* is one of the city's main shopping streets. Take the third right into Ben Hillel, cross over main King George V Street and you will be standing in front of Felafel & Shwarma King, which makes supposedly the best falafels in the city.

A balcony in the historic neighbourhood of Nakhalat Shiva ④

For hotels and restaurants in this area see pp257–8 and pp272–4

Mahane Yehuda

Continue west along **Agrippas Street** ⑥, passing on the right a passage that leads to top restaurant Arcadia. This has traditionally been a poor area with cheap rents that have proved attractive to recent immigrants, hence all the signs in Cyrillic. Agrippas is also the southern boundary of **Mahane Yehuda Market** ⑦, the city's colourful prime source of fresh produce, from fruit and vegetables to fish and meat *(see p148)*.

Exit the market back onto Jaffa Road, now returning east. Pass by a building on your right that has a doorway flanked by two lions on pillars – the former residence of the British Consul, 1863–90 – before arriving at a major junction marked by a small monument of a mortar on a plinth; this is a **Davidka** ⑧, a weapon that played a large role in the 1948 War. The Hebrew inscription is from the Old Testament Book of Isaiah and reads, "For I will defend this city to save it".

Fork left at the monument to follow historic **Ha-Neviim Street** ⑨ *(see p124)*, which during the 19th century was one of Jerusalem's main avenues. It is lined by some notable buildings, including at No. 64 a fine house once occupied by the English Victorian painter William Holman Hunt and, at No. 58, Thabor House, designed and once occupied by the German Conrad Schick, one of the city's foremost early architects.

The garden terrace at Ticho House, open daily for lunch ⑪

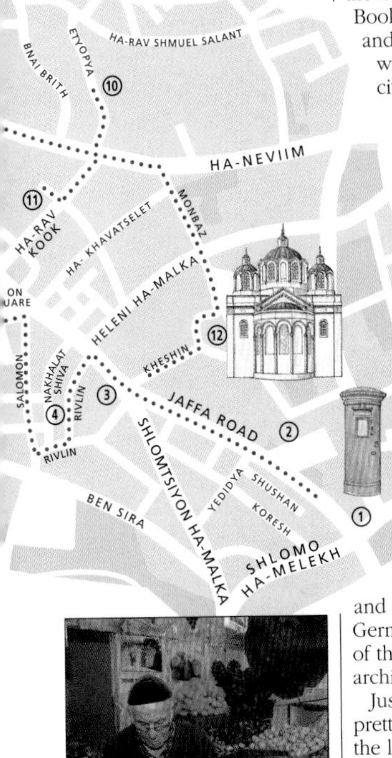

A stall in one of the covered lanes of Mahane Yehuda Market ⑦

Just past Thabor House, a pretty, high-walled lane on the left leads to the **Ethiopian Church** ⑩, a modest basilica with an interior painted in nursery blues and pinks, and filled with glittery, golden icons and smoky incense.

Return to Ha-Neviim and cross over to head south down Ha-Rav Kook Street looking for the signs for **Ticho House** ⑪ *(see p123)*. This is an historic Arab residence that has been turned into a lively cultural centre hosting art exhibitions and regular jazz, folk and classical recitals; it also has a pleasant garden terrace.

Decorative panel, Ethiopian Church ⑩

Returning to Ha-Neviim, take the next right and walk straight over the roundabout; the end point of the walk is visible ahead in the form of the three Muscovite-styled domes of the **Cathedral of the Holy Trinity** ⑫ *(see p124)*. Consecrated in 1872, the church was built to cater to Russian pilgrims, who at the time far outnumbered pilgrims from any other country. From here, it's just a short step back to Jaffa Road and the start of the walk.

TIPS FOR WALKERS

Starting point: *Jaffa Road.*
Length: *2 miles (3 km).*
Best time to walk: *Any time, but avoid Friday afternoon and Saturday, when everything is closed.*
Stopping-off points: *In addition to the places mentioned in the walk, there are dozens of food stalls around Mahane Yehuda Market, including some selling "meorav Yerushalmi", literally "Jerusalem meats", a mix of chopped livers, kidneys, hearts and beef, fried and served in pockets of bread. At the end of the walk, there are two good cafés at the junction of Heleni Ha-Malka and Jaffa Road, and many more cafés and restaurants in Nakhalat Shiva, which is just across Jaffa Road.*

A 90-Minute Walk around East Jerusalem

East Jerusalem is the Palestinian Arab part of the city. It lies north of the Old City and east of the main north–south road Derekh Ha-Shalom, swelling over the Mount of Olives and down the other side. The main street is Salah ed-Din Street, which is visited as part of this walk. High-profile tourist sights are few, but it is a vibrant area with many points of interest, including Christian pilgrimage sights and the Holy Land's most atmospheric old hotel.

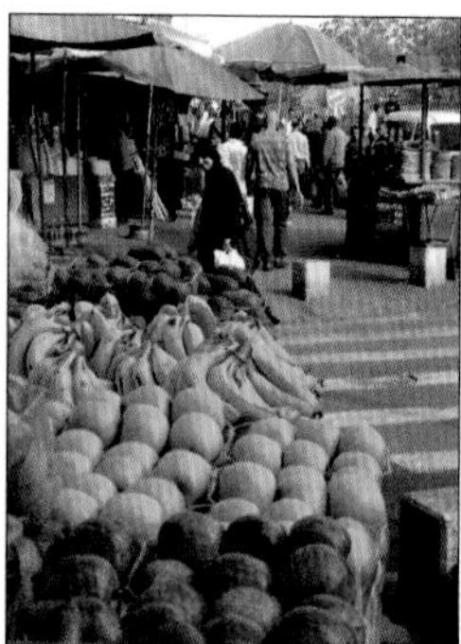

A fruit stall on the corner of Nablus Road

Nablus Road

The walk starts at **Damascus Gate** ① *(see p66)*, the largest and one of the busiest of the Old City gates. Taking advantage of the perpetual crowds, small traders spread their wares on sheets around the amphitheatre-like space in front of the gate so that it operates as a small makeshift market. Cross the busy road that runs parallel with the city walls to the junction with Nablus Road, which is also busy with street traders selling breads and fruit. Some of these traders stand in the shadow of **Schmidt's Girls' College** ②, part of the St Paul's Hospice complex, designed in fine Germanic style by the same architect responsible for Mount Zion's Church of the Dormition *(see p116)*.

Walk north up Nablus Road and shortly you come to an alley enclosed between high walls off to the right: this leads to the **Garden Tomb** ③ *(see p126)*. The claims for it as the burial place of Jesus Christ have been dismissed by archaeologists, but that does not seem to deter the coach-loads of Christian pilgrims who flock here each day to engage in open-air prayer sessions in what is, admittedly, a lovely garden setting. Stroll on, passing on your left the Arab bus station for services to West Bank towns and Ramallah. At the next traffic junction, marked by the modest little Sadd and Said Mosque, continue north as Nablus Road becomes a narrow, leafy lane squeezed beside the fortified bulk of the local US Consulate. On your right at No. 14 is **Palestinian Pottery** ④, founded on this site back in 1922 by the Balians, one of three Armenian families brought over by the British authorities from Kuthaya, Turkey, to renovate the ceramic tiles on the Dome of the Rock. Ring the bell to enter and visit the showrooms and a small museum on the history of ceramics in Jerusalem. You can also watch the craftspeople at work hand-painting designs onto the pottery prior to firing.

An elderly Palestinian

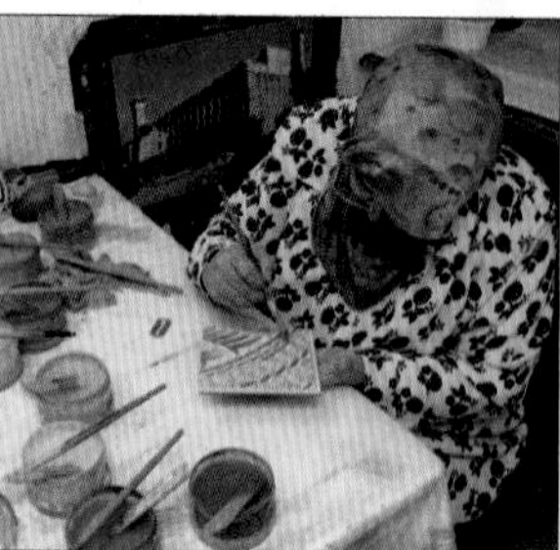

Handpainting a ceramic tile at the Palestinian Pottery workshop ④

Further along, on the left, are several fine examples of late 19th- and early 20th-century buildings, including a villa that houses the East Jerusalem offices of the British Council. On the right is the high wall that rings St George's Cathedral, which is visited later in the walk.

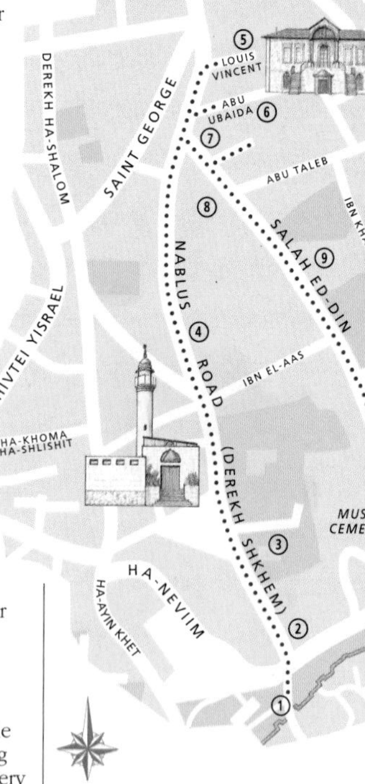

KEY

··· Walk route

Nablus Road now joins with Salah ed-Din Street, but continue on, taking the second right, Louis Vincent Street, a short cul-de-sac leading to the **American Colony Hotel** ⑤ *(see p127)*. Originally built (1865–76) as a home by a wealthy Arab merchant, the building was subsequently sold to pilgrims from Chicago, hence the name, before later becoming a hospice and then a hotel. It boasts a beautiful courtyard café and equally welcoming cellar bar. Opposite the main entrance to the hotel, beside an attractive little giftshop, steps leads up to the excellent Bookshop at the American Colony Hotel *(see p149)*.

Lobby area of the historic American Colony Hotel ⑤

Salah ed-Din Street

Return the way you came, taking a quick detour left down Abu Obeida Street to take a look at **Orient House** ⑥, an elegant 1897 villa that served as the headquarters of the Palestinian Authority in Jerusalem until it was shut down by the Israeli government in 2001.

Back on Salah ed-Din Street an easily missed, plain door in a wall gives access to the **Kings' Tombs** ⑦ *(see p127)* – actually the tomb of a single queen, dating from the 1st century AD. From here, cross over the street to the main gate of **St George's Cathedral** ⑧ *(see p126)* and buzz for admittance. Visitors are usually free to wander the gardens and courts of what is a surprisingly large compound. It is worth finding your way into the cathedral for its admirably restrained interior, which contains the royal arms formerly displayed in Government House during the time of British rule and deposited here when the Mandate came to an end in 1948. Services are still held throughout the week, although the language of mass these days is Arabic.

South of the cathedral, **Salah ed-Din Street** ⑨ becomes a busy high street with a clutter of low-rise shops, moneychangers, pharmacies and snack joints. Although vibrant, the scene is very visibly poorer than the corresponding main streets over in West Jerusalem. At its southern end Salah ed-Din Street terminates opposite the city walls and **Herod's Gate** ⑩, which to the Arabs is the far more poetic Bab el-Zahra, or "Flower Gate". At this point you can enter the Old City; or turn left and follow the walls down to the very worthwhile **Rockefeller Museum** ⑪ *(see p127)* and its archaeological finds from the Holy Land; or bear right and follow Sultan Suleyman Street, past rows of small clothes and jewellery shops, and eateries, back to the Damascus Gate area.

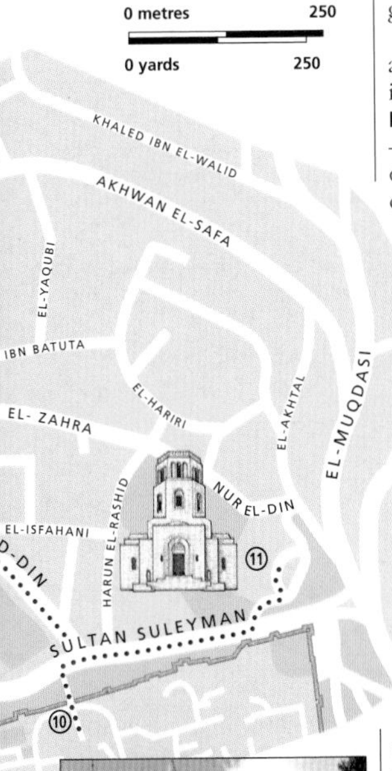

The Gothic bell tower of St George's Cathedral ⑧

Decorative sarcophagus at the Rockefeller Museum ⑪

TIPS FOR WALKERS

Starting point: *Damascus Gate.*
Length: *1.5 miles (2.4 km).*
Palestinian Pottery: *Open 9am–4pm Mon–Sat.*
Stopping-off points: *The American Colony Hotel serves lunch in the courtyard garden or indoors in Val's Brasserie Lounge. Café Europe, at 9 El-Zahra Street, just off Salah ed-Din, offers good value Western-style cuisine, including ham and eggs, in premises that resemble an English tearoom.*

Shops and Markets

When it comes to shopping, the main attractions in Jerusalem are the souks (markets) of the Old City. In comparison with the great bazaars of Istanbul or Cairo, Jerusalem's souks are perhaps a little small, and the array of goods on offer is largely limited to souvenir items such as T-shirts and religious articles, but they still reward exploration. There is better shopping elsewhere, however, notably in the modern centre of West Jerusalem, where you'll find high-street shopping and malls, and areas of interesting boutiques: see Where to Shop, below. For more information on methods of payment and bargaining, see pp284–5.

Sacks of spices at a shop on the Old City's Souk Khan el-Zeit Street

Religious souvenirs are popular throughout the Old City

OPENING HOURS

Shops in the Muslim Quarter of the Old City and in East Jerusalem are open daily except for Friday morning – Friday being the Islamic holy day. Many shops and stalls in the souks of the Old City are also closed all day Sunday, as many of the shop owners are Christian. Shops in the Jewish Quarter of the Old City and throughout West Jerusalem are open Sunday to Thursday from around 9am to 7pm, Friday from 9am to 3pm, and closed Saturday. Beware of local religious holidays *(see pp36–9)*: during the holy month of Ramadan Muslim shops close 30 minutes to one hour before sunset. All Jewish-owned businesses close for Jewish holidays.

WHERE TO SHOP

Away from the Old City, visit King George V Street around the intersection with Jaffa Road for general high-street shopping. For boutique shopping, visit nearby Ben Hillel and Bezalel Streets. For the most diverse selection of interesting shops you need to take a taxi south to Emek Refa'im Street in the German Colony (it is just five minutes from the King David Hotel/ YMCA), which boasts a mile-long stretch of chic boutiques and cafés.

MARKETS

The streets in the Muslim and Christian Quarters of the Old City form a single large market, or souk. In the traditional Middle Eastern manner, different areas specialise in specific wares. David Street, for example, which runs east from the Jaffa Gate area, is almost entirely devoted to tourist trinkets and is the place to buy Christian-themed kitsch. Christian Quarter Road, off David Street, is more upmarket and, in addition to more religious souvenirs, also sells items such as richly coloured Palestinian rugs, covers and dresses. Many of the shops in the Muristan *(see pp90–91)* specialise in leather, while the Via Dolorosa is strong on religious items. Most diverse of all is Souk Khan el-Zeit, where stores sell everything from CDs and clothes to live chickens and honey-drenched Arabic pastries.

West Jerusalem has an excellent covered central market in **Mahane Yehuda**, which runs between Agrippas Street and Jaffa Road. Many stalls sell fruit and vegetables, but there are also fishmongers, butchers, sellers of dairy produce, olives, nuts and dried fruits, plus clothing stalls. There are a handful of cafés and even a couple of small jewellery and designer apparel boutiques. The market is open Sunday to Thursday from 9am to 8pm, and Friday 9am to one hour before Shabbat.

ANTIQUES

In Jerusalem (and Israel in general), unlike other parts of the Holy Land, you may buy antiques and objects from excavations, but to take them

A typical antiques shop in the Christian Quarter of the Old City

out of the country you must obtain a permit from the Israeli Antiquities Authority *(see pp284–5)*. Only certain shops are authorised to deal in antiques of this kind; buy from a non-accredited source and there is a chance that you may be buying looted goods. **Tzadok** in West Jerusalem is an authorised specialist that often has items for sale garnered from recent digs. Founded in 1938, **Baidun** is one of the better known antique dealers along the Via Dolorosa. It sells pieces from the Chalcolithic era to early Islamic times. There are many antique stores along this street but it is advisable to check a store is authorized before you commit to buying anything.

BOOKS

Israel's oldest and largest bookstore chain is **Steimatzky**, founded in Jerusalem in 1925. It still has several branches in the city (including on Jaffa Road, Ben Yehuda Street and King George V Street), all of which sell English-language newspapers and magazines, fiction and non-fiction, and books about Jerusalem and Israel. However, the best selection on the history and politics of the city, and the Middle East in general, is found at the **Bookshop at the American Colony Hotel**. It also carries a well chosen selection of general English-language literature.

CERAMICS

Distinctive items of pottery are sold in shops throughout the Old City but for the best quality visit **Palestinian Pottery** *(see p146)*. Its showrooms are filled with displays of the company's trademark hand-painted cups, bowls, tiles and vases, with prices starting from a few dollars.

In West Jerusalem, the narrow lanes of Nakhalat Shiva are full of pottery-stocked gift stores, including the **Guild of Ceramists Gallery Shop**, which has different collections of unique pieces by a variety of Israeli artisans.

Distinctive items of handpainted ceramics at Palestinian Pottery

JEWELLERY

Israeli-jewellery designer **Michal Negrin**, whose whimsical designs are sold in her own-brand boutiques across the world, has several stores in Jerusalem; the most central of these is located in Nakhalat Shiva. **Goldtime** is another respected local chain store with several branches in Jerusalem. For more one-off and highly decorative designs visit **Puenta**, which is also in Nakhalat Shiva.

RELIGIOUS ARTICLES

For Christian religious items there is a plethora of shops along the Old City's David Street and the Muristan area of the Christian Quarter, specialising in crucifixes, rosaries and biblical scenes crafted from olive wood. Shops selling exquisitely crafted items of Judaica are found all throughout the Old City's Jewish Quarter, particularly on the ancient Cardo, which is where you'll find **Ot Ezra** and **Chabad**. In the new city they cluster on King David Street, near the King David Hotel and YMCA. Try **Tzadok**, which carries everything from simple silver candlesticks to *chalah* dishes and *menorahs*, and *shofars*, the traditional Jewish trumpet made from a ram's horn.

DIRECTORY

MARKETS

Mahane Yehuda
120 Jaffa Road, West Jerusalem.

ANTIQUES

Baidun
20 Via Dolorosa, Muslim Quarter, Old City.
Map 4 D2.
Tel *(02) 626 1469.*
www.baidun.com

Tzadok
18 King David Street, West Jerusalem.
Map 1 B4.
Tel *(02) 625 8039.*

BOOKS

Bookshop at the American Colony Hotel
23 Nablus Road, East Jerusalem.
Map 1 C1.
Tel *(02) 627 9731.*

Steimatzky
33 Jaffa Road, West Jerusalem.
Map 1 A3.
Tel *(02) 625 3654.*

CERAMICS

Guild of Ceramists Gallery Shop
27 Yoel Salomon Street, Nakhalat Shiva, West Jerusalem.
Map 1 A3.
Tel *(02) 624 4065.*

Palestinian Pottery
14 Nablus Road, East Jerusalem.
Map 1 C2.
Tel *(02) 628 2826.*
www.palestinianpottery.com

JEWELLERY

Goldtime
8 King George V Street, West Jerusalem.
Map 1 A3.
Tel *(02) 625 5883.*

Michel Negrin
12 Yoel Salomon Street, Nakhalat Shiva, West Jerusalem.
Map 1 A3.
Tel *(02) 622 3573.*
www.michalnegrin.com

Puenta
21 Yoel Salomon Street, Nakhalat Shiva, West Jerusalem. **Map** 1 A3.
Tel *(02) 624 0383.*

RELIGIOUS ARTICLES

Chabad
Cardo, Jewish Quarter, Old City. **Map** 3 C4.
Tel *(02) 627 2217.*

Ot Ezra
8 Cardo, Jewish Quarter, Old City. **Map** 3 C4.
Tel *(02) 628 8166.*

Entertainment

For a relatively small city, Jerusalem offers a wide range of high-quality entertainment, especially in the fields of theatre and classical music. It enjoys several months of dynamic artistic and cultural activity a year, focused on summer and the Christmas season. Every May and June there is the Israel Festival, the country's most important cultural jamboree, and in April/May there is the Jerusalem Arts Festival. The Jerusalem Film Festival is in July and there is an annual Jewish Film Festival. For information on what's on, consult the daily *Jerusalem Post* or the free monthly *Time Out*, available at hotels and tourist offices.

The Armenian Tavern, a lone drinking spot in the Old City

BARS & PUBS

Apart from a small but characterful bar in the corner of the **Armenian Tavern** restaurant, just south of the Citadel, there is nowhere to drink in the Old City. You need to go to West Jerusalem and, specifically, the district of narrow lanes known as Nakhalat Shiva. This small neighbourhood has become the centre of nightlife in the city, with dozens of bars, whose patrons spill outside in the warmer months. Among them, **Barood** stands out for its superb selection of spirits and liqueurs, including shelves of absinthes, schnapps and home-made flavoured vodkas. Nearby **Stardust** is the place for terrific music on the sound system, sports on the big screen, and a happy hour that lasts four hours.

Also in West Jerusalem, just off King George V Street is **Link**, a bar-restaurant with a pleasant garden terrace. One block north and west in the premises of the Bezalel Art School, **Mona** is another good bar-restaurant, beloved of the city's secular population for being one of the few places open on Shabbat (the Jewish day of rest).

Predominantly Muslim East Jerusalem is naturally thin on venues serving alcohol, but a drink at the **Cellar Bar** of the American Colony Hotel is a signature experience every bit as essential as a stroll along the Via Dolorosa. It's the place to meet UN officials, international correspondents, NGO workers and Palestinian entrepreneurs. Otherwise, the **Kan Zaman** garden restaurant at the Jerusalem Hotel serves Palestinian beers, wine and *nargilehs* (waterpipes).

CHILDREN

The **Jerusalem Biblical Zoo** *(see p138)* brings together all the animals that the Bible mentions as living in the Holy Land. It is beautifully designed and kids love it. The **Bloomfield Science Museum** is devoted to acquainting children with science via lots of interactive exhibits. It's fun for adults too. In the Liberty Bell Gardens (Ha-Pa'amon), just south of the Bloomfield Gardens *(see p121)* is the **Train Theater**, with a permanent repertoire of puppetry, plays and annual productions. The park itself is also very child friendly, with basketball courts, ping-pong tables and a rollerblade rink.

CINEMA

Jerusalem's cinemas screen both local Israeli films plus international and Hollywood hits. Non-Hebrew films are usually screened in the original language with subtitles. For mainstream fare, the best bet is the **Globus Malcha** cinema complex in the Malcha Mall in the south-west of the city. The **Jerusalem Cinematheque**, on the slopes of the Hinnon Valley just outside the Old City walls, screens seasons of classics and retrospectives, as well as recent world cinema releases. Every July it hosts the Jerusalem Film Festival. **Lev Smador** in the German Colony is another quality art-house cinema, specialising in European and independent films. The nearby **Third Ear** is a courtyard book, DVD and CD store that also screens films in its small auditorium.

MUSIC

The **Henry Crown Concert Hall** at the Jerusalem Theatre is the major venue for classical performances and home to

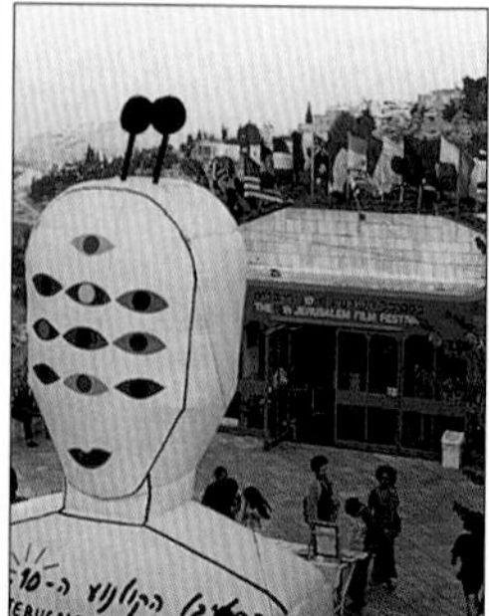

Creative advertising for the Cinematheque Film Festival

the Jerusalem Symphony Orchestra. Organ and choral concerts are held regularly at the **Church of the Dormition** *(see p116)* on Mount Zion, while the **YMCA** and **Ticho House** host regular classical recitals by soloists and ensembles, and regular folk evenings. In East Jerusalem, the **Kan Zaman** restaurant has Friday night performances of classical Arabic music.

ROCK, POP & JAZZ

The city's premier live music venue is **Yellow Submarine**, which features nightly acts performing blues, jazz, rock and folk. It is in an industrial district south of the centre, but it's only a short taxi ride from the Jaffa Road area. For world and ethnic music and festivals head to **Confederation House** on Emile Botta Street. When the occasional big name plays town, the venue is the **Sultan's Pool** on Hebron Road, a now-dry ancient reservoir, which, when not in use, resembles an abandoned quarry, just outside the city walls.

THEATRE & DANCE

The **Jerusalem Theatre** is the city's largest and most active cultural centre. In addition to the main Sherover Theatre, it has three other concert spaces and is a busy venue for both local and foreign productions. Smaller, but housed in a beautifully renovated old Ottoman structure, the **Khan Theatre** has two performance spaces, kept busy with a lively programme of international productions.

The **Gerard Bahar Performance Centre**, which is just west of central King George V Street, hosts regular theatre and dance events (it's the home of the respected Vertigo and Kombina dance companies), as well as occasional music concerts. **Ha-Ma'abada** (The Lab) is a beautifully designed modern performance space that is used for avant-garde theatre and dance. Over in East Jerusalem you'll find **El-Hakawati Palestinian National Theatre**, featuring performances in Arabic which are often of a political nature.

DIRECTORY

BARS & PUBS

Armenian Tavern
79 Armenian Patriarchate Road, Armenian Quarter, Old City. **Map** 3 B4. **Tel** *(02) 627 3854.*

Barood
31 Jaffa Street, Nakhalat Shiva, West Jerusalem. **Map** 1 A3. **Tel** *(02) 625 9081.*

Cellar Bar
American Colony Hotel, 2 Louis Vincent Street, off Nablus Road, East Jerusalem. **Map** 1 C1. **Tel** *(02) 627 9777.*

Kan Zaman
Jerusalem Hotel, Nablus Road, East Jerusalem. **Map** 1 C1. **Tel** *(02) 628 3282.*

Link
3 Hama'alot Street, West Jerusalem. **Tel** *(02) 625 3446.*

Mona
12 Shmuel Ha-Nagid, West Jerusalem. **Tel** *(02) 622 2283.*

Stardust
6 Rivlin Street, Nakhalat Shiva, West Jerusalem. **Map** 1 A3. **Tel** *(02) 622 2196.*

CHILDREN

Bloomfield Science Museum
Hebrew University, Givat Ram, West Jerusalem. **Tel** *(02) 654 4888.*

Jerusalem Biblical Zoo
Manahat, West Jerusalem. **Tel** *(02) 675 0111.*

Train Theater
Liberty Bell Park, West Jerusalem. **Map** 1 B5. **Tel** *(02) 561 8514.* **www**.traintheater.co.il

CINEMA

Globus Malcha
Malcha Mall, Manahat, West Jerusalem. **Tel** *(02) 678 8448.*

Jerusalem Cinematheque
11 Hebron Road, West Jerusalem. **Map** 1 B5. **Tel** *(02) 565 4333.*

Lev Smador
4 Lloyd George Street, German Colony, West Jerusalem. **Tel** **5155.*

Third Ear
8 Emek Refa'im Street, German Colony, West Jerusalem. **Tel** *(02) 563 3093.*

MUSIC

Church of the Dormition
Mount Zion, Old City. **Map** 1 C5. **Tel** *(02) 565 5330.*

Henry Crown Concert Hall
Jerusalem Theatre, 20 David Marcus Street, Talbiye, West Jerusalem. **Tel** *(02) 560 5757.* **www**.jso.co.il

Kan Zaman
See Bars & Pubs.

Ticho House
9 Ha-Rav Kook Street, West Jerusalem. **Map** 1 A2. **Tel** *(02) 624 4168.*

YMCA
26 King David Street, West Jerusalem. **Map** 1 A4. **Tel** *(02) 569 2692.*

ROCK, POP & JAZZ

Confederation House
12 Emile Botta Street, Yemin Moshe. **Map** 1 B4. **Tel** *(02) 624 5206.*

Yellow Submarine
13 Ha-Rechavim Street, Talpiot, West Jerusalem. **Tel** *(02) 679 4040.*

THEATRE & DANCE

Gerard Bahar Performance Centre
11 Bezalel Street, Nakhla'ot, West Jerusalem. **Tel** *(02) 625 1139.*

Ha-Ma'abada (The Lab)
28 Hebron Road, West Jerusalem. **Map** 1 B5. **Tel** *(02) 629 2000.* **www**.maabada.org.il

El-Hakawati Palestinian National Theatre
El-Nuzha Street, East Jerusalem. **Map** 1 C1. **Tel** *(02) 628 0957.*

Jerusalem Theatre
20 David Marcus Street, Talbiye, West Jerusalem **Tel** *(02) 560 5757.* **www**.jerusalem-theatre.co.il

Khan Theatre
2 David Remez Square, West Jerusalem. **Tel** *(02) 671 8281.*

JERUSALEM STREET FINDER

The map references that are given throughout the Jerusalem chapters of this guide refer to the maps on the following pages. References are also given in the listings for hotels *(see pp256–8)* and restaurants *(see pp272–4)*. Some of the many small streets and alleys may not be named on the maps. Many streets and monuments have two or even three names: one in Hebrew, one in Arabic and, occasionally, a commonly used English-language form, too. What we call Damascus Gate is also known as Shaar Shkhem to Israelis and Bab el-Amud to Arabs. In this guide and on the following maps, where there is a sufficiently well-recognized English name, we have used it; otherwise, we have used the Arabic names for predominantly Arab areas (for example, the Muslim Quarter of the Old City) and Hebrew names for Jewish areas. Spellings in this guide may vary from those you see on street signs.

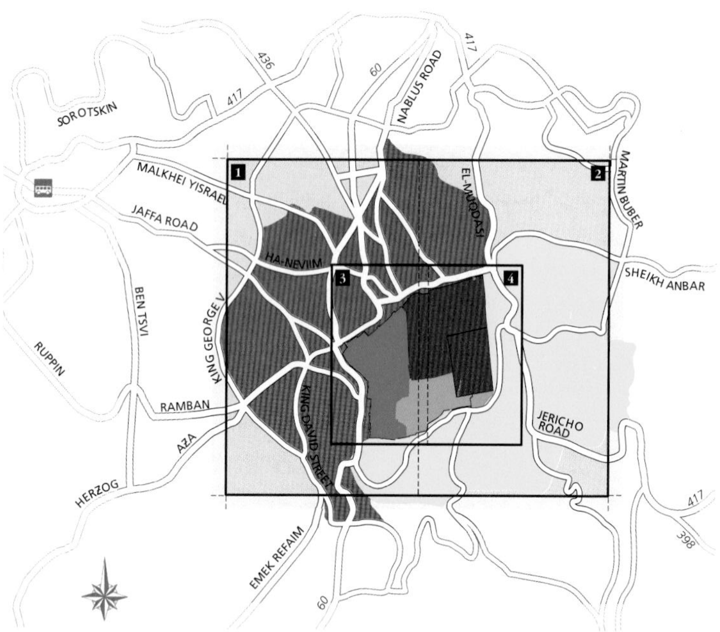

KEY TO JERUSALEM STREET FINDER

- Major sight
- Other sight
- Other important building
- Bus station
- Light Rail stop
- Taxi rank
- Parking
- Tourist information
- Synagogue
- Church
- Mosque
- Route of Via Dolorosa
- IV Station of the Cross
- Police station
- Post office
- Hospital with casualty unit
- City wall
- Covered street
- 25» Street number

SCALE OF MAP ABOVE

0 metres 1000

0 yards 1000

SCALE OF MAPS 1 – 2

0 metres 250

0 yards 250

SCALE OF MAPS 3 – 4

0 metres 100

0 yards 100

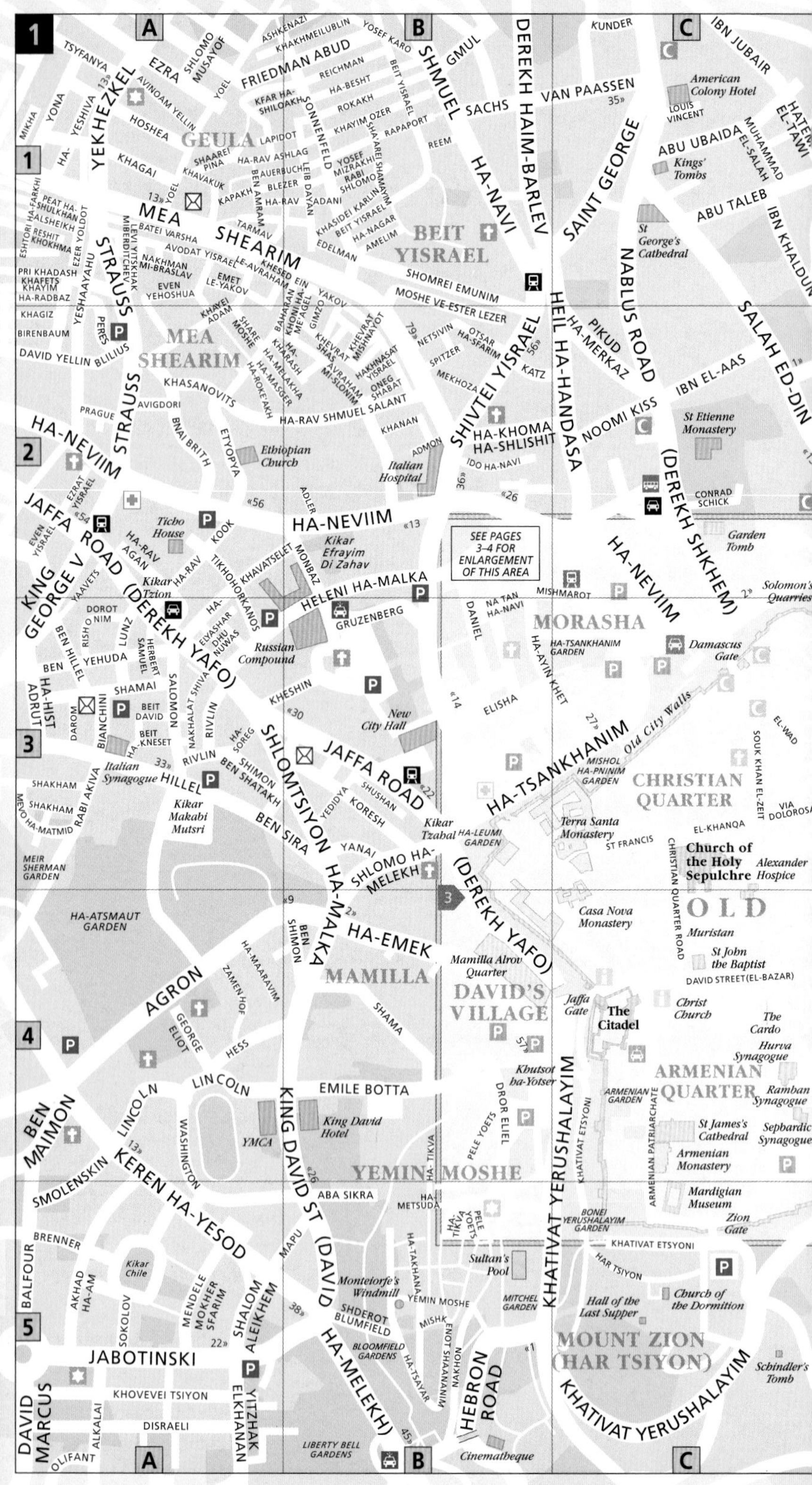
1
A
B
C
1
2
3
4
5
GEULA
MEA SHEARIM
BEIT YISRAEL
MORASHA
MAMILLA
DAVID'S VILLAGE
YEMIN MOSHE
CHRISTIAN QUARTER
ARMENIAN QUARTER
OLD
MOUNT ZION (HAR TSIYON)
SEE PAGES 3–4 FOR ENLARGEMENT OF THIS AREA
American Colony Hotel
Kings' Tombs
St George's Cathedral
St Etienne Monastery
Garden Tomb
Solomon's Quarries
Damascus Gate
Old City Walls
Ethiopian Church
Italian Hospital
Ticho House
Kikar Efrayim Di Zahav
Kikar Tzion
Russian Compound
New City Hall
Italian Synagogue
Kikar Makabi Mutsri
Kikar Tzahal
Terra Santa Monastery
Church of the Holy Sepulchre
Alexander Hospice
Casa Nova Monastery
Muristan
St John the Baptist
Mamilla Alrov Quarter
Jaffa Gate
The Citadel
Christ Church
The Cardo
Hurva Synagogue
Ramban Synagogue
Sephardic Synagogue
St James's Cathedral
Armenian Monastery
Mardigian Museum
Zion Gate
Church of the Dormition
Hall of the Last Supper
Schindler's Tomb
Khutsot ha-Yotser
YMCA
King David Hotel
Kikar Chile
Sultan's Pool
Montefiore's Windmill
Cinematheque
HA-ATSMAUT GARDEN
MEIR SHERMAN GARDEN
HA-TSANKHANIM GARDEN
MISHOL HA-PNININ GARDEN
HA-LEUMI GARDEN
ARMENIAN GARDEN
BONEI YERUSHALAYIM GARDEN
MITCHEL GARDEN
BLOOMFIELD GARDENS
LIBERTY BELL GARDENS
YEKHEZKEL
EZRA
SHLOMO MUSAYOF
FRIEDMAN ABUD
SHMUEL HA-NAVI
DEREKH HAIM-BARLEV
VAN PAASSEN
SACHS
GMUL
SAINT GEORGE
IBN JUBAIR
HATEM EL-TAWI
ABU UBAIDA
MUHAMMAD EL-SALAH
ABU TALEB
IBN KHALDUN
NABLUS ROAD
SALAH ED-DIN
MEA SHEARIM
STRAUSS
SHOMREI EMUNIM
MOSHE VE-ESTER LEZER
SHIVTEI YISRAEL
HEIL HA-HANDASA
PIKUD HA-MERKAZ
IBN EL-AAS
NOOMI KISS
(DEREKH SHKHEM)
HA-KHOMA HA-SHLISHIT
HA-RAV SHMUEL SALANT
HA-NEVIIM
JAFFA ROAD (DEREKH YAFO)
KING GEORGE V
HELENI HA-MALKA
SHLOMTSIYON HA-MALKA
JAFFA ROAD
HA-TSANKHANIM
(DEREKH YAFO)
SHLOMO HA-MELEKH
HA-EMEK
AGRON
HILLEL
BEN SIRA
EMILE BOTTA
LINCOLN
KING DAVID ST (DAVID HA-MELEKH)
KEREN HA-YESOD
BEN MAIMON
SMOLENSKIN
JABOTINSKI
KHOVEVEI TSIYON
DISRAELI
DAVID MARCUS
YITZHAK ELKHANAN
SHALOM ALEIKHEM
HEBRON ROAD
KHATIVAT YERUSHALAYIM
KHATIVAT ETSYONI
HAR TSIYON
ARMENIAN PATRIARCHATE
CHRISTIAN QUARTER ROAD
DAVID STREET(EL-BAZAR)
VIA DOLOROSA
SOUK KHAN EL-ZEIT
EL-WAD
EL-KHANQA
ST FRANCIS
CONRAD SCHICK
MISHMAROT
ABA SIKRA
SHDEROT BLUMFIELD
YEMIN MOSHE
GEORGE ELIOT
WASHINGTON
BALFOUR
BRENNER
SOKOLOV
JAFFA ROAD
OLIFANT
ALKALAI

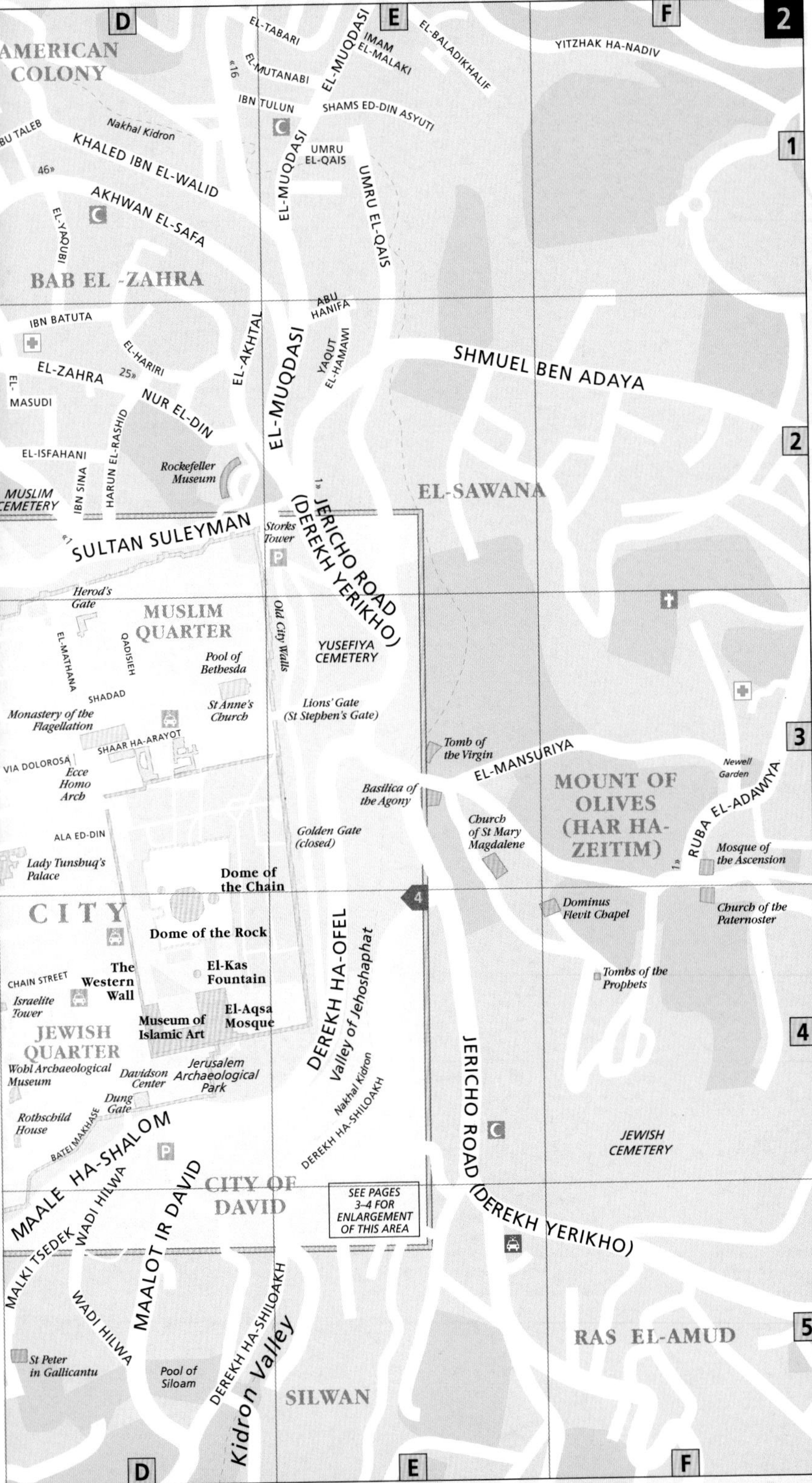
2
D
E
F
AMERICAN COLONY
BAB EL -ZAHRA
EL-SAWANA
MUSLIM QUARTER
CITY
JEWISH QUARTER
CITY OF DAVID
MOUNT OF OLIVES (HAR HA-ZEITIM)
RAS EL-AMUD
SILWAN
EL-TABARI
EL-MUQDASI
IMAM EL-MALAKI
EL-BALADIKHALIF
YITZHAK HA-NADIV
EL-MUTANABI
IBN TULUN
SHAMS ED-DIN ASYUTI
Nakhal Kidron
KHALED IBN EL-WALID
AKHWAN EL-SAFA
UMRU EL-QAIS
EL-YAQUBI
IBN BATUTA
ABU HANIFA
YAQUT EL-HAMAWI
EL-HARIRI
EL-ZAHRA
EL-MASUDI
NUR EL-DIN
EL-AKHTAL
SHMUEL BEN ADAYA
EL-ISFAHANI
HARUN EL-RASHID
IBN SINA
MUSLIM CEMETERY
Rockefeller Museum
SULTAN SULEYMAN
Storks Tower
JERICHO ROAD (DEREKH YERIKHO)
Herod's Gate
EL-MATHANA
QADISIEH
Old City Walls
YUSEFIYA CEMETERY
Pool of Bethesda
St Anne's Church
SHADAD
Lions' Gate (St Stephen's Gate)
Monastery of the Flagellation
SHAAR HA-ARAYOT
VIA DOLOROSA
Ecce Homo Arch
Tomb of the Virgin
EL-MANSURIYA
Newell Garden
Basilica of the Agony
RUBA EL-ADAWIYA
Church of St Mary Magdalene
Golden Gate (closed)
ALA ED-DIN
Lady Tunshuq's Palace
Dome of the Chain
Mosque of the Ascension
Dominus Flevit Chapel
Church of the Paternoster
Dome of the Rock
DEREKH HA-OFEL
Valley of Jehoshaphat
The Western Wall
El-Kas Fountain
CHAIN STREET
Tombs of the Prophets
Israelite Tower
El-Aqsa Mosque
Museum of Islamic Art
Wohl Archaeological Museum
Davidson Center
Jerusalem Archaeological Park
Nakhal Kidron
DEREKH HA-SHILOAKH
Dung Gate
Rothschild House
BATEI MAKHASE
MAALE HA-SHALOM
WADI HILWA
MAALOT IR DAVID
JEWISH CEMETERY
SEE PAGES 3–4 FOR ENLARGEMENT OF THIS AREA
JERICHO ROAD (DEREKH YERIKHO)
MALKI TSEDEK
WADI HILWA
DEREKH HA-SHILOAKH
Kidron Valley
St Peter in Gallicantu
Pool of Siloam
1
2
3
4
5

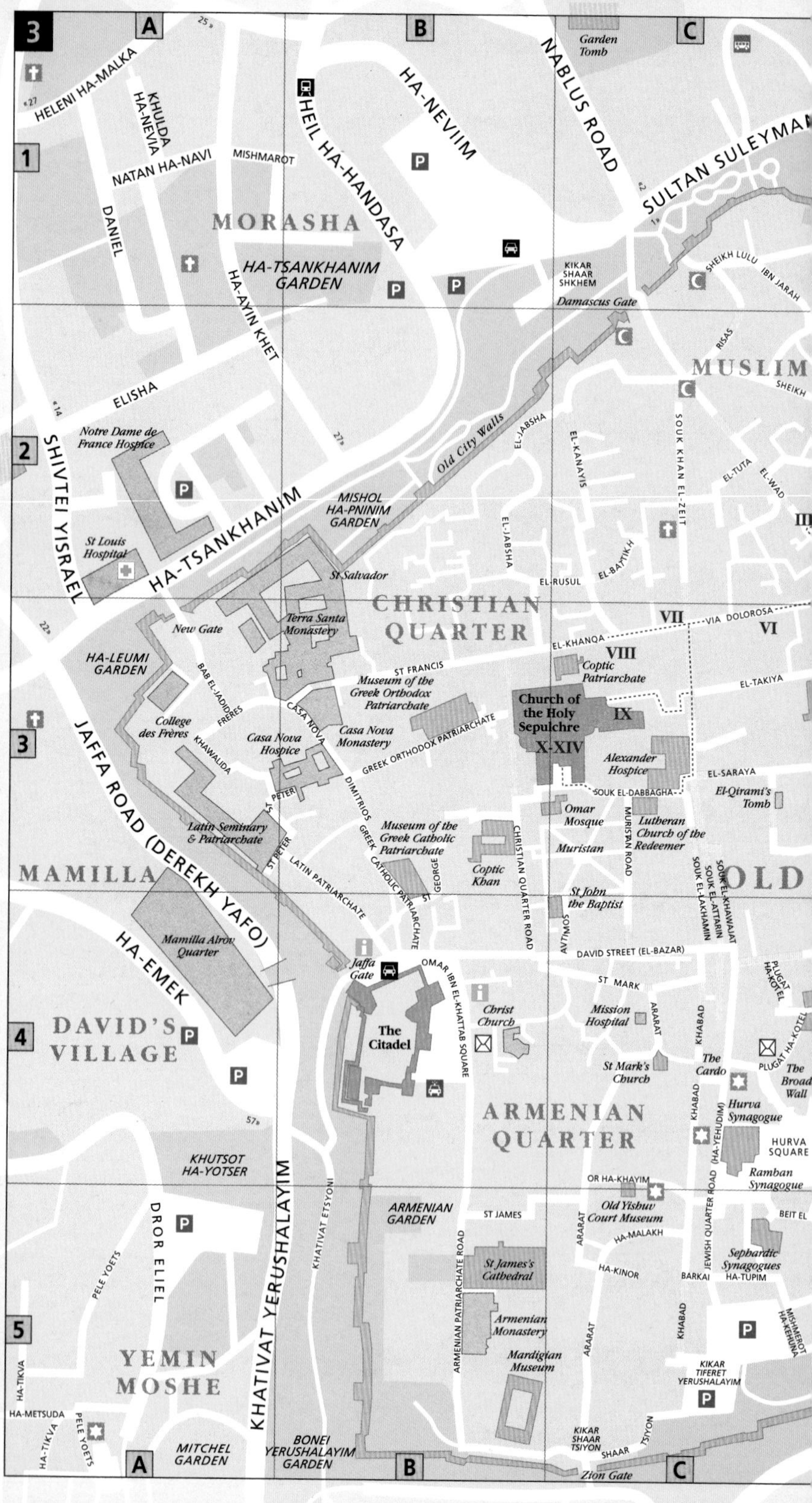

3
A
B
C
1
2
3
4
5
Garden Tomb
HELENI HA-MALKA
KHULDA HA-NEVIA
HEIL HA-HANDASA
HA-NEVIIM
NABLUS ROAD
SULTAN SULEYMAN
NATAN HA-NAVI
MISHMAROT
DANIEL
MORASHA
HA-TSANKHANIM GARDEN
HA-AYIN KHET
KIKAR SHAAR SHKHEM
Damascus Gate
SHEIKH LULU
IBN JARAH
RISAS
MUSLIM
SHEIKH
ELISHA
Notre Dame de France Hospice
Old City Walls
EL-JABSHA
EL-KANAYIS
SOUK KHAN EL-ZEIT
EL-TUTA
EL-WAD
SHIVTEI YISRAEL
HA-TSANKHANIM
MISHOL HA-PNININ GARDEN
St Louis Hospital
St Salvador
EL-BAYTIKH
EL-RUSUL
CHRISTIAN QUARTER
VII
VIA DOLOROSA
VI
New Gate
Terra Santa Monastery
EL-KHANQA
VIII
HA-LEUMI GARDEN
BAB EL-JADID
ST FRANCIS
Coptic Patriarchate
Museum of the Greek Orthodox Patriarchate
EL-TAKIYA
FRERES
CASA NOVA
Church of the Holy Sepulchre
IX
College des Frères
Casa Nova Hospice
Casa Nova Monastery
GREEK ORTHODOX PATRIARCHATE
X-XIV
KHAWALDA
Alexander Hospice
JAFFA ROAD (DEREKH YAFO)
ST PETER
DIMITRIOS
SOUK EL-DABBAGHA
EL-SARAYA
El-Qirami's Tomb
Omar Mosque
Latin Seminary & Patriarchate
Museum of the Greek Catholic Patriarchate
CHRISTIAN QUARTER ROAD
MURISTAN ROAD
Lutheran Church of the Redeemer
Muristan
GREEK CATHOLIC PATRIARCHATE
SOUK EL-LAKHAMIN
SOUK EL-ATTARIN
SOUK EL-KHAWAJAT
MAMILLA
LATIN PATRIARCHATE
ST GEORGE
Coptic Khan
OLD
St John the Baptist
Mamilla Alrov Quarter
AVTIMOS
DAVID STREET (EL-BAZAR)
HA-EMEK
Jaffa Gate
OMAR IBN EL-KHATTAB SQUARE
ST MARK
PLUGAT HA-KOTEL
Christ Church
Mission Hospital
ARARAT
KHABAD
DAVID'S VILLAGE
The Citadel
St Mark's Church
The Cardo
The Broad Wall
ARMENIAN QUARTER
Hurva Synagogue
HURVA SQUARE
KHUTSOT HA-YOTSER
KHATIVAT YERUSHALAYIM
KHATIVAT ETSYONI
OR HA-KHAYIM
JEWISH QUARTER ROAD (HA-YEHUDIM)
Ramban Synagogue
ARMENIAN GARDEN
ST JAMES
Old Yishuv Court Museum
BEIT EL
DROR ELIEL
HA-MALAKH
PELE YOETS
ARMENIAN PATRIARCHATE ROAD
St James's Cathedral
HA-KINOR
BARKAI
Sephardic Synagogues
HA-TUPIM
Armenian Monastery
MISHMEROT HA-KEHUNA
YEMIN MOSHE
Mardigian Museum
KIKAR TIFERET YERUSHALAYIM
HA-TIKVA
HA-METSUDA
BONEI YERUSHALAYIM GARDEN
MITCHEL GARDEN
KIKAR SHAAR TSIYON
SHAAR TSIYON
Zion Gate

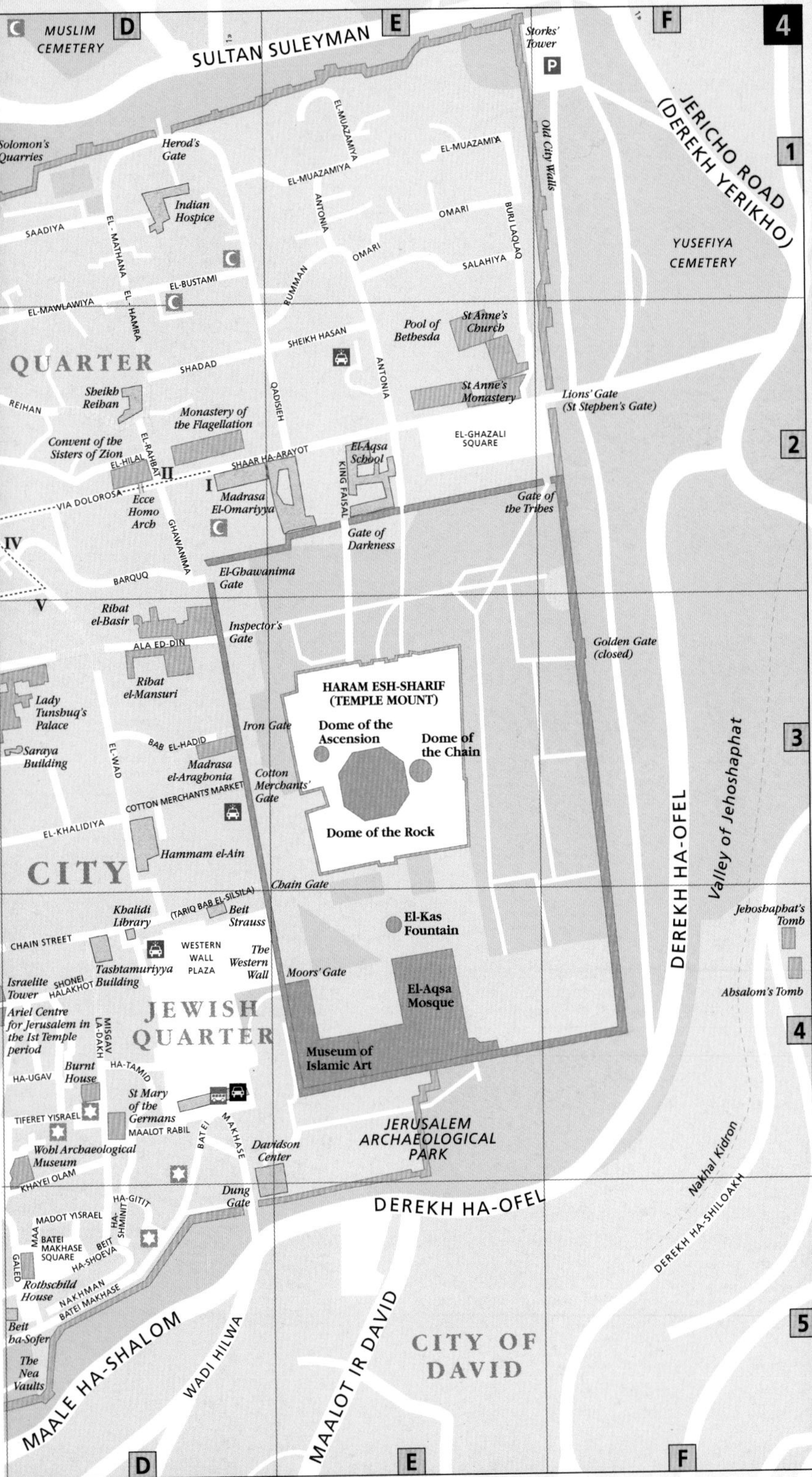

MUSLIM CEMETERY
D
E
F
4
SULTAN SULEYMAN
Storks' Tower
JERICHO ROAD (DEREKH YERIKHO)
Solomon's Quarries
Herod's Gate
EL-MUAZAMIYA
Old City Walls
Indian Hospice
ANTONIA
OMARI
BURJ LAQLAQ
SAADIYA
EL-MATHANA
YUSEFIYA CEMETERY
SALAHIYA
EL-BUSTAMI
RUMMAN
EL-MAWLAWIYA
EL-HAMRA
Pool of Bethesda
St Anne's Church
SHEIKH HASAN
QUARTER
SHADAD
Sheikh Reiban
QADISIEH
St Anne's Monastery
Lions' Gate (St Stephen's Gate)
REIHAN
Monastery of the Flagellation
Convent of the Sisters of Zion
EL-RAHBAT
EL-HILAL
EL-GHAZALI SQUARE
SHAAR HA-ARAYOT
El-Aqsa School
KING FAISAL
VIA DOLOROSA
Ecce Homo Arch
Madrasa El-Omariyya
GHAWANIMA
Gate of the Tribes
Gate of Darkness
BARQUQ
El-Ghawanima Gate
Ribat el-Basir
Inspector's Gate
Golden Gate (closed)
ALA ED-DIN
Ribat el-Mansuri
HARAM ESH-SHARIF (TEMPLE MOUNT)
Lady Tunshuq's Palace
Iron Gate
Dome of the Ascension
Dome of the Chain
Saraya Building
BAB EL-HADID
EL-WAD
Madrasa el-Araghonia
Cotton Merchants' Gate
COTTON MERCHANTS MARKET
Valley of Jehoshaphat
DEREKH HA-OFEL
EL-KHALIDIYA
Dome of the Rock
Hammam el-Ain
CITY
Chain Gate
(TARIQ BAB EL-SILSILA)
Khalidi Library
Beit Strauss
El-Kas Fountain
Jehoshaphat's Tomb
CHAIN STREET
WESTERN WALL PLAZA
The Western Wall
Tashtamuriyya Building
Moors' Gate
Israelite Tower
SHONEI HALAKHOT
El-Aqsa Mosque
Absalom's Tomb
JEWISH QUARTER
Ariel Centre for Jerusalem in the Ist Temple period
MISGAV LA-DAKH
Museum of Islamic Art
HA-TAMID
HA-UGAV
Burnt House
St Mary of the Germans
TIFERET YISRAEL
MAALOT RABIL
BATEI MAKHASE
JERUSALEM ARCHAEOLOGICAL PARK
Wohl Archaeological Museum
Davidson Center
KHAYEI OLAM
Nakhal Kidron
Dung Gate
HA-GITIT
DEREKH HA-OFEL
DEREKH HA-SHILOAKH
MADOT YISRAEL
HA-SHMINIT
MAA
BATEI MAKHASE SQUARE
BEIT HA-SHOEVA
GALED
Rothschild House
NAKHMAN BATEI MAKHASE
Beit ha-Sofer
MAALE HA-SHALOM
WADI HILWA
MAALOT IR DAVID
CITY OF DAVID
The Nea Vaults
1
2
3
5
I
II
IV
V

ISRAEL, PETRA & SINAI REGION BY REGION

Israel, Petra & Sinai at a Glance

The Holy Hand is rich in historical sights far beyond its biblical associations. In Petra it has one of the most unusual and magical ruined cities in the world, and the Roman-era remains at sites such as Jerash in Jordan and Beth Shean in northern Israel are similarly stunning. The scenery that the visitor encounters while travelling can also be dramatic, especially in the region of the Dead Sea (a geographic marvel in itself) and in the Sinai peninsula. Off the coast of Sinai, the Red Sea conceals underwater scenery every bit as spectacular as that on dry land.

Waterfront at Jaffa, a virtual suburb of Tel Aviv and a favourite place for city-dwellers to dine at weekends

THE DEAD SEA AND THE NEGEV DESERT
(See pp186–205)

Beautiful sandstone cloisters at the Church of the Nativity in Bethlehem

THE RED SEA AND SINAI
(See pp236–249)

St Catherine's Monastery, Sinai, one of the world's oldest continuously functioning monasteries

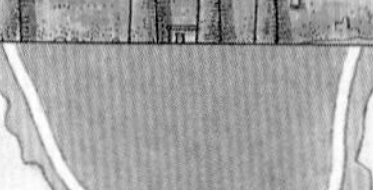

◁ Spectacular desert scenery at Wadi Rum in Western Jordan

View from the shore of the Sea of Galilee, rich in associations with the miracles and teachings of Jesus Christ

The ruined main street of Jerash, the best-preserved Roman city in the Holy Land

The mountaintop fortress of Masada on the Dead Sea, the most visited site in Israel after Jerusalem

The incredible shaping of the landscape in the carved rock façades of Petra

0 kilometres 50

0 miles 50

THE COAST AND GALILEE

A fertile corridor squeezed between the sea and the desert, this is the Promised Land of the Old Testament. The green hills and fresh waters of Galilee provided the setting for many episodes in the early life and ministry of Christ. Beside all its religious associations this is very much a secular paradise too, the heartland of modern Israel and a sun-drenched scenic magnet for tourists.

The wealth of ancient sites along this stretch of coast bears witness to the fact that for centuries this has been an important land corridor connecting Africa, Europe and Asia. The great empires of ancient Egypt to the south and Assyria and Babylon to the east met here in trade and battle. Later, the Romans exploited this coastline with the laying of a great highway, the Via Maris, and Herod built a magnificent port in Caesarea *(see p176)*, one of the grandest and most important in the eastern Mediterranean. Ports such as this formed the nuclei of the Latin Kingdoms when the Crusaders came conquering in the Middle Ages. The Muslim Arabs eventually drove out the Christian knights but their legacy remains in some superb muscular architecture, especially at Akko, which retains one of the most charming old towns in the whole of the Holy Land.

When in the 19th century the first major waves of Jewish immigrants began arriving, it was on the fertile coastal plains and rolling hills of Galilee that they chose to settle. They planted wheat and cotton in the fields, orange groves and vineyards on the slopes, and cities overlooking the sea. The capital they founded, Tel Aviv, has become a vibrant centre of culture and commerce, while Haifa, attractively tumbling down Mount Carmel to the sea, is a thriving economic powerhouse. Inland Galilee remains rural and idyllic, equally pleasing to pilgrims on the trail of Christ and to seekers after relaxation and the picturesque.

The harbour at Akko, stronghold of the Crusaders and one of the Holy Land's best preserved old cities

◁ A vaulted street in Jaffa, an important ancient port now part of metropolitan Tel Aviv

Exploring the Coast and Galilee

Northern Israel is arguably the most attractive region in the Holy Land. The coast has long white sandy beaches, while Galilee is a landscape of rolling green hills, forested valleys and clear freshwater lakes. The Golan even has mountains that are capped with snow for part of each year. Places of interest include the hilltop Jewish holy town of Safed, Nazareth, traditionally held to be where Jesus spent his childhood, and many fine archaeological sites, including Crusader castles and Roman towns. With such a concentration of beauty spots and picturesque vistas, this is an area ideally explored by car.

SIGHTS AT A GLANCE

Akko pp178–9 4
Belvoir Castle 13
Beth Alpha 14
Beth Shean 15
Caesarea 2
Capernaum 10
Golan Heights 8
Haifa and Mount Carmel 3
Megiddo 5
Nazareth 6
Safed 7
Sea of Galilee pp182–3 9
Tabgha 11
Tel Aviv pp168–73 1
Tiberias 12

Herod the Great's port of Caesarea, now an impressive set of ruins beside the sea

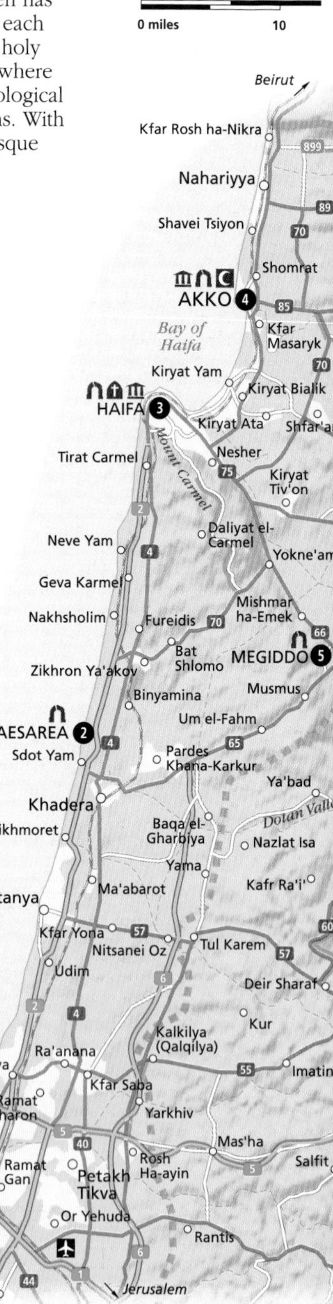

GETTING AROUND

Jerusalem and Tel Aviv are linked by a good motorway. Buses depart roughly every 15 minutes and the journey takes less than an hour. Northbound services along the coastal highway from Tel Aviv to Caesarea and Haifa are only slightly less frequent. Trains link Tel Aviv and Jerusalem, and there is a coastal line from Tel Aviv to Nahariya.

For additional map symbols *see back flap*

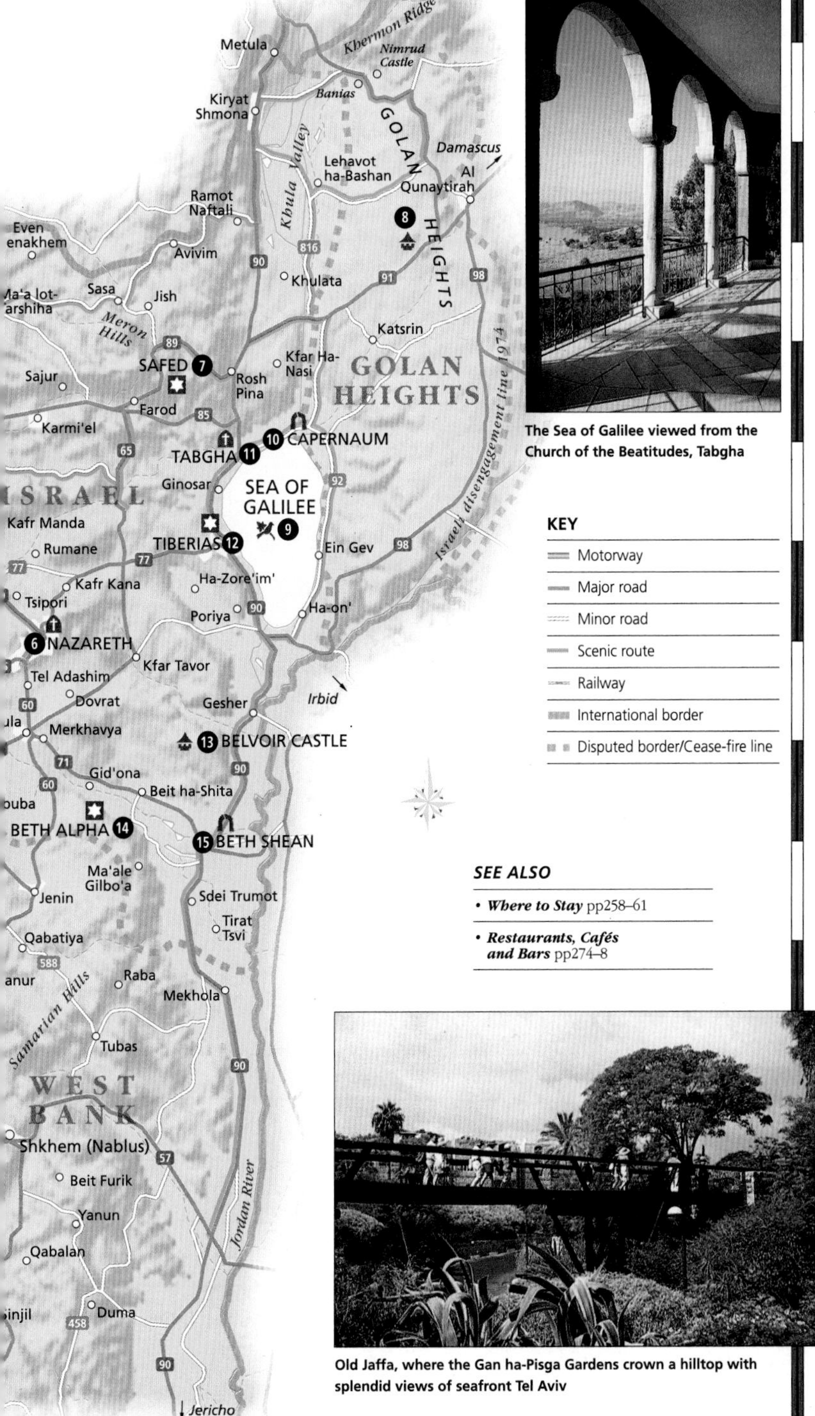

The Sea of Galilee viewed from the Church of the Beatitudes, Tabgha

SEE ALSO

- ***Where to Stay*** pp258–61
- ***Restaurants, Cafés and Bars*** pp274–8

Old Jaffa, where the Gan ha-Pisga Gardens crown a hilltop with splendid views of seafront Tel Aviv

Tel Aviv ❶

Tel Aviv represents the modern face of the Jewish state – a brash, confident centre of commerce and contemporary culture. It is also a true Mediterranean resort city, with a long, sandy beach fringed by cafés, bars and shops. Away from the seafront are gracious palm-filled avenues, lined with elegant buildings in the international Bauhaus style *(see p171)*. All this has been created since 1909, when the Jewish National Fund purchased land among the dunes north of the old Arab port of Jaffa *(see pp174–5)* on which to build a new city, to be called Tel Aviv ("Hill of the Spring").

Mosaic flooring at the Eretz Israel Museum in northern Tel Aviv

Beit Hatfutsot (Museum of the Jewish People)

University Campus, Gate 2, Klausner Street, Ramat Aviv. ***Tel*** *(03) 745 7800.* ◯ *10am–4pm Sun–Tue, Thu; 10am–6pm Wed; 9am–1pm Fri.* **www**.bh.org.il

When it opened in 1978, this was regarded as one of the world's most innovative museums. It is still worth setting aside several hours to visit. Instead of showing historical artifacts, it uses thematically arranged dioramas, interactive displays and short videos to illustrate aspects of life in the Jewish diaspora, past and present, throughout the world, and the influence of Jewish arts and literature on other cultures. One of the highlights is a display of beautifully made scale models of synagogues from various countries. The permanent collection is supplemented by temporary exhibitions. There is also a genealogy centre, where Jews from around the world can trace their lineage.

Beit Hatfutsot exhibit

Eretz Israel Museum

2 Haim Levanon, Ramat Aviv. ***Tel*** *(03) 641 5244.* ◯ *10am–4pm Sun–Wed, 10am–8pm Thu, 10am–2pm Fri, Sat.* **www**.eretzmuseum.org.il

Built around the site of Tel Qasile, where excavations have revealed layers of human habitation dating back to 1200 BC, this museum depicts the history and culture of the land of Israel. It comprises a number of themed pavilions, all containing permanent exhibitions. One has a very fine collection of ancient and Islamic-era glass; others are devoted to coins, ancient pottery, Judaica, copper mining, postal history and philately, and to ancient crafts. Additionally there's a square with a collection of beautiful mosaic floors from early synagogues, churches and mosques; an old olive oil press; a reconstructed flour mill; and a 1925 fire engine given by the city of New York to Tel Aviv's volunteer fire brigade in 1947.

Historical Jewish personages – part of a display at Beit Hatfutsot

Old Port

North of the centre, at the point at which the Yarkon River empties into the Mediterranean, Tel Aviv's port (known as the Namal) was developed in the late 1930s to lessen Jewish dependence on the Arab

For hotels and restaurants in this region see pp258–9 and pp275–8

SIGHTS AT A GLANCE

Beachfront Promenade ④
Beit Hatfutsot (Museum of the Jewish People) ①
Bialik Street ⑧
Dizengoff Street ⑤
Eretz Israel Museum ②
Manshiye ⑫
Neve Tzedek ⑬
Old Port ③
Rabin Square ⑥
Rothschild Avenue ⑪
Shalom Tower ⑩
Tel Aviv Museum of Art ⑦
Yemenite Quarter ⑨

Key to Symbols *see back flap*

port of Jaffa. It was decommissioned in 1965, when bigger facilities were created in Ashdod to the south, and lay neglected for around 30 years until the site was revitalised in the 1990s. It is now a lively area of bars, cafés, restaurants, nightclubs and shops. There is even an antiques market on Saturdays. Many of the businesses are on the boardwalk facing the sea; many also have a view of the disused power plant just across the river, which serves as the venue for the Ha'aretz Art Festival every autumn.

A café on the boardwalk in the fashionable Old Port area

Beachfront Promenade

A white-sand beach stretches right along the seafront of central Tel Aviv, backed by a long promenade, modern hotels and Miami-style condominiums. It is possible to walk all the way from the Old Port in the north down to Jaffa in the south *(see pp174–5)* along the promenade. At its northern end this takes the form of a big, rolling wooden deck, which in parts gently undulates like sand dunes. This is a favourite area for fishermen and for wedding couples, who have their photographs taken with the Mediterranean Sea as a backdrop.

Further south, in the vicinity of **Independence Park** (Gan Ha-Atzmaut), there's a small children's playground. Beside this, a section of beach is screened off for the use of Orthodox Jews (men and women on different days).

The city centre stretch of beach is dominated by the huge, pink **Opera Towers**, with shops and restaurants at street level, and a distinctive stepped profile. The beach here is crowded all summer with sun-seekers and, after dark, with open-air concert- and disco-goers. Strong sea currents mean that you should swim only where you see white flags. Red flags mean that it is dangerous; black flags that it is forbidden.

VISITORS' CHECKLIST

Road Map B3. *390,000. Ben Gurion, 22 km (14 miles) SE. Arlosoroff Station, Arlosoroff Rd, tel. *5770. New Central Bus Station, Levinsky St, (03) 639 4444 (local buses), (03) 694 8888 (long-distance buses). 46 Herbert Samuel Rd (03) 516 6188. Beach Festival (Jul & Aug). daily.*

The beachfront parade in central Tel Aviv, part of a well-maintained promenade that stretches the length of the city

Exploring Tel Aviv

North central Tel Aviv is where the money is. Visit Basel Street for chic cafés and boutiques. The real heart of the city, however, lies south of Ben Gurion Avenue, which is named for Israel's first prime minister *(see p53)*; his former home at No. 17 is now a museum. The main streets run north–south and are Ben Yehuda Street and Dizengoff Street (see below), both of which run almost the whole length of the city centre. South again is the Yemenite Quarter and the districts of Manshiye and Neve Tzedek, which are some of the oldest parts of Tel Aviv.

Dizengoff Square with a performing fountain at its centre

Dizengoff Street

The city's main shopping street is named after Tel Aviv's first mayor, Meir Dizengoff. It is at its liveliest around the junction with Frishmann Street, where there are plenty of street cafés with pavement seating and a large branch of the Israeli chain bookstore Steimetzky's. Also here is the **Bauhaus Center**, which is dedicated to raising awareness of Tel Aviv's unique architectural heritage *(see p171)*. To this end, the Center runs two-hour English-language tours at 10am each Friday visiting some of the city's Bauhaus buildings.

One block south of the Bauhaus Center is **Dizengoff Square**, an irregularly shaped concrete platform raised above a traffic underpass. It sports a drum-like fountain by Israeli artist Yaakov Agam that has water jets programmed to perform hourly light and music shows. At the weekend, the square is host to a flea market. On the east side are two beautifully renovated Bauhaus buildings, one of which is now the **Hotel Cinema Eden**; it's possible to take the elevator up to the fifth-floor roof terrace to enjoy the city views.

Bauhaus Center

99 Dizengoff Street. ***Tel*** *(03) 522 0249.* *10am–7:30pm Sun–Thu; 10am–2:30pm Fri, noon–7:30pm Sat.* **www**.bauhaus-center.com

Rabin Square

A large, rectangular plaza in the eastern part of central Tel Aviv, Rabin Square is overlooked by **City Hall**, a brutal concrete block that is only slightly softened by having its windows painted in different colours. The square is a venue for demonstrations, celebrations and concerts. It was at one such gathering – a peace rally on 4 November 1995 – that the then Israeli Prime Minister Yitzhak Rabin was assassinated. The basalt stones of the **Rabin Memorial** on Ibn Givrol Street, beside City Hall, occupy the very spot where he was shot. Nearby is a wall covered with graffiti drawn by mourning citizens and now preserved behind glass.

At the centre of the square is another memorial, the **Monument of Holocaust and Resistance**, a huge glass and iron structure erected in the 1970s and designed by well-known and often controversial Israeli artist Yigal Tumarkin.

There are some good shops on the west side of the square, notably Tola'at Seferim, a bookshop with a pleasant café, and Mayu, a youthful fashion boutique. Across on the east side is Brasserie, an excellent Art Deco, French-style restaurant.

Tel Aviv Museum of Art

27 Ha-Melekh Shaul Avenue.
Tel *(03) 607 7000.* *10am–4pm Mon, Wed & Sat; 10–10pm Tue & Thu; 10am–2pm Fri.*
www.tamuseum.com

Israel's most important collection of 19th- and 20th-century art includes works representing the major trends of modernism: Impressionism (Degas, Renoir, Monet), Post-Impressionism (Van Gogh, Gauguin, Cézanne), Cubism (Braque, Leger, Metzinger) and Surrealism (Miró), as well as key pieces by Pablo Picasso. Other works range from 17th-century Flemish to modern Israeli. In addition to the permanent collections, there are excellent temporary exhibtions. A ticket also covers entrance to the **Helena Rubenstein Pavilion** on Habima Square, where additional contemporary art shows are held.

Modern large-scale sculpture outside the Tel Aviv Museum of Art

Tel Aviv's Bauhaus Architecture

Tel Aviv has the world's largest assemblage of buildings in the International Modern style, also known as Bauhaus. Altogether there are some 4,000 examples within the city. These buildings, largely erected in the 1930s and 1940s, were designed by immigrant architects trained in Europe, particularly in Germany, home of the modernist Bauhaus School between 1919 and 1933. The simplicity and functionality of the style, which aimed to unify art with technology, was considered highly appropriate to the socialist ideals of Zionism that underpinned the founding of the new city.

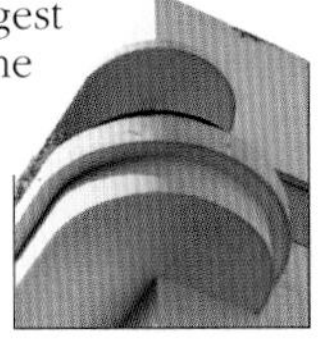

Rounded balconies on a Bauhaus building

In 2003, Tel Aviv's unique and bountiful Bauhaus legacy was recognised by the United Nations cultural agency UNESCO, who declared the "White City" on the Mediterranean a World Heritage Site.

Horizontals Characteristics of Bauhaus architecture include asymmetrical façades with "ribbons" of windows running horizontally. Balconies are often curbed and have overhanging ledges to provide shade for the rooms below.

Verticals The sole vertical element in the typical Bauhaus building is provided by the internal stairwell; this appears on the façade as a ladder-like arrangement of windows.

Ships Some of the most striking buildings were inspired by the superstructure of the ships that brought the Jewish immigrants to Palestine. Windows shaped like maritime portholes are a common feature.

Where to look The highest concentration of Bauhaus buildings is on Rothschild Boulevard and neighbouring Ahad Ha'am Street. The Bauhaus Centre, on Dizengoff Street *(see p170)* is a source of books and information on the subject, as well as a place to find some unusual souvenirs.

Rounded forms Although initially Bauhaus buildings were completely rectilinear, later architects began to introduce more rounded forms. This was decried by purists who regarded curves as heretical because of their supposed impracticality: "How do you hang a picture on a curved wall?" they asked.

Nahum Gutman's colourful, mosiac-covered fountain on Bialik Street

Bialik Street

Bialik is one of the city's most historic streets. At No. 14 is the **Rubin Museum**, the former residence of one of Israel's most famous painters, Reuven Rubin (1893–1974). It now contains a permanent collection of 45 of his works, as well as a historical archive of his life. Changing exhibits feature other Israeli artists. A few doors along, **Bialik House** (Beit Bialik) is the former home of Haim Nahman Bialik (1873-1934), Israel's national poet. The house has been kept as it was during Bialik's time, and includes a library and paintings by some of Israel's best-known artists.

At the end of the street is a striking mosaic-covered fountain by Nahum Gutman *(see p173)*.

A little south of Bialik, **Bezalel Street** is home to a street market famed for cut-price fashion. South again, **Sheinkin Street** was a centre of alternative culture in the 1980s. That is no longer the case, but it still boasts many independent shops and cafés.

Rubin Museum

14 Bialik Street. ***Tel** (03) 525 5961.* *10am–3pm Mon, Wed, Thu; 10am–8pm Tue; 11am–2pm Sat.* **www**.rubinmuseum.org.il

Bialik House

22 Bialik Street. ***Tel** (03) 525 4530.* *11am–5pm Mon–Thu, 10am–2pm Fri & Sat.* *(book ahead).*

Yemenite Quarter

A masterplan for Tel Aviv was drawn up by Scottish urban planner Sir Patrick Geddes at the request of Mayor Dizengoff in 1925. This influenced the growth of the city for decades to come. The Yemenite Quarter (Kerem Ha-Temanim), however, predates the Geddes plan, and its maze of small streets contrasts sharply with the orderly layout of the rest of the city. The architecture also predates the arrival of the Bauhaus style that characterises much of the rest of Tel Aviv. Here, buildings instead employ motifs from Classical, Moorish and Art Nouveau styles. This is most apparent on **Nakhalat Binyamin Street**, which boasts many curious, if slightly faded, examples of this eclectic architecture. The street is especially worth visiting on Tuesdays and Fridays, when it hosts a busy craft market. This is also one of the busiest nightlife streets, in particular the area around the junctions with Rothschild Avenue and Lilienblum Street.

Street performer on Nakhalat Binyamin

The other local landmark is **Carmel Market** (open 9am–6pm Sun–Thu, 9am–3pm Fri), which is on Ha-Carmel Street and is the city's largest and busiest open-air market. It begins near the junction with Allenby Street with stalls selling cheap clothing and household items, before switching to fresh fish, meat, fruit and vegetables, spices and herbs, breads and biscuits, and nuts and seeds. Many of the side streets off Ha-Carmel specialise in different food produce.

Shalom Tower

9 Ahad Ha'am Street. ***Tel** (03) 517 7304.* *9am–6pm Sun-Thu; 9am–2pm Fri.*

One block west of Nakhalat Binyamin Street, this austere, 1960s office building sits on the former site of Israel's first secular Hebrew school. At the time of its construction, the tower was the tallest structure in Israel (it is now surpassed by the radio tower near the Tel Aviv Museum). There are impressive mosaics in the lobby area, and original street lamps at ground level. Also of interest in the tower are several small exhibitions that contain models of Tel Aviv and multimedia presentations, plus an art gallery. A number of attractive restaurants and cafés make this a pleasant spot to while away an afternoon.

Rothschild Avenue

This is one of Tel Aviv's most elegant old thoroughfares, lined with palm trees and some of the city's finest examples of Bauhaus buildings *(see p171)*.

Twice-weekly craft market on Nakhalat Binyamin in the Yemenite Quarter

Attractive Hassan Bek Mosque, founded by a local governer

Independence Hall (Beit Ha-Tanakh) at No. 6 was once the residence of the first mayor, Meir Dizengoff. This is also where Ben Gurion declared the independence of Israel on 14 May 1948. The museum's Hall of Declaration remains as it was on that day, with original microphones on the table and a portrait of Herzl, the Zionist leader. Nearby 23 Allenby Street is now the **Haganah Museum**. The Haganah was the clandestine pre-1948 military organisation that later became the Israeli army.

Independence Hall
16 Rothschild Boulevard. ***Tel*** *(03) 517 3942.* *9am–1:45pm Sun–Fri.*

Haganah Museum
23 Rothschild Boulevard. ***Tel*** *(03) 560 8624.* *8am–4pm Sun–Thu.*

Manshiye

Manshiye is the coastal neighbourhood that acts as a buffer between the twin municipalities of Tel Aviv and Jaffa *(see pp174–5)*. Its most distinguished landmark is the little **Hassan Bek Mosque** on the main seafront road, built in 1916 by a governor of Jaffa of the same name. During the 1948 War, Arab soldiers used the mosque's minaret as a firing position; this is one of the episodes recorded in the nearby **Etzel Museum 1947–1948**, which is dedicated to the Israeli defence forces and their role in this particular conflict. Historical documents, photos, newspaper clippings and weapons are exhibited in a purpose-built, black-glass structure in attractive **Charles Clore Park** on the seafront. The park is a venue for many of the city's big open-air events, including the annual Love Parade.

Etzel Museum 1947–1948
15 Goldman Street. ***Tel*** *(03) 517 2044.* *8am–4pm Sun–Thu.*

Neve Tzedek

Neve Tzedek is where Tel Aviv began. The settlement was founded on empty sandy flats in the late 1880s by a group of Jewish families keen to escape overcrowding in the port of Jaffa. Today, the area retains the feel of a small village, with narrow lanes lined by high walls and a strange mix of architectural styles. Decades of neglect are currently being reversed by an energetic programme of renovation and restoration.

At the heart of the district is the **Suzanne Dellal Centre** for dance and drama. It boasts four performance halls in a building that was once a local school. The main courtyard, with orange trees and tiled murals, is a popular place to meet and relax.

Nearby, the **Rokach House Museum** occupies the former home of Shimon Rockach, one of the founding fathers of Neve Tzedek. Inside, photos and documents illustrate the daily life of the community at the end of the 19th century.

A few doors away, the **Nahum Gutman Museum** is dedicated to another of Israel's best-known artists, a Russian-born painter who was also admired for his children's books. As well as displaying a small collection of Gutman's work, the galleries are used for temporary exhibitions.

Suzanne Dellal Center
6 Yehieli Street.
Tel *(03) 510 5656.*

Rokach House Museum
36 Shimon Rockach Street. ***Tel*** *(03) 516 8042.* *10am–2pm Fri & Sat.*
www.rokach-house.co.il

Nahum Gutman Museum
21 Shimon Rokach Street.
Tel *(03) 516 1970.* *10am–4pm Sun–Wed; 10am–8pm Thu; 10am–2pm Fri; 10am–5pm Sat.*
www.gutmanmuseum.co.il

The history of Neve Tzedek in tiled murals at the Suzanne Dellal Centre

Street-by-Street: Old Jaffa

Artists' Quarter mural

According to the Bible, Jaffa (then called Joppa) was founded in the wake of the great flood by Noah's son Japheth. Archaeologists have unearthed remains dating back to the 20th century BC, establishing Jaffa as one of the world's oldest ports. However, with the growth of Tel Aviv, Jaffa, which had flourished under the Ottomans, went into decline. Following Jewish victory in the 1948 War it was absorbed into the new city to the north. The core of the old town has since been revived as an attractive arts, crafts and dining centre.

The seafront of Old Jaffa, with its warehouses reborn as restaurants

Archaeological Museum
Housed in an elegant 18th-century local government building, this museum holds finds from digs in the area.

Ha-Pisga open-air amphitheatre is used for concerts during the summer.

To Flea Market

To Clock Tower

The Mahmoudiya Mosque dates from 1812 and remains in use by the local Muslim community.

A 19th-century *sabil* (fountain)

To the Promenade

MIFRAZ SHLOMO

HA-ALIYAH HA-SHNIYA

Napoleonic cannons

The Sea Mosque was the mosque of local fishermen.

Clock Tower
Built in 1901 to mark the 25th anniversary of the then Turkish sultan, the clock tower has since been heavily restored and now serves as a symbol of modern Jaffa.

Gan ha-Pisga
Ha-Pisga garden lies on top of the ancient 'tel' (mound) of Jaffa. An observation area, marked by the curious Statue of Faith, *offers good views across to Tel Aviv.*

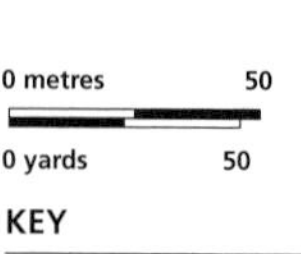

0 metres 50

0 yards 50

KEY

— — — Suggested route

For hotels and restaurants in this region see pp258–9 and pp275–8

★ Artists' Quarter
A compact area of old Arab houses and narrow stone-flagged alleys, in recent times this has been transformed into residences, studios and galleries for artists and craftspeople.

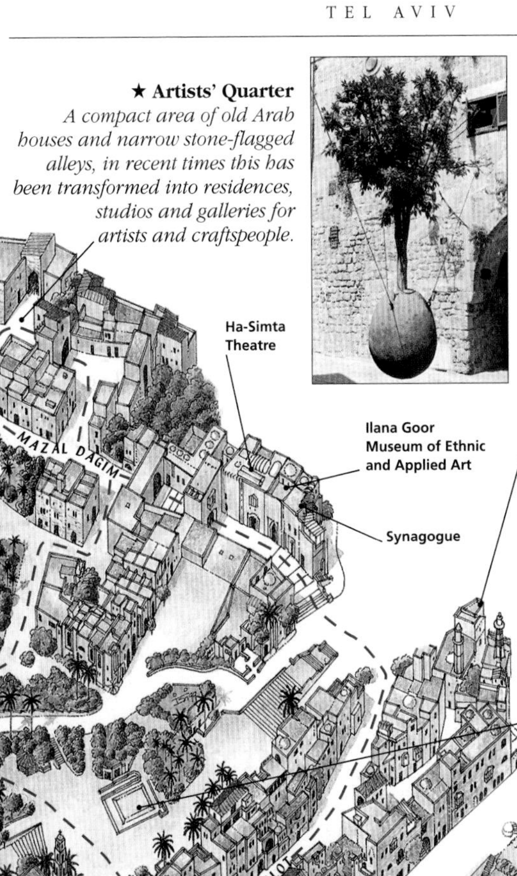

Ha-Simta Theatre

Ilana Goor Museum of Ethnic and Applied Art

Synagogue

VISITORS' CHECKLIST

2 km (1 mile) S of central Tel Aviv. *Kedumim Square, (03) 518 4015.* *10am–6pm.* **www**.oldjaffa.co.il **Archaeological Museum** 10 Mifraz Shlomo St. ***Tel*** *(03) 682 5375.* *call for opening hours.*

The House of Simon the Tanner is traditionally held to be where the apostle Peter once stayed (Acts 9: 43).

★ Kedumim Square
Underneath the picturesque main square of Old Jaffa is the Visitors' Centre, with exposed Roman-era exhibits and a light and sound show about the old city.

St Michael's Church
Dating from the 19th century, this small Greek Orthodox church has recently been renovated.

The Monastery of St Nicholas, built around 1667, still serves Jaffa's Armenian community.

The Wishing Bridge was renovated in 2005. It is said to bring true the wish of anyone crossing it if they touch the bronze statue of their zodiac sign while looking at the sea.

Monastery of St Peter
Built in Latin American Baroque style, this Roman Catholic monastery and church was dedicated in 1891. It stands on a site formerly occupied by a Crusader citadel.

STAR SIGHTS

★ Artists' Quarter

★ Kedumim Square

The impressive Roman aqueduct at Caesarea

Caesarea ❷

Road map B2. 🚌 76 and 77 from Khadera. ℹ *(04) 617 4444.*

At the height of his power, in 29–22 BC, Herod the Great *(see pp43–5)* built a splendid city over the site of an ancient Phoenician port and dedicated it to Augustus Caesar, the Roman emperor. The splendour of this city is attested to by the lavish description of it by Flavius Josephus in his book *The Jewish War*. Until the many recent excavations, this had been seen by many scholars as wild exaggeration.

This period of prosperity lasted in Caesarea until AD 614, after which its history became more unstable. During the early 12th century and the Crusades, Caesarea again became an important city, and was used once more as a port. By the late 13th century however, it had been destroyed by the Mamelukes and was left to be reclaimed by the sand, with only a small Arab village remaining. The importance of these great hidden ruins was not realized until the 1940s, and now Caesarea is one of Israel's major archaeological sites.

Most of the main sights lie in the **Caesarea National Park**. If entering from the south, you will first see the huge Roman theatre. With seats for 4,000 spectators, it has been restored, and hosts summer concerts. A short distance to the west, on a small coastal promontory, a group of half-submerged walls indicate the site of Herod's palace. Further inland are the neglected ruins of one of the largest hippodromes in the Roman Empire.

On the coast by the inner harbour is the Crusader citadel, still surrounded by walls which date back to around AD 1250. Enclosing this whole area are the ruins of the much larger Crusader city walls. Within these ruins lies the unique **Underwater Archaeological Park**. The four diving complexes at this new park enable divers to see the techniques used to build the ancient port, as well as remnants of wrecked ships.

North of the ancient city is the extraordinary Roman aqueduct dating from the Herodian period. Extending for 17 km (11 miles), it carried water from the foothills of Mount Carmel to Caesarea. A short way to the south of the site, the **Caesarea Museum** has interesting artifacts from the Roman city.

Caesarea National Park
◯ *8am–4pm daily.*

Underwater Archaeological Park
Caesarea Harbour. ***Tel*** *(04) 626 5898.*
◯ *6am–dark daily.*

Caesarea Museum
Kibbutz Sdot Yam. ***Tel*** *(04) 636 4367.* ◯ *daily.*

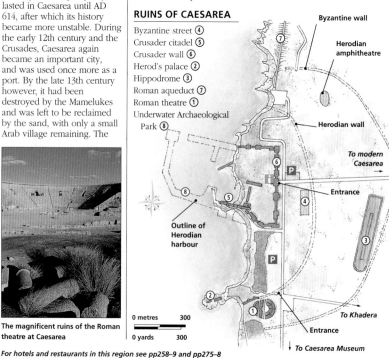

The magnificent ruins of the Roman theatre at Caesarea

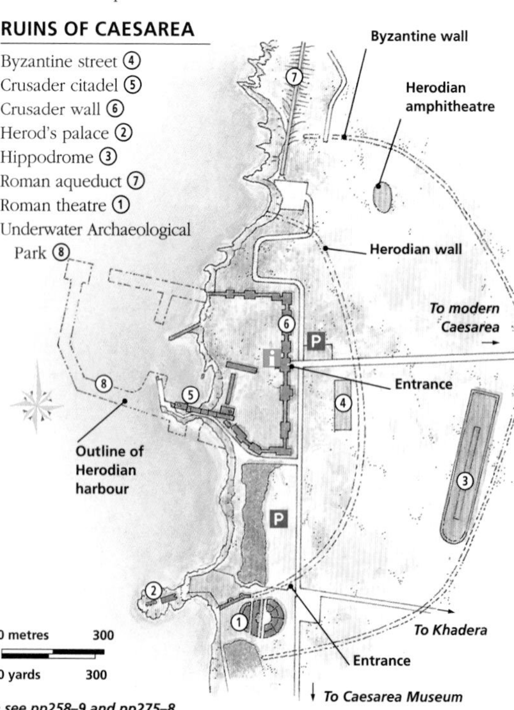

Haifa and Mount Carmel ❸

The city of Haifa lies on the Mediterranean coast at the foot of Mount Carmel. Israel's third largest city, it is a major industrial centre. Away from the busy port, steep slopes rise up the mountain, providing quiet, attractive suburbs for the wealthy. A small trading port for most of its history, Haifa was conquered by the Crusaders in the early 12th century *(see pp48–9)*, and later fortified under Ottoman rule. In the late 19th century it became an important refuge for Jewish immigrants. Between 1918 and 1948 Haifa was taken over by the British in the occupation of Palestine. Today it is a mixed, non-religous city, and the only one where buses run on Saturdays.

The spectacular Baha'i Temple and gardens in Haifa

National Museum of Science and Technology

Old Technion, 12 Balfour St. ***Tel*** *(04) 861 4444.* *Daily.*
www.mustsee.co.il

The former Technology Institute in the city centre is one of Haifa's most important buildings. Founded by German immigrants in the early 1900s, it was Israel's first institute of higher education. Renovated many times, it is now home to the National Museum of Science and Technology, which has many interesting interactive exhibits, exploring the latest innovations in Israeli science.

Baha'i Shrine and Gardens

Ha-Ziyonut St. ***Tel*** *(04) 831 3131.* *daily (shrine: am only).* www.bahaigardens.org.il

On the edge of the city centre towards Central Carmel is Haifa's most striking landmark, the impressive golden-domed Baha'i Shrine. Standing imperiously on the hillside, it is surrounded by a splendidly manicured garden, and is the headquarters of the Baha'i faith. Its followers believe that no religion has a monopoly on the truth, and aim to reconcile the teachings of all holy men. The ornate shrine houses the tomb of the Bab, the herald of Bahaulla. Bahaulla (1817–92) is the central figure of the Baha'i faith and is considered by his disciples to have been the most recent of God's messengers.

Central Carmel

South of the temple, Central Carmel spreads up the slopes of the mountain. A largely wealthy residential area, it manages to resist the onslaught of traffic and busy modern life. Its many parks, cafés, and stylish bars make it a relaxing detour.

VISITORS' CHECKLIST

Road map B2. *290,000.* *48 Ben Gurion St, (04) 853 5606.*

Bat Galim

Northwest of Central Carmel is the popular coastal area of Bat Galim. Close to the city centre, its beach and busy seafront promenade have made it a favourite with tourists. For those wanting more extensive beaches, however, try the attractive Carmel Beach. This is 6 km (4 miles) to the south, away from the busy city.

Carmelite Monastery

Stella Maris St. ***Tel*** *(04) 833 7758.* *daily.*

On much of the upper slopes of Mount Carmel are wide stretches of vegetation, the remnants of an ancient forest. On these slopes, to the southwest of Bat Galim, is the Stella Maris Carmelite Monastery, which can be reached by cable car or on foot. Built in an area that for centuries was frequented by hermits, this was a place of worship near where the Carmelite order was founded. The beautiful church here dates from the early 1800s.

Elijah's Cave

201 Allenby St. ***Tel*** *(04) 852 7430.* *Sun–Fri (Fri: am only).*

Below the monastery is Elijah's Cave, where Elijah is said to have lived and meditated before defeating the pagan prophets of Baal on Mount Carmel. Today it is a synagogue with a Torah Ark and a niche in the ceiling where visitors can place notes.

Dome of the Stella Maris Carmelite Monastery

Akko ❹

Khan el-Umdan clocktower

Outside of Jerusalem Akko (the historic Acre) has the most complete and charming old town in all of the Holy Land. Its origins date back to the Hellenistic period, but the form in which it survives today was set by the Arabs and their Crusader foes. After the Crusaders took Jerusalem in 1099, they seized Akko as their main port and lifeline back to Europe. Lost at one point to the Muslim armies under Saladin, it was regained by Richard I "the Lionheart". For most of the 13th century, with Jerusalem in the hands of the Muslims, Akko was the Crusaders' principal stronghold. As the Christian armies steadily lost ground, it was the last bastion to fall. Akko's fortunes were revived under a series of Ottoman governors, one of whom, Ahmed Pasha el-Jazzar, successfully defended the city against an invasion by Napoleon in 1799.

The harbour at Akko, in continuous use since Hellenistic times

Exploring Akko

Crusader Akko was destroyed by the victorious Arab armies in 1291 and what can be seen today is largely an 18th-century Turkish town built on the site of the old. The defensive **walls** are rebuildings of the original Crusader walls, fragments of which are still discernible. The warren-like street pattern is interrupted by three great khans, or merchants' inns: the **Khan el-Umdan** (Khan of the Columns) with its distinctive clocktower; the **Khan el-Faranj** (Khan of the Franks or Foreigners); and the **Khan a-Shuarda** (Khan of the Martyrs). While the khans date from the Ottoman era they echo the fact that in Crusader times Akko had autonomous quarters given over to the merchants of Italy and Provence. Such was the rivalry between these colonies that at one point open warfare erupted between the Venetians and Genoese, who fought a sea battle off Akko in 1256. The khans are no longer in commercial use but Akko does have a lively **souk**, selling fruit, vegetables and household items. You'll also find plenty of fresh fish, which you can see being brought ashore at the town's picturesque harbour early each morning.

There is also the Ethnographic Museum, which has a beautiful collection that illustrates life in Galilee from the 19th century to the beginning of the 20th century.

Mosque of El-Jazzar

El-Jazzar St. ***Tel*** *(04) 991 3039.*
daily. during prayers.

Akko lay semi-derelict for more than 400 years after its destruction in 1291. Its rebirth came with the rule of the emir Dahr el-Amr and his successor, Ahmed Pasha El-Jazzar ("the Butcher"), both of whom governed the city for the Ottomans in the second half of the 18th century. El-Jazzar, in particular, was a prolific builder. Among his legacy is the Turkish-style mosque (built 1781) that bears his name and continues to dominate the old town skyline. Its courtyard contains recycled columns from the Roman ruins of Caesarea and, at the centre, a small, elegant fountain used for ritual ablutions. By the mosque are the sarcophagi of El-Jazzar and his son, while underneath are the remains of a Crusader church that El-Jazzar had transformed into a cistern to collect rainwater.

Crusader City

El-Jazzar St. ***Tel*** *(04) 995 6706.*
winter: 8:30am–4pm daily; summer: 8:30am–5pm daily.

When the Ottoman governors rebuilt Akko they did so on top of the ruins of the Crusader city. The Crusader-era street level lies some 8 m (25 ft) below that of today. Part of it has been excavated revealing a subterranean wealth of well-preserved examples of 12th- and 13th-century streets and buildings. There are some amazingly grand Gothic

Akko's dominant landmark, the Turkish-style Mosque of el-Jazzar

For hotels and restaurants in this region see pp258–9 and pp275–8

Gothic-arched halls of the former Crusader city in Akko

knights' halls, built around a broad courtyard. An extensive network of drainage channels has also been excavated. South of the couryard is a large refectory with huge columns; in two corners you can still see carved lilies that may indicate building work done in the period of Louis VII of France, who arrived at Akko in 1148. One of Akko's other well-known visitors was Marco Polo and it is quite possible that he dined in this very room. Below the refectory is a network of underground passageways that lead to an area known as El-Bosta (from the Arabic for "post office", which is what the Turks used this space for); divided by columns into six sections, it was originally the crypt of St John's Church.

Citadel

Off Ha-Haganah St. ***Tel*** *(04) 995 6707.* *Sun–Fri (Fri: am only).*

Akko's Citadel was built by the Turks in the 18th century on top of Crusader foundations. During the British Mandate it served as a prison for Jewish activists and political prisoners, some of whom were executed in the gallows room. These events are commemorated in the Citadel's **Museum of Underground Prisoners**.

Hammam el-Pasha

Off El-Jazzar St.
Tel *(04) 995 1088.*
daily (Fri: am only).

This is not a museum as such, but a Turkish bathhouse dating to 1780 and the rule of El-Jazzar (hence the name of Hammam el-Pasha, meaning "Bathhouse of the Governor"). It was in use until as recently as the 1940s and remains in an excellent state of repair. The floors and walls are composed of panels of different coloured marble, and the fountain in the "cold room" (where patrons would relax after bathing) retains most of its beautiful majolica decoration. A sound and light show introduces visitors to the history of Akko and the life of a typical bathhouse attendant.

Fountain from the Hammam el-Pasha

VISITORS' CHECKLIST

Road map B2. *46,000.* *Ha-Arbaa St.* *David Remez St, (04) 856 4444.* *El-Jazzar St, (04) 995 6706.* *daily.* *Fringe Theatre Festival (Sep–Oct).* **www**.akko.org.il

THE OLD CITY OF AKKO

Citadel ④
Crusader City ②
El-Jazzar's Wall ⑤
Hammam el-Pasha ③
Khan el-Faranj ⑦
Khan a-Shuarda ⑧
Khan el-Umdan ⑥
Lighthouse ⑩
Mosque of El-Jazzar ①
Souk ⑨

0 metres 50
0 yards 50

Key to Symbols *see back flap*

Aerial view of the ruined hilltop city of Megiddo

Megiddo 5

Road map B2. Route 66, 35 km (22 miles) SE of Haifa. ***Tel*** *(04) 659 0316.* *from Haifa & Tiberias.* *8am–4pm (winter: 3pm) daily (closes 1 hour earlier Fri).*

This ancient town at the head of the Jezreel valley was the scene of so many battles that the Book of Revelation in the New Testament says that it is where the final battle between Good and Evil will take place at the end of the world. The biblical name of "Armageddon" derives from "Har Megedon", or mountain of Megiddo.

The settlement controlled the main communication routes between the East and the Mediterranean, and in the 3rd millennium BC it was already a fortified city. In 1468 BC its Canaanite fortress was destroyed by the troops of the Egyptian pharaoh Thutmose III, and became an Egyptian stronghold. Megiddo was subsequently conquered and again fortified, possibly by Solomon, and in the 8th century BC came under Assyrian rule, after which it fell slowly into decline.

Extensive excavation of the spectacular mound (or "tel") has, over the years, revealed 20 successive settlements, each built over the other. The visible remains include defensive walls, a temple, an enormous grain silo and the foundations of many buildings. On the eastern side of the "tel" is an old reservoir, at the base of which a tunnel leads to a spring that lies outside the city walls. Visitors can go through the tunnel at the end of their tour of the site.

In 2005, the site joined UNESCO's World Heritage list, reflecting its historical importance and powerful influence on later civilizations.

Nazareth 6

Road map B2. *75,000.* *Casa Nova St, (04) 601 1072.* **www**.nazarethinfo.org

Lying on the rise between the Jordan Valley and the Jezreel plain, Nazareth consists of two parts. The old town is inhabited by Christian and Muslim Palestinians, and contains all of the major sights. To the north is Nazareth Illit, a large Jewish district founded in 1957 by colonists as part of the plan to settle all Galilee.

Mosaic of Joseph, Basilica of the Annunciation, Nazareth

Famous as the site of the Annunciation and the childhood of Jesus, Nazareth has had a colourful history. The village suffered at the hands of the Romans during the Jewish Revolt of AD 66 *(see p43)*, then flourished under the Byzantines, and later became an important Christian site with the Crusader conquest of the Holy Land in 1099. After the resurgence of Muslim power in the 12th and 13th centuries, Christians found it increasingly dangerous to visit. Improving relations by the 18th century allowed the Franciscans to acquire the Basilica, and they have maintained a Christian presence here ever since. Today the town is a pilgrimage site, with its many Christian churches attracting large numbers of visitors. Recent restoration projects and modern hotel developments have helped Nazareth to cope with the crowds. Unfortunately though, such high levels of tourism have done little to preserve the city's magical atmosphere. The old town is still fascinating however, with much of its traditional architecture remaining. The souk, the heart of local life, is a maze of narrow alleys where you can find a wide range of unusual goods.

Built in 1969 over the ruins of the original Byzantine church, and the successive Crusader one, the **Basilica of the Annunciation** is the major focal point in Nazareth. A bold, modern church, its large dome towers over the town. The crypt includes the Cave of the Annunciation, where the angel Gabriel is said to have appeared to Mary. A peaceful garden leads to **St Joseph's**, a small church, rebuilt in 1914 on what is thought to be the site of Joseph's home and workshop.

Environs

The main attraction of the ruined fortified town of **Tsipori** (Sepphoris), northwest of Nazareth, is its splendid 3rd-century AD mosaics. The hilltop site includes a Roman theatre that seated 5,000, the remains of a Crusader citadel and sections of the ancient water supply. Tsipori is also famous as being the supposed birthplace of the Virgin Mary.

On **Mount Tabor**, 10 km (6 miles) east of Nazareth, is a beautiful basilica, built here in 1924 to commemorate the Transfiguration (Mark 9: 9–13). It lies within the ruins of a 12th-century Muslim fortress.

Tsipori
Route 79, 3 km (2 miles) NW of Nazareth. ***Tel*** *(04) 656 8272.* *8am–4pm (summer: 5pm) daily (closes 1 hour earlier Fri).*

Safed 7

Road map C2. *26,000.* *100 Ha-Palmach St, (04) 680 1465.* **www**.zhr.org.il

The highest town in Israel, Safed is also one of the four holy cities of the Talmud, together with Jerusalem, Hebron and Tiberias. In the Middle Ages Safed became a popular meeting place for many groups of Sephardic Jews who had been driven out of Spain in the course of the Christian Reconquest. Religious schools were founded and many interpreters of the Kabbalah lived in the town. To this day Safed has remained an important centre of Jewish religious studies.

Safed covers a number of small hilltops, with its attractive old town centre located around the slopes of Gan ha-Metusda, once the site of a Crusader citadel. The old quarters of the town centre are best explored on foot, via their narrow streets and steep stairways. The Synagogue Quarter has many interesting Kabbalist synagogues including those of Itzhak Luria, Itzhak Abuhav and Joseph Caro. The former Arab Quarter (which became Jewish in 1948) is now home to a large colony of artists and is known as the Artists' Quarter. In the narrow streets and alleys between the area's picturesque houses, artists display their paintings and sculptures.

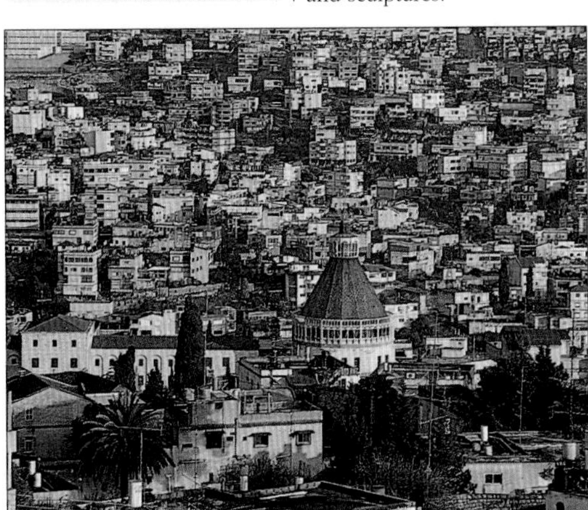

Old quarter of Nazareth, dominated by the Basilica of the Annunciation

Banias Falls, Golan Heights

Golan Heights 8

Road map. C2 *to Katsrin.* *(04) 696 2885.* **www**.tour.golan.org.il

This region of long-running historical conflict has nevertheless got much to recommend it. A high fertile plateau, dominated by Mount Hermon, it borders Israel, Syria, Jordan and Lebanon. This unique geography, aside from making it strategically important, also makes it a spectacular place to visit, with incredible vistas all around.

A major source of the Jordan River, one of the most popular places to visit is **Banias**, 15 km (9 miles) east of Kiryat Shmona. Here a large spring cascades downstream to the attractive Banias Falls nearby. **Nimrud Castle**, a short way to the northeast, originates from biblical times, though it owes its present shape to the rule of the Mameluke sultan Baybars I (1260–77). Nine of the defensive towers remain, along with much of the outer wall, a keep, and the moat.

In the south of the Golan is the administrative capital of **Katsrin**. Founded as an Israeli settlement in 1974, the town itself is unremarkable, but is a good base for exploring the beautiful countryside around. This is ideal hiking country, and the spectacular **Yehudiya Reserve** to the south of Katsrin is well worth a visit.

Nimrud Castle
26 km (16 miles) E of Kiryat Shmona. ***Tel*** *(04) 694 9277.* *daily.*

Sea of Galilee 9

Israel's chief source of water, the Sea of Galilee (Lake Tiberias/Kinneret) lies 212 metres (696 feet) below sea level and is fed and drained by the Jordan River. It is 21 km (13 miles) long, and 9 km (6 miles) wide, and since biblical times has been famous for its abundance of fish. Many of Jesus's disciples were fishermen here, and he did much of his preaching by its shores. Today, this beautiful area is one of Israel's most popular tourist centres, with a mix of fascinating historical and religious sites, and a varied selection of hotels and outdoor activities.

Statue of Saint Peter, Tiberias

Speedboating on the Sea of Galilee, one of many water sports available

KEY

- Major road
- Minor road
- Ferry
- Excursion boat
- Water sports
- Camping site
- Beach
- Viewpoint

Mount of the Beatitudes *(see p184)*

To Safed

Church of the Primacy of St Peter *(see p184)*

Church of the Multiplication of the Loaves and the Fishes *(see p184)*

Kibbutz Ginosar is home to a fishing boat from Jesus's time, found here in 1986 *(see p29)*.

Capernaum
Tabkha
Kibbutz Ginosar
Migdal
HAR ARBEL
90
TIBERIAS
77
To Nazareth
HAR MENORIM
Poriya
Kibbutz Kinneret

Tiberias
The largest town on the Sea of Galilee, Tiberias is a popular resort with many hotels, bars and restaurants. The busy lakeside offers beaches and water sports.

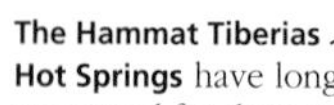

The Hammat Tiberias Hot Springs have long been renowned for their curative properties, and are said to date from the time of Solomon.

Kibbutz Kinneret's cemetery, with great views of the sea, is resting place to many spiritual leaders of the Zionist movement.

Yardenet Baptism Site
The Jordan River has always been an important Christian site since Christ was supposedly baptized here. At Yardenet, large crowds of pilgrims gather to be baptized in the river themselves.

A View of the Sea of Galilee
This view is taken from the hills above the northeastern shore.

VISITORS' CHECKLIST

Road map C2. *from Tel Aviv and Jerusalem.* *19 Habanim St, Tiberias, (04) 672 5666.* *for groups only from Tiberias to Kibbutz Ein Gev, phone to check times (04) 665 8008.* *Holyland Sailing for Pilgrim Groups, Tiberias (all year round), (04) 672 3006; Lido Kinneret Sailing Co, Tiberias (all year round), (04) 672 1538.* *Kibbutz Ein Gev Music Festival (Apr), Galilee Song Festival (May).*

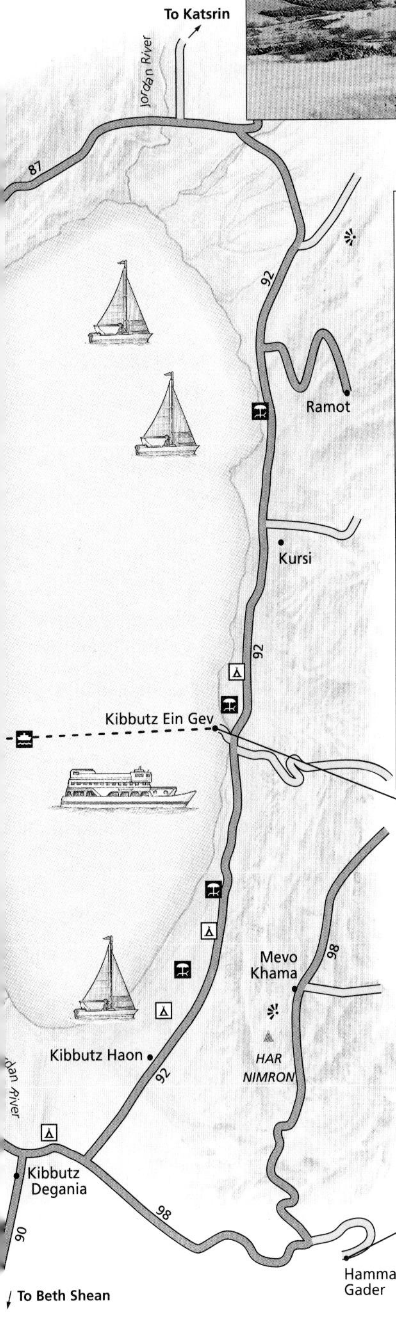

THE FIRST KIBBUTZ – DEGANIA

Conceived by Eastern European Jews, the first kibbutz was founded at Degania in 1909. The guiding ideals behind Israel's kibbutzim are self-sufficiency and equality, with everyone working for the common good. Rural farming communities, they are highly productive, and hold their own plenary meetings to decide on community matters. There are now two kibbutzim here, with the original called Degania Alef (A). By the main gate to the kibbutz is a Syrian tank, stopped here by the kibbutzniks when they famously defeated an entire armoured column during the 1948 war.

Typical kibbutz house at Degania

To Katsrin
Jordan River
87
92
Ramot
Kursi
Kibbutz Ein Gev
Mevo Khama
98
HAR NIMRON
Kibbutz Haon
Jordan River
Kibbutz Degania
90
To Beth Shean
Hammat Gader

Kibbutz Ein Gev is renowned for its fish restaurants, good beaches and its annual international music festival.

Hammat Gader Alligator Farm
The large alligator farm at Hammat Gader is open to the public. The town is also famous for its ancient Roman hot springs, which have now been largely restored. You can still bathe in their relaxing waters.

0 kilometres 4
0 miles 2

Capernaum ⑩

Road map C2. Route 87, 12 km (7.5 miles) N of Tiberias. *from Tiberias.* ***Tel*** *(04) 672 1059.* *daily.*

Capernaum, on the northern shoreline of the Sea of Galilee, was an important Roman town and one of the focal points of Christ's teachings in Galilee. It was also home to a number of his Disciples, including Simon Peter. In Capernaum's fascinating archaeological precinct there are surviving houses from the period, as well as a church, built over the ruins of what is said to have been **Simon Peter's house**. There are also the remains of a synagogue that has been dated to the 4th century AD.

Carved relief, Church of the Multiplication

Tabgha ⑪

Road map C2. Route 87, 10 km (6 miles) N of Tiberias. *from Tiberias to junction of routes 90 and 87.*

Just to the southwest of Capernaum, Tabgha (Ein Sheva) is one of the most important sites of Christ's ministry in Galilee, where he did much of his preaching. Heading from the bus-stop, a short way along Route 87 you will come to the **Church of the Multiplication of the Loaves and the Fishes**. Built in the 1980s, it boasts the remains of a 5th-century Byzantine basilica and fragments of splendid mosaics. This original church was built over the supposed spot from which Christ fed 5,000 followers with five loaves and two fish.

Nearby to the east, on the lakeside, is the **Church of the Primacy of Peter**. A black basalt Franciscan chapel, it is built on the site where Jesus Christ is said to have appeared to the Apostles after his Resurrection. The area has various other ruins, including a 4th-century chapel.

On top of the hill behind, known as the Mount of the Beatitudes, is the modern **Church of the Beatitudes**. The hill is so called, because it is thought that here, overlooking the lake, Christ gave his Sermon on the Mount. This famously began with his blessings or "beatitudes".

The modern Church of the Beatitudes near Tabgha

Tiberias ⑫

Road map C2. *39,500.* *Archaeological Garden, Rehov ha-Banim, (04) 672 5666.*

The busy town of Tiberias (Tverya) is the largest on the shores of the Sea of Galilee. It was founded during Roman times by Herod Antipas, who dedicated it to the Emperor Tiberius and moved the regional capital here from Tsipori. The town has been home to many notable scholars and rabbis, and became one of Israel's holy cities, along with Jerusalem, Hebron and Safed. The **Tomb of Maimonides**, the great medieval Jewish philosopher, can be found on Ben Zakai Street.

Today, Tiberias is a popular tourist centre, with an attractive lakeside setting and in an ideal location for exploring Galilee. The town has a lively atmosphere, especially along the busy lakeside promenade. Just off the promenade is **St Peter's Church**, built originally by the Crusaders. The current church has a boat-shaped nave, reflecting St Peter's life as a fisherman.

Tiberias is also known for its curative hot springs, of which there are several to visit in the town. There are also some public beaches to the north of town, and the popular **Gai Beach Water Park** is 1 km (half a mile) to the south of Tiberias.

Gai Beach Water Park
Sederot Eliezer Kaplan. ***Tel*** *(04) 670 0700.* *daily.* *Nov–Mar.*

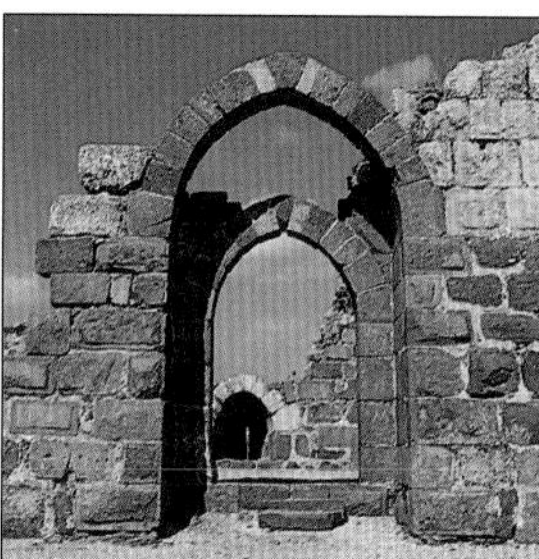

Ruined arches at Belvoir Castle

Belvoir Castle ⑬

Road map C2. Off Route 90, 27 km (17 miles) S of Tiberias. *to Beth Shean, then taxi.* ***Tel*** *(04) 658 1766.* *8am–4pm (winter: 3pm) daily.*

The ruined Crusader fortress of Belvoir, in the Kokhav ha-Yarden nature reserve, offers incomparable views of the Jordan Valley. The impressive fortress is surrounded by two huge walls, the outer one pentagonal and the inner one square. Built by the Knights Hospitallers in 1168, Belvoir was besieged many times by Saladin. It capitulated only in 1189 after a siege of more than a year, with the Muslim leader sparing both the fortress and its defenders' lives, in recognition of their great courage. Belvoir was finally destroyed by troops from Damascus in the 13th century. The area around the fortress is dotted with modern sculpture.

For hotels and restaurants in this region see pp258–60 and pp275–6

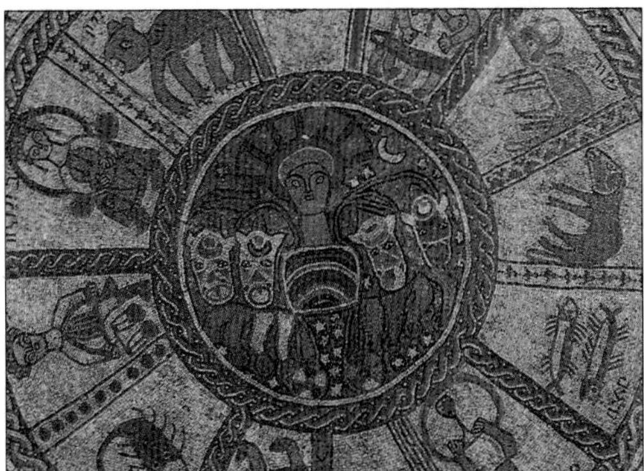

Detail from 6th-century mosaic at Beth Alpha, showing signs of the zodiac

Beth Alpha ⓮

Road map C2. Off Route 71, 11 km (7 miles) W of Beth Shean. ***Tel*** *(04) 653 2004.* *8am–5pm (winter: 4pm) daily.*

The remnants of this 6th-century synagogue were found by chance in 1928 by colonists from the nearby Hefzi-Bah kibbutz. The ruined walls give an idea of the original basilica-shaped building, but the main interest is the magnificent mosaic floor, which has survived largely intact. The upper part of the floor depicts the Ark of the Covenant, with cherubs, lions and religious symbols. The large central patterns represent the zodiac and symbols of the seasons. These show the continuing importance of pagan beliefs at the time, and the need for Judaism to try to accommodate them. The lower part relates the story of Abraham and the sacrifice of his son Isaac.

Beth Shean ⓯

Road map C2. *18,000.* *from Tiberias.*

The best-preserved Roman-Byzantine town in Israel, Beth Shean lay on the old trade routes between Mesopotamia and the Mediterranean. First inhabited 5,000 years ago during the Canaanite era, it later became the main city in the region during the period of Egyptian occupation *(see p41)*. Falling to the Philistines in the 11th century BC, it then became part of Solomon's kingdom. After the conquest of Alexander the Great it was renamed Scythopolis, and became a flourishing Hellenistic city. The Roman conquest in the 1st century BC saw Scythopolis further prosper as one of the ten city states of the Decapolis. It later retained its economic importance under the Byzantines, also becoming a major centre of Christianity. An economic collapse, then an earthquake in AD 749, eventually left only a small remaining Jewish community.

The archaeological sites at Beth Shean are in two areas. The main site comprises the Roman-Byzantine city, and the archaeological mound, or "tel". These are both within the Beth Shean National Park, 1 km (half a mile) north of the town. The jewel of this site is the Roman theatre, one of the best preserved in Israel, and once capable of seating 7,000. The old Byzantine baths have surviving mosaic and marble decoration, and tall columns from the ruined temples are equally impressive. The *tel* offers a good overview of the site, and consists of 16 or more superimposed towns. It is difficult however to understand the details of its complex archaeology.

The other site focuses on the ruined Roman amphitheatre, a short way to the south. Used for gladiatorial contests, it was connected to the main town by a paved street. Some of this street survives today, paved with huge blocks of basalt.

Beth Shean National Park
Tel *(04) 658 7189.* *daily.*

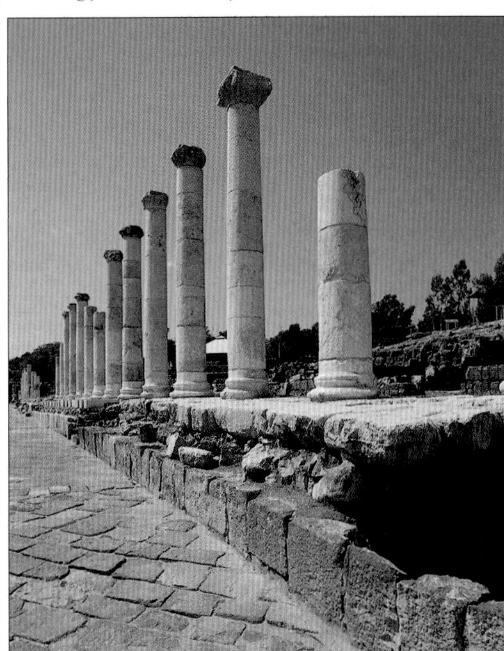

Ruined colonnade along an old Byzantine street, Beth Shean

THE DEAD SEA AND THE NEGEV DESERT

In this the most arid and inhospitable region of the Holy Land, even the waters of its great lake are incapable of supporting life, hence the "Dead Sea". But in times past, the harsh remoteness of the hills and desert was prized by reclusive communities and rebels, and so the area is dotted with ancient ruins charged with biblical significance.

Today, the Dead Sea is no longer so remote – just a 20-minute ride from Jerusalem on an air-conditioned bus. Tourists flock to its shores to test its incredibly buoyant waters. The lowest body of water in the world, it has such a high salt content it is impossible to sink. Its mineral-rich mud is also claimed to have therapeutic qualities and a string of lakeside spas do good business out of the black, sticky silt. Away from the water, high up on the rocky hillsides are the caves in which the Dead Sea Scrolls were discovered, while on a mountain top to the south is Herod the Great's fortress of Masada, one of the most stunning attractions in all Israel.

Where the Dead Sea ends, the Negev Desert begins. Here, the only signs of life, apart from the odd convoy of tourists exploring canyons and craters, are a few groups of Bedouin *(see p249)* tenaciously clinging to traditional nomadic ways.

Over the centuries, there have been many attempts to cultivate the desert. More than 2,000 years ago, the Negev was the final stage for caravans on the spice and incense route from India and southern Arabia to the Mediterranean; the Nabataeans who controlled the route perfected irrigation and cultivation techniques and established flourishing cities, such as Ovdat *(see p202)*. More recently, Israel has initiated programmes for the economic development of the region in the form of desert kibbutzim.

In spite of this desire to tame the desert, more and more people these days come in search of all that remains wild and undeveloped. In this respect, the Negev still has much to offer.

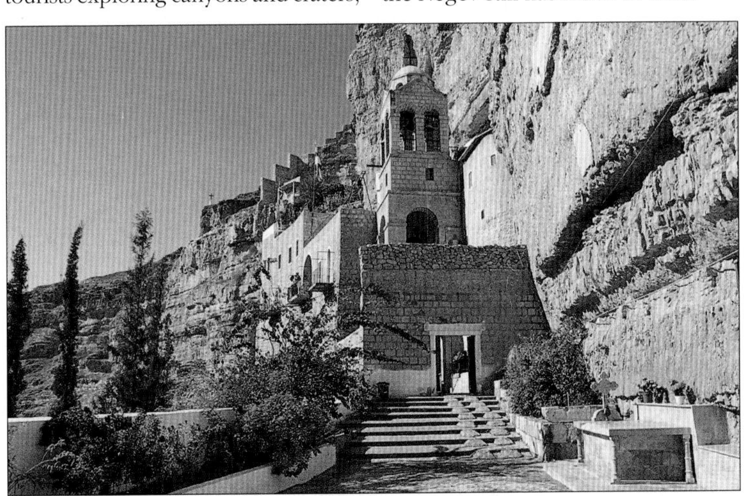

The secluded retreat of St George's Monastery, hidden in a desert canyon near Jericho

◁ **Masada on the Dead Sea, one time rebel fortress, now the most visited archaeological site in Israel**

Exploring the Dead Sea and the Negev Desert

All the sites as far south as Masada can be visited in a series of day trips from Jerusalem. Heading south beyond Masada or Beersheva and into the Negev Desert is more of an undertaking. There are only two main routes through this vast wedge of sun-baked wilderness: along the border with Jordan on Route 90; or straight down the centre of the country via Ein Ovdat and Mitspe Ramon. This latter route is by far the more interesting.

Tel Aviv
Bat Yam
MEDITERRANEAN SEA
Rekhovot
Gedera
Ashdod
Nitsanim
Ashkelon
Qiryat Gat
Gaza
GAZA STRIP
Sderot
Sa'ad
Netivot
Re'im
Rahat
Khan Yunis
Magen
Rafah (Rafi'akh)
Ofakim
BEERSHEVA 15
Kerem Shalom'
Tse'elim
ISRAEL
Agur Sands
Mash'abei Sade
Nitsana
Sde Boker
EIN OVDAT 12
OVDAT 13
Mitspe Ramon
MAKHTESH 16
Negev

The mountain-top fortress of Masada, conveniently visited as a day trip from Jerusalem

SIGHTS AT A GLANCE

Beersheva 15
Bethlehem pp193–5 6
The Dead Sea 9
Eilat 19
Ein Gedi 8
Ein Ovdat 12
Hebron 14
Herodion 5
Jericho 2
Khai Bar Yotvata Wildlife Reserve 17
Makhtesh Ramon 16
Mar Saba Monastery 4
Masada pp200–1 10
Nebi Musa 3
Ovdat 13
Qumran 7
Sodom 11
St George's Monastery 1
Timna National Park 18

0 kilometres 25
0 miles 20

Ein Gedi, where waterfalls and greenery provide respite from the heat and dust

For additional map symbols *see back flap*

Bethlehem and surrounding hills, viewed over the roof of the Church of the Nativity

GETTING AROUND

The easiest way of getting from Jerusalem to Jericho, Bethlehem and Hebron is a shared taxi from Damascus Gate *(see p310)*. You can also take bus No. 163 from Jaffa Road in Jerusalem to Bethlehem. From Bethlehem you can take a taxi on to the Herodion or Mar Saba. For longer trips, the Israeli bus company Egged serves all Dead Sea and Negev locations *(see pp312–13)*. For those who only want to visit the Negev Desert, there are direct flights to Eilat with Arkia *(see pp308–9)*.

The volcano-like mound of the Herodion, a 1st-century BC hilltop fortress

KEY

- Motorway
- Motorway under construction
- Major road
- Minor road
- Scenic route
- Railway
- International border
- Disputed border/Cease-fire line

SEE ALSO

- ***Where to Stay*** pp261–2
- ***Restaurants, Cafés and Bars*** pp278–9

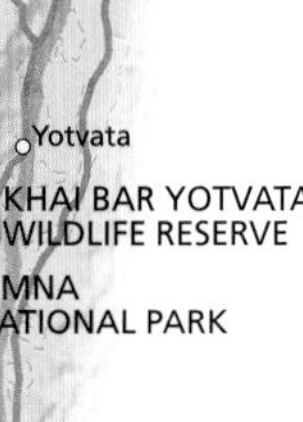

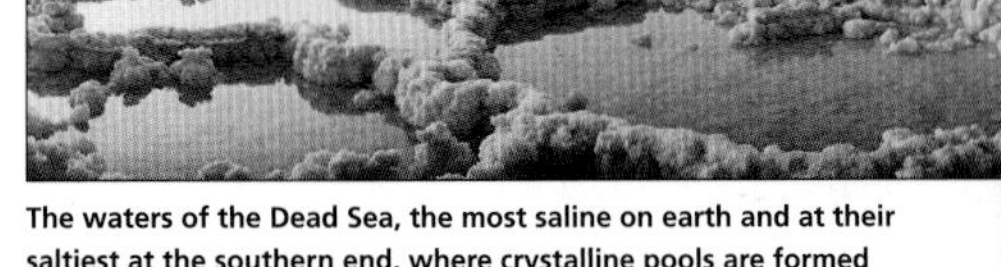

The waters of the Dead Sea, the most saline on earth and at their saltiest at the southern end, where crystalline pools are formed

St George's Monastery ❶

Road map C3. Route 1, 27 km (17 miles) E of Jerusalem. ***Tel*** *(050) 534 8892.* *from Jerusalem.* *7am–4pm daily.*

One of the finest hikes in the region is rewarded by the spectacle of St George's Monastery, an ancient retreat hollowed out of the sheer rock wall of a deep and narrow gorge. The monastery was founded in AD 480 around a cluster of caves where, according to tradition, St Joachim learned from an angel that Anne, his sterile wife and mother-to-be of the Virgin Mary, had conceived.

In AD 614 invading Persians massacred the monks and destroyed the monastery. It was partially reoccupied by the Crusaders in the Middle Ages but only fully restored at the end of the 19th century. Some attractive 6th-century mosaics remain, and there is a Crusader-era church with a shrine containing the skulls of the martyred monks.

The monastery can be reached in 20 minutes on foot via a signposted track off the old Jerusalem–Jericho road. From a starting point on the modern road hikers can take a more scenic path to the monastery which follows along the full length of the Wadi Qelt gorge.

St George's Monastery, built into the cliff face of Wadi Qelt

Jericho, regarded as perhaps the world's oldest city

Jericho ❷

Road map C3. *17,000.* *or taxi from Jerusalem.* *daily.* *Jericho Festival (Feb).*

It is best to check with the authorities first before visiting the city to make sure it is safe for tourists as unrest has returned to the region.

Claimed to be the world's oldest city and with rich biblical associations, Jericho lies just a few miles north of the Dead Sea, 258 m (846 ft) below sea level, in the middle of the Judaean desert. It owes its existence to the Ain es-Sultan spring (the biblical Elisha's Spring), the same one that, 10,000 years ago in the late Mesolithic period, attracted a semi-nomadic population of hunter-gatherers to first settle here.

Islamic-era mosaic from Hisham's Palace

According to the Bible, Jericho was the first town captured by the Israelites under the leadership of Joshua. The Book of Joshua tells how, in order to possess the land promised to them by God, the Israelites brought down the city walls with a tremendous shout and a trumpet blast (Joshua 6). During Roman times Mark Antony made a gift of the oasis town to Cleopatra of Egypt, who, in turn, leased the place to Herod the Great. Being at a lower altitude than Jerusalem, Jericho is notably warmer, and Herod wintered in a palace here, as had the Hasmonean rulers before him.

The Bible's New Testament mentions several visits to Jericho by Jesus, who healed two blind men and lodged at the home of the tax collector Zacchaeus (Luke 19: 1-10). Near the centre of town there is still the centuries-old sycamore tree up which Zacchaeus was said to have climbed in order to see Jesus.

Repeated Bedouin raids led to the decline of Jericho around the 12th century, and it wasn't until the 1920s that the town's former irrigation network was restored and the area was brought to bloom again. In 1948, the town took in more than 70,000 Palestinian refugees. The camps have since gone, and Jericho is now administered by the Palestinian National Authority.

Other attractions include **Tel Jericho** (also known as Tell es-Sultan), the sun-baked earthen mound that represents something like 10,000 years of continuous settlement. Most striking of all is a large stone tower with great thick walls that dates back as far as 7,000 BC.

A cable car service connects Tell es-Sultan with the Greek Orthodox **Monastery of the Temptation** 2 km (1 mile) to the north. Like St George's in Wadi Qelt, this holy retreat has a spectacular location, perched high up on a cliff face. The views from its terraces are breathtaking. The monastery dates back to the 12th century and is supposedly built around the grotto where the Devil appeared to tempt Jesus away from his 40-day fast (Matthew 4: 1–11).

Hisham's Palace (Qasr Hisham) is an early Islamic hunting lodge built in AD 724 for the Omayyad caliph Hisham. It lies in ruins, destroyed centuries ago by an earthquake, but it is worth a visit if only for a gorgeous floor mosaic depicting a lion hunting gazelles grazing under a broad leafy tree.

Tel Jericho
daily.

Monastery of the Temptation
***Tel** (02) 232 2827. Mon–Sat.*

Hisham's Palace
***Tel** (02) 232 2522. daily.*

Nebi Musa, regarded by Muslims as the burial place of Moses

Nebi Musa ❸

Road map C4. Route 1, 10 km (6 miles) S of Jericho. *to Jericho, then taxi.*

Although the claim is heavily disputed, Muslims revere the desert monastery of Nebi Musa as the burial place of Moses. There has been a mosque on the site since 1269, built under the patronage of the Mameluke emir Baybars. In 1470–80 a two-storey hospice was added to accommodate visiting pilgrims. However, the attractive whitewashed structures of the present day date from around 1820 and the days of Ottoman rule. The disputed cenotaph of Moses, covered with a traditional Islamic green drape, occupies the spartan, domed tomb chamber of the mosque.

Although the five-day festival of feasting and prayer that used to occur here each year now no longer happens, many Muslims still desire to be laid to rest in the large cemetery that covers the hills around the complex.

Mar Saba Monastery ❹

Road map C4. Off Route 398, 17 km (11 miles) E of Bethlehem. ***Tel** (02) 277 3135. Bethlehem, then taxi. 8am–5pm daily. Ring bell. No women allowed.*

Located out in the wilds of the Judaean desert, Mar Saba is one of the dozens of retreats built in this area from the 5th century on by hermits seeking an austere life of solitude and prayer. This particular monastery was founded in AD 482 by St Saba, a monk born in Cappadocia, Turkey, whose preachings were said to have impressed the Byzantine emperor Justinian. Despite a massacre of the monks by the Persians in the 7th century (the skulls are preserved in a chapel), the monastery survived to bloom in the 8th and 9th centuries, when its thick defensive walls housed up to 200 devotees.

Although only around 20 monks now live in Mar Saba, it remains a functioning desert monastery. As seen today, topped by bright blue domes, the complex largely dates to 1834, when it was rebuilt following a major earthquake.

An ornate canopy in the monastery's main church supposedly shelters the remains of St Saba, which were returned to the Holy Land only in 1965 having being carried off by the Crusaders and kept in Venice for seven centuries. The church walls are hung with icons and a lurid fresco depicting Judgment Day.

Unfortunately, women are not allowed to enter the monastery, but the views of Mar Saba from a neighbouring tower (which women are permitted to climb) are alone worth the trouble of a visit.

The distinctive blue domes of the gorge-top monastery of Mar Saba

The hilltop Herodion with sweeping views of the landscape

Herodion ❺

Road map B4. Route 356, 12 km (7 miles) SE of Bethlehem. ***Tel*** *(050) 5 505 007.* *Bethlehem, then taxi.* *8am–4pm (Fri: 3pm) daily.* *on Sat but call ahead.*

Dominating the desert landscape south of Bethlehem is the volcano-like mound of the Herodion, named for Herod the Great. He had this circular fortified palace built in 24–15 BC for entertaining, and to mark the defeat of his rival, Antigonus. It was long thought this might also have been his mausoleum, but despite extensive excavations no tomb has been found.

During the Second Revolt in AD 132–5 the Herodion became the headquarters of the Jewish leader Bar-Kokhba. In expectation of a Roman attack, the rebels turned its cisterns into a network of escape tunnels.

Around the 5th century, the site became a monastery with cells and a chapel, where you can still see carved Christian symbols. Also identifiable are a massive round tower and three semicircular ones, ruins of the palace baths, the *triclinium* (dining room) and fragments of mosaics, all dating from Herod's time.

At the foot of the mound are the remains of the Lower Herodion, with the dry imprint of a large pool that, in Herod's day, served as a reservoir and centrepiece for ornamental gardens.

Bethlehem ❻

Perched on a hill at the edge of the Judaean desert, Bethlehem is in biblical tradition the childhood home of David, who was named king here as he tended his father's sheep. It is also the birthplace of Jesus Christ and a major site of pilgrimage since the construction of the Church of the Nativity in the 4th century AD. The town flourished until Crusader times, but the following centuries witnessed a great reduction in population, reversed only after the 1948 war with the arrival of thousands of Palestinian refugees.

Getting to Bethlehem

The best way to reach Bethlehem is to go via Rachel's Crossing checkpoint. To get to Rachel's Crossing from Jerusalem either catch bus No. 163 from Jaffa Road outside the Central Bus Station, or take a shared taxi *(see p310)* from Damascus Gate, or take a taxi. You will then have to walk through the checkpoint and take another taxi on the other side. Hire cars may be driven from Israel into Bethlehem, but check that your insurance policy includes the Palestinian territories before setting out.

Exploring Bethlehem

Since 1995 Bethlehem has been under the control of the Palestinian National Authority, which has initiated a programme of economic recovery and tourism. Despite the huge number of pilgrims and chaotic urban growth, Bethlehem has retained a certain fascination, especially in the central area around Manger Square and in the souk just to the west. The souvenir shops are filled with kitsch religious objects but also sell fine carved olive-wood crib scenes that local craftsmen have produced for centuries. No visitor should miss the **Church of the Nativity** *(see pp194–5)* on Manger Square. Built in the fourth century over the supposed site where Jesus Christ was born, the church is one of the holiest Christian sites.

The prominent **Mosque of Omar** was built in 1860 and is the only Islamic place of worship in the town centre, despite the fact that Muslim residents now outnumber Christians in Bethlehem.

St Catherine's Church

Manger Square. ***Tel*** *(02) 274 2425.* *daily.*

Connected to the Church of the Nativity, St Catherine's faces a heavily-restored, Crusader-period cloister *(see p174)*. The church was built by Franciscans in the 1880s on the site of a 12th-century Augustinian monastery, which had replaced a 5th-century monastery associated with St Jerome. On the right side of the nave, stairs descend to the grottoes of the Holy Innocents, St Joseph and St Jerome, which connect to the Grotto of the Nativity. These were used as burial places by Christians as early as the 1st century AD and contain the tombs of St Jerome and St Paula.

The church spires and towers of Bethlehem, birthplace of Jesus Christ

The Virgin Mary and Child, a relationship celebrated at the Milk Grotto

The Milk Grotto

Milk Grotto Street. **Tel** *(02) 274 3867.* *daily.*

This grotto is considered sacred because tradition has it that the Holy Family took refuge here during the Massacre of the Innocents, before their flight into Egypt. While Mary was suckling Jesus, so the story goes, a drop of milk fell to the ground, turning it white. Both Christians and Muslims believe that scrapings from the stones in the grotto help to boost the quantity of a mother's milk and also enhance fertility.

The present building was put up by the Franciscans in 1872 on the site of a 4th-century church.

Baituna Al-Talhami

Paul VI Street. *Mon–Sat (Thu: am only).*

In an old Palestinian house on the town's main street, the Arab Women's Union has created this small but interesting craft museum. One room is given over to the embroidery typical of Palestinian women's dress and to silver jewellery, which normally represented a family's fortune. The *diwan* (living room) is furnished with rugs, musical instruments and oil lamps. The kitchen contains old copper utensils and an oven. Examples of traditional handstitched embroidery are usually available to buy.

VISITORS' CHECKLIST

Road map B3.. *40,000* *Hebron Road.* *Manger Square, (02) 274 1581.* *daily.* *Almond Blossom Festival (Feb), Olive Harvest (Oct), Midnight Mass (24 Dec).*

Rachel's Tomb

Hebron Road.

On the road to Jerusalem, just before the border checkpoint between Israel and the Palestinian territory, is the tomb of Rachel, wife of Jacob and the mother of two of his twelve sons. The tomb can only be accessed from the Israeli side, via Hebron Road. It is the third most holy site in Judaism, and is also sacred to Muslims. The actual "tomb" consists of a rock covered by a velvet drape with eleven stones on it, one for each of the eleven sons of Jacob who were alive when Rachel died in childbirth. The structure around the tomb was built in the 12th century by the Crusaders and altered many times in the centuries that followed, including in 1860 by Moses Montefiore *(see p51)*. The site is visited by Jewish women who come to pray that they will conceive.

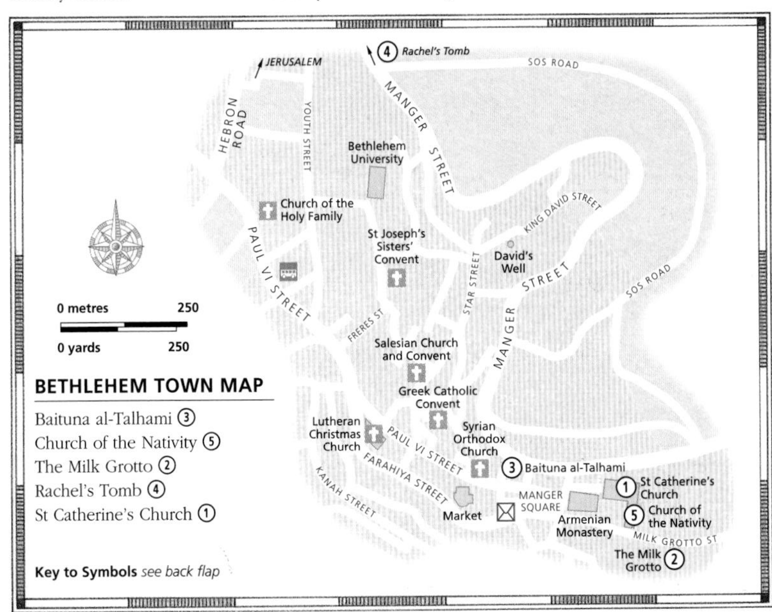

BETHLEHEM TOWN MAP

Baituna al-Talhami ③
Church of the Nativity ⑤
The Milk Grotto ②
Rachel's Tomb ④
St Catherine's Church ①

Key to Symbols *see back flap*

Church of the Nativity

Statue of Mary

The first evidence of a cave here being venerated as Christ's birthplace is in the writings of St Justin Martyr around AD 160. In 326, the Roman emperor Constantine ordered a church to be built and in about 530 it was rebuilt by Justinian. The Crusaders later redecorated the interior, but much of the marble was looted in Ottoman times. In 1852 shared custody of the church was granted to the Roman Catholic, Armenian and Greek Orthodox churches, the Greeks caring for the Grotto of the Nativity.

Plaza in front of the Church of the Nativity, with the plain façade in the distance

Stairs to main church

★ Grotto of the Nativity
The grotto is the church's focal point. A silver star is set in the floor over the spot where Christ is said to have been born.

St Catherine's Church ***(see p193)***

Altar of the Adoration of the Magi (Manger Altar)

Nave
The wide nave survives intact from Justinian's time, although the roof is 15th-century, with 19th-century restorations. Fragments of high-quality mosaics decorate the walls.

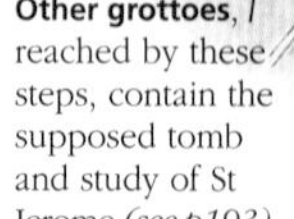

Other grottoes, reached by these steps, contain the supposed tomb and study of St Jerome *(see p193)*.

Statue of St Jerome

Cloister of St Catherine's Church
Incorporating columns and capitals from the 12th-century Augustinian monastery that previously stood here, this attractive, peaceful cloister was rebuilt in Crusader style in 1948.

Painted Columns
Thirty of the nave's 44 columns carry Crusader paintings of saints, and the Virgin and Child, although age and lighting conditions make them hard to see. The columns are of polished, pink limestone, most of them reused from the original 4th-century basilica.

VISITORS' CHECKLIST

Manger Square, Bethlehem.
Tel *(02) 274 2440.* *summer: 6:30am–7:30pm daily; winter: 5:30am–5pm daily. Grottoes closed Sun am.*

★ Mosaic Floor
Trap doors in the present floor, here and to the left of the altar, reveal sections of mosaic floor surviving from the 4th-century basilica.

Wall mosaics, made in the 1160s, once decorated the entire church.

The narthex was originally a single, long porch with three large doors leading into the church and three onto the street.

★ Door of Humility
The Crusader doorway, marked by a pointed arch, was reduced to the present tiny size in the Ottoman period to prevent carts being driven in by looters. A massive lintel above the arch indicates the door's even larger original size.

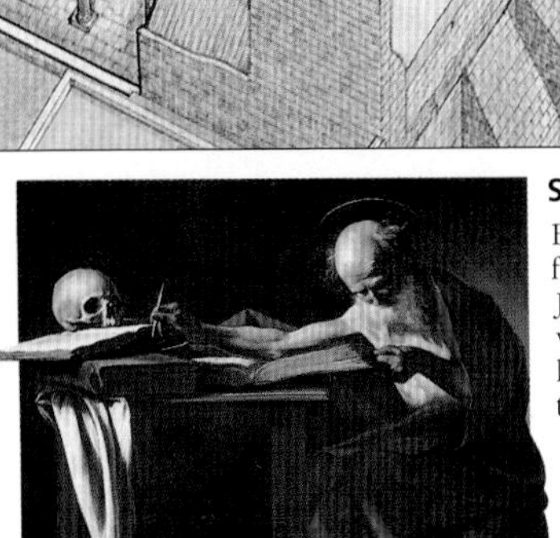

***St Jerome Writing* (c.1604) by Caravaggio**

ST JEROME

Born at Stribo (not far from Venice), St Jerome (c.342–420) was one of the most learned scholars of the early Christian Church. He travelled widely and, in 384, settled in Bethlehem, where he founded a monastery. Here, he completed a new version of the Bible *(see p24)*, inspired by the pope's suggestion that a single book should replace the many differing texts in circulation. His great work later became known as the Vulgate. Tradition places the saint's study and tomb next to the Grotto of the Nativity.

STAR FEATURES

★ Grotto of the Nativity

★ Mosaic Floor

★ Door of Humility

Caves at Qumran, where the hot, dry, desert climate helped to preserve the Dead Sea Scrolls

Qumran ❼

Road map C4. Route 90, 20 km (12 miles) S of Jericho. ***Tel*** *(02) 994 2235.* *from Jerusalem.* *8am–5pm Sat–Thu (winter: 4pm), 8am–3pm Fri (winter: 2pm).*

Qumran is known chiefly as the place where the Dead Sea Scrolls were discovered. From 150 BC to AD 68 this remote site was the home of a radically ascetic and reclusive community, often identified with the Essenes. According to their school of thought, the arrival of the Jewish Messiah was imminent, and they prepared for this event with fasting and purification through ritual ablutions. These activities were rudely brought to a halt through conflict with the Romans.

The Essenes largely vanished from history until 1947 when a Bedouin shepherd boy looking for a lost goat happened upon a cave full of jars. These jars were found to contain a precious hoard of 190 linen-wrapped scrolls that had been preserved for 2,000 years. Following much study by academics some of the scrolls are now on view in a purpose-built hall at the Israel Museum *(see pp136–7)*.

Visitors to Qumran watch a short film on the Essenes and view a small exhibition on the community before being directed to the archaeological site at the foot of the cliffs. Signs indicate the probable uses of different areas of the vaguely defined remains. From the site you can see the caves above where the scrolls were found. You can scramble up to the caves for a fine view, but you need to allow about two hours and carry a substantial supply of water.

Ein Gedi ❽

Road map C4. Route 90, 56 km (35 miles) S of Jericho. *from Jerusalem.*

Ein Gedi is famous as a lush oasis in an otherwise barren landscape. Several springs provide plentiful water to support a luxuriant mix of tropical and desert vegetation. The site is mentioned in the Bible for its beauty (Song of Songs: 1–14) and as a refuge of David who was fleeing from King Saul (I Samuel: 24).

Protected as **Ein Gedi Nature Reserve**, the oasis is a haven for desert wildlife such as ibexes and rock hyraxes, which look like large rodents, while the more remote areas are the abode of the desert leopard. Two gorges, belonging to the Nakhal David and Nakhal Arugot rivers, are at the core of the reserve; these are crossed by a network of paths. The shortest walking tour takes about an hour and ends at the spectacular Shulamit Falls. A short way from the reserve's entrance are the ruins of a 5th-century BC synagogue with mosaics and inscriptions in Hebrew and Aramaic.

Ein Gedi is also a popular spot with Dead Sea bathers *(see p197)*. For a more luxurious experience, the **Ein Gedi Health Spa**, a further 3 km (2 miles) to the south, has hot sulphur baths and private access to the Dead Sea.

Ein Gedi Nature Reserve
Highway 90, Dead Sea ***Tel*** *(08) 658 4285.* *daily.*

Ein Gedi Health Spa
Highway 90, Dead Sea
Tel *(08) 659 4726.* *daily.*

Trail sign for one of the gorges in the Ein Gedi Nature Reserve

The Dead Sea ❾

The Dead Sea (which is actually a lake, not a sea) lies half in Israel, half in Jordan. It is 76 km (47 miles) from north to south and less than 16 km (10 miles) across. At 411 m (1,348 ft) below sea level, it is also the lowest point on earth. The water is so mineral-laden that it is around 26 per cent solid. The therapeutic qualities of the water and its mud have been touted since ancient times, and spas are dotted along its shores. However, the Dead Sea is endangered. Its water level has gone down 12 m (40 ft) since the beginning of the 20th century because its main source, the Jordan River, has been over-exploited for irrigation purposes.

VISITORS' CHECKLIST

30 km (18 miles) E of Jerusalem. *from Jerusalem for Qumran, Ein Gedi, Masada and Neve Zohar; from Amman for Amman Beach, Dead Sea Panorama and Wadi Mujib Nature Reserve.* **Dead Sea Panorama *Tel*** *(05) 349 1133.* **Museum** *9am–5pm daily.* **Wadi Mujib Nature Reserve *Tel*** *(06) 463 3589.* compulsory. www.rscn.org.jo

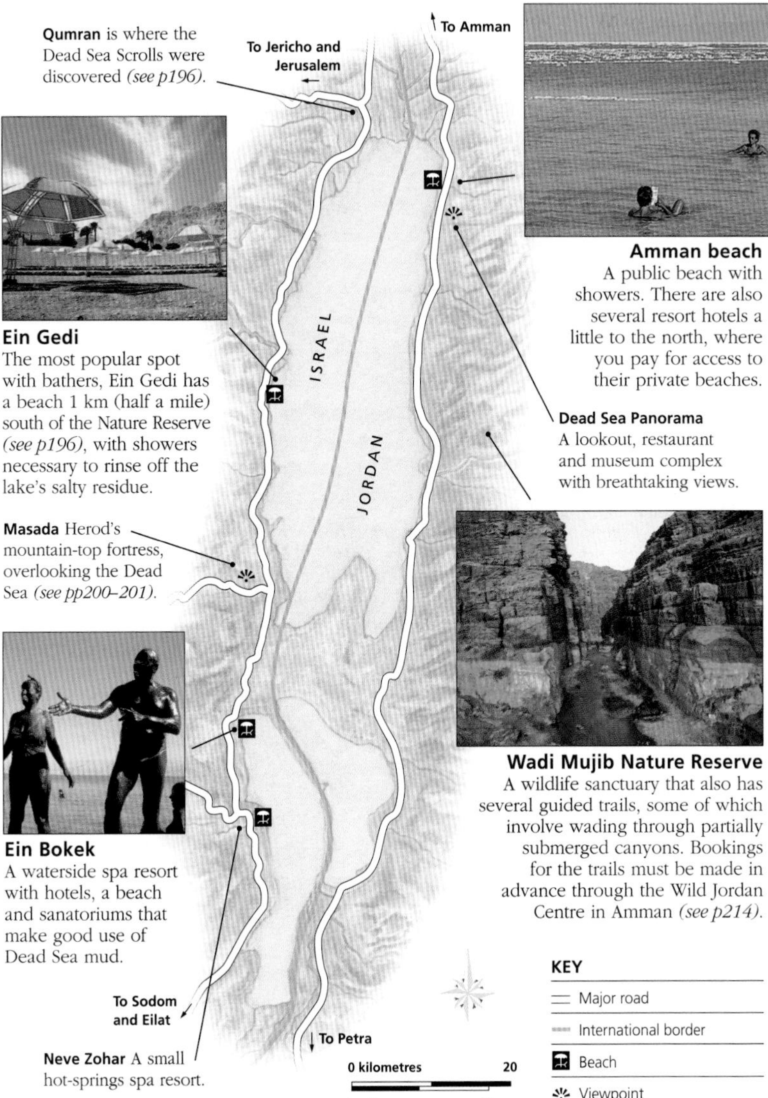

Qumran is where the Dead Sea Scrolls were discovered *(see p196)*.

Ein Gedi
The most popular spot with bathers, Ein Gedi has a beach 1 km (half a mile) south of the Nature Reserve *(see p196)*, with showers necessary to rinse off the lake's salty residue.

Masada Herod's mountain-top fortress, overlooking the Dead Sea *(see pp200–201)*.

Ein Bokek
A waterside spa resort with hotels, a beach and sanatoriums that make good use of Dead Sea mud.

Neve Zohar A small hot-springs spa resort.

Amman beach
A public beach with showers. There are also several resort hotels a little to the north, where you pay for access to their private beaches.

Dead Sea Panorama
A lookout, restaurant and museum complex with breathtaking views.

Wadi Mujib Nature Reserve
A wildlife sanctuary that also has several guided trails, some of which involve wading through partially submerged canyons. Bookings for the trails must be made in advance through the Wild Jordan Centre in Amman *(see p214)*.

Shimmering white salt deposits on the southern shores of the Dead Sea ▷

Masada ⑩

This isolated mountain-top fortress about 440 m (1,300 ft) above the banks of the Dead Sea was fortified as early as the 1st or 2nd century BC and then enlarged and reinforced by Herod the Great, who added two luxurious palace complexes. On Herod's death the fortress passed into Roman hands but it was captured in AD 66 during the First Revolt by Jews of the Zealot sect. After the Romans had crushed the rebels in Jerusalem, Masada remained the last Jewish stronghold. Held by less than 1,000 defenders, it was under Roman siege for over two years before the walls were breached in AD 73.

Cable Car
The cable car operates daily between 8am and 4pm; otherwise it is a strenuous 45–60-minute climb up the twisting Snake Path.

★ Hanging Palace
Part of the large Northern Palace complex, the Hanging Palace was Herod's private residence. It was built on three levels; the middle terrace had a circular hall used for entertaining, the lower had a bathhouse.

The Water Gate is at the head of a winding path to reservoirs below.

Calidarium
Masada's hot baths are one of the best preserved parts of the fortress. The columns remain on which the original floor was raised to allow hot air to circulate underneath and heat the room.

Synagogue
Possibly built by Herod, this synagogue is thought to be the oldest in the world. The stone seats were added by the Zealots.

STAR FEATURES

★ Hanging Palace

★ Western Palace

For hotels and restaurants in this region see pp261–2 and pp278–9

Cistern
At the foot of the mountain Herod built dams and canals that collected the seasonal rainwater to fill cisterns on the northeast side of the fortress. This water was then carried by donkey to the cisterns on top of the rock, such as this one in the southern part of the plateau.

VISITORS' CHECKLIST

Road map C4. Off Route 90, 18 km (11 miles) S of Ein Gedi. *from Jerusalem or Eilat.* ***Tel*** *(08) 658 4207. 8am–5pm (winter: 4pm) Sat–Thu; sites close 1 hr early Fri & hols.* **Sound and Light Show:** *Mar–Oct.* ***Tel*** *(08) 995 9333 for reservations.*

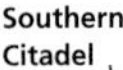

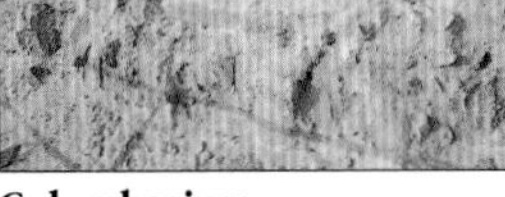

Columbarium
This is a small building with niches for funerary urns; it is thought the urns held the ashes of non-Jewish members of Herod's court.

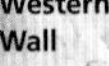

★ Western Palace
Used for receptions and the accommodation of Herod's guests, the Western Palace was richly decorated with mosaic floors and frescoes adorning the walls.

The Roman ramp is now the western entrance to the site.

THE ROMAN SIEGE OF MASADA (AD 72–73 OR AD 73–74)

According to a 1st-century account by historian Flavius Josephus, the Roman legions laying siege to Masada numbered about 10,000 men. To prevent the Jewish rebels from escaping, the Romans surrounded the mountain with a ring of eight camps, linked by walls; an arrangement that can still be seen today. To make their attack, the Romans built a huge earthen ramp up the mountainside. Once this was finished, a tower was constructed against the walls. From the shelter of this tower the Romans set to work with a battering ram. The defenders hastily erected an inner defensive wall, but this proved little obstacle and Masada fell when it was breached. Rather than submit to the Romans the Jews inside chose to commit mass suicide. Josephus relates how each man was responsible for killing his own family. "Masada shall not fall again" is a swearing-in oath of the modern Israeli army.

Roman catapult missiles

Remains of one of the Roman base camps viewed from the fortress top

Sodom ⓫

Road map C4. Route 90, 50 km (31 miles) S of Ein Gedi. *from Jerusalem.*

Biblical tradition holds that the city of Sodom lay on the southern shore of the Dead Sea (Genesis 19). Its sinful inhabitants, along with those of neighbouring Gomorrah, angered God, and he destroyed the cities with "brimstone and fire". Archaeologists now favour Bab ed-Dhra in Jordan as the likely site, but the name Sodom remains attached to a spot on the Israeli side of the Dead Sea. There is nothing to visit but nearby are the two spas of **Ein Bokek** and **Neve Zohar**, famous for their therapeutic centres, and a public beach with fresh-water showers *(see p197)*.

Inland and 9 km (6 miles) south of Neve Zohar is **Mount Sodom**, a mountain composed largely of rock salt. A well-marked path goes up to the top, from where you can enjoy incomparable views of the Dead Sea and the Moab mountains in Jordan. You can also go up by car: take the dirt road that heads west off route 90 just north of the unattractive Dead Sea Works plant. Another sign-posted scenic hiking route leads to what is known as the **Flour Cave**. The cave gets its name from the white crumbly chalk coating that covers the interior and the clothing of all who visit.

A typically barren Dead Sea landscape near Sodom

Spring-fed pool at Ein Ovdat in the shade of canyon walls

Ein Ovdat ⓬

Road map B5. Route 40, 52 km (32 miles) S of Beersheva. ***Tel*** *(08) 655 5684.* *from Jerusalem.* *daily. Summer: 8am–4:45pm (3:45pm Fri & hols); winter: 8am–3:45pm (2:45pm Fri & hols); last entry 1 hr before closing.*

At Ein Ovdat a white-walled gorge gouged 200 m (656 ft) deep into the desert floor shades two icy-cold pools. The larger of the pools is fed by a waterfall with its source in the rock face high above. Archaeologists have found traces of human presence in this area that date back perhaps 35,000 years, suggesting that the springs were known in antiquity.

A well-marked trail through the gorge begins at a roadside viewpoint 2 km (1 mile) south of the turn-off for Kibbutz Sde Boker. The trail ends with a set of rough rock-cut steps ascending the cliffs; the views from these down the gorge are spectacular. A path leads to a roadside car park 7 km (4 miles) south of the viewpoint.

Ovdat ⓭

Road map B5. Route 40, 60 km (37 miles) S of Beersheva. ***Tel*** *(08) 655 1511.* *from Jerusalem.* *daily. Summer: 8am–5pm (2pm Fri & hols); winter: 8am–4pm (3pm Fri & hols); last entry 1 hr before closing.*

Located on a flat hilltop, the ancient town of Ovdat was built by the Nabataeans in the 2nd century BC as a stopover on the trade route between Egypt and Asia Minor. It continued to prosper under the Byzantines, and most of what you see today dates from the 4th or 5th century, including the remains of houses, baths and two churches. The smaller of these has its original apse and bishop's throne; a white line divides the original and reconstructed parts. The views across the desert are excellent. Below the hill you can make out evidence of the network of dams built by the Nabataeans to channel rain-water towards the dry land, enabling them to plant vine-yards and fruit orchards. Ovdat was abandoned after the Persian invasion of 620. The Visitors' Centre has an exhibition of archaeological finds from the ancient site.

Partially reconstructed Byzantine-era ruins at Ovdat

Hebron ⓮

Road map B4. *120,000.* *daily.*

Nestled among hills 40 km (25 miles) south of Jerusalem, Hebron is one of the most densely populated towns in the West Bank. Its fame rests on its glassmaking, which began in the Middle Ages and has always been managed by one single family.

This coloured glassware can be found for sale in another of Hebron's major attractions, its medieval Arab souk, which contains some imposing Crusader-era vaulted passageways.

However, Hebron is a town undermined by troublesome political tensions. It is divided

The Tomb of the Patriarchs, mosque and burial site of Sarah, Isaac and Jacob

into two zones: the greater area is governed by the Palestinian Authority, but the town centre is occupied by Jewish settlers. Large numbers of Israeli soldiers maintain a constant peace-keeping presence. Friction between the two communities dates back to a 1929 pogrom in which the Arabs massacred Hebron's centuries-old Jewish community. After the Six Day War of 1967 the centre of town was resettled by militant Jewish colonists. Tension continues to erupt into occasional violence. For your personal safety, ask about the situation before making a trip to Hebron.

Hebron is regarded as a sacred place by the Jewish, Christian and Muslim religions alike; it was here they believe that Abraham buried his wife Sarah, in the Cave of Machpelah, purchased from the Hittite Ephron. The cave then became his own tomb and later that of his descendants Isaac and Jacob.

Around 20 BC Herod the Great sealed the cave and built a great hall over it. Under Byzantine rule the structure was turned into a church and then, after the Arab conquest of 638, a mosque. The invading Crusaders attempted to reclaim the site for Christianity and built much of the present-day construction, but it was completed by Saladin as a mosque. In the 13th century the Mameluke ruler Baybars forbade non-Muslims from entering the building.

After the 1967 war the mosque remained Muslim, but access was granted to Jews as well. Today, the complex, known as the **Tomb of the Patriarchs** (Haram al-Khalil in Arabic), is divided into a Jewish synagogue and a Muslim mosque, each with its own entrance. It remains a bone of contention between the faiths; in 1994 Jewish colonist Baruch Goldstein entered the mosque and killed 29 Muslim worshippers.

Tomb of the Patriarchs
Tel (02) 960 5602. 8am–4pm Sun–Thu except during prayer times.

Beersheva ⓯

Road map B4. *200,000. 1 Hebron Rd, Beer Abraham, (08) 623 4613. Bedouin market Thu.*

The so-called capital of the Negev is a city that has grown rapidly and chaotically. In the Old Testament it is famous as the place where Abraham made a pact with Abimelech for the use of a well for his animals (Genesis 21: 25–33). Beersheva means "well of the covenant". For centuries it remained little more than a Bedouin well until the Turks transformed the site into an administrative centre (which was the object of a valiant cavalry charge by the Australians in World War I).

Since the Israelis captured Beersheva in 1948, it has attracted many immigrants to become the country's fourth largest city.

There is an attractive grouping of an Ottoman-era mosque and Governor's House in the town centre, but the most interesting thing about Beersheva is the **Bedouin market**. This is held on the edge of town every Thursday from dawn and attracts hundreds of nomads. Besides the livestock and everyday objects bought by the locals, visitors can also buy traditional Bedouin handicrafts such as jewellery and copperwork.

Just outside town is **Tel Beersheva**, a city founded at the end of the 11th century BC and fortified around the time of Solomon. It was destroyed in the 9th century by the Egyptians but was rebuilt, remaining a bulwark of the southern frontier of Judaea until it was razed to the ground by the Assyrians. Remains include a 10th-century BC city gate and a Roman fortress. There is also a museum of Bedouin life.

Tel Beersheva
6 km (4 miles) NE of Beersheva.
Tel (08) 646 7286. daily.

Bedouin selling sheep at Beersheva's Thursday market

Makhtesh Ramon 16

Road map B5. Route 40, 80 km (50 miles) S of Beersheva. *from Beersheba.* **Visitors' Centre *Tel*** *(08) 658 8691. 8am–5pm Sun–Thu, 8am–4pm Fri (last entry 1 hour before closing).*

Makhtesh Ramon is Israel's most spectacular natural phenomenon: a crater some 40 km (25 miles) long, 9 km (5 miles) wide, with a depth of 300 m (1,300 ft). It is the largest of three craters in the Negev Desert, which scientists believe were formed more than half a million years ago by a combination of tectonic movement and erosion.

Traffic between Beersheva and Eilat has to cross Makhtesh Ramon, negotiating switchback roads that wind down to the crater floor and back up again. Nabataean caravans also travelled this way between Petra and Ovdat, and the ruins of an ancient caravanserai stand at the centre of the depression.

On the crater's rim is the town of Mitspe Ramon, the main base for exploring this part of the desert. The town's Visitors' Centre has exhibits on the geology of the great crater and its flora and fauna. It also has hiking maps – but make sure to take plenty of water if you go trekking here. In Mitspe Ramon you can also arrange to tour the crater by camel or jeep.

Spectacular geological scenery at Timna National Park

Khai Bar Yotvata Wildlife Reserve 17

Road map B6. Route 90, 35 km (22 miles) N of Eilat. ***Tel*** *(08) 637 6018. from Eilat. daily. Summer: 8:30am–4pm; winter: 8:30am–5pm (Fri & Sat: 4pm). Obligatory with departures every hour.*

A caracal, one of the biblical species at Khai Bar

Khai Bar was founded with the aim of reintroducing some of the creatures named in the Bible, which have since vanished from the Negev. Most of the animals roam freely, safari-park style, in a 40-sq km (15-sq mile) territory in the Arava Valley. Visits can be made only by jeep in the company of a ranger guide. Native species in the reserve (not all of which receive biblical mention) include scimitar-horned oryxes, wild Somali donkeys, ostriches and the addax antelope with their curved horns. A Predator Centre houses wildcats, caracals (desert lynxes), foxes, leopards and hyenas in spacious enclosures.

Timna National Park 18

Road map B6. Route 90, 28 km (18 miles) N of Eilat. ***Tel*** *(08) 631 6756. from Eilat. 8am–4pm Sat–Thu, 8am–3pm Fri).* **www**.timna-park.co.il

Ancient remains indicate working mines at Timna as far back as 3000 BC, and the Egyptians were mining copper here around 1500 BC. They left two temples dedicated to the goddess Hathor, protectress of mines. A hieroglyphic inscription in one of the temples mentions pharaoh Rameses III offering a sacrifice to Hathor. The mines continued to be worked under the Nabataeans and Romans before being abandoned. With the added attraction of some curious mushroom-shaped rock formations created by wind erosion, the area has been preserved as a national park. An underground passage gives access to the ancient mines, and you can see Egyptian graffiti representing ibexes and hunters armed with bows and arrows.

Modern sculptures set in the natural splendour of Makhtesh Ramon

Eilat ⓳

Road map B7. *50,000.* *8 Beit ha-Gesher Street.* *8:30am–5pm Sun–Thu, 8am–1pm Fri (08) 630 9111.*

Lying at the end of the Gulf of Aqaba, on a stretch of Israel's 12 km (7 mile) long southern coast, Eilat is the only Israeli town on the Red Sea. The town is filled with hotels and tourist villages, and is a centre for diving and trips into the desert. Eilat is similar in many ways to Aqaba, which faces it from 6 km (4 miles) away on the other side of the Gulf. Along with an equally stunning location, Eilat also shares a similar history to Aqaba. Now separated by political boundaries however, it is Eilat that has prospered the most. With the United Nations partition of Palestine in 1947, Israel was ceded this small stretch of coastline, and Eilat has since developed rapidly, both as a port and as a popular holiday resort, with excellent tourist facilities.

The bottom of the Red Sea is the main attraction here. If you don't want to dive to admire this multicoloured ecosystem, there are glass-bottomed boats as well as the "Yellow Submarine". This large 23-m (75-ft) long submersible leaves from Coral World, and cruises out over the reef, descending to a depth of around 60 m (200 ft).

Coral Island, south of Eilat in the Gulf of Aqaba

The large **Coral World Underwater Observatory** is an oceanographic complex where you can get a close-up view of the marvellous marine life here. It contains 25 tanks with more than 500 species of fish, sponges, corals and invertebrates. The most interesting displays are those with the larger creatures such as sharks and sea turtles. The main spectacle though is at the underwater observatory itself, which is 6 m (20 ft) underwater and gives a spectacular live view of the local marine life through its large glass windows.

Divers and expert swimmers will be delighted at **Dolphin Reef**, where small groups led by an instructor can actually swim with the dolphins and observe their behaviour as they play, swim and hunt.

The salt marshes just north of Eilat are the feeding grounds of many species of migratory birds travelling between Africa and Eurasia every spring and autumn. The **International Birdwatching Centre**, at Kibbutz Eilot, has an interpretation centre, and organizes guided birdwatching tours. In season, the skies are filled with thousands of flamingos, storks and herons, as well as eagles, hawks and buzzards.

By boat you can go to the fabulous reefs off **Coral Island** (or Pharaoh's Island), which lies just across the Egyptian border. Regular trips are run for divers, but those wishing to land and visit the 12th-century Crusader fortress that dominates the island will need to find a tour that can arrange a group visa.

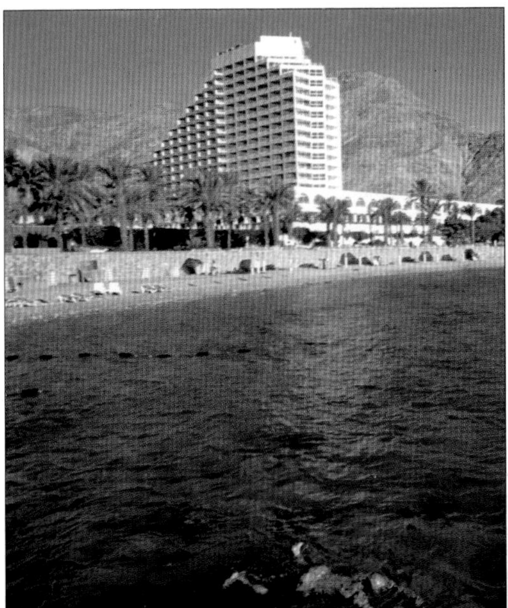

Swimming in the perfectly clear waters off the beach at Eilat

Coral World Underwater Observatory
Coral Beach. ***Tel*** *(08) 636 4200.* *8:30am-4pm daily.* **www**.coralworld.com/eilat

Dolphin Reef
Southern Beach. ***Tel*** *(08) 630 0100.* *daily.* **www**.dolphinreef.co.il

International Birdwatching Centre
Kibbutz Eilot, 2 km (1 mile) N of Eilat. ***Tel*** *(08) 633 5339.* *Oct-Jun: Sun-Thu (am only).*

PETRA AND WESTERN JORDAN

While most visitors to Jordan come for the sole purpose of seeing the magnificent rock-cut city of Petra, many depart greatly impressed by the gracious and hospitable locals. Besides these attractions, Western Jordan has many fascinating archaeological sites from prehistoric, Roman, Byzantine and Crusader times.

Only partitioned off from Palestine in 1923 and made fully independent in 1946, the nation of Jordan has a maturity that belies its youth. That the kingdom is viewed as an anchor in the often turbulent sea of Middle Eastern politics is due, in large part, to the efforts of the late King Hussein (1953–99) who worked solidly to establish and maintain peace in the region. The extreme warmth and friendliness of the population is an expression of the stability Hussein secured for his country. Day-to-day patterns of life in Jordan are also shaped by a relaxed and tolerant interpretation of Islam. Tourists who have just visited neighbouring Israel may well appreciate the laissez-faire nature of the Jordanian people.

Although Jordan has an area of about 92,000 sq km (36,000 sq miles), around nine-tenths of this is desert. Consequently, the population of approximately 5.5 million is concentrated in the northwest on a plateau above the Jordan Valley. Watered by the Jordan River and surrounded by mountains, this little pocket enjoys a lush greenhouse-like climate and is entirely devoted to agriculture. But south of Amman the fertile plains abruptly end and give way to the vast stony desert that extends all the way down to the Red Sea. Largely shunned by the local populace, this is the region that visitors come to see. This is where you find the craggy sandstone landscapes out of which Petra was carved. Further south is Wadi Rum with its great cinemascope sandy oceans that provided a dramatic backdrop for the exploits of Lawrence of Arabia.

Perfectly suited to the Jordanian terrain, the camel, pictured here at Mount Nebo

◁ **The Treasury at Petra, arguably the single most spectacular sight in the whole Middle East**

Exploring Petra and Western Jordan

Though possessing few sites itself, Jordan's modern capital, Amman, makes a very comfortable base from which to explore the northwest of the country. The Arab fortress at Ajlun, the Roman ruins at Jerash, the Byzantine mosaics of Madaba, and further mosaics along with splendid views at Mount Nebo, are all within an hour's drive. If you can spare the time and secure the use of a car (self-drive or a taxi hired by the day), then Amman is certainly worth a couple of days. The Crusader castles of Kerak and Shobak are perhaps best visited while heading south, en route to the site that truly epitomizes the magic of the region, Petra. While it is possible to see the major attractions in just one day, Petra more than repays repeated visits: multiple-day passes are available. Accommodation is easy to find in the neighbouring town of Wadi Musa. Be sure also to leave enough time for the surreal rockscapes of Wadi Rum.

The impressive stone sweep of the colonnaded Oval Plaza at Jerash

SIGHTS AT A GLANCE

GETTING AROUND

Most major tourist destinations can be reached by good, modern roads. There are two main routes south – take the King's Highway (Route 49) for Mount Nebo, Madaba, Kerak and Shobak, and the Desert Highway (Routes 15 and 53) to head directly to Petra and Wadi Rum. It is possible to fly between Amman and Aqaba and an inexpensive bus service connects all areas of the country. For many people, however, coach tours are the most comfortable way to get about.

SEE ALSO

• ***Where to Stay*** pp262–3

• ***Restaurants, Cafés and Bars*** pp279–80

Wadi Rum, where sandstone mountains rise sheer from the desert floor

65

AQABA 11

For additional map symbols *see back flap*

The Roman theatre in the shadow of modern central Amman

KEY

- Motorway
- Major road
- Minor road
- Four-wheel-drive track
- Scenic route
- Railway
- International border

The Royal Tombs at Petra, a site that ranks with the Pyramids as a surviving wonder of the ancient world

Umm Qais ❶

Road map C2.100 km (62 miles) NW of Amman. *7am–sunset daily.*

Umm Qais is the site of the ancient Graeco-Roman city of Gadara. The ruins lie in lush hill country overlooking the Golan Heights and the Sea of Galilee. The city is well known from the Bible for Jesus's miracle of the Gadarene Swine, when he cast out demons into pigs (Matthew 8: 28–34). Since 1974, archaeologists have uncovered many impressive Roman remains, including a colonnaded street, a theatre and a mausoleum.

Ajlun ❷

Road map C3. 50 km (31 miles) W of Amman. *(02) 642 0115.* **Fortress** *8am–5pm daily (winter: 4pm).*

The market town of Ajlun is dominated by the fortress of **Qalat ar-Rabad**, a superb example of Arab military engineering. Built in 1184–5, partly in response to Crusader incursions in the region, it was later used by the Ottomans up until the 18th century. At a height of more than 1,200 m (4,000 ft), it offers fantastic views over the Jordan Valley.

Environs
About 30 km (19 miles) northwest of Ajlun is **Pella**. Water, fertile land and, later, its location on two major trade routes were drawing settlers here well before 3000 BC. Its Roman-Byzantine ruins are today's attraction.

The Arab fortress at Ajlun, built to stem the Crusaders' advance

View of Jerash's Cardo, Agora (market place) and unusual Oval Plaza

Jerash ❸

Road map C3. 50 km (31 miles) N of Amman. *from Amman.* ***Tel*** *(04) 635 1272. Oct–Apr: 8am–4pm Mon–Thu, 9am–4pm Fri–Sun; May–Sep: 8am–7pm Mon–Thu, 9am–4pm Fri–Sun. Jordan Festival (late Jul–early Aug).* ***Tel*** *(06) 566 0156.*

Excavations of Jerash, known as Gerasa in Classical times, began in the 1920s, bringing to light one of the best preserved and most original Roman cities in the Middle East. It was during the Hellenistic period of the 3rd century BC that Jerash became an urban centre and a member of the loose federation of Greek cities known as the Decapolis *(see p42)*. From the 1st century BC Jerash drew considerable prestige from the semi-independent status it was given within the Roman province of Syria. It prospered greatly from its position on the incense and spice trade route from the Arabian Peninsula to Syria and the Mediterranean. Jerash lost its autonomy under Trajan, but his annexation of the Nabataean capital Petra *(see pp220–31)* in AD 106 brought the city even more wealth. By AD 130 ancient Gerasa was at its zenith. Having become a favourite city of Hadrian *(see p43)*, it flourished both economically and socially. After a period of decline in the 3rd century, it enjoyed a renaissance as a Christian city under the Byzantines, notably in the reign of Justinian (AD 527–65). The Muslims took over the city in 635, and it was badly damaged by a series of earthquakes in the 8th century. The final blow to the city was dealt by Baldwin II of Jerusalem in 1112 during the Crusades *(see pp48–9)*.

Detail of floor mosaic in St George's Church

The city is reached through **Hadrian's Arch**, built in honour of the Roman emperor. Alongside is the **Hippodrome**, where Gerasa's chariot races and other sporting events took place, and a little way down the track is the **South Gate**, part of the 4th-century AD city wall. To its left, and on a prominent rise is first the **Temple of Zeus**, and then the **South Theatre**, which nowadays is used as a venue for the Jordan Festival *(see p37)*. The most unusual feature of the Roman city is the **Oval Plaza** (1st century AD) which, with its asymmetrical shape, is a unique monument from the Roman world. The plaza, 80 m by 90 m (262 ft by 295 ft), is enclosed by 160 Ionic columns. Beneath its stone paving runs a complex drainage system. From here, going north, is the **Cardo**, a spectacular paved street about 600 m (660 yards) long, which was lined with the city's major buildings, shops and residences. Chariot tracks are visible in the stones. To the left lies the **Agora**, the city's main food market, which had a central fountain. At the Tetrapylon (crossroads) the Cardo meets a second major street, the **South Decumanus**, which runs east–west. Further along on the left side of the Cardo is the 2nd-century **Nymphaeum**, a lavish public fountain. One of its basins has a design of four fish kissing. Nearby is the impressive **Temple of Artemis**, the patron goddess of the city in Greek and Roman times.

Close to the Temple are the remains of several Byzantine churches. The largest is usually referred to as the **Cathedral**. There is also a complex of three churches, dedicated to **SS Cosmas and Damian, St John the Baptist** and **St George**, which dates back to AD 526–33 and has fine mosaic floors. Further along the Cardo, to the right, is the **Propylaeum Church** with the remains of an ornate plaza in front, while next to it are the ruins of an **Omayyad Mosque**. Beyond lie the unexcavated **West Baths**, which preserve a splendid domed ceiling. At the **North Tetrapylon**, once marked by a dome resting on four arches, the road to the left leads to the small **North Theatre**.

Allow at least half a day to see the ruins, and finish off with the **Museum**, displaying sarcophagi, statuary and coins.

Temple of Zeus (2nd century AD)

Irbid
Modern town
Museum
South Gate
Byzantine city wall
Entrance
Visitors' Centre and restaurant
Amman
0 metres 200
0 yards 200

KEY TO THE RUINS OF JERASH

Agora (7)
Cardo (6)
Cathedral (9)
Hadrian's Arch (1)
Hippodrome (2)
North Tetrapylon (16)
North Theatre (17)
Nymphaeum (12)
Omayyad Mosque (14)
Oval Plaza (5)
Propylaeum Church (13)
SS Cosmas and Damian, St John the Baptist and St George (10)
South Decumanus (8)
South Theatre (4)
Temple of Artemis (11)
Temple of Zeus (3)
West Baths (15)

The reconstructed South Gate, the 4th-century AD entrance to Jerash

Amman ❹

Like Jordan itself, Amman is a modern creation, but one whose roots run deep into history. The hills of Downtown hosted the biblical capital of the Ammonites and the Roman city of Philadelphia before the Omayyad Arabs built a palace on the same well-defended hill-top. In the modern age, Amman only began to prosper in the early 1920s when Emir Abdullah made it the capital of Trans-Jordan. Today, it is a bustling, modern and forward-looking Arab city of over two million people.

King Hussein Mosque, built on the site of a 7th-century mosque

Exploring Central Amman

Amman's most interesting district for the visitor is the recently renovated Downtown, with its bustling markets and fascinating Roman ruins.

More than anything, Amman is a town of hills (*jebels*) and, of these, the most historically important is Jebel el-Qalaa, which rises north of Downtown. This is the site of the Citadel, a Roman temple and the main city museum.

Head of a Semite chief

Downtown

The backstreet souks (markets) around El-Malek Faisal, El-Hashemi and Quraysh streets form the commercial hub of Amman. Shops here stock everything from marinated olives to gold jewellery, while pastry stalls, falafel stands, and aromatic coffee and spice grinders also compete for the attention of passers-by. There are also several interesting souvenir stalls on El-Hashemi Street.

The central **King Hussein Mosque**, built in 1924 on the site of a mosque erected in AD 640 by the caliph Omar, is the best attended in the city. Also nearby is the **Roman Nymphaeum**, built in AD 191 as a complex of pool and fountain, and dedicated to the nymphs. Jordan's Department of Antiquities is currently excavating the Nymphaeum, and the site should have been restored to something like its original condition by 2010. A new National Museum is also due to open in Downtown in 2010–11.

Citadel

Jebel el-Qalaa. ***Tel*** *(06) 463 8795.*
winter: 8am–4pm Sat–Thu, 9am–4pm Fri; summer: 8am–6pm daily.

For thousands of years Jebel el-Qalaa has served as the fortified heart of Amman. The Ammonite capital of Rabbath Ammon was situated here but most of the remains visible today are part of what was an Omayyad Palace, completed around AD 750 and destined to last for only 30 years. The large complex includes an impressive audience hall, a colonnaded street, Byzantine basilica, large cistern and the residence of Amman's local governor. The southern Roman Temple of Hercules, with its towering columns and ornately carved stonework, was built at the same time as the city's Roman Theatre *(see p213)* and offers fine views over the city.

Ruins of the Temple of Hercules at the Citadel

SIGHTS AT A GLANCE

Archaeological Museum ③
Citadel ②
Darat el-Funun ⑥
Downtown ①
Folklore Museum & Museum of Popular Traditions ⑤
King Abdullah Mosque ⑦
Roman Theatre ④
Royal Automobile Museum ⑧
Wild Jordan Centre ⑨

Chequered *keffiyehs* – traditional Jordanian men's headwear

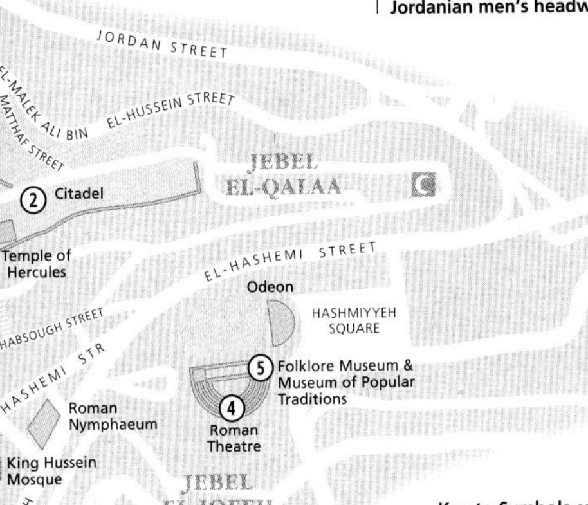

Key to Symbols *see back flap*

VISITORS' CHECKLIST

Road Map C3. *2,125,000.* *Ministry of Tourism, El-Mutanabbi Street, Jebel Amman (Third Circle), (06) 464 2311.*

Archaeological Museum

Jebel el-Qalaa. ***Tel*** *(06) 463 8795.* *summer: 8am–6pm Sat–Thu, 9am–4pm Fri; winter: 8am–4pm Sat–Thu, 9am–4pm Fri.*

This small museum at the Citadel records over 8,000 years of Middle Eastern history. Finds include Neolithic skulls and elephant bones from the Jordan Valley, a collection of copper-plated Dead Sea scrolls *(see p137)* and several Nabataean artifacts from Petra *(see pp220–31)*. The very modern-looking bug-eyed statues from Ain Gazal are over 8,500 years old. Look out also for the impressive doorway transported here from the Arab castle of Qasr el-Tuba in the Eastern Desert. Local finds include the graceful statue of Athena, from the nearby Roman Theatre, and the head of Tyche, the town god.

Roman Theatre

El-Hashemi Street. *summer: 8am–6pm Sat–Thu, 9am–6pm Fri; winter: 8am–4pm Sat–Thu, 9am–4pm Fri.*

Amman's most obvious remnant from the past is its impressive Roman Theatre, dating from around AD 170 and with a seating capacity of around 6,000. It's a fine place to sit, meet the locals and take in the city. The back rows of the theatre were added later and carved out of an existing necropolis. At the foot of the theatre are a Corinthian colonnade and the old Odeon (a small theatre or meeting hall). The nearby Hashemite Square is a popular hangout for local Jordanian families.

Folklore Museum & Museum of Popular Traditions

El-Hashemi Street. ***Tel*** *(06) 465 1742.* *summer: 8am–6pm Sat–Thu, 9am–6pm Fri; winter: 8am–4pm Sat–Thu, 9am–4pm Fri.*

The vaults below the Roman Theatre house these two modest but interesting museums. The Folklore Museum has some traditional costumes, a Bedouin tent, fine examples of the rababa (a one-stringed musical instrument) and traditional coffee grinders. The second museum displays Circassian and Armenian silver jewellery, traditionally given to the bride on her wedding day, plus amulets made from Turkish coins and symbols representing the hands of Fatima. There are some fine mosaics from Jerash *(see p210–11)* and the baptism site of Wadi el-Kharrar.

The Roman Theatre, built during the reign of Emperor Marcus Aurelius

Exploring Amman

Although the majority of Amman's places of interest are concentrated in the neighbouring Downtown and Jebel el-Qalaa districts, it is well worth exploring further afield. Just west of the centre, Jebel Amman is the city's main hill, and is home to the Wild Jordan Centre and the landmark King Abdullah Mosque. West again, the upscale districts of Abdoun and, stretching to the north, Shmeisani boast the majority of Amman's shops and restaurants. The city is quite spread out, so taxi is the best way to get around.

The hilly landscape of the modern city of Amman

Darat el-Funun

Nimer bin Adwan Street, Jebel el-Webdeh. ***Tel*** *(06) 464 3251.* *10am–7pm Sat–Wed, 10am–8pm Thu.* **www**.daratalfunun.org

This art gallery, pleasant café and small garden dotted with archaeological remains offer a tranquil escape from the nearby Downtown bustle. The rotating exhibits of contemporary art, regular lectures and occasional music concerts make this the best place to tap into Amman's thriving arts scene. The main gallery is housed in a 1920s villa, next to the charming remains of a 6th-century Byzantine church, itself built on the site of a Roman temple. Above the church is the house in which TE Lawrence is said to have written sections of *The Seven Pillars of Wisdom.*

King Abdullah Mosque

Suleyman el-Nabulsi Street, Jebel el-Webdeh. *8am–11am & 12:30–2pm Sat–Thu, 9am–10am Fri.*

Amman's most impressive Islamic monument is the King (El-Malek) Abdullah Mosque, completed in 1990 and dedicated by King Hussein to his grandfather. The soaring central blue dome covers the largest religious space in the city – the prayer hall can hold up to 7,000 worshippers. The cavernous, octagonal interior is decorated with fine Quranic calligraphy and several huge chandeliers. Remove your shoes when you enter the mosque. Women should wear a headscarf (provided). The attached small **Islamic museum** contains coins and examples of Islamic decorative arts.

Royal Automobile Museum

King Hussein Park. ***Tel*** *(06) 541 1392.* *10am–7pm Wed, Thu, Sat–Mon, 11am–7pm Fri.* **www**.royalautomuseum.jo

The former King Hussein was passionate about automobiles. This museum, 5 km (3 miles) northwest of the city centre, exhibits around 70 classic cars and motorcycles from his own personal collection. These range from a 1916 Cadillac to an array of more modern Lotus, Ferrari and Porsche sporting models, all driven by the King. Also on display is the Mercedes-Benz jeep that carried the casket in his funeral procession in 1999.

Wild Jordan Centre

Othman bin Aafarn Street, Jebel Amman. ***Tel*** *(06) 463 3589.* **www**.rscn.org.jo

Jordan's innovative Royal Society for the Conservation of Nature (RSCN) runs this cutting-edge centre, which focuses on Jordan's natural heritage. The Wild Jordan Nature shop stocks products made by RSCN-operated development initiatives throughout Jordan, including natural handmade olive oil soaps from Ajlun, worked silver from Dana and Mujib, Bedouin-made candles from Feynan and hand-painted ostrich eggs from the Eastern Desert. The excellent café serves tasty and healthy lunches, and the terrace, in particular, affords fantastic views over Downtown.

This is also the place for information on ecotourism excursions to Jordan's many nature reserves; possibilities include hiking and canyoning in Wadi Mujib *(see p197)*, and the chance to see Arabian oryx in the wild at the Shaumari Wildlife Reserve.

The distinctive blue dome that caps the striking King Abdullah Mosque

Detail of a mosaic from the Memorial Church of Moses on Mount Nebo

Mount Nebo ❺

Road map C3. 10 km (6 miles) NW of Madaba. *from Madaba then a 4-km (2.5-mile) walk, or taxi.* *7am–7pm daily (Oct–Apr: 5pm).*

This mountain rises at the end of the long chain skirting the Dead Sea, and offers spectacular views of the Jordan River and Dead Sea 1,000 m (3,300 ft) below. It was from here that Moses saw the Promised Land just before he died (Deuteronomy 34: 1–5).

In the early 4th century a sanctuary, mentioned by the pilgrim nun Egeria *(see p32)*, was built on Mount Nebo (Fasaliyyeh in Arabic) to honour Moses, probably over the remains of a more ancient construction. During the Byzantine period, the church was transformed into a fine basilica with a sacristy and new baptistry. Monastic buildings were added later.

Since 1933, reconstruction work has been carried out on the church, now known as the **Memorial Church of Moses**. Mosaics inside include a remarkable example in the Old Baptistry depicting farmers, hunters and an assortment of animals surrounded by geometric decoration. A Greek inscription dates it to AD 531. Next to the New Baptistry, a mosaic cross from the original church stands on a modern altar. Outside, the foundations of the monastery can be seen.

Madaba ❻

See pp216–7.

Kerak ❼

Road map C4. *19,000.* *El-Mujamma Street, (03) 235 4263.*

The town of Kerak, on top of a hill with a sheer drop on three sides, is dominated by a magnificent Crusader citadel. Kerak was an important city (and for a time the capital) of the Biblical kingdom of Moab. For this reason, the **castle** is also sometimes known as Krak des Moabites.

It was built in 1142 by the Frankish lord of Oultrejourdain, Payen le Bouteiller, to whom the territory had been ceded by King Baldwin II of Jerusalem in 1126. It was the pearl in the chain of fortifications that ran between Jerusalem and Aqaba, and replaced Shobak as the centre of Oultrejourdain. Under Reynald de Châtillon it resisted assaults by Saladin's troops in 1183 and 1184, but finally fell after a siege in 1188.

Arab repairs and additions in white limestone contrast with the Crusader parts built in dark, volcanic tufa. The upper courtyard, containing a much-damaged Crusader chapel, provides an exceptional viewpoint. Steps lead down to vast, dimly-lit, vaulted rooms and corridors below ground. The lower courtyard gives access to a small **Archaeological Museum** displaying locally excavated artifacts.

Castle
El-Mujamma St. *daily.*

Archaeological Museum
Tel *(03) 235 1862.* *9am-5pm daily.*

Shobak ❽

Road map C5. 60 km (37 miles) S of Tafila. ***Tel*** *(03) 213 2138.* *to Shobak village, then taxi.* *daily.*

Shobak, isolated on a rocky, conical hill in rough, barren surroundings at 1,300 m (4,265 ft) above sea level, is perhaps the most impressively sited castle in Jordan. It was called Krak de Montréal, or Mons Regalis, and was the first outpost (1115) built beyond the Jordan River by King Baldwin I of Jerusalem to guard the road from Egypt to Damascus. It resisted many sieges until 1189, when it fell to Saladin's troops.

The towers and walls are well preserved and decorated with carved inscriptions dating from 14th-century Mameluke renovations, but the inside is ruinous. Near the gatehouse, a well with over 350 dangerously slippery, spiral, rock-cut steps descends to a spring.

The impressive and well-preserved Crusader fortress at Kerak

Madaba ❻

Road map C4. *75,000.* *from Amman.* *Hussein bin Ali St, (05) 325 3563.*

According to the Old Testament the Moabite city of Madaba was one of those conquered by the tribes of Israel. After changing hands several times it flourished under Roman dominion and by the 4th century AD it had become an important centre of Christianity with its own bishop. The town weathered invasions by the Persians and Muslims but declined under the Mamelukes and was abandoned during the 16th century. It was not reoccupied until the late 19th century.

The main attraction is the fabulous mosaic map housed in **St George's Church** in the town centre. An icon of the Virgin Mary in the church is believed by Christians to incorporate a miraculous blue "helping hand". An **Archaeological Park** encompasses the remains of several more 6th-century churches, all with impressive mosaics, including one depicting scenes from the legend of Adonis and Aphrodite. The **Church of the Apostles** on the southern edge of town has a mosaic depicting the sea goddess Thetis surrounded by fish and sea monsters.

St George's Church
8:30am (10:30am Fri & Sun)–6pm daily.

Archaeological Park
daily.

St George's Church, also known as the Church of the Map

The Madaba Mosaic Map

Mosaic gazelle from the map

In the late 19th century clashes with the Muslim community led to a group of Christians from Kerak voluntarily moving to the long-uninhabited site of ancient Madaba. They were permitted to build new churches only on the sites of old ones. In 1884, while clearing such a site, the mosaic map was uncovered. It was incorporated into the new St George's Church but was badly damaged in the process. It wasn't until ten years later that scholars recognized the great historic value of the mosaic, which was probably made during the reign of the Emperor Justinian (AD 527–65).

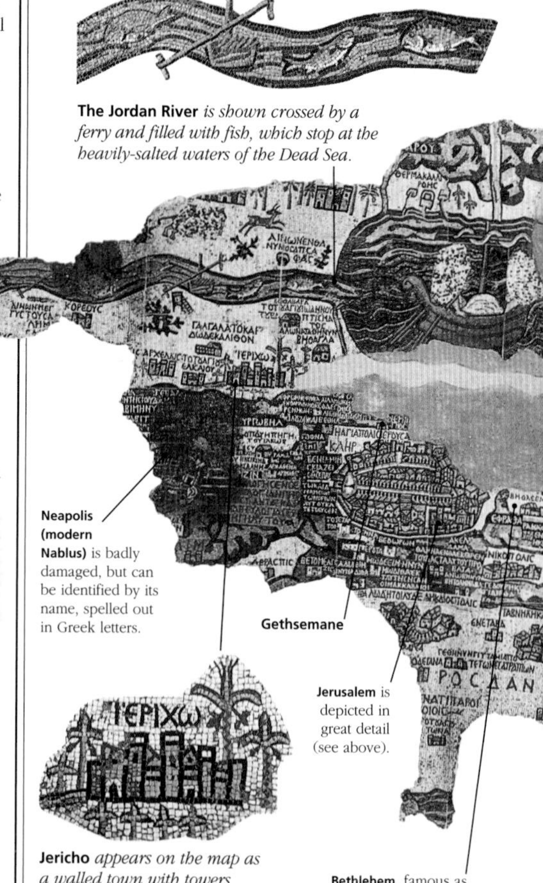

The Jordan River *is shown crossed by a ferry and filled with fish, which stop at the heavily-salted waters of the Dead Sea.*

Neapolis (modern Nablus) is badly damaged, but can be identified by its name, spelled out in Greek letters.

Gethsemane

Jerusalem is depicted in great detail (see above).

Jericho *appears on the map as a walled town with towers, agreeing with the evidence found at the site of Tel Jericho.*

Bethlehem, famous as the birthplace of Jesus, is shown as a small village dominated by the Church of the Nativity.

The Madaba map, visited by up to a thousand visitors a day

JERUSALEM AS DEPICTED ON THE MAP

In the 6th century, Jerusalem was still essentially the Roman city of Aelia Capitolina with its walls and gates, and the main streets of the Cardo Maximus and the Decumanus. Identifiable landmarks include Damascus Gate and the Church of the Holy Sepulchre, as well as the long-vanished Nea Basilica *(see p82)* and Damascus Gate column.

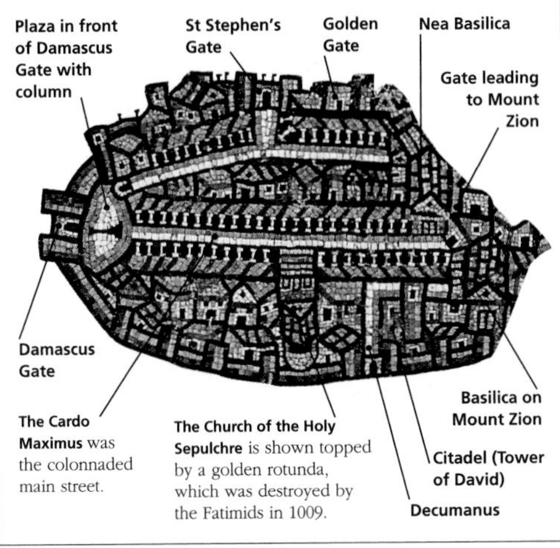

Kerak sits on top of a high mountain.

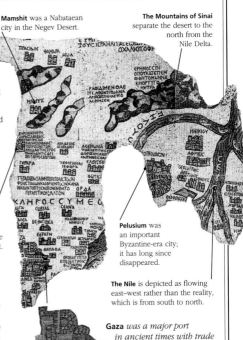

Mamshit was a Nabataean city in the Negev Desert.

The Mountains of Sinai separate the desert to the north from the Nile Delta.

The Dead Sea is shown with two boats carrying salt and grain. The sailors have been hacked out, probably by iconoclasts who objected to the representation of living beings in art.

Beersheva, although existing only in part, can be identified by the text – and by its accurate location in the western Negev Desert.

Ashdod, an ancient port on the Mediterranean, remains an important deep-water harbour.

Pelusium was an important Byzantine-era city; it has long since disappeared.

The Nile is depicted as flowing east–west rather than the reality, which is from south to north.

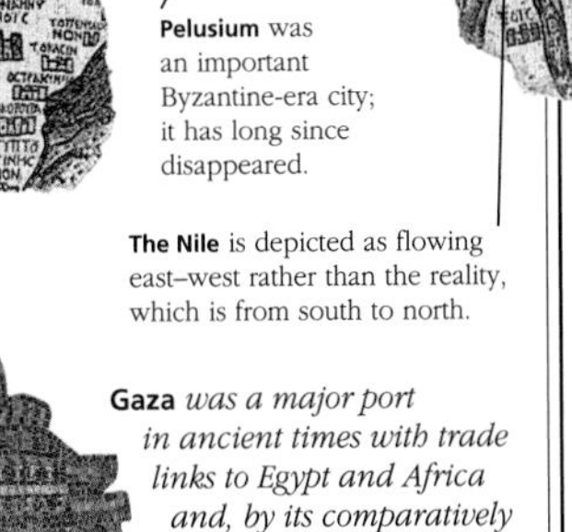

Gaza *was a major port in ancient times with trade links to Egypt and Africa and, by its comparatively large size, the map accords it great importance.*

WHAT THE MAP SHOWS

The map is oriented east–west rather than north–south, with Palestine on the left and Egypt's Nile Delta on the extreme right. The cities and villages are located remarkably accurately for the time, and they are represented in plan form, corresponding to a large degree to modern cartography.

Petra ❾

Petra is one of the world's most impressive and atmospheric archaeological sites. Its marvellously preserved rock-hewn tombs and temples once encircled a thriving metropolis. There has been human settlement here since prehistoric times, but before the Nabataeans *(see p227)* came, Petra was just another desert watering hole. Between the 3rd century BC and the 1st century AD, they built a superb city and made it the centre of a vast trading empire. In AD 106 Petra was annexed by Rome. Christianity arrived in the 4th century, the Muslims in the 7th and the Crusaders briefly in the 12th. Thereafter Petra lay forgotten until 1812 when rediscovered by JL Burckhardt *(see p223)*.

The City of Petra
The city's main street leads to the Temenos Gate, entrance to the sacred precinct of Qasr el-Bint, Petra's most important temple (see pp228–9).

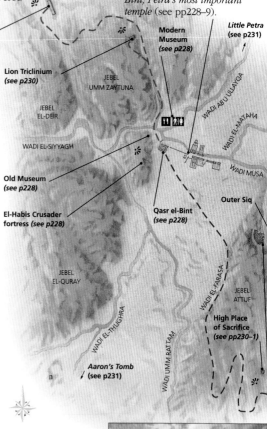

★ The Monastery
The imposing façade of the Monastery, or El-Deir, is 47 m (154 ft) wide and 40 m (131 ft) high. This magnificent Nabataean temple may later have served as a church (see pp230–31).

Modern Museum *(see p228)*

Little Petra **(see p231)**

Lion Triclinium ***(see p230)***

JEBEL UMM ZAYTUNA

JEBEL EL-DEIR

WADI ABU ULLAYQA

WADI EL-MATAHA

WADI EL-SIYYAGH

WADI MUSA

Old Museum ***(see p228)***

El-Habis Crusader fortress ***(see p228)***

Qasr el-Bint ***(see p228)***

Outer Siq

JEBEL EL-QURAY

WADI EL-FARASA

JEBEL ATTUF

High Place of Sacrifice ***(see pp230–1)***

WADI EL-THUGHRA

WADI UMM RATTAM

Aaron's Tomb **(see p231)**

VISITING PETRA

- It is worth spending more than a day here. There are passes for 1–4 days.
- Cars allowed up to ticket gate but not beyond.
- Horses may be hired to take you the 900 m (half a mile) to Siq entrance.
- Two-seater horse-drawn carts go from the ticket office to the Treasury. From there Petra can be covered on foot or camel.
- Basic food and drinking water available in Petra.
- Wear sunhat and high-factor sunscreen.
- Avoid wandering off main walk routes without guide and water supply.
- A new visitors' centre is planned near the Siq.

The Theatre
Carved into the mountainside by the Nabataeans, probably in the 1st century AD, this theatre follows the standard Roman design of the time. It was large enough to seat up to 7,000 people (see p225).

★ The Royal Tombs
These monumental façades sculpted into the mountain at the eastern end of the Petra basin create an awe-inspiring panorama when viewed from a distance (see pp226–7).

VISITORS' CHECKLIST

Road map C5. Wadi Musa, 260 km (160 miles) S of Amman. *to Wadi Musa from Amman, Aqaba.* *6am–sunset daily.* *passes sold for 1, 2, 3 or 4 days.* *Ask at the Visitors' Centre.* *Petra Visitors' Centre, (03) 215 6044 or 215 6060 (6:30am–5pm daily). Do not photograph Bedouin without their permission.* **Museum** *9am–4:30pm daily (summer: 5:30pm).*

★ The Siq
Access to Petra is through this deep ravine, formed when a split in the mountain was swept clear by water from the Wadi Musa (see pp222–3).

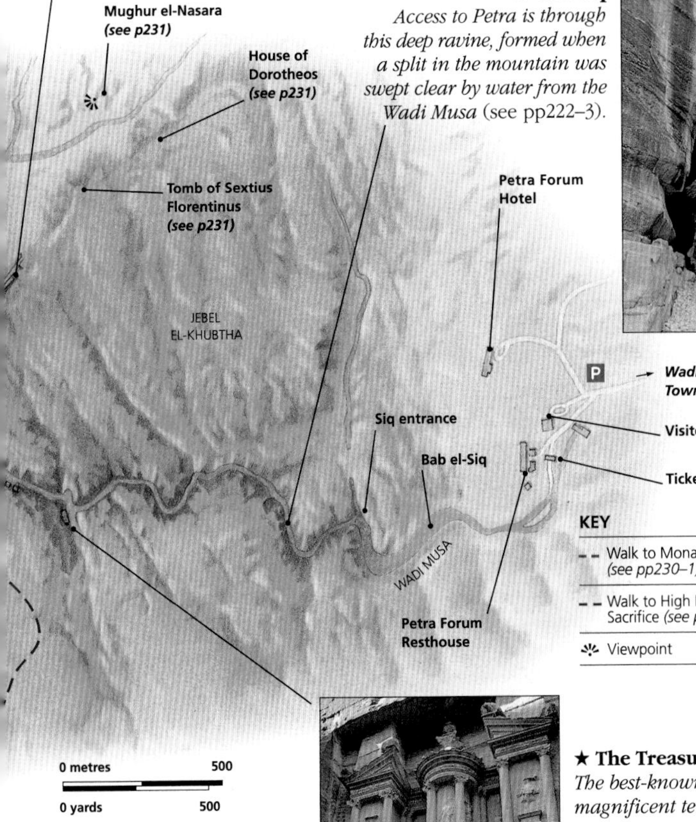

Mughur el-Nasara
(see p231)

House of Dorotheos
(see p231)

Tomb of Sextius Florentinus
(see p231)

JEBEL EL-KHUBTHA

Petra Forum Hotel

Wadi Musa Town

Visitors' Centre

Ticket gate

Siq entrance

Bab el-Siq

WADI MUSA

Petra Forum Resthouse

KEY

- Walk to Monastery *(see pp230–1)*
- Walk to High Place of Sacrifice *(see pp230–1)*
- Viewpoint

0 metres 500

0 yards 500

★ The Treasury
The best-known of all Petra's magnificent temples, deliberately positioned at the end of the Siq for maximum impact, the 1st-century BC Treasury takes its name from Bedouin folklore. They believed that the Khasneh el-Faroun (Treasury of the Pharaoh) was the magical creation of a great wizard who had deposited treasure in its urn (see p224).

STAR SIGHTS

- ★ The Siq
- ★ The Treasury
- ★ The Royal Tombs
- ★ The Monastery

The Siq: the Ancient Entrance to Petra

To reach the Siq, the narrow Gorge that leads into Petra, you must first walk 900 m (half a mile) along the wide valley known as the Bab el-Siq. This prelude to Petra has many tantalizing examples of the Nabataeans' appetite for sculpting monuments out of mountainsides. The entrance to the Siq is marked by the remains of a monumental arch. It is the start of a gallery of intriguing insights into the Nabataeans' past. These include water channels cut into the rock, Nabataean graffiti, carved niches with worn outlines of ancient deities, Nabataean paving stones, and eerie flights of steps leading nowhere. As the Siq descends, it closes in and at its deepest, darkest point unexpectedly opens out on Petra's most thrilling monument – the Treasury *(see pp224–5)*.

Djinn Blocks
In Arab folklore these carved blocks, of which Petra has 26, house djinn *(spirits). They may have been tower tombs.*

Obelisk Tomb and Bab el-Siq Triclinium
Two rock-cut tombs on the way to the Siq stand one above the other. They seem to be one complex but are, in fact, separate. The upper, probably earlier, Obelisk Tomb shows Egyptian inspiration. The lower structure, known as the Bab el-Siq Triclinium (funerary dining chamber), is a superb illustration of the Nabataean Classical style (see p225).

A votive niche, to one side of the remains of the monumental arch supports, was reached by steps.

FROM THE TICKET GATE, THROUGH THE SIQ, TO THE TREASURY

It is about 1.5 km (nearly one mile) from the ticket gate to the end of the Siq. The route follows the course of a wadi which runs through the Siq and into the city. As the Siq descends, almost imperceptibly, it becomes deeper and narrower. At its narrowest point, the walls are only one metre apart.

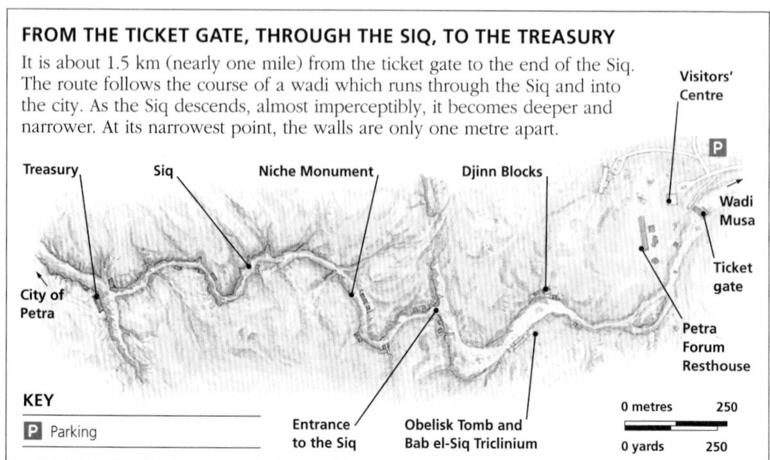

Nabataean Pavements
The Siq was probably paved by the Nabataeans in the 1st century AD. Substantial stretches of this paving can still be seen. Next to the most extensive stretch is the Niche Monument (see below).

Water Channels
The water channels were part of a sophisticated system of water conservation and flood prevention devised by the Nabataeans.

Entrance to the Siq
In ancient times, the Siq was entered via a monumental arch. It fell in 1896, leaving only traces of its supporting structures.

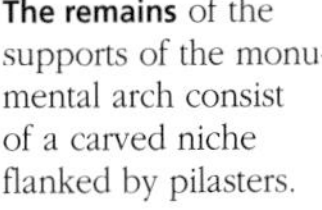

The remains of the supports of the monumental arch consist of a carved niche flanked by pilasters.

The Niche Monument
Carved into a free-standing rock, a quarter of the way along the Siq, is a small Classical shrine. Within the niche are two Djinn blocks, one of which has eyes and a nose.

View of the Treasury
The first breathtaking glimpse of the Treasury is when its pink-hued, finely chiselled façade suddenly appears through a chink in the dark, narrow walls of the Siq. It is a moment filled with powerful contrasts.

JOHANN LUDWIG BURCKHARDT

In 1812, after lying hidden for more than 500 years to all except local Arabs, Petra was rediscovered by an explorer called Johann Ludwig Burckhardt. The son of a Swiss colonel in the French army, he was an outstanding student with a thirst for adventure. In 1809 he was contracted by a London-based association to explore the "interior parts of Africa". Three years later, after intense study of Islam and Arabic, he disguised himself as a Muslim scholar, took the name Ibrahim ibn Abdullah and set out for Egypt. On his way through Jordan, however, he was lured by tales of a lost city in the mountains. To get there, he had to persuade a guide to take him. Using the pretence that he wanted to offer a sacrifice to the Prophet Aaron, he became the first modern Westerner to enter Petra.

Burckhardt in the disguise he assumed to enter Petra

From the Treasury to the Theatre

Set deep in the rock and protected by the valley walls, the magnificent 1st-century BC Treasury creates a formidable first impression of Petra. As its design had no precedent in the city, it is thought that architects from the Hellenistic Near East were brought in to create it. From the Treasury the path leads into the Outer Siq, lined on both sides with tombs of all sizes, some half buried by risen ground levels. At the end of the Outer Siq, in the midst of this great necropolis, is the Classical Theatre. Started by the Nabataeans and possibly added to by the Romans, it was a project requiring advanced engineering skills.

The Outer Siq
From the Treasury to the Theatre tombs display a range of intermediate design styles. One, freestanding, uniquely combines Classical features with a crowstep used as a battlement.

Treasury Tholos
The central figure may be the Petran fertility goddess El-Uzza. Bullet marks in the tholos and urn have been made over the years by Bedouin attempting to release hidden treasure.

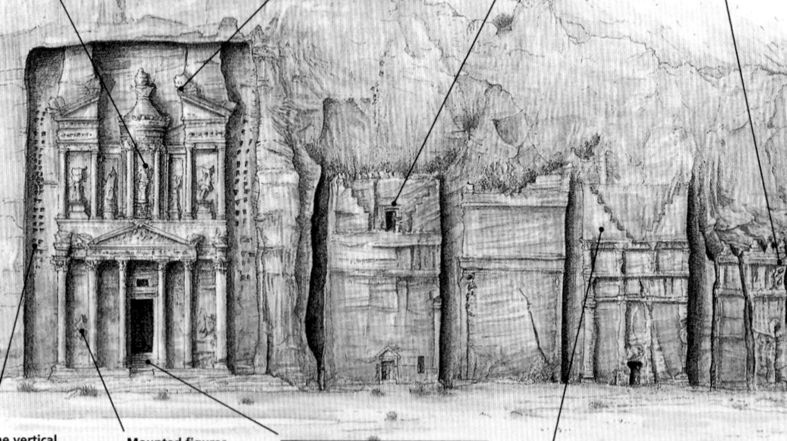

Eagle, Nabataean male deity symbol

"Attic" burial chambers were a device to protect the dead from animals and tomb robbers.

The vertical footholds may have been to aid the sculptors.

Mounted figures of Castor and Pollux, sons of Zeus, flank the portico.

The single-divide crowstep was a design devised by the Nabataeans to complement the Classical cornice.

THE OUTER SIQ

The artwork above shows some of the major constructions on the left-hand side of the Outer Siq as you walk from the Treasury to the Theatre. In reality, of course, the route bends and twists and on both the left and right sides are a great number of other tombs and features of architectural interest that could not be included.

Treasury Interior
A colossal doorway dominates the outer court (left) *and leads to an inner chamber of 12 sq m (14 sq yards). At the back of the chamber is a sanctuary with an ablution basin, suggesting that the Treasury was in fact a temple.*

For hotels and restaurants in this region see pp262–3 and pp279–80

THE ARCHITECTURE OF PETRA

The Nabataeans were adventurous architects, inspired by other cultures but always creating a distinctive look. The multiple crowstep can be seen as a design of the first settlers, whereas complex Nabataean Classical buildings reflect a later, cosmopolitan Petra. However, the dating of façades is very difficult, as many examples of the simple "early" style appear to have been built during the Classical period or even later.

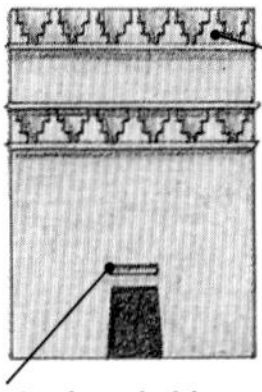

Multiple crowstep

Slot for primitive plaster pediment

This early design, *seen in the Streets of Façades, was probably Assyrian-inspired. Fragments of the once brightly painted plaster pediments have been found.*

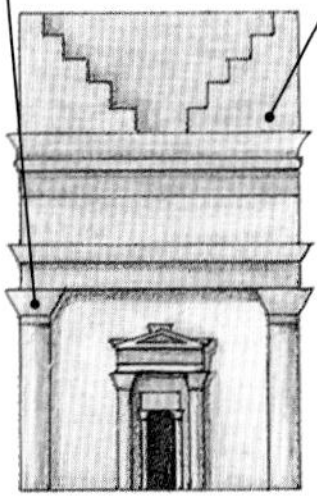

Nabataean concave "horned" capitals, resting on "cushions"

Single-divide crowstep, lending height

This intermediate style, *seen frequently in Petra, replaced multiple crowsteps with a huge single-divide crowstep, adding Classical cornices and pillars and Hellenistic doorways. This style continued well into the 1st century AD.*

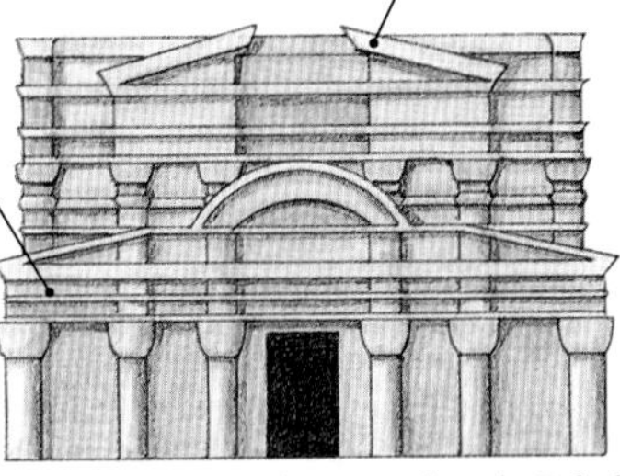

Nabataean Classical *designs, such as the Bab el-Siq Triclinium* (above), *are complex, possibly experimental fusions of Classical and native styles.*

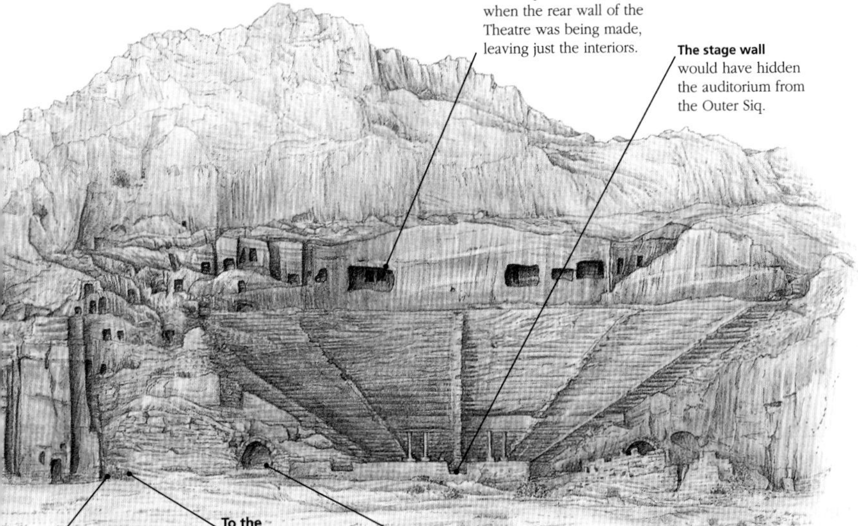

Tomb façades were cut away when the rear wall of the Theatre was being made, leaving just the interiors.

The stage wall would have hidden the auditorium from the Outer Siq.

Stairway to High Place of Sacrifice ***(see p231)***

To the Streets of Façades

Theatre Vaults

For access there were tunnels either side of the stage. Inside (right) *these were dressed with painted plaster or marble.*

Streets of Façades

Carved on four levels, these tightly packed tombs may include some of Petra's oldest façades. Most are crowned with multiple crowsteps.

The Royal Tombs

Carved into the base of El-Khubtha mountain, a short detour to the right at the point where the Outer Siq opens out on to Petra's central plain, are the Urn, Corinthian and Palace Tombs. They are collectively known as the Royal Tombs, their monumental size suggesting they were built for wealthy or important people, possibly Petran kings or queens. These tombs and their neighbours are also remarkable for the vivid striations of colour rippling through their sandstone walls, an effect heightened in the warm glow of the late afternoon sun. Particularly striking are the Silk Tomb and the ceiling inside the Urn Tomb.

Panoramic view of the Royal Tombs from the direction of the ruined city

Palace Tomb
The largest of all the Royal Tombs, the Palace Tomb had a grandiose façade on five levels which was taller than the rock into which it was carved. The upper levels, since collapsed, had to be built up using large blocks of stone.

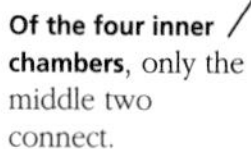

Of the four inner chambers, only the middle two connect.

Corinthian Tomb
There is no doubt that this was an important tomb in its day, but its design has baffled archaeologists because of its lack of symmetry. The doorways, each in a different style, are a clear illustration of this.

THE ROYAL TOMBS
First in the sequence of Royal Tombs is the towering Urn Tomb *(far right)*, reached by a stairway. Its name refers to a relatively tiny urn on top. Further along is the badly eroded Corinthian Tomb, which seems to be modelled largely on the Treasury, and beyond that the Palace Tomb, thought to be based on Nero's Golden House in Rome.

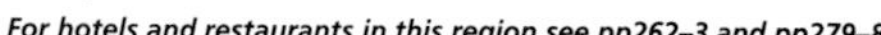

THE NABATAEANS

The Nabataeans were a people whose original homeland lay in north-eastern Arabia and who migrated westward in the 6th century BC, settling eventually in Petra. As merchants and entrepreneurs, they grasped the lucrative potential of Petra's position on the spice and incense trade routes from East Asia and Arabia to the Mediterranean. By the 1st century BC they had made Petra the centre of a rich and powerful kingdom extending from Damascus in the north to Leuke Kome in the south and had built a city large enough to support 20–30,000 people. Key to their success was their ability to control and conserve water. Conduits and the remains of terracotta piping can be seen along the walls of the Outer Siq – part of an elaborate system for channelling water around the city. The Romans felt threatened by their achievements and took over the city in AD 106. Although the Nabataeans ceased to be an identifiable political group, Petra continued to thrive culturally for a time. In the end the transfer of trade from land to sea and two devastating earthquakes in the 4th and 8th centuries AD brought about the city's demise.

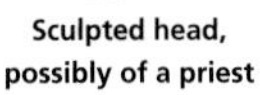
Sculpted head, possibly of a priest

Greek *(left)* and Nabataean pottery vessels found at Petra

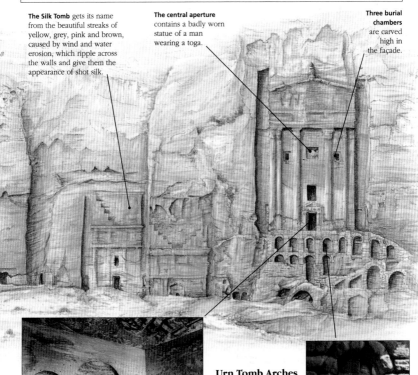

The Silk Tomb gets its name from the beautiful streaks of yellow, grey, pink and brown, caused by wind and water erosion, which ripple across the walls and give them the appearance of shot silk.

The central aperture contains a badly worn statue of a man wearing a toga.

Three burial chambers are carved high in the façade.

Urn Tomb Interior

In AD 447 the Urn Tomb was turned into a church and two of the four recesses in the back wall were combined to make an apse. A Greek inscription records the consecration.

Urn Tomb Arches

Two levels of arches support the large terrace in front of the Urn Tomb. Their appearance earned them a place in Bedouin folklore as sinister dungeons underneath a law court.

The City of Petra

Just past the theatre, the Outer Siq opens out into a wide plain. The ruins of the city of Petra are in the middle of this vast basin and the path alongside the Wadi Musa leads down to the site. Today, fragmented remains of the main street and a few nearby buildings are almost all that is left of the great city that once filled the valley. The grand Roman-style Cardo would have been Petra's main artery, fringed with markets and leading to the city's most sacred temple, the Qasr el-Bint. This building, like all the important buildings around the Cardo, would have been lavishly decorated. Traces of ornate plasterwork and marble veneer can still be seen on its walls and steps.

View of the ancient city of Petra from a point just past the Theatre

Modern Museum
Among the exhibits are a marble basin with lioness handles found in Petra Church and a small carved plaque of the Nabataean goddess al-Uzza (left) *found in the Great Temple.*

The Old Museum is in a rock-cut tomb, built, unusually for Petra, with windows. It houses a collection of statuary.

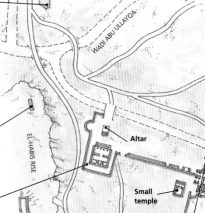

El-Habis Crusader Fortress
As its name suggests, this small fortress was built by the Crusaders. While they were here they also used the Qasr el-Bint as a stable.

Qasr el-Bint el-Faroun
The name "Palace of the Pharaoh's Daughter" was a colourful invention of Bedouin mythology. The 1st-century BC building was probably Petra's main temple, the huge slab of stone at the foot of the steps being an altar to the sun god Dushara, chief deity of the Nabataean pantheon.

Temenos Gate
The imposing entrance to the sacred precinct of Qasr el-Bint had freestanding columns in front of its three massive, possibly metal-clad wooden doors. It probably dates from after the Roman annexation. The carvings of animal deities on its capitals are a Nabataean slant on an otherwise Classical design.

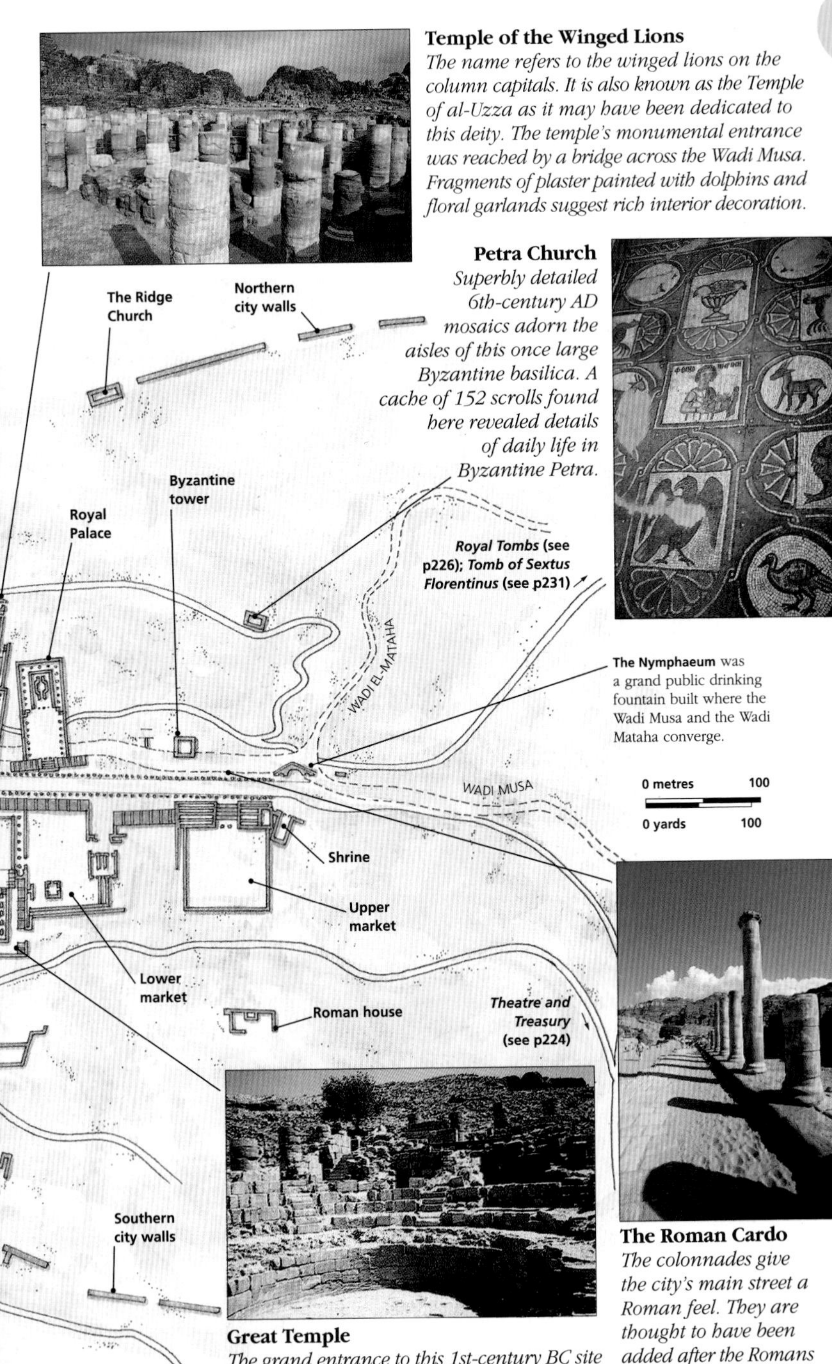

Temple of the Winged Lions
The name refers to the winged lions on the column capitals. It is also known as the Temple of al-Uzza as it may have been dedicated to this deity. The temple's monumental entrance was reached by a bridge across the Wadi Musa. Fragments of plaster painted with dolphins and floral garlands suggest rich interior decoration.

Petra Church
Superbly detailed 6th-century AD mosaics adorn the aisles of this once large Byzantine basilica. A cache of 152 scrolls found here revealed details of daily life in Byzantine Petra.

The Nymphaeum was a grand public drinking fountain built where the Wadi Musa and the Wadi Mataha converge.

The Roman Cardo
The colonnades give the city's main street a Roman feel. They are thought to have been added after the Romans annexed Petra in AD 106. The street has been partly restored by Jordan's Department of Antiquities.

Great Temple
The grand entrance to this 1st-century BC site led into a colonnaded lower precinct laid with hexagonal paving stones. Under the floor were extensive water ducts. Great stairways swept up to a 600-seat auditorium, of uncertain function. The decor was red and white stucco.

Other Sites Around Petra

Many of Petra's most famous sights can be visited in half a day. However, having come so far, it would be a pity not to explore more of this unique capital of a vanished civilization. A full day is enough to do the basic route from the ticket gate to the ancient city *(see p228)*, taking in the Royal Tombs *(see p226)*, and to include a walk to either the Monastery or the High Place of Sacrifice. Two days will enable you to do the basic route, both excursions and leave you with time to explore the area around the Tomb of Sextius Florentinus. Of the more distant sights, Little Petra can be visited in a day, while two days should be allowed for Aaron's Tomb.

High Place of Sacrifice: the round altar with the main altar behind

Façade of the Lion Triclinium

Walk to the Monastery

Just beyond the Qasr el-Bint *(see p228)* a path crosses the Wadi Musa. It leads past the Forum Restaurant to the start of an arduous but thoroughly worthwhile climb to one of Petra's most awe-inspiring and best-preserved monuments – the Monastery. The path, which cuts through the wadi, is paved in parts and features more than 800 rock-cut steps. The afternoon, when the sun is not directly in front, is the best time to do this walk.

A short detour off the main route, indicated by a Department of Antiquities signpost, leads to the **Lion Triclinium**. This monument, with the peculiar keyhole effect in the façade, caused by erosion, has blurred leonine representations of the goddess al-Uzza guarding its entrance. Its largely Classical façade has unusually ornate Nabataean features, such as "horned" capitals with floral scrollwork. After this, the path to the Monastery rises steeply. There are occasional flights of steps through the winding and narrowing gorge, and several interesting carved monuments along the way. Finally, the path slips between two boulders, and drops on to a wide, once-colonnaded, rock-cut terrace. Immediately to the right is the **Monastery**, Petra's most colossal temple, dedicated to the deified king, Obodas 1, who died in 86 BC. Although it resembles the Treasury *(see p224)*, it was never as ornate, even when statues adorned its niches. Its simple, powerful architecture, thought to date from the 1st century AD, is seen by many as the quintessential Nabataean Classical design *(see p225)*. The interior has one large chamber with an arch-topped niche where the altar stood. It came to be known as the Monastery because of the many Christian crosses carved on its walls.

The Monastery's massive tholos crowned with an urn resting on Nabataean "horned" capitals

Walk to the High Place of Sacrifice

Midway between the Treasury and the Theatre, a rock-cut stairway, marked at the start by several djinn blocks *(see p222)*, leads to the top of Jebel Attuf mountain. It is here, at 1,035 m (3,000 ft), that one of the best preserved of Petra's many places of sacrifice is located. The ascent, while gradual, requires stamina and a good head for heights, and is best attempted in the early morning. The first part of the summit is a large terrace with two 6-m (20-ft) stone obelisks, possibly fertility symbols. The second, reached by a northwards scramble past the ruins of a small Nabataean building, is another plateau. Here, just beyond a rock-cut cistern, is the **High Place of Sacrifice**. In the centre of a large courtyard is a low offering table. Steps at the far end lead up to the main altar, which has a rectangular indentation in the top. The adjacent round altar has a basin with a carved channel, quite possibly for draining the blood of animal and human sacrifices. The nearby cisterns may have been used for ritual ablutions.

The path winding down the other side of Jebel Attuf into the Wadi Farasa valley is a spectacular stepped descent, sometimes with sheer drops. The first thing you see, carved into the rock face, is the **Lion Monument**, representing the goddess al-Uzza. It was originally a fountain, perhaps for pilgrims to the High Place, with water pouring from the lion's mouth. Water channels and the shape of the lion's head and legs can still be seen.

Thereafter, the path becomes a series of steps leading to the delightfully secluded **Garden Triclinium**. The tomb takes its name from the surrounding greenery. On top of the tomb is a large cistern. Further along, to the left, is the **Tomb of the Roman Soldier**, so called because of the remains in one of the façade niches of a figure wearing the uniform of a high-ranking Roman officer. Although Classical, the façade has Nabataean "horned" capitals on top of the pillars. Opposite is the façadeless **Triclinium**, thought have been part of the Roman Soldier Tomb complex. It has the only carved interior in Petra and its niches, fluted half columns and cornice are spectacularly enhanced by the amazing bands of colour running through the walls and ceiling.

Beautifully carved interior of the Triclinium, unusual for Petra

Further down the track is the relatively plain **Broken Pediment Tomb**, named after its most striking feature. Nearby is the elegant **Renaissance Tomb**, with the three urns above its arched entrance. Similar in style to the Tomb of Sextius Florentinus, it may date from the same period. Past this point the Wadi Farasa widens and the descent ends in the main valley, not far from the Qasr el-Bint *(see p228)*.

Aaron's Tomb

This site is venerated by Muslims, Christians and Jews as the place where Moses's brother Aaron was buried. The white dome of the shrine can be seen from the High Place of Sacrifice, which may be a close enough viewing for most people. The journey there involves a three-hour ride on horseback and a hard three-hour climb to the top of Petra's highest peak – Jebel Haroun. For those determined to go, a guide and adequate supplies are essential.

The lonely mountaintop shrine of Aaron's Tomb, Petra's holiest place

Tomb of Sextius Florentinus

Beyond the Palace Tomb *(see p226)*, along a track skirting the cliff, stands the **Tomb of Sextius Florentinus**. Despite its badly eroded north-facing façade, the beautiful and unusual details of its design are clearly visible. Above its entrance is a Latin inscription listing the positions held by Florentinus up to his last post as Governor of Arabia in AD 127. Further north is the **Carmine Façade** with its vivid striations of red, blue and grey. Continuing alongside the Wadi Mataha brings you to a rock-cut complex known as the **House of Dorotheos** because of two Greek inscriptions found here. On the other side of the wadi is a cluster of homes and tombs known as **Mughar el-Nasara**, including the fine **Tomb with Armour**. Local Christians were probably responsible for the many crosses etched into the walls.

Tomb of Sextius Florentinus, Roman governor of the province of Arabia

Little Petra

This northern suburb of Petra, Siq el-Berid, has come to be known as Little Petra because it is like a miniature version of the main city. Situated 8 km (5 miles) north of Wadi Musa town, it is most easily reached by taxi. The journey on foot, north along the Wadi Abu Ullayqa, which starts just past the Qasr el-Bint, is hard, but rewarding. A guide is essential.

Detail from ceiling of the Painted House

Little Petra seems to have been a largely residential settlement, as relatively few tombs have been discovered here. It may well have been where Petra's wealthy merchants had their homes. Just outside its Siq-like entrance, which was once controlled by a gate, are a large cistern and a Classical temple. The gorge, shorter than the one leading into Petra, contains a simple temple. As you emerge from the quiet of the gorge into the town, the incredible profusion of façades is overwhelming, with houses, temples and cisterns carved into every exposed rock face. Flights of steps shoot off in all directions, evoking images of a bustling urban centre. One of Little Petra's main attractions is the **Painted House** with its plaster ceiling and walls delightfully decorated with flowers, vines, bunches of grapes, Eros with his bow and Pan playing his pipes.

Wadi Rum ⑩

Thamudic rock graffiti

The desert landscape of Wadi Rum is one of the most awe-inspiring sights in the entire Middle East. Huge ochre-coloured rock pinnacles, weathered into bulbous, outlandish shapes, rise up 600 m (2,000 ft) from the flat valley floors, like islands in a sea of red sand. Hundreds of hiking and climbing routes wind their way up and around the many peaks. This area was once on a major trade route, and evidence of settlement here includes ruins of a temple built by the Nabataeans *(see p227)* and carvings and inscriptions left by the later Thamud people. Today the region is still inhabited by semi-nomadic Bedouin tribes.

Aqaba
Petra
JEBEL HUBEIRA
JEBEL LEYYAH
WADI LEYYAH
WADI RUM
JEBEL RUM
JEBEL UMM ISHRI
Nabataean Temple
Rum
JEBEL UMM EJIL
WADI RUMMAN
WADI UM
Abu Aina campsite
JEBEL QATTAR
Aqaba

★ Lawrence's Spring
Not far from Rum village, this tranquil spring was described by TE Lawrence as "a paradise just 5 feet square". A Nabataean-built water channel can be seen nearby.

Rum Village
The main settlement is a rapidly growing Bedouin village. The Rest House on the outskirts offers spartan accommodation and simple meals.

Khazali Canyon
This steep defile is dotted with Thamudic inscriptions. It is possible to scramble 200 m into the canyon, starting on a ledge on the right-hand side.

0 kilometres 4

0 miles 2

KEY

- Road
- Walk
- Hike/scramble
- Four-wheel-drive/camel track

STAR SIGHTS

★ Lawrence's Spring

★ Jebel Umm Fruth Rock Bridge

For hotels and restaurants in this region see pp263–3 and pp279–80

Rock Map at Jebel Amud
In a cave 20 km (12 miles) north-east of Rum is a rock marked with indentations and lines. It is thought by some to be a topographical map of the area, dating from around 3000 BC.

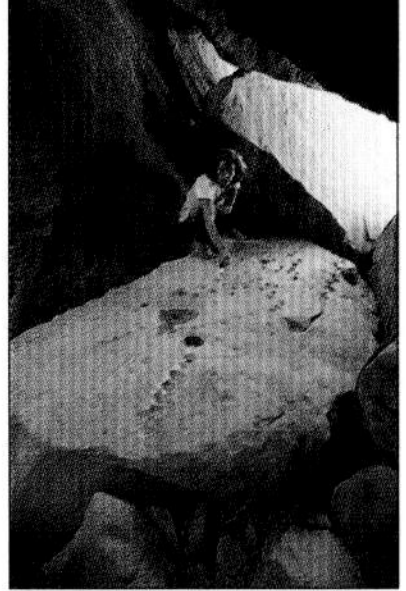

VISITORS' CHECKLIST

Road map C7. 30 km (19 miles) SE of the Desert Highway (Route 53). Turn off 45 km (28 miles) N of Aqaba. *advisable for visiting the desert. Jeeps, camels and guides available at the Rest House or in Rum village.* **Rest House** ***Tel*** *(03) 201 8867.*

Seven Pillars of Wisdom
This spectacular peak, also known as Jebel Makhras, is named after TE Lawrence's famous book, not, as is often suggested, vice versa. Wadi Siq Makhras, just to the south, provides hiking access to Wadi Umm Ishrin and beyond.

Jebel Barrah
This large outcrop, seen here at its northern end, flanks beautiful Barrah Canyon, which is a stunning hike best negotiated from the south.

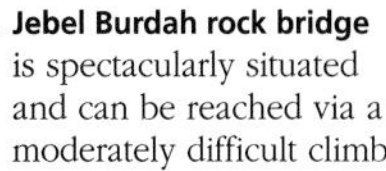

Jebel Burdah rock bridge is spectacularly situated and can be reached via a moderately difficult climb.

★ Jebel Umm Fruth Rock Bridge
This dramatic natural phenomenon is one of several rock bridges in the area. It rises straight from the desert floor and can be climbed and crossed without difficulty.

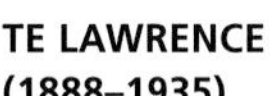

TE LAWRENCE (1888–1935)

Lawrence of Arabia, the most famous British hero of World War I, earned his nickname for his exploits fighting alongside the Arab tribes that revolted against Turkish rule in 1915. Sent to Mecca in 1916 to liaise with leaders of the revolt, he then led many Arab guerrilla operations in the desert, including attacks on the Hejaz Railway, some launched from Wadi Rum. He also took part in the capture of Aqaba and the advance on Damascus. *The Seven Pillars of Wisdom*, his account of the Arab Revolt, contains lyrical descriptions of the dramatic scenery around Wadi Rum.

Exploring Wadi Rum

There are essentially two main ways to explore the desert of Wadi Rum: through a combination of jeep and hiking, or by camel trekking. Jeeps allow you to travel further and faster, but the more traditional means of transport will bring you much closer to the stillness of the desert. Either way, make sure you carry lots of water and avoid travelling during the midday heat, especially in summer. For contact details of companies organising Wadi Rum expeditions, see page 309.

Jeeps, the best way to cover large distances quickly in Wadi Rum

Jeep Tours

A wide range of jeep tour options is posted at the main reception gate, 7 km (4.5 miles) before Rum village. If you have not pre-arranged a trip, you will be allocated a driver here. It is possible to join up with other travellers to share the cost of a jeep. There are two main areas to explore: the main southern section of Wadi Rum and the less-visited northern scenery closer to the village of Diseh.

The most popular destinations include the striking red sand dunes of Jebel Umm Ulaydiyya, the small oasis known as "Lawrence's Spring" and the narrow *siq* (gorge) of Khazali Canyon.

Other noteworthy attractions include the Nabataean inscriptions and petroglyphs of Anfaishiyya, the natural rock bridge of Jebel Umm Fruth and the various "sunset sites", which are all ideal places to witness the changing afternoon colours of the desert rocks.

Hiking

Many of the best trips offer a combination of jeep travel and hiking. The 5-km (3-mile) stroll through the towering walls of Barrah Canyon is a favourite option. Some hikes require a guide, such as the excellent hour-long scramble up to the Jebel Burdah rock bridge and the exciting half-day hike through labyrinthine Rakhabat Canyon.

Most trips require jeep transport to get you to the start of the hike. The only walk you can really do by yourself is from the visitor centre east to Makharas Canyon and back; take a guide if you are unsure of your route-finding skills.

Hikers taking a break with their Bedouin guides

Tourists on a camel-trek through the canyons of the Wadi

Adventure Activities

An excellent alternative to making arrangements on the spot is to arrange a more active itinerary in advance with one of Wadi Rum's excellent Bedouin guides. Most can arrange jeep and overnight trips but you'll need a specialist for climbing or canyoning. Overnight trips that combine a jeep excursion, camel ride and some rock scrambling are very popular.

Camel trekking is fun but the pace is slow and can be highly uncomfortable after a couple of hours. Still, it is undeniably the best way to get a feel for the desert in classic "Lawrence" fashion. The three-day ride from Wadi Rum south to Aqaba *(see p235)* is a challenging adventure.

Horse riding is possible on the periphery of the park, as is mountain biking over the desert flats.

It is well worth fitting in an overnight at a Bedouin camp during your visit. The larger fixed camps can be touristy but are fun nonetheless. The smaller ones shift location regularly and offer a more authentic, but also more basic, experience. The food is generally excellent; you may get to try *mensaf* (a Bedouin dish of lamb and rice) or, if you are lucky, a "Bedouin barbeque" – meat slow-cooked in a desert oven called a *zerb*. Reclining by an open fire, gazing at the stars and sipping a mint tea in the stillness of the desert is perhaps the quintessential Wadi Rum experience.

Aqaba ⓫

Road map: B7. *62,000.* *El-Koornish St (next to the Fort), (03) 201 3363.*

The only Jordanian outlet to the sea, Aqaba is a very important commercial port town. The relentless stream of heavy trucks going to and coming from Amman along the Desert Highway is clear evidence of this.

South of the town however, away from the busy port, the crystal clear waters are home to fabulous coral reefs. These are the main reason for Aqaba's popularity with visitors, as they offer some of the best scuba diving in the world. Closer to the shore, many other types of water sports also help to provide escape from the extreme summer heat. Large sandy beaches stretch out along the coast, bounded by modern hotels, and the steep mountains behind form a spectacular natural backdrop.

Aqaba's long and glorious past also provides it with some notable archaeological sites to visit. It is thought to be close to the site of biblical Ezion-Geber, the large port which is said to have been built by King Solomon. Its existence has, however, yet to be proved.

The town's deep freshwater springs ensured that Aqaba became a popular caravan stop for merchants travelling between Egypt, the Mediterranean coast and Arabia. By the 2nd century BC, the now prosperous town had fallen under the control of the Nabataeans *(see p227)*. Such prosperity saw it conquered by the Romans in AD 106, and later the Muslims in AD 630. Under Muslim control, Aqaba became an important stage on the pilgrimage to Mecca, and the Muslims built the fortified town of **Ayla** nearby to the north. After suffering a major earthquake in 748, the town was rebuilt, and thrived with an increasing sea trade. Following another earthquake in 1068 however, and then the Crusader conquests of the 12th century, the city was finally abandoned. You can visit the ruins at the Ayla digs, next to the coastal Corniche road. Much of the foundations of walls, towers and a series of buildings still remain. The **Archaeological Museum**, next to the tourist office, features material from the digs, as well as illustrating the history of Aqaba.

Ruins of the old fortified Islamic town of Ayla, in modern Aqaba

The other main archaeological site in Aqaba is the **Mameluke Fort**, set between the palm trees on La Côte Verte. Built in the 16th century, its portal now bears the coat-of-arms of the Hashemites, placed there after Lawrence of Arabia's troops conquered the port during World War I. The fort also served as a caravanserai for hundreds of years, and some restored rooms pay testament to this more peaceful role.

Sign to Aqaba Aquarium

By going west past the industrial port and just beyond the ferry passenger terminal you will come to the small Aqaba Marine Science Station **Aquarium**. This contains a collection of the most important species of the varied flora and fauna in the Gulf of Aqaba, including moray eels and deadly stonefish. It also displays information on the campaign to protect the Red Sea.

Sailing boats anchored in the Gulf of Aqaba

Archaeological Museum
El-Koornish St (next to Fort). ***Tel*** *(03) 201 9063.* *8am–5pm daily.*

Mameluke Fort
La Côte Verte. ***Tel*** *(03) 201 9063.* *daily.*

Aquarium
South Coast (near ferry terminal). ***Tel*** *(03) 201 5145.* *7:30am–3:30pm daily.*

THE RED SEA AND SINAI

Once coveted by Egypt's pharaohs for its reserves of turquoise, copper and gold, Sinai is now equally prized by tourists for its white, palm-fringed sands and the limpid waters of the Red Sea, rich with marine life. Its close association with key episodes from the Old Testament also makes the Sinai's mountainous interior an area of deep religious significance for Jews, Muslims and Christians alike.

The Sinai Peninsula forms a triangle between the gulfs of Aqaba and Suez, two finger-like extremities of the Red Sea. Although the whole of Sinai is Egyptian territory, Israel and Jordan also have small stretches of Red Sea coast at Eilat and Aqaba, respectively.

The word "Sinai" probably derives from "Sin", the moon god worshipped in Egypt under the pharaohs. But the region is better known through the Bible as the "great and terrible wilderness" negotiated by Moses and his people in their epic 40-year journey from Egypt to the Promised Land. It's here that God supposedly first spoke to Moses through the medium of a burning bush and here, on Mount Sinai, that Moses received the Ten Commandments. The peninsula has been crossed by countless armies, including most recently that of the Israelis, who held the region from 1967 to 1982 when it was returned to Egypt under the terms of the Camp David peace treaty. In the years since then tourism has boomed as southern Sinai and the peninsula's eastern coast have been developed with all-inclusive resorts, such as Sharm el-Sheikh. But the wilderness is far from tamed. Inland Sinai remains virtually uninhabited with barren mountains sheltering hidden oases such as Feiran, with its thousands of date palms. More dramatic still are the underwater landscapes of the Red Sea, where vast coral reefs provide a home for more than 1,000 species of marine life, making for one of the world's richest dive sites.

Divers filming at Eilat's Dolphin Reef

◁ Central Sinai inland of Nuweiba, dramatic but accessible only by four-wheel-drive or camel

Exploring the Red Sea and Sinai

Most visitors head for where the mountains and desert meet the clear cool waters of the Red Sea; specifically, Eilat, Aqaba and, most picturesque of all, the Sinai peninsula's east coast. Its string of modern resorts, while uninteresting in themselves, are set against a backdrop of extraordinary natural beauty. Nuweiba, Dahab, Naama Bay and Sharm el-Sheikh are the largest and most well-developed tourism centres, but there are many smaller, more private beach retreats. St Catherine's Monastery can be visited as a day trip.

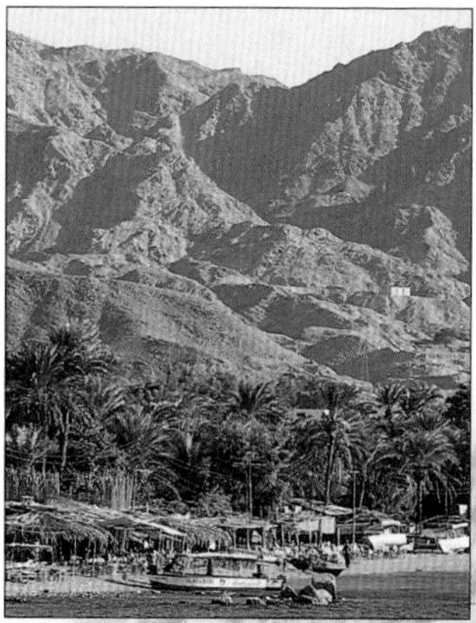

SIGHTS AT A GLANCE

Dahab 3
Feiran Oasis 8
Mount Sinai 7
Nuweiba 2
Ras Muhammad National Park 5
St Catherine's Monastery pp246–9 6
Sharm el-Sheikh 4
Taba 1

Aqaba, with a typical Red Sea scene of beach, palms and looming mountains

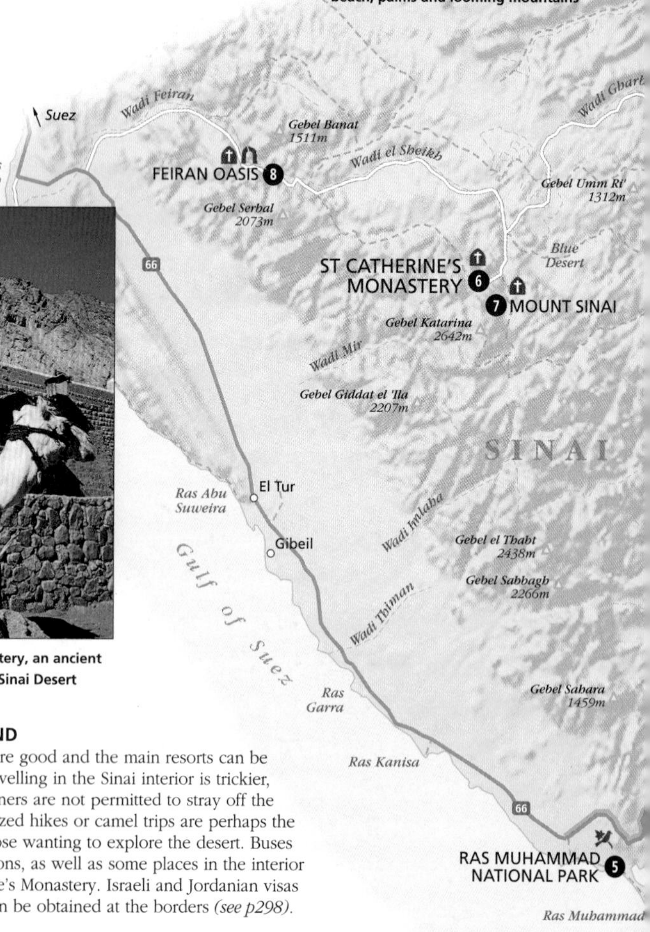

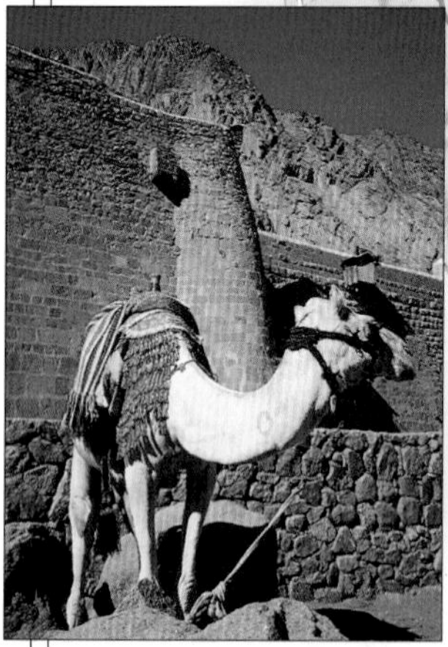

St Catherine's Monastery, an ancient walled retreat in the Sinai Desert

GETTING AROUND

The coastal roads are good and the main resorts can be reached by car. Travelling in the Sinai interior is trickier, especially as foreigners are not permitted to stray off the main roads. Organized hikes or camel trips are perhaps the best options for those wanting to explore the desert. Buses serve coastal locations, as well as some places in the interior such as St Catherine's Monastery. Israeli and Jordanian visas and Sinai passes can be obtained at the borders *(see p298)*.

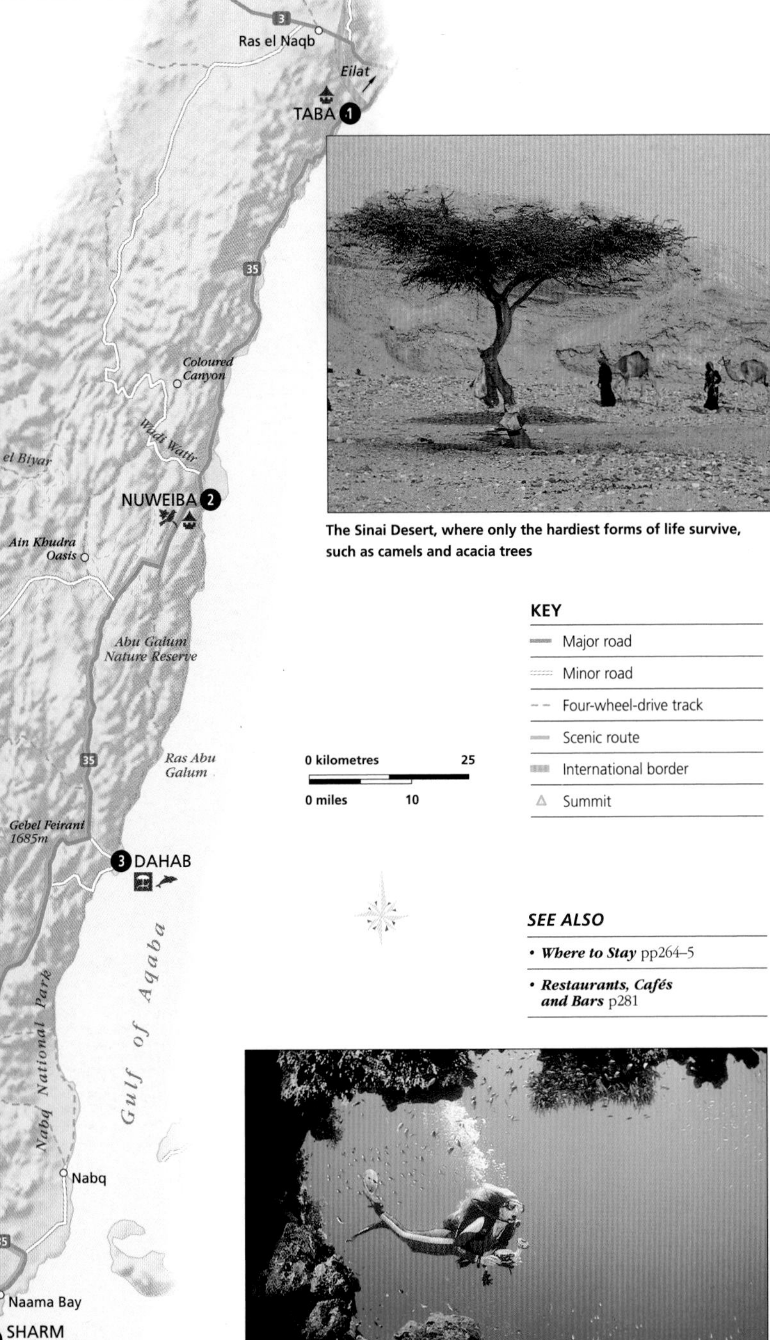

The Sinai Desert, where only the hardiest forms of life survive, such as camels and acacia trees

The underwater scenery and marine life of the Red Sea, which is every bit as stunning as the desert and mountain landscapes above

SEE ALSO

- ***Where to Stay*** pp264–5
- ***Restaurants, Cafés and Bars*** p281

The Coral Reefs of the Red Sea

The Coral Reef is one of the richest ecosystems on earth. Visitors to the Red Sea cannot but marvel at the contrast between the barren, almost lifeless desert and the explosion of marine life on the coastal reefs. The waters are so clear that even from the surface you can appreciate the huge diversity of species inhabiting the reefs. Scuba divers can use the facilities of the many diving centres along the coast *(see pp292–5)*. Remember that a reef is an extremely fragile and threatened environment and divers should look but not touch.

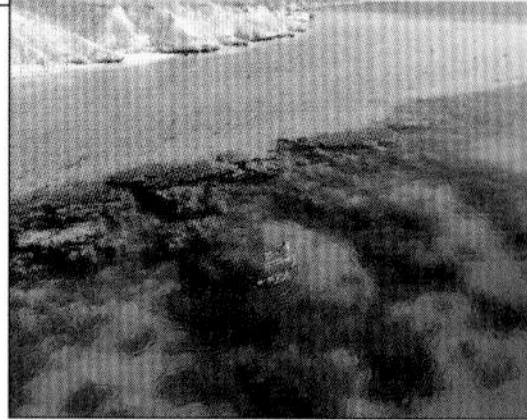

View of lagoon and the shallow waters covering the reef-top

The edge of the reef is the best place for snorkellers to appreciate its wealth of marine life.

The lagoon teems with small colourful fish, including the fry of species found on the reef beyond.

The clown fish *protects itself from the sea anemone's stinging tentacles with a layer of mucus, using its host as a refuge from predators and for laying its eggs.*

Moray eel, emerging from its reef-wall lair

School of flag basslets, a very common species in the Red Sea

Manta rays *are harmless plankton-eaters. Growing up to 6 m (20 ft) across, they are most common in open water or where there are strong currents.*

Alcyonarians, brightly coloured soft corals

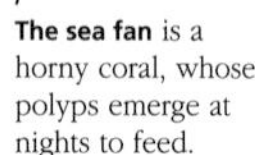

The sea fan is a horny coral, whose polyps emerge at nights to feed.

Jacks are usually seen in large schools in open water, but large solitary individuals will visit the reef.

CORALS, THE ARCHITECTS OF THE REEF

Corals are animals, colonies of polyps, which require very precise conditions of water temperature and sunlight to grow. They take many forms – from hard rock-like corals, such as *Acropora* species, to the horny gorgonians which project from the reef into the current to feed on micro-organisms, to various soft corals. Most reefs are built over many thousands of years from the skeletons of hard corals.

Gorgonians filtering the water for plankton

An *Acropora* growing in still, shallow water

Feathery red plume of Klunzinger's soft coral

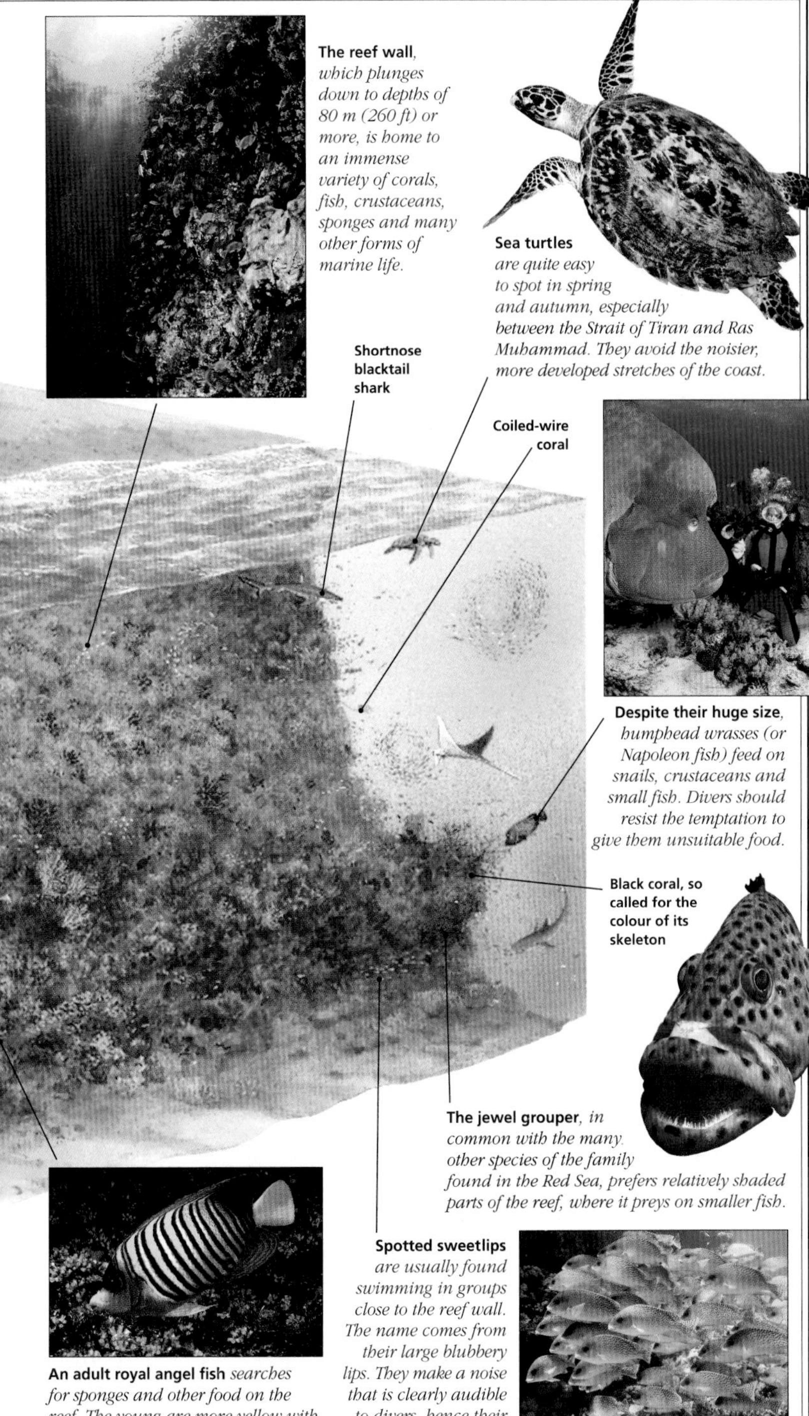

The reef wall, *which plunges down to depths of 80 m (260 ft) or more, is home to an immense variety of corals, fish, crustaceans, sponges and many other forms of marine life.*

Sea turtles *are quite easy to spot in spring and autumn, especially between the Strait of Tiran and Ras Muhammad. They avoid the noisier, more developed stretches of the coast.*

Despite their huge size, *humphead wrasses (or Napoleon fish) feed on snails, crustaceans and small fish. Divers should resist the temptation to give them unsuitable food.*

The jewel grouper, *in common with the many other species of the family found in the Red Sea, prefers relatively shaded parts of the reef, where it preys on smaller fish.*

An adult royal angel fish *searches for sponges and other food on the reef. The young are more yellow with a large eyespot on the dorsal fin.*

Spotted sweetlips *are usually found swimming in groups close to the reef wall. The name comes from their large blubbery lips. They make a noise that is clearly audible to divers, hence their other name – grunters.*

Carvings on the Haggar Maktub, in the desert near Nuweiba

Taba ❶

Road map F5.

Since Israel returned ownership of the Sinai peninsula to Egypt in 1982, the small coastal town of Taba has served as a border post between the two countries. A pleasing stretch of beach is overlooked by a five-star hotel. Just under 20 km (12 miles) to the south is the new resort of **Taba Heights**, which boasts some of the most luxurious hotels in Sinai, as well as an 18-hole golf course and a marina. There are views from the resort across the Red Sea to the Israeli, Jordanian and Saudi Arabian coastlines.

Between Taba and Taba Heights, just offshore is **Pharaoh's Island** (or Coral Island), which is dominated by an impressive Crusader fortress. Tickets for the boat across to the island are available from the Salah ed-Din Hotel on the coastal road.

Nuweiba ❷

Road map F6.
from Aqaba (Jordan).

Nuweiba lies midway along the Gulf of Aqaba at the side of a promontory and consists of two distinct districts. To the south is the luxuriant Nuweiba Muzeina oasis, which for centuries was a port for pilgrims going to Mecca. It now has many hotels and tourist villages. To the north is Nuweiba el-Tarabin, named after the Bedouin tribe that lives here. You can visit the ruins of the large **Tarabin fortress**, built in the 16th century by the Mameluke sultan Ashraf el-Ghouri. The Nuweiba area is rich in beaches, and diving and snorkelling sites.

Bedouin with his camel, outside Nuweiba

Environs

Nuweiba makes a convenient starting point for trips to the Sinai interior. One of the most fascinating is to the **Coloured Canyon**, a narrow sculpted gorge created by water erosion. Its sandstone walls have taken on many hues of yellow, red and ochre due to the slow process of oxidation of the ferrous minerals in the rocks. The canyon opening can be reached by car from the Ain Furtaga oasis, about 15 km (9 miles) from Nuweiba on the road west, and thence by following the Wadi Nekheil track.

Another fascinating trip uses a jeep track from Ain Furtaga through the immense Wadi Ghazala to **Wadi Khudra**. Midway along the track you will come to the Ain Khudra oasis, a lovely patch of palms and tamarisks seemingly wedged between the high, near vertical, red walls of the canyon. If you continue a little further along the trail you will come to the solitary Haggar Maktub (Rock of Inscriptions). Since the Nabataean period, pilgrims going to Sinai have left graffiti carved on the rock.

Heading south from Nuweiba Muzeina along the coast leads to the **Abu Galum Nature Reserve**. A maze of narrow wadis penetrates the interior, with an abundance of plants and wildlife, such as foxes, ibexes and hyraxes. The beach at Ras Abu Galum is usually deserted except for a few Bedouin fishermen.

Dahab ❸

Road map F6.

In Arabic the word *dahab* means "gold", and the name derives from the sand on the beautiful beaches. The crown of palm trees, the beaches and the light blue sea make this one of the most popular localities in Sinai. It has grown up around the old Bedouin village of Assalah, which still survives today. The many camping sites, simple hotels and beachside restaurants attract an array of mainly independent travellers who lend a raffish air to the town.

Many also visit for the world-class diving sites around Dahab. Among the

Raccoon butterflyfish with diver, off the coast of Dahab in the Gulf of Aqaba

Four Seasons Resort, one of numerous luxury hotels in Sharm el-Sheikh

most famous and dangerous are the "Canyon" and the "Blue Hole". Almost entirely surrounded by reef, the Blue Hole drops to a depth of 80 m (260 ft) only a few metres off the shore. Although many sites are for expert scuba divers only, there are still plenty of others suitable for beginners or snorkellers.

Sharm el-Sheikh ❹

Road map E7. *Tourist Office, Sharm el-Sheikh, (069) 366 4721.*

Until the latter half of the 20th century, the most famous resort in Sinai was only a military airport. Situated on the western side of the Strait of Tiran, Sharm became famous when Egyptian president Nasser decided to block Israeli access to the Red Sea, thus provoking the 1967 war. Under Israeli occupation of Sinai, the first hotels were built and began to attract tourists, especially expert scuba divers. The Sharm el-Sheikh bay is still a military port, but the neighbouring Sharm el-Maiya bay has hotels, shops and small restaurants. Most of the tourist development, however, has focused on **Naama Bay**, a few kilometres to the north. This is the place that most people actually mean when they talk about Sharm el-Sheikh. It has a long beach with a host of luxury hotels and diving centres. Boats take snorkellers as well as scuba divers out to the open sea. Here, in the Strait of Tiran, you can observe manta rays, sharks, dolphins and, occasionally, sea turtles. For those wanting to stay above water, tourists are taken in glass-bottomed boats to observe the coral reef from above. Other attractions include all manner of water sports, plus camel treks, quad biking and excursions inland.

Another spectacular sight is the long reef under the cliffs to the west of the **Ras Umm Sidd** lighthouse. Reachable from land, here you can admire a forest of gorgonians, huge Napoleon fish and, sometimes, barracuda.

Environs

A 29-km (18-mile) journey by jeep along the coast road north of Sharm el-Sheikh brings you to the 600-sq km (232-sq mile) **Nabq National Park**. This coastal park on the edge of the desert boasts crystal-clear lagoons and the most northerly mangrove forest in the world, which extends for 4 km (2.5 miles) along the shoreline. The hardy mangroves are able to live in salt water, making this is an extremely important environment, linking land to sea. It is used as a feeding ground by migratory birds, including storks, herons and many species of birds of prey.

Ras Muhammad National Park ❺

Road map E7. 20 km (12.5 miles) S of Sharm el-Sheikh. *to Sharm el-Sheikh, then taxi.* *daily.*

On the southern tip of the Sinai peninsula, where the waters of the Gulf of Suez and the Gulf of Aqaba converge, is a park instituted in 1983 to protect the incredibly varied coastal and marine environment. It includes extensive coral reefs, a lagoon, mangroves and a rugged desert coastline, and there is a series of well-marked trails leading to the most interesting spots. Among the most beautiful of these is the Ras Muhammad headland, the southernmost point in Sinai. Formed from fossilized corals, the headland is surrounded by beautiful reefs. The diving sites are very varied, with both reefs and wrecks to explore. There are also long, sandy beaches and a clifftop "Shark Observatory".

Gazelle at Ras Muhammad National Park

Entrance to Ras Muhammad National Park

Diver exploring coral reef in the Red Sea, surrounded by glittering shoal of sweeper fish ▷

St Catherine's Monastery ❻

A community of Greek orthodox monks has lived here, in the shadow of Mount Sinai, almost uninterruptedly since the monastery was founded in AD 527 by Byzantine emperor Justinian. It replaced a chapel built in 337 by St Helena, mother of Emperor Constantine, at the place where tradition says that Moses saw the Burning Bush. The monastery was named after St Catherine only in the 9th or 10th century, after monks claimed to have found her body on nearby Mount Catherine.

Library
The collection of priceless early Christian manuscripts is second only to that in the Vatican Library in Rome.

★ Icon Collection
Most of the monastery's 2,000 icons, such as this one of St Theodosia, are kept here, in the Icon Gallery. A selection is always on public view in the Basilica.

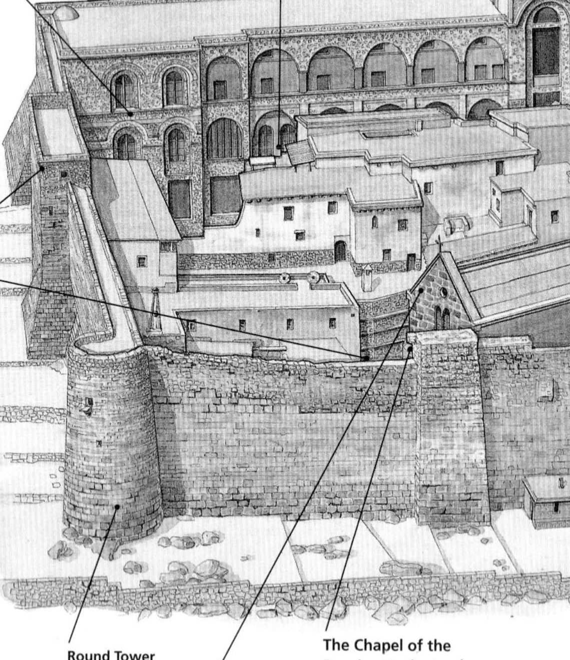

The Walls of Justinian, built in the first half of the 6th century, are part of the complex's original structure.

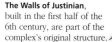

The Burning Bush
This spiny evergreen is said to be from the same stock as the bush from which Moses heard God's voice, instructing him to lead his people out of Egypt to the Promised Land.

Round Tower

The Chapel of the Burning Bush stands where it is claimed the miraculous bush seen by Moses originally grew.

★ Basilica of the Transfiguration
This magnificently decorated church owes its name to the 6th-century Mosaic of the Transfiguration in the apse. It can be glimpsed behind the gilded iconostasis that dates from the early 17th century.

STAR SIGHTS

- ★ Basilica of the Transfiguration
- ★ Icon Collection

For hotels and restaurants in this region see pp264–5 and p281

Bell Tower
This was built in 1871. The nine bells were donated by Tsar Alexander II of Russia and are nowadays rung only on major religious festivals.

VISITORS' CHECKLIST

Road map E6. *Sinai, 90 km (56 miles) W of Dahab and Nuweiba.* ✈ *10 km (6 miles) NE of monastery.* 🚌 *from Taba, Nuweiba or Dahab to St Catherine's Village (El-Milga), then taxi 3.5 km (2 miles). Petrol available at monastery.* ◯ *9am–noon Mon–Thu, Sat.* ● *Greek Orthodox hols. Admission free, but offerings welcome.*

The Mosque was created in 1106 by converting a chapel originally dedicated to St Basil.

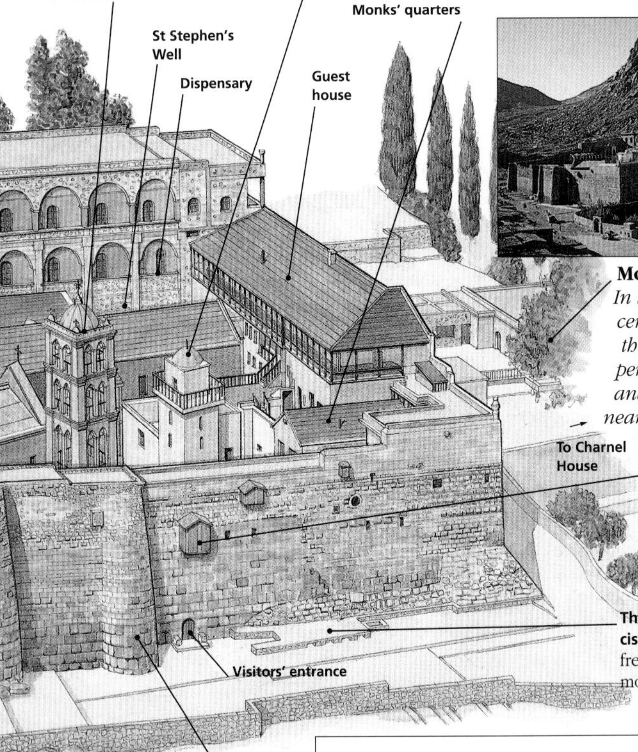

Monastery Gardens
In the orchard lies the cemetery, from which the monks' bones are periodically exhumed and transferred to the nearby Charnel House.

The elevated entrance, reached by a pulley system, used to be the only access.

The underground cistern was dug to store fresh water from the monastery's springs.

Well of Moses
One of the monastery's main water sources, this is also known as the Well of Jethro, as Moses is said to have met his future wife, Jethro's daughter, here.

ST CATHERINE OF ALEXANDRIA

St Catherine is one of the most popular of early Christian female saints. Her legend, not recorded before the 10th century, recounts that she was a virgin of noble birth, martyred in Alexandria in the early 4th century. After being tortured on a spiked wheel (hence the Catherine wheel), she was beheaded. Her body was then transported by angels to Sinai, where it was found, uncorrupted, some six centuries later by the local monks.

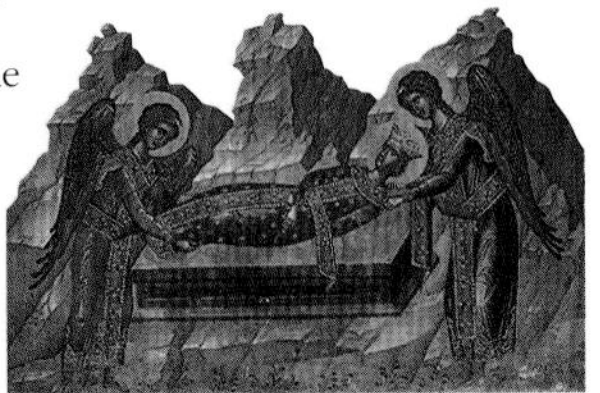

Detail from icon showing angels setting down the body of St Catherine in Sinai

Exploring St Catherine's Monastery

Fortified by massive curtain walls, the monastery lies at the head of Wadi el-Deir (Valley of the Monastery), surrounded by high, red granite mountains. It is inhabited by about 20 Greek Orthodox monks, who follow the rule of St Basil, and the only buildings normally open to visitors are the Basilica and the Charnel House. Despite this and the constant crowds of pilgrims and tourists, the remote location in the heart of Sinai and spectacular, rugged scenery are awe-inspiring. For the reasonably fit, there are well-marked paths to the top of Mount Sinai and other nearby peaks.

Rock steps leading to the Gate of Confession on Mount Sinai near St Catherine's Monastery

Inside the monastery

Entry nowadays is through a small postern in the curtain wall, whose impressive thickness varies from 1.8–2.7 m (6–9 ft). Some sections of wall survive from the monastery's origins in the 6th century, but large-scale rebuilding took place in the 14th century, after an earthquake, and in 1800, on Napoleon's orders.

The monastery's Basilica was built in AD 527 with three aisles in typical Byzantine style. Eleventh-century, carved wooden doors open into the narthex (porch), where some of the monastery's splendid icons, all painted on wood, are displayed. The collection is exceptional for its size and quality, and because it contains the only examples of Byzantine painting to have survived the Iconoclast era (726–843). Among them are a *St Peter* (5th–6th century), a *Christ in Majesty* (7th century), both in encaustic painting, and the *Ladder of Paradise* (7th century).

Carved cedar doors, made in the 6th century, lead into the central nave, which contains 12 columns topped by grey granite capitals and hung with icons showing the saints of the months of the year. The marble floor and coffered ceiling are 18th century. The iconostasis, dating from 1612, is by a Cretan monk, Jeremiah the Sinaite. The large figures represent Christ, the Virgin Mary and Saints Michael, Nicholas, Catherine and John the Baptist.

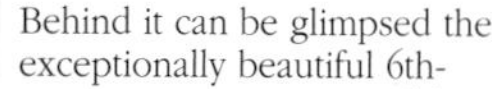

Behind it can be glimpsed the exceptionally beautiful 6th-century Mosaic of the Transfiguration decorating the roof of the apse. It shows Christ surrounded by Elijah, Moses and the Disciples John, Peter and James. In the apse, (often closed), on the right, is a marble coffin containing the remains of St Catherine.

Coptic Cross in monastery wall

The Chapel of the Burning Bush, behind the apse and also usually closed to the public, is the holiest part of the monastery. It was built on the site where God is thought to have appeared to Moses for the first time (Exodus 3: 2–4). Tradition says that the bush itself *(see p246)* was moved outside when the chapel was built.

The library has over 3,000 manuscripts in Greek, Coptic, Syriac, Arabic, Georgian, Armenian and Old Slavonic. The oldest is the 5th-century *Codex Syriacus*, one of the earliest existing copies of the Gospels.

St Catherine's has, uniquely for a Christian monastery, a mosque within its walls. It was built for the Bedouin who worked in the monastery and also as a way of avoiding attacks by the Muslims.

Outside the walls

In the gardens *(see p247)* are the monks' cemetery and the Chapel of St Triphonius. The latter's crypt holds the Charnel House containing the bones of deceased monks. The robed skeleton is that of Stephanos, a 6th-century guardian of the path to Mount Sinai.

Moses receiving the tablets inscribed with the Ten Commandments from God, 6th-century wall painting, St Catherine's Monastery

For hotels and restaurants in this region see pp264–5 and p281

Chapel of the Holy Trinity on the summit of Mount Sinai

Mount Sinai ❼

Road map E6. Sinai, 90 km (56 miles) W of Dahab and Nuweiba.

According to tradition, Mount Sinai (Gebel Musa, the Mountain of Moses) is the Biblical Mount Horeb, where Moses spent 40 days and received the Ten Commandments (Exodus 24). Two paths climb to the 2,286-m (7,500-ft) summit from behind the monastery, both requiring three hours' walking. The route said to have been taken by Moses is the most tiring as it consists of 3,700 rock steps called the Steps of Repentance. There are several votive sites along the way.

A cypress-shaded plain, 700 steps below the summit, is the so-called Amphitheatre of the Seventy Elders of Israel, where those who accompanied Moses stopped, leaving him to go to the top alone. It is also called Elijah's Hollow, as Elijah is said to have heard the voice of God here. It contains **St Stephen's Chapel** and is where people spending the night on the mountain are asked to sleep. This is also where the second, longer but easier, path joins the first. Camels can be hired to this point, but the final 700 steps have to be done on foot.

On the summit is the small **Chapel of the Holy Trinity** (often closed). It was built in 1934 on the ruins of a 4th–5th-century church and is said to be where God spoke to Moses from a fiery cloud. Nearby is a small, 12th-century mosque and the cave where Moses spent the 40 days. The summit offers grandiose views, but is often crowded. If you join the many who go up to see the sunrise or sunset, take a flashlight and warm clothes.

The mountain lies at the heart of the St Catherine Protectorate, a conservation area recognised as a Unesco World Heritage site. The area is ideal for trekking. One of the longer hikes is to the top of Mount Catherine (Gebel Katarina), Egypt's highest peak. Angels supposedly transported St Catherine of Alexandria's body here, away from her torturers' wheel. Hikers can pick up informative booklets to trails in the area at the Protectorate Office in the village of El-Milga, 3.5 km from St Catherine's Monastery. All treks must be done with a Bedouin guide, which is also arranged through the office.

Feiran Oasis ❽

Road map E6. Sinai, 60 km (37 miles) W of St Catherine's Monastery.

This is the largest and most fertile oasis in Sinai, verdant with date palms, tamarisks and cereal fields. Just south of the Bedouin village of adobe houses is a small, modern convent built with stone from the Byzantine bishop's palace which formerly stood here.

The oasis was the earliest Christian site in Sinai. Many chapels already existed here when, in 451, it became the seat of a bishopric. This governed St Catherine's Monastery until the 7th century, when Feiran's bishop was deposed for heresy and the city fell into ruin. Excavations have revealed its fortified walls, several churches and many other buildings. Feiran is said to be the place where Joshua defeated the Amalekites (Exodus 17).

Shaded gardens surrounding the convent in the Feiran Oasis

THE BEDOUIN OF THE SINAI PENINSULA

In Arabic the word *bedu* means "desert dwellers" and refers specifically to the nomadic tribes that live in Saudi Arabia, the Negev and Sinai. For centuries the Bedouin have lived in close contact with nature, depending for their livelihood on the breeding of sheep, goats and camels. Those in Sinai descend from the peoples who arrived from the Arabian Peninsula from the 14th to the 17th century. The last 20 years of the 20th century have seen a drastic change in their customs and traditions. Today, about 25,000 Bedouin live in Sinai. Many are still nomadic livestock breeders, while others live in permanent camps in wood and corrugated-iron dwellings, making their living as guides, desert tour operators, or by working in large hotels on the coast.

TRAVELLERS' NEEDS

WHERE TO STAY

Jerusalem offers an impressive range of accommodation: from the luxury of the King David and the American Colony hotels, to the plain but welcoming hospices of the various Christian communities, which cater for pilgrims and tourists alike. You will find even more varied accommodation in the rest of the Holy Land. Across Israel, kibbutz hotels offer moderately priced accommodation with good facilities and attractive country settings. Field schools are located near many of the country's nature reserves, and have cheaper, more basic rooms. By the Dead Sea there are many hotels and health resorts, while by the Red Sea and along the Sinai coast, large tourist villages offer water sports and diving. Those who want to cater for themselves will also find many options at a range of prices, from rented villas and apartments to the many excellent youth hostels and camp sites. The listings on pages 256–65 give details on a selection of accommodation to suit every budget.

Doorman at the King David Hotel

The towering Sheraton Hotel, Tel Aviv

GRADING AND FACILITIES

There is no official hotel grading system in Israel and no plan to introduce one, although hotels in Jordan do have their own rating system, with the best (4–5 stars) being comparable to a standard international hotel. Most of the Israeli hotels lie within the medium to high price range, with excellent levels of service and amenities. Rooms are normally equipped with air conditioning, televisions and minibars, with other facilities often including fitness centres, pools and business suites. Most hotels also have bars and restaurants, as well as a dining area where a large buffet-style breakfast is served.

For disabled travellers, many hotels have wheelchair access, and bathrooms and other facilities which have been specially adapted. The largest hotels in the Jewish areas are also equipped to satisfy the needs of practising Jews. These are classed as kosher hotels, and they observe the main Jewish religious laws, especially those concerning Shabbat and *Kashrut*. Many have synagogues and automatic lifts which can be used during the Shabbat rest.

Larger hotels and tourist villages, such as those by the Red Sea, offer private beaches, scuba diving and a range of water sports; while the Dead Sea hotels, often more akin to health resorts, are ideal for those in need of pampering, with their therapeutic hot spas.

PRICES

Compared to Western standards, hotel prices in Israel and Jordan are usually rather high, although the same level of accommodation and service will cost you significantly less in Sinai. Hotel rates fluctuate widely, depending on the season and the various Christian, Muslim and Jewish holidays, so make sure to verify the price before booking. The price of a room almost always includes breakfast, but not other extras. In Israel the room price also includes local taxes, although you can avoid the 17 per cent VAT by paying in foreign currency or on credit card. US dollars, especially, are taken almost everywhere, and all major credit cards are accepted.

In Jordan and Sinai the situation is slightly different. In the large hotels and tourist villages in Sinai all costs over and above the basic room price are subject to double taxation if paid together with the final bill, or on credit card. You can avoid this by paying in cash at the time. Also, listed room rates in Sinai and Jordan exclude tax, which can be as much as 23 per cent, so make sure that you know the final cost. Credit cards are accepted in both Sinai and Jordan, but when using cash, note that while most major currency is taken in Sinai, you can only use dinars in Jordan.

BOOKING A HOTEL

During certain periods of the year, such as Christmas and Easter, or during Jewish holidays – Passover, Rosh ha-Shanah, Yom Kippur, Sukkoth and Hannukah *(see pp36–9)* – finding accommodation can be a real problem, especially in Jerusalem. In Israel as a whole, you may also have difficulty finding a room during the hottest months of

◁ **Diners at a restaurant overlooking the harbour at Old Jaffa**

A reception room at the luxurious American Colony Hotel, Jerusalem

July and August, as this is the busiest time of year, with many Israelis also taking their own holidays.

It is, therefore, always wise to book well in advance, and the **Israel Hotel Association**, the **Kibbutz Hotel Chain**, field schools, youth hostels and some local bed-and-breakfast associations all have centralized booking services, which are often accessible via the Internet and e-mail. The same also applies to many independent hotels and guest houses. If you do need to make arrangements yourself over the phone, most hotel staff can speak good English.

KIBBUTZ HOTELS

These hotels were first established as a source of supplementary income for the largely agricultural kibbutzim, and are completely separate from the very basic type of accommodation offered to those on kibbutz working holidays *(see p293)*. Located mostly in the country, they are ideally placed for visitors wanting a relaxing country break or a base near some of the region's archaeological attractions. Here again there is no grading system: accommodation ranges from very plain lodgings on working kibbutzim, offering bed and breakfast, to more comfortable (albeit informal) hotel complexes with restaurants, swimming pools and other facilities. Most of the hotels are members of the **Kibbutz Hotel Chain** (KHC), the largest hotel group in Israel. As well as providing accommodation, they also organize package tours, adventure breaks, organized nature tours and fly-drive holidays. These can often be good options, as, owing to their often remote locations, many kibbutz hotels are not served by public transport, and may only be convenient if travelling by car.

Kibbutz hotels are very popular among the Israelis for their own vacations, especially during the Jewish holidays and in July and August. It is consequently difficult to find accommodation during these times, unless you book well in advance. Prices usually range between NIS 300–700 for a double room and breakfast, depending on the type of kibbutz and the season.

SELF-CATERING

In Jerusalem and throughout the rest of Israel you can find a wide selection of property to rent, from smart city apartments to luxury country homes. The cost can vary considerably, depending on the type of property you require, but if you are a large family or party, then it can often work out very reasonably when compared to the same length of stay in a hotel. One of the biggest agents dealing with rented holiday homes in Israel is **Homtel**.

CHRISTIAN HOSPICES AND GUEST HOUSES

This type of accommodation, mainly in Jerusalem and near the holy sites, is a popular and inexpensive alternative to hotels. Clean and unashamedly basic, they are often centrally located, and for many are an ideal place to stay for a few nights. You don't have to be a practising Christian to lodge at the Christian hospices, but at times the house rules can be quite strict (you must leave the room early in the morning and the doors are locked at 10–11pm). For unmarried couples it may also be difficult to find a double room. Many guest houses have over the years become bona fide hotels, with their own special charm and character. In this case, prices are slightly higher, although they are still good value when compared to the large hotels.

Enjoying the view of Jerusalem's Old City from the terrace at the King David Hotel

Holiday-makers relaxing on one of the beautiful beaches at Eilat, on the Red Sea coast

YOUTH HOSTELS

For those on a tight budget youth hostels are ideal, and often the cheapest places to stay in Israel. They have no age limits either, so you will find a mixture of people staying at them, from young backpackers to many older travellers. There are plenty of hostels to choose from, with around 32 **Israel Youth Hostel Association (IYHA)** hostels, affiliated to Hostelling International, as well as a large number of independent ones.

Hostels in Israel are located in the major tourist areas – Jerusalem, Tel Aviv, Eilat and Galilee – and also throughout the rest of the country. Most offer single, double, and family rooms as well as the more usual dormitories, with prices starting between NIS 40 and NIS 120 per person. Israeli hostels are generally modern, with basic facilities and clean, simple accommodation. The price includes linen, and in the Israel Youth Hostel Association hostels it also includes breakfast. In the independent hostels you can pay for the room only, and be entirely self-catering.

If you plan to stay at IYHA hostels for any length of time, you may want to pay for membership. While this is not compulsory, it does entitle you to preferential rates, and may be more cost-effective.

As well as providing basic accommodation, the IYHA also offers package tours. A range of different itineraries includes full dinner, bed-and-breakfast at a choice of hostels, and passes for public transport and national parks. They also organize fly-drive packages, which can be a cheap and easy way of seeing the country if you want to follow your own, more flexible, holiday schedule.

FIELD SCHOOLS

There are 24 Field Study Centres in Israel, run by the **Society for the Protection of Nature in Israel (SPNI)**. These are located in the vicinity of some of Israel's major natural reserves, and were established as a way of promoting a better understanding of the country's natural environment and history through organized educational holidays, lecture programmes and summer schools. This is still their main focus, and their varied selection of organized holidays revolves around the region's diverse history, archaeology, geology, flora and fauna.

If you would prefer to visit these areas on your own, these centres will also often offer accommodation at a daily rate. The rooms are simple but clean, and all include a private bathroom and air conditioning. Most bedrooms sleep between four and six people, although some doubles and a few dorm rooms are also available. If you are paying on a room-only basis, the cost is generally less than NIS 190 per person, although prices for the organized holidays can vary significantly depending on the type of itinerary. Booking in advance is obligatory, and the SPNI's centralized booking office can also reserve rooms at some of the kibbutz hotels located in the natural reserves and parks.

Sunbathing by the Dead Sea

CAMPING

There are campsites across Israel for those wanting to spend time under canvas and visit more remote places. Details can be obtained from the **Society for the Protection of Nature in Israel (SPNI)** or from tourist information offices. Prices start from NIS 12 per person, increasing at sites with better facilities. These may include launderettes, electricity points, shops, bars and swimming pools. Some places will also hire out tents or trailer homes.

Campsites in Jordan and Sinai are much less common, with fewer facilities. They are found only in some of the more popular national parks and at some Red Sea resorts.

In Israel, camping rough is also quite common, but choose a secluded public area and leave the site tidy if you want to avoid problems. Places such as the West Bank and Gaza Strip are totally no-go areas, as are all military and border zones. If in doubt, check first. Also be very aware of your possessions and personal safety, especially if in a remote area and alone. Make sure that you have protection against mosquitoes, and check thoroughly for other unwanted guests, such as scorpions.

Camping in the woods near the Sea of Galilee

JORDAN AND SINAI

Parts of Sinai and Jordan offer the full spectrum of accommodation. Amman, in particular, has a full complement of international five-star chain hotels, including a Four Seasons, Grand Hyatt, Kempinski, Marriott, Le Meridien and Sheraton, plus a healthy budget scene in the Downtown district. The choice is also broad at Wadi Musa (for Petra), although given the large number of visitors, it is wise to book well in advance, especially in March/April and September/October (peak times). Elsewhere in Jordan the choice is greatly diminished, although the country is small enough that most sights can be visited from either Amman or Petra.

Major credit cards are accepted in all mid-range and top-end hotels in Jordan. A sales and service tax of up to 26 per cent is commonly added to bills.

Sinai resorts such as Dahab, Nuweiba and Sharm el-Sheikh offer many top-class resort hotels, many with prime beachfront locations, some boasting beautiful architecture and all offering a full range of facilities, from multiple bars and restaurants to dive and water sports centres. Such is the abundance of accommodation that a little Internet research can sometimes throw up some bargain room rates. Peak seasons are during the Muslim feasts of Eid el-Fitr and Eid el-Adha *(see p38)*, around Christmas and especially New Year, and during July and August; at such times you need to book ahead.

There are no proper hostels in Jordan or Sinai with an official national association, but there are many cheap hotels and dormitories that serve the same purpose. Dahab and other smaller Sinai resorts often have simple bamboo-constructed huts for rent on the beach – these are especially popular with budget travellers.

DIRECTORY

BOOKING A HOTEL

Israel Hotel Association
29 Hamered Street,
PO Box 50066,
Tel Aviv, Israel.
Tel *(03) 517 0131.*
Fax *(03) 510 0197.*
infotel@israelhotels.org.il
www.Israelhotels.org.il

KIBBUTZ HOTELS

Kibbutz Hotel Chain (KHC)
41 Montefurie,
Beit Nesuah,
Tel Aviv, Israel 65201.
Tel *(03) 560 8118.*
Fax *(03) 560 7710.*
khc_rsv@kibbutz.co.il
www.kibbutz.co.il

SELF CATERING

Good Morning Jerusalem
9 Coresh Street,
Jerusalem,
Israel 94144.
Tel *(02) 623 3459.*
Fax *(02) 625 9330.*
gmjer@netvision.co.il
www.accommodation.co.il

Homtel
Home Association of Jerusalem,
PO Box 7547,
Jerusalem,
Israel 91074.
Tel *(02) 645 2198.*
www.bnb.co.il

CHRISTIAN HOSPICES AND GUEST HOUSES

Christian Information Centre
Jaffa Gate
(opposite David Tower).
Tel *(02) 627 2692.*
Fax *(02) 628 6417.*
www.cicts.org

YOUTH HOSTELS

Israeli Youth Hostel Association (IYHA)
Jerusalem International Convention Centre,
PO Box 6001, Jerusalem,
Israel 91060.
Tel *(02) 655 8406.*
Fax *(02) 655 8432/8431.*
www.iyha.org.il

FIELD SCHOOLS

Society for the Protection of Nature in Israel (SPNI)
13 Heleni Hamalka St,
Jerusalem,
Israel.
Tel *(02) 624 4605 (shop), (03) 638 8688 (rooms).*
www.teva.org.il

CAMPING

Society for the Protection of Nature in Israel (SPNI)
3 Hasfela St,
Tel Aviv,
Israel 66183.
Tel *(03) 638 8674.*
www.teva.org.il

Choosing a Hotel

The hotels in this guide have been selected across a wide price range for the excellence of their facilities, location or character. This section lists hotels in Jerusalem by area, with price ranges given in US dollars. For Jerusalem map references, see the Street Finder on pages 156–9; for further afield see back endpaper map.

PRICE CATEGORIES
Prices categories are per night for two people occupying a standard double room, with tax, breakfast and service included:
$ Under $65
$$ $65–$100
$$$ $100–$175
$$$$ $175–$250
$$$$$ Over $250

THE MUSLIM QUARTER

Ecce Homo Convent — $$

41 Via Dolorosa, 97626 ***Tel*** *(02) 627 7293* ***Fax*** *(02) 628 2224* ***Rooms*** *120* ***Map*** *4 D2*

Superbly situated on the Via Dolorosa, this hospice, built in 1856, is just a few minutes' stroll from Jerusalem's holiest sites. There are impressive Roman-era ruins underfoot and just outside the door, and a magnificent view of the Old City from the roof. The rooms are very modest but clean. **www.eccehomoconvent.com**

Austrian Hospice — $$$

Via Dolorosa 37 ***Tel*** *(02) 626 5800* ***Fax*** *(02) 627 1472* ***Rooms*** *31* ***Map*** *3 C2*

The historic Austrian Hospice of the Holy Family, inaugurated in 1863, serves as both a guesthouse and a cultural centre. Just off the bustling Via Dolorosa, this island of calm has simple but attractively furnished rooms, a Viennese cafe, a garden and breathtaking views of the old city's roofscape. **www.austrianhospice.com**

THE CHRISTIAN AND ARMENIAN QUARTER

Casa Nova — $$

10 Casa Nova St, 97600 ***Tel*** *(02) 627 1441* ***Fax*** *(02) 626 4370* ***Rooms*** *89* ***Map*** *3 B3*

Just two blocks up the slope from the Holy Sepulchre, this simple Franciscan hospice, built in 1866, maintains a high quality of cleanliness and comfort. Its location and the good value it offers make it popular with pilgrimage groups so book well ahead. **www.custodia.org/casanovaj**

Maronite Monastery Hospice — $$

25 Maronite Convent St, 97111 ***Tel*** *(02) 628 2158* ***Fax*** *(02) 627 2821* ***Rooms*** *27* ***Map*** *3 B4*

Situated two blocks inside Jaffa Gate, this hospice, also known as Foyer Mar Maroun, is virtually across the street from the Citadel. Ensconced in a centuries-old building that is one of the most beautifully-kept in the Armenian Quarter, its rooms are spotless and well-maintained. **www.maronitejerusalem.org**

Christ Church Guest House — $$$

Omar Ibn el-Khattab Square, 97604 ***Tel*** *(02) 627 7727* ***Fax*** *(02) 628 2999* ***Rooms*** *23* ***Map*** *3 B4*

Run by an evangelical Anglican organization founded in 1809, this hospice just inside Jaffa Gate was constructed in the 19th century on foundations that go back to Roman times. Rooms are small and plain but comfortable, and there's a good range of services. Very popular, especially during holiday periods. **www.cmj-israel.org**

THE MOUNT OF OLIVES AND MOUNT ZION

Mount of Olives — $$

53 Mount of Olives Rd ***Tel*** *(02) 628 4877* ***Fax*** *(02) 626 4427* ***Rooms*** *55* ***Map*** *2 F3*

Set atop Jerusalem's highest hill, this welcoming, family-run hotel is just a short stroll from the Mount of Olives' many Christian, Jewish and Muslim sites, including the Mosque of the Ascension, which is next door. Eleven of the clean, quiet rooms have panoramic views of the Old City. **www.mtolives.com**

Seven Arches — $$$

Ruba el-Adawiya St, Mount of Olives Plaza ***Tel*** *(02) 626 7777* ***Fax*** *(02) 627 1319* ***Rooms*** *196* ***Map*** *2 F4*

Offering one of the most spectacular city panoramas in the world, this large, modern hotel, built in classic 1960s style, sits on top of the Mount of Olives. The rooms are comfortable and the service courteous. The Old City is a 15-minute walk down the hill, past half-a-dozen major Christian sites. **www.7arches.com**

Key to Symbols *see back cover flap*

MODERN JERUSALEM

Agron Guest House $$

6 Agron St, 94265 ***Tel*** *(02) 594 5522* ***Fax*** *(02) 622 1124* ***Rooms*** *55* ***Map*** *1 A4*

Sandwiched between the smart Rehavia neighbourhood and West Jerusalem's lively commercial heart, this hostel is a 10-minute walk from the Old City. The Conservative Jewish Movement has a religious and cultural centre next door. Rooms are institutional but comfortable. Reception is closed on the Sabbath. **www.iyha.org.il**

Azzahra $$$

13 El-Zahra St, 97200 ***Tel*** *(02) 628 2447* ***Fax*** *(02) 628 3960* ***Rooms*** *15* ***Map*** *2 D2*

A small, family-run hotel on a quiet alleyway near East Jerusalem's commercial centre. Just a few blocks northeast of Damascus Gate, this place is known for its friendly atmosphere and service. Offers good value for money and has a well-regarded Middle Eastern restaurant. **www.azzahrahotel.com**

Notre Dame $$$

12 Ha-Tsanhanim St, 91204 ***Tel*** *(02) 627 9111* ***Fax*** *(02) 627 1995* ***Rooms*** *150* ***Map*** *1 B3*

Built between 1885 and 1904 to house French Catholic pilgrims, this neo-Romanesque complex is now a Vatican-run ecumenical, cultural and pilgrimage centre. The guest rooms are comfortable and offer good value. It is situated across the street from the Old City's New Gate, which leads into the Christian Quarter. **www.notredamecenter.org**

Palatin $$$

4 Agripas St, 94301 ***Tel*** *(02) 623 1141* ***Fax*** *(02) 625 9323* ***Rooms*** *28* ***Map*** *1 A2*

On a pedestrianized street in the heart of West Jerusalem's commercial centre, around the corner from bustling King George St, this hotel has been run by the same family since it was built in 1936. It is close to Mahaneh Yehuda food market. The rooms are modest but modern, and one has a balcony. **www.palatinhotel.com**

St Andrew's Scottish Guest House $$$

1 David Remez St, 91086 ***Tel*** *(02) 673 2401* ***Fax*** *(02) 673 1711* ***Rooms*** *20* ***Map*** *1 B5*

This delightful hospice has large, simple bedrooms and a somewhat colonial atmosphere, which is not surprising since the all-stone building dates from the late 1920s. Situated a short walk from the cafés and restaurants of Emeq Refaim St. There is disabled access to one room. **www.scotsguesthouse.com**

St George $$$

8 Salah ed-Din St, 95908 ***Tel*** *(02) 627 7232* ***Fax*** *(02) 628 2575* ***Rooms*** *47* ***Map*** *1 C1*

Situated just two blocks north of Herod's Gate and the Muslim Quarter, this hotel is in the heart of East Jerusalem's commercial centre. It was opened by Jordan's late King Hussein in 1965 and still has a 1960s feel, especially in the lobby. Rooms overlooking Salah ed-Din St can be noisy. **www.hotelstgeorge-jer.com**

YMCA Three Arches $$$

26 King David St, 94101 ***Tel*** *(02) 569 2692* ***Fax*** *(02) 623 5192* ***Rooms*** *55* ***Map*** *1 A4*

Housed in a richly decorated, landmark building dedicated in 1933, the YMCA is just across the street from the King David Hotel. It has a unique Mandate-era ambience and offers superb views of the city. Rooms are unexciting but comfortable. Has good sports facilities and offers excellent value. Non-smoking building. **www.ymca3arch.co.il**

Jerusalem $$$$

Nablus Road, 97200 ***Tel*** *(02) 628 3282* ***Fax*** *(02) 628 3282* ***Rooms*** *14* ***Map*** *1 C2*

Situated just north of Damascus Gate and housed in a 19th-century Arab house, this place has been run by the same family since 1960. It has spacious rooms with high ceilings, all elaborately furnished in traditional Oriental style. There's a delightful, vine-shaded garden terrace and the service is excellent. **www.jrshotel.com**

Mount Zion $$$$

17 Hebron Rd, 93546 ***Tel*** *(02) 568 9555* ***Fax*** *(02) 673 1425* ***Rooms*** *137* ***Map*** *1 B5*

Built in 1882 as a British-run hospice, this hotel has comfortable, well-furnished rooms with lots of character. Many afford fine views of Mount Zion, the Hinnom Valley, the Judean Desert and, on clear days, the hills of the Hashemite Kingdom of Jordan. Guests can stroll through the delightful gardens at the back of the hotel. **www.mountzion.co.il**

American Colony $$$$$

23 Nablus Rd, 97200 ***Tel*** *(02) 627 9777* ***Fax*** *(02) 627 9779* ***Rooms*** *86* ***Map*** *1 C1*

This hotel was founded in 1902 by the American family that still own it. Long the preferred hang-out of journalists and diplomats, the fabled American Colony pampers visitors with classic Arabian architecture, a flowery Turkish courtyard, lush gardens and plenty of Oriental charm. **www.americancolony.com**

King David $$$$$

23 King David St, 94101 ***Tel*** *(02) 620 8888* ***Fax*** *(02) 620 8880* ***Rooms*** *237* ***Map*** *1 B4*

A favourite with American Presidents, the historic King David, built of pink sandstone in 1931, is famed for its King Solomon-style lobby, grassy gardens and Mandate-era atmosphere. Rooms have classic styling; the pricier ones afford stunning views of the Old City. Amenities include a tennis court and two kosher restaurants. **www.danhotels.com**

Mamilla Hotel
$$$$$

11 King Solomon St, 94182 ***Tel*** *(02) 548 2200* ***Fax*** *(02) 548 2201* ***Rooms*** *194* ***Map*** *1 B5*

Jerusalem's most fashionable venue, this exquisite, contemporary hotel is loved by locals as well as guests. Unwind at either the spectacular rooftop restaurant, the trendy mirror bar, the lobby café or the organic snack and juice bar. Close to Alrov Mamilla Avenue, one of Jerusalem's beautiful shopping and entertainment strips. **www.mamillahotel.com**

The David Citadel Hotel
$$$$$

7 King David St, 94101 ***Tel*** *(02) 621 1111* ***Fax*** *(02) 621 1000* ***Rooms*** *384* ***Map*** *1 B4*

The David Citadel is known for luxury, elegance and professional service. It is a favourite with international politicians and celebrities as well as families, and is ideally located – overlooking the walls of the Old City and just a short stroll from Jaffa Gate and Mamilla Avenue. **www.thedavidcitadel.com**

FURTHER AFIELD

A Little House in Bakah
$$$

1 Yehuda St, Bakah, 93627 ***Tel*** *(02) 673 7944* ***Fax*** *(02) 673 7955* ***Rooms*** *35*

About 2 km (1 mile) due south of the King David Hotel, this welcoming place has an arched and colonnaded façade, modest rooms and a garden restaurant. The Sherover Promenade in Talpiyot and the Emeq Refaim St restaurant zone are close by. **www.jerusalem-hotel.co.il**

Neve Shalom/Wahat al-Salam Hotel
$$$

Neve Shalom/Wahat al-Salam, 99761 ***Tel*** *(02) 999 3030* ***Fax*** *(02) 991 7412* ***Rooms*** *39*

Located in a quiet, rural community in which Jews and Palestinians live together and work towards peaceful coexistence, this hotel's name means "oasis of peace". It has modest rooms in single-storey buildings and is situated about 30 km (19 miles) west of Jerusalem, south of Latrun, so guests need to have a car. **www.nswas.com/hotel**

Notre Dame de Sion Guest House
$$$

23 Ha-Oren St, Ein Karem, 95744 ***Tel*** *(02) 641 5738* ***Fax*** *(02) 643 7739* ***Rooms*** *28*

About 8 km (5 miles) west of the Old City in a delightful natural setting, this peaceful, stone-built B&B was built as a French convent in the mid-1800s. Still run by nuns, it has simple, spacious rooms, a cafeteria and a large, peaceful garden (closed Sun) that lends itself to contemplation. Rooms have neither phone nor TV. **www.sion-ein-karem.org**

Yitzhak Rabin Youth Hostel & Guest House
$$$

1 Nahman Avigad St, Givat Ram, 91390 ***Tel*** *(02) 678 0101* ***Fax*** *(02) 679 6566* ***Rooms*** *77*

A striking and modern establishment whose facilities are positively deluxe by hostel standards, making it an excellent value option for families. Situated near the Israel Museum, the Givat Ram campus of the Hebrew University, the Valley of the Cross and the Knesset, it is served by buses 17 and 18 from the Central Bus Station. **www.iyha.org.il**

Ambassador
$$$$

Nablus Rd, Sheikh Jarrah, 97200 ***Tel*** *(02) 541 2222* ***Fax*** *(02) 582 8202* ***Rooms*** *115*

Situated in the attractive East Jerusalem neighbourhood of Sheikh Jarrah, with its many consulates, this comfortable hotel is about 2 km (1 mile) north of the Old City and about 1 km (half a mile) west of Mount Scopus. It has a vine-shaded terrace and tasteful rooms. Limited disabled access. **www.jerusalemambassador.com**

Ramat Rachel
$$$$

Kibbutz Ramat Rachel, 90900 ***Tel*** *(02) 670 2555* ***Fax*** *(02) 673 3155* ***Rooms*** *164*

On the southern outskirts of Jerusalem, this hotel is on the grounds of Kibbutz Ramat Rachel, which was founded in the 1920s. The large, grassy gardens and some of the rooms afford fine views of the Judean Desert. The hotel has a first-rate swimming, fitness and spa complex (additional fee to use fitness and spa facilities). **www.ramatrachel.co.il**

THE COAST AND GALILEE

BEIT ALFA (BETH ALFA) Beit Alfa Guest Rooms
$$$

Kibbutz Beit Alfa, 10802 ***Tel*** *(04) 653 3026* ***Fax*** *(04) 653 3882* ***Rooms*** *39* ***Map*** *C2*

At the foot of Mount Gilboa, this kibbutz guesthouse is on the grounds of Kibbutz Beit Alfa, known for its ancient synagogue, Japanese garden, mini-zoo and herd of dairy cows. The modest rooms, in low-rise, kibbutz-style buildings, come with a microwave and a fridge. **www.beit-alfa.com**

BEIT SHEAN Beit Shean Guest House
$$$

126 Menahem Begin Ave, 11741 ***Tel*** *(04) 606 0760* ***Fax*** *(04) 606 0766* ***Rooms*** *62* ***Map*** *C2*

Housed in an impressive, ultra-modern stone building and endowed with functional but comfortable facilities, this hostel is an excellent base for visits to the Beit Shean antiquities, the Beit Alfa synagogue and Belvoir Castle. A humourous wall mural with thousands of tiny figures illustrates daily life in ancient Beit Shean. **www.iyha.org.il**

Key to Price Guide *see p256* **Key to Symbols** *see back cover flap*

CAESAREA Dan Caesarea $$$$$

Caesarea, 30600 **Tel** *(04) 626 9111* **Fax** *(04) 626 9122* **Rooms** *114* **Map** *B2*

Surrounded by some of the most beautiful coastal scenery in Israel, this quiet hotel is next to Israel's only 18-hole golf course. The superb Roman ruins of Caesarea are a five-minute drive away. The hotel offers a wide variety of sports activities, including scuba diving and sailing. Some rooms have sea views. **www.danhotels.com**

CARMEL FOREST Carmel Forest Spa Resort $$$$$

Near Kibbutz Beit Oren, 31900 **Tel** *(04) 830 7888* **Fax** *(04) 830 7886* **Rooms** *126* **Map** *B2*

Overlooking the Mediterranean and amid the natural beauty of Mount Carmel, this luxury spa-hotel offers a wide variety of spa and health treatments, as well as cooking, wine and yoga courses. Amenities include a genuine Turkish hammam and two swimming pools. Guests must be 16. Cellphones can be used only in rooms. **www.isrotel.co.il**

GOLAN HEIGHTS Golan Field School $$

Katsrin, 12900 **Tel** *(04) 696 1234* **Fax** *(04) 696 5033* **Rooms** *33* **Map** *C2*

On the edge of the Golan's largest town, this hostel is run by the Society for the Protection of Nature in Israel. It has spartan rooms situated in a grassy campus, and is an ideal base for visiting the Golan countryside, including Yehudiya Nature Reserve, and the area's Roman-era Jewish archaeological sites. **www.aspni.org**

GOLAN HEIGHTS Alaska Inn $$$

Metulla, 10292 **Tel** *(04) 699 7111* **Fax** *(04) 699 7118* **Rooms** *49* **Map** *C1*

Overlooking Lebanon's Ayun Valley from a spot 20 km (12 miles) west of Banias Spring on the Golan Heights, the village of Metulla is "Israel's Switzerland". The hotel is an ideal base for exploring the northern Golan and the far north of Galilee. Local attractions include the Canada Centre ice-skating rink. **www.alaskainn.co.il**

HAIFA Dan Panorama $$$$

107 HaNassi Ave, 34632 **Tel** *(04) 835 2222* **Fax** *(04) 835 2235* **Rooms** *266* **Map** *B2*

Occupying two high-rise towers perched high on top of Mount Carmel, this deluxe hotel offers breathtaking Mediterranean panoramas and affords easy access to the charms of Central Carmel. Pricier rooms look out on Haifa Bay. Situated very near the upper station of the Carmelit funicular railway and the zoo. **www.danhotels.co.il**

KFAR PEKI'IN Peki'in Youth Hostel & Family Guesthouse $$

Kfar Peki'in, 24914 **Tel** *(02) 594 5677* **Fax** *(04) 957 4116* **Rooms** *50* **Map** *C2*

This modern, well-appointed hostel is situated in a Druze village in the Galilee hills, 7 km (4 miles) southeast of Maalot-Tarshiha, with views of Mount Meron. Walking tours and hosting by local families help guests get a sense of Druze life and culture. Served by Egged bus 44 from Nahariya and bus 271 from Haifa. **www.iyha.org.il**

NAHARIYA Carlton Nahariya $$$

23 HaGaaton Blvd, 22444 **Tel** *(04) 900 5511* **Fax** *(04) 982 3771* **Rooms** *200* **Map** *B2*

Situated 10 km (6 miles) north of Akko on the café-lined main street of the seaside resort of Nahariya, this hotel is just steps from the beach. The comfortable, airy rooms have a modern, Mediterranean ambience. Amenities include spa treatments, a sun terrace with sea views and free bicycles. **www.carlton-hotel.co.il**

NAHSHOLIM Nahsholim Seaside Resort $$$$

Nahsholim, 30815 **Tel** *(04) 639 9533* **Fax** *(04) 639 7614* **Rooms** *128* **Map** *B2*

Set on a bay along Israel's most gorgeous strip of Mediterranean coastline, this kibbutz-run tourist village is a short walk from the delightful Dor Nature Reserve and a string of fish ponds that attract flocks of migrating birds, making it a great location for bird watchers, if not the local fish farmers. **www.nahsholim.co.il**

NAZARETH Plaza Hotel $$$

2 Hermon St, Upper Nazareth, 17502 **Tel** *(04) 602 8200* **Fax** *(04) 602 8222* **Rooms** *184* **Map** *B2*

The Plaza is a modern, 10-storey hotel known for its convenient facilities, efficient service and central Lower Galilee location rather than its character. It makes a good base for car trips to Nazareth, Megiddo, Tsipori, the Jezreel Valley and Beit Alfa. Amenities include a sauna and a Jacuzzi. **www.israelhotels.org.il**

ROSH PINA Auberge Shulamit $$$

David Shuv St, Rosh Pina, 12000 **Tel** *(04) 693 1494* **Fax** *(04) 693 1495* **Rooms** *4* **Map** *C2*

In one of the oldest and most charming Jewish villages in Galilee, this old basalt house, built in the 1930s, is decorated with exquisite taste and attention to detail. The rooms have Jacuzzi baths. The breakfasts are superb and the fine restaurant affords a breathtaking view. Weekends are booked up long in advance. **www.shulamit.co.il**

ROSH PINA Mizpe Hayamim $$$$$

On Hwy 89 between Safed and Rosh Pina, 12000 **Tel** *(04) 699 4555* **Fax** *(04) 699 9555* **Rooms** *99* **Map** *C2*

Midway between Rosh Pina and Safed, on a hillside perch overlooking the Sea of Galilee and Golan, this ultra-luxurious resort is a great place to be pampered. Surrounded by orchards and gardens, its exquisitely designed rooms exude French style. The spa offers some uniquely Galilean treatments. **www.mizpe-hayamim.com**

SAFED Ruth Rimonim Inn $$$$

Tet-Zayin St, Safed, 13110 **Tel** *(04) 699 4666* **Fax** *(04) 692 0456* **Rooms** *77* **Map** *C2*

In the heart of Safed, the ruins of a 17th-century inn have been transformed into one of Galilee's most attractive hostelries. The atmospheric rooms, built partly of stone, have a very local flavour, and most come with fine views of Mount Meron. There are also spa facilities. **www.rimonim.com**

SEA OF GALILEE Nof Ginosar $$$

Kibbutz Ginosar, 14980 ***Tel*** *(04) 670 0300* ***Fax*** *(04) 679 2170* ***Rooms*** *161* ***Map*** *C2*

On the lakeside, 10 km (6 miles) north of Tiberias, this sprawling, grassy place has a fine beach and is a superb choice for a stay by the Sea of Galilee. Rooms are simple, pleasant and comfortable. Next to the hotel are 75 guest houses, where rooms are cheaper than in the main hotel. **www.ginosar.co.il**

SEA OF GALILEE Vered HaGalil Guest Farm $$$

Off Hwy 90 between Tiberias and Rosh Pina, 12385 ***Tel*** *(04) 693 5785* ***Fax*** *(04) 693 4964* ***Rooms*** *18* ***Map*** *C2*

This quiet, family-run ranch, which doubles as a horseback riding school, offers individual stone-and-wood cottages and cabins, each with lots of woody furnishings and a veranda. The rustic restaurant serves up American fare. The stables supply horses for guided riding tours of the area. **www.veredhagalil.co.il**

SEA OF GALILEE YMCA Peniel-by-Galilee $$$

On Hwy 90 north of Tiberias, 14101 ***Tel*** *(04) 672 0685* ***Fax*** *(04) 672 5943* ***Rooms*** *13* ***Map*** *C2*

In a superb lakefront location, this guesthouse is ensconsed in a lovely stone building with lots of character and is awash in greenery. The ornate Middle Eastern-style lobby and the lovely chapel are a feast for the eyes. Rooms are simply furnished; the best ones come with great lake views. Lakewater swimming pool. **www.ymca-galilee.co.il**

SEA OF GALILEE Rimonim Galei Kinneret $$$$$

1 Eliezer Kaplan St, Tiberias, 14209 ***Tel*** *(04) 672 8888* ***Fax*** *(04) 679 0260* ***Rooms*** *120* ***Map*** *C2*

Right on the shore of the Sea of Galilee, this luxurious spa hotel offers classy, up-to-date facilities and a wide variety of health and beauty treatments. Rooms are decorated in a contemporary style and many come with delightful watery views. Activity options include water-skiing. **www.rimonim.com**

SEA OF GALILEE Scots Hotel $$$$$

1 Gdud Barak St, Tiberias, 14100 ***Tel*** *(04) 671 0710* ***Fax*** *(04) 671 0711* ***Rooms*** *69* ***Map*** *C2*

Full of character, this former hospital was founded in the late 1800s and is still owned by the Church of Scotland. Sitting on the lakefront and surrounded by a lovely garden, it is an oasis of tranquility in the town centre. The modern and very attractive facilities include a spa and, in summer, a swimming pool. **www.scotshotels.co.il**

SHLOMI Shlomi Youth Hostel $$

Shlomi, 22832 ***Tel*** *(04) 980 8975* ***Fax*** *(04) 980 9163* ***Rooms*** *100* ***Map*** *B2*

Clean and quiet, if a bit old-fashioned, this unpretentious hostel makes a good budget base for exploring the far northwest of Galilee, including Akko and the coast around Nahariya. Served by Egged buses 22 and 23 from Nahariya. **www.iyha.org.il**

TEL AVIV HaYarkon 48 Hostel $$

48 HaYarkon St, 63305 ***Tel*** *(03) 516 8989* ***Fax*** *(03) 510 3113* ***Rooms*** *17 & dormitories* ***Map*** *B3*

Just two blocks from the city's broad, sandy beach and a five-minute walk from the Carmel Market, this bright yellow hostel is welcoming and cheery. Many of the private rooms come with balconies. Dorm beds are cheap. Amenities include bicycle parking and a pool table. **www.hayarkon48.com**

TEL AVIV Old Jaffa Hostel & Guest House $$

13 Amiad St, 68139 ***Tel*** *(03) 682 2370* ***Rooms*** *23* ***Map*** *B3*

Just around the corner from Jaffa's famous flea market and Clock Tower, this atmospheric hostel occupies an Ottoman-era residence with sky-high ceilings and a mellow rooftop lounge offering views of the Mediterranean. Dorm beds are cheap. Breakfast is not included but is available in nearby cafés. **www.telaviv-hostel.com**

TEL AVIV Dizengoff Square Apartments $$$

89 Dizengoff St & 4 Dizengoff Circle, 64396 ***Tel*** *(03) 524 1151* ***Fax*** *(03) 523 5614* ***Rooms*** *32* ***Map*** *B3*

Occupying two Bauhaus-style buildings at Dizengoff Circle, the ever-lively focal point of central Tel Aviv, this place offers tastefully furnished apartments with kitchenettes. Larger suites come with sofas and a kitchen table. Excellent value for money in a great location. **www.hotel-apt.com**

TEL AVIV Dizengoff Suites Hotel $$$

39 Gordon St, 63461 ***Tel*** *(03) 523 4363* ***Fax*** *(03) 527 3524* ***Rooms*** *20* ***Map*** *B3*

On the corner of lively Dizengoff St, midway between the beach and Rabin Square, this family-run place has suites with kitchenettes, fridges and, in most cases, balconies. Breakfast can be bought at the stylish, Italian-style café-restaurant on the ground floor. **www.dizengoffsuites.co.il**

TEL AVIV Cinema $$$$

1 Zamenhof St, 64373 ***Tel*** *(03) 520 7100* ***Fax*** *(03) 520 7101* ***Rooms*** *82* ***Map*** *B3*

In a former cinema right on Dizengoff Circle, this stylish hotel has a film-themed lobby that is as evocative of 1930s elegance as its curvaceous, Bauhaus-style façade. Some rooms have balconies and/or kitchenettes. Amenities include a sun roof, sauna and Jacuzzi. **www.atlas.co.il**

TEL AVIV Sheraton City Tower $$$$

14 Shalom Zissman St, Ramat Gan, 52521 ***Tel*** *(03) 754 4444* ***Fax*** *(03) 754 4445* ***Rooms*** *170* ***Map*** *B3*

At the gateway to Ramat Gan's skyscraper Diamond Exchange District, this business-oriented hotel, on floors nine to 17 of a 40-storey tower, is across the Ayalon Expressway from the Central Tel Aviv Train Station, linked by rail with Akko, Haifa, Beersheva and Jerusalem. Amenities include an outdoor Jacuzzi and a spa. **www.sheraton.co.il**

Key to Price Guide *see p256* **Key to Symbols** *see back cover flap*

TEL AVIV Alexander $$$$$

3 Habakuk St, 63505 ***Tel*** *(03) 545 2222* ***Fax*** *(03) 546 9346* ***Rooms*** *67* ***Map*** *B3*

This all-suites hotel sits at the northern entrance to Metzitzim beach, just two blocks south of the Tel Aviv Port nightlife area and a few minutes on foot from HaYarkon Park and the Yarkon River Estuary. The spacious suites come with a kitchenette and a work area, and can accommodate up to seven people. **www.alexander.co.il**

TEL AVIV Dan Panorama $$$$$

Charles Clore Park, 68012 ***Tel*** *(03) 519 0190* ***Fax*** *(03) 517 1777* ***Rooms*** *480* ***Map*** *B3*

Overlooking a grassy, seafront park midway between Dizengoff Circle and Old Jaffa, this 18-storey hotel is across the street from a broad, sandy beach and just three blocks from the chic boutiques and cafés of Neve Tzedek. All rooms have sea views and balconies. Only one room has facilities for wheelchair users. **www.danhotels.com**

TEL AVIV Sheraton Tel Aviv $$$$$

115 HaYarkon St, 63573 ***Tel*** *(03) 521 1111* ***Fax*** *(03) 523 3322* ***Rooms*** *314* ***Map*** *B3*

One of the city's most luxurious hotels, this venerable, high-rise hotel overlooks the beachfront and is just steps from several beachside cafés. The public areas are an excellent example of modern Israeli interior design. All rooms are spacious and stylish, and have balconies with Mediterranean panoramas. **www.sheraton-telaviv.co.il**

TEL AVIV Tel Aviv Hilton $$$$$

Independence Park, 63405 ***Tel*** *(03) 520 2222* ***Fax*** *(03) 527 2711* ***Rooms*** *598* ***Map*** *B3*

The most luxurious and best equipped of the city's hotels, the Hilton sits on a cliff overlooking the Mediterranean. Rooms are spacious and come with marble bathrooms and wonderful sea views. Activity options include sailing, windsurfing and cycling. The beach is a two-minute walk down the hill. **www.hilton.com**

YEHIAM Teva be-Yehiam $$$

Kibbutz Yechiam, 25125 ***Tel*** *(04) 985 6057 or (050) 444 4362* ***Fax*** *(04) 952 4567* ***Rooms*** *60* ***Map*** *B2*

In the Galilee hills 23 km (14 miles) northeast of Akko, this kibbutz hotel is built next to the ruins of a Crusader castle. It is a very quiet spot, with lots of greenery and lawns, but the speciality here is arranging challenging outdoor activities for groups, such as kayaking, speedboating, jeep tours and cliff rapelling. **www.rimoney-hagalil.com**

THE DEAD SEA AND THE NEGEV DESERT

DEAD SEA Tsell Harim Hotel $$$

Ein Bokek, 86930 ***Tel*** *(08) 668 8111* ***Fax*** *(08) 668 8100* ***Rooms*** *160* ***Map*** *C4*

This low-rise, beachfront complex, centred around an outdoor swimming pool, has modest but serviceable rooms. When it is too hot outside you can take refuge in the indoor pool, filled with Dead Sea water. Amenities include a Finnish sauna and a rooftop solarium; black mud baths are free. **www.tsell-harim.co.il**

DEAD SEA Magic Nirvana Club $$$$

Neve Zohar, 86910 ***Tel*** *(08) 668 9444* ***Fax*** *(08) 668 9400* ***Rooms*** *300* ***Map*** *C4*

On the shores of the Dead Sea, this 10-storey club hotel has indoor and outdoor pools, palm-dotted lawns, plenty of activities and rooms with views across the water to the Mountains of Edom in Jordan. Prices include meals and activities. **www.fattal.co.il**

EILAT Eilat Field School $$

Coral Beach, 88000 ***Tel*** *(08) 637 2021* ***Fax*** *(08) 637 1771* ***Rooms*** *50* ***Map*** *B7*

About 3 km (2 miles) south of town, this basic hostel offers bed and breakfast in low-rise buildings surrounded by lawns. It is opposite the Coral Reef Nature Reserve, which has the area's best snorkelling. Experienced guides offer hiking tours of nearby mountain areas. The Coral World Underwater Observatory is a short walk away. **www.aspni.org**

EILAT Eilat Youth Hostel & Guest House $$

7 Arava Road, 88101 ***Tel*** *(08) 637 0088* ***Fax*** *(08) 637 5835* ***Rooms*** *105* ***Map*** *B7*

Just a short walk from Eilat's seafront promenade, this hostel's creative architecture incorporates some cleverly shaded public spaces. Rooms are spartan, but clean and serviceable. The rooftop deck affords great views of the city, the deep blue Red Sea and the mountains of Aqaba. **www.iyha.org.il**

EILAT Ambassador $$$

Coral Beach, 88103 ***Tel*** *(08) 638 2222* ***Fax*** *(08) 638 2209* ***Rooms*** *247* ***Map*** *B7*

Near a quiet strip of sandy coastline about 3 km (2 miles) south of Eilat, next to the Coral Reef Reserve, this hotel is centred around an expansive swimming pool. You are never far from grass and greenery, and shaded areas for relaxing. The diving club offers a wide range of approved scuba courses. **www.isrotel.co.il**

EILAT Orchid $$$

Southern Beach, 88000 ***Tel*** *(08) 636 0360* ***Fax*** *(08) 637 5323* ***Rooms*** *180* ***Map*** *B7*

Overlooking the Coral World Underwater Observatory from a landscaped hillside about 5 km (3 miles) south of town, this Thai-inspired village has wooden buildings, spacious rooms and a tropical feel. Amenities include free bicycles and snorkelling equipment. The beach is close by and free shuttles give lifts to the town centre. **www.orchidhotel.co.il**

EILAT Reef $$$

Coral Beach, 88103 **Tel** *(08) 636 4444* **Fax** *(08) 636 4488* **Rooms** *79* **Map** *B7*

Situated on the beach about 3 km (2 miles) south of the city centre, not far from the Coral World Underwater Observatory, this is one of Eilat's smaller, more accessibly-priced hotels. It has an outdoor Jacuzzi, balcony-equipped rooms and lots of water sports options, including diving, snorkelling and windsurfing. **www.reefhoteleilat.com**

EILAT Eilat Princess $$$$$

Taba Beach, 88000 **Tel** *(08) 636 5555* **Fax** *(08) 637 6333* **Rooms** *420* **Map** *B7*

Situated 5 km (3 miles) south of Eilat near the Taba border crossing to Egypt, this opulent hotel affords fine views across the Red Sea to the mountains of Jordan and Saudi Arabia. Amenities include swimming pools, tennis courts, a spa, a heath centre and cuisine from around the world. **www.eilatprincess.com**

EIN GEDI Ein Gedi Youth Hostel $$

Ein Gedi, 86980 **Tel** *(08) 658 4165* **Fax** *(08) 658 4445* **Rooms** *66* **Map** *C4*

Set around a quiet courtyard, the location is ideal for guests who want to combine hiking in the nearby oases with relaxing in the Dead Sea. It is situated next to the Nahal David Nature Reserve and served by Egged buses 486 and 487 from Jerusalem. Some rooms have balconies with sea views. **www.iyha.org.il**

EIN GEDI Ein Gedi Guesthouse $$$$

Kibbutz Ein Gedi, 86980 **Tel** *(08) 659 4220* **Fax** *(08) 658 4328* **Rooms** *150* **Map** *C4*

Overlooking the Dead Sea and the lush Ein Gedi oasis, this kibbutz guesthouse has comfortable, cheerful rooms with creative furnishings inspired by the local desert landscape. Superb hiking, a Dead Sea beach and the Ein Gedi Spa are nearby. **www.ein-gedi.co.il**

JERICHO Jericho Resort Village $$$

Near Hisham Palace **Tel** *(02) 232 1255* **Fax** *(02) 232 2189* **Rooms** *104* **Map** *C3*

Set amid spacious, oasis-like grounds, this modern hotel-resort, on the northern outskirts of town on the road leading up the Jordan Valley, has attractive rooms and bungalows for up to six people. Amenities include swimming pools, restaurants and courts for tennis, basketball and sand volleyball. **www.jerichoresorts.com**

KHAI BAR YOTVATA WILDLIFE RESERVE Kibbutz Lotan Guesthouse $$$

Kibbutz Lotan, 88855 **Tel** *(08) 635 6935* **Fax** *(08) 635 6927* **Rooms** *20* **Map** *B6*

Surrounded by the mountains of the southern Arava Desert, this guesthouse is run by the environmentally conscious Kibbutz Lotan, affiliated with the Reform Jewish Movement, and is part of their Centre for Ecotourism and Birdwatching. It is situated 15 km (9 miles) north of the Khai Bar Nature Reserve and 50 km (31 miles) north of Eilat.

MASSADA Massada Youth Hostel $$

Massada, 86935 **Tel** *(08) 995 3222* **Fax** *(08) 658 4650* **Rooms** *88* **Map** *C4*

A deluxe option as far as youth hostels go, this appealing place, situated next to the cable car station, has spotless, modern rooms, most with bunk beds, and a lovely swimming pool. The cafeteria offers plentiful, adequate food. The hostel is served by Egged buses 444 and 486 from Jerusalem. **www.iyha.org.il**

MITZPE RAMON Ramon Inn $$$$

1 Ein Akev St, 80600 **Tel** *(08) 658 8822* **Fax** *(08) 658 8151* **Rooms** *96* **Map** *B5*

A great base for exploring the remote, highland reaches of the southern Negev, including spectacular Makhtesh Ramon, this attractive establishment is architecturally in harmony with the desert. Some apartments have space for up to five people. There is a covered swimming pool and two saunas. **www.isrotel.co.il**

PETRA AND WESTERN JORDAN

AJLOUN Al-Jabal Hotel $

Al Qala' St **Tel** *(02) 642 0202* **Fax** *(02) 642 0991* **Rooms** *20* **Map** *C3*

Set among pine trees on the road up from Ajloun town towards the Crusader-era hilltop castle, this clean, simple hotel offers courteous service, decent rooms (many with superb views) and a sense of isolation. It is quiet out-of-season, but often packed with families in summer so booking is advisable. **www.jabal-hotel.com**

AMMAN San Rock International $$

Sa'eed Abu Japer St, Jabal Amman, 11191 **Tel** *(06) 551 3800* **Fax** *(06) 551 3600* **Rooms** *105* **Map** *C3*

Popular with Western tour groups, this is one of Amman's best mid-range hotels. It is situated near 6th Circle, in a part-residential, part-commercial district on the western edge of the city centre, which cuts down driving time to and from the airport. Decor is a little tired, but service is notably good. **www.sanrock-hotel.com**

AMMAN Shepherd $$

Zaid bin al-Harith St, Jabal Amman, 11181 **Tel** *(06) 463 9197* **Fax** *(06) 463 9197* **Rooms** *48* **Map** *C3*

This decent lower-priced hotel is situated in a characterful location between the 1st and 2nd circles, within walking distance of the shops and cafés of Rainbow St and the restaurants and craft outlets around 2nd Circle. Rooms are pleasant – ask for one at the back to avoid street noise. Excellent value for money. **www.shepherd-hotel.com**

Key to Price Guide *see p256* **Key to Symbols** *see back cover flap*

AMMAN Hisham $$$

Mithqal al-Fayez St, Jabal Amman, 11183 ***Tel*** *(06) 464 4028/2720* ***Fax*** *(06) 464 7540* ***Rooms*** *25* ***Map*** *C3*

Peaceful and discreet, this modest, family-run hotel is situated between the 3rd and 4th circles, on a leafy corner in the heart of the diplomatic quarter, just behind the French Embassy. It has a long history and a good deal of character. The rooms are comfortable, and service is outstanding: genial, attentive and accommodating.

AMMAN Marriott $$$$

Issam al-Ajlouni St, Shmeisani, 11190 ***Tel*** *(06) 560 7607* ***Fax*** *(06) 567 0100* ***Rooms*** *293* ***Map*** *C3*

This top-class chain hotel is regularly voted by journalists and business-people as one of the finest hotels in the Middle East. Located in a smart district, it features every luxury, from first-class restaurants and large, elegantly appointed rooms to personal service and a full range of business facilities. **www.marriott.com**

AMMAN InterContinental $$$$$

Between 2nd and 3rd circles, Jabal Amman, 11180 ***Tel*** *(06) 464 1361* ***Fax*** *(06) 464 5217* ***Rooms*** *475* ***Map*** *C3*

One of Amman's, and Jordan's, longest-established hotels has been entirely renovated to international luxury quality. A premier venue for top-level congresses, and a favoured venue for visiting journalists, it offers large, well-appointed rooms, good sports facilities and outstanding restaurants. **www.intercontinental.com**

AQABA Coral Bay $$

The Royal Diving Club, South Beach, 77110 ***Tel*** *(03) 201 7035* ***Fax*** *(03) 201 7097* ***Rooms*** *69* ***Map*** *B7*

A comfortable hotel attached to the Royal Diving Club, located on its own private, west-facing beach about 18 km (11 miles) south of Aqaba town centre. Rooms are airy and pleasant, and you get access to pools, a restaurant, a beach bar and, of course, the exquisite coral reefs immediately offshore for snorkelling and diving. **www.rdc.jo**

AQABA Mövenpick $$$$

King Hussein St, North Beach, 77110 ***Tel*** *(03) 203 4020* ***Fax*** *(03) 203 4040* ***Rooms*** *235* ***Map*** *B7*

A first-class luxury resort hotel, on a prime plot straddling the beach road alongside the town centre: one of the four swimming pools occupies an overbridge linking the main hotel building with its seafront villas and beaches. The hotel design has Arabic influences, and the dining options are lavish. **www.movenpick.com**

AQABA InterContinental $$$$$

King Hussein St, North Beach, 77110 ***Tel*** *(03) 209 2222* ***Fax*** *(03) 209 3318* ***Rooms*** *255* ***Map*** *B7*

This beautiful resort hotel faces south over Aqaba's bay and is within easy walking distance of the town centre. The opulent decor includes acres of marble in the public areas and lavish gardens around the hotel's pools and sandy beaches. Rooms are large and modern, many with balconies. The restaurants are outstanding.

DANA Dana Guest House $$

Dana Village ***Tel*** *(03) 227 0497/0498* ***Fax*** *(03) 227 0498* ***Rooms*** *9* ***Map*** *C5*

The Dana Guest House sits in idyllic countryside and is run superbly well by Jordan's Royal Society for the Conservation of Nature (RSCN). It looks out over a pristine, silent landscape of mountains and valleys – heaven for birdwatchers. Most rooms have a balcony and one is ensuite (the rest share bathrooms). Book rooms and meals well ahead.

DEAD SEA Kempinski Ishtar $$$$

Dead Sea Rd, Sweimeh, 11180 ***Tel*** *(05) 356 8888* ***Fax*** *(05) 356 8800* ***Rooms*** *345* ***Map*** *C4*

This stylish resort hotel on the Dead Sea has world-class facilities, with chalet-style villa suites dotted around eight pools, including a magical infinity pool, a variety of restaurants, the largest spa in the Middle East and a great beach. Design is cool and contemporary and the service is memorably good. **www.kempinski-deadsea.com**

DEAD SEA Mövenpick Resort & Spa $$$$$

Sweimeh, Dead Sea Rd, 11180 ***Tel*** *(05) 356 1111* ***Fax*** *(05) 356 1122* ***Rooms*** *340* ***Map*** *C4*

One of Jordan's flagship Dead Sea resort hotels, boasting unusual taste and character. Design plays a central role, from the Damascene-style hard-carved wooden ceiling in the lobby bar, to the spacious guest rooms, which are housed in low-rise, two-storey villas of local stone and stucco. Pools, beach facilities, restaurants and a spa add to the attraction.

PETRA Petra Moon $

Wadi Musa, 71810 ***Tel*** *(03) 215 6220* ***Fax*** *(03) 215 4547* ***Rooms*** *17* ***Map*** *C5*

Excellent low-budget hotel located a short walk up the hill behind the Mövenpick Hotel, barely five minutes from the ticket gate into Petra. Rooms are well-kept (all ensuite), and service is pleasant and attentive. There is no air conditioning, but that is rarely an issue outside July and August. **www.petramoonhotel.com**

PETRA Petra Palace $$

Wadi Musa, 71810 ***Tel*** *(03) 215 6723* ***Fax*** *(03) 215 6724* ***Rooms*** *160* ***Map*** *C5*

This comfortable hotel benefits from a location on the "Tourist Road" strip of shops and restaurants, near the Petra ticket gate and is easy to reach after a hard day of walking. Rooms, which are bright and tidy, are therefore priced slightly high, but the atmosphere is pleasant and there is a good bar too. **www.petrapalace.com.jo**

PETRA Mövenpick $$$$$

Wadi Musa, 71810 ***Tel*** *(03) 215 7111* ***Fax*** *(03) 215 7112* ***Rooms*** *183* ***Map*** *C5*

This is a spectacularly well-designed and well-appointed five-star hotel. The location is unbeatable, directly at the entrance to Petra, a few metres from the visitor centre and ticket gate. The interior is stunning, featuring intricate Arabesque designs and a soaring atrium. Rooms are very comfortable. **www.movenpick.com**

Choosing a Hotel in the Red Sea and Sinai

This section lists hotels in the Red Sea and Sinai area. The price ranges are given in Egyptian pounds. For key to symbols and map references, see back endpaper.

PRICE CATEGORIES
Prices categories are per night for two people occupying a standard double room, with tax, breakfast and service included:
£ Under LE 150
££ LE 150– LE 250
£££ LE 250–LE 650
££££ LE 650–LE 1,500
£££££ Over LE 1,500

THE RED SEA AND SINAI

DAHAB Bishbishi Camp £

Mashraba, Dahab **Tel** *(069) 3640 727* **Rooms** *40* **Map** *F6*

The camp is ideal for young travellers keen for the experience of living close to nature rather than having luxurious surroundings and lots of modern conveniences. It comprises a series of beachside bamboo-style huts equipped with the essentials. Some have ceiling fans to help ease the heat of the day.

DAHAB Jasmine Pension £

Mashraba, Dahab **Tel** *(069) 3640 852* **Fax** *(069) 3640 885* **Rooms** *17* **Map** *F6*

The Jasmine Pension is an attractive, inexpensive alternative to the many camps that can be found in Dahab. The rooms are basic but comfortable, and are equipped with their own bathrooms and fans. The complex's own restaurant may not be luxurious, but it serves good, hearty food. **www.jasminepension.com**

DAHAB Blue Beach club £££

Lighthouse, Dahab **Tel** *(069) 3640 411* **Fax** *(069) 3640 413* **Rooms** *36* **Map** *F6*

The attractive rooms that form the Blue Beach Club have great views of the surrounding neighbourhood of Asilah, Dahab, in one direction and the beach in the other. Room facilities include a fan and fridge. This quiet hotel is ideal for couples looking for a relaxing base from which to explore the area. **www.bluebeachclub.com**

DAHAB Club Red £££

Mashraba, Dahab **Tel** *(069) 3640 380* **Fax** *(069) 3640 380* **Rooms** *18* **Map** *F6*

The Club Red is a no-frills hotel that attracts young people, especially divers, because of its close proximity to the sea and good dive facilities. Discounts are offered to divers. Some rooms can be shared to keep costs down, while others come complete with fans and adjoining bathrooms. **www.club-red.com**

DAHAB Nesima Hotel £££

Mashraba Asilah, Dahab **Tel** *(069) 3640 320* **Fax** *(069) 3640 321* **Rooms** *51* **Map** *F6*

The Nesima Hotel has one of the most popular diving centres in the Asilah area, and is known for its good food served in a traditional-styled restaurant. Its rooms are well-presented, with many featuring domed ceilings and sea views. Its pool overlooks the sea. The hotel welcomes people with disabilities. **www.nesima-resort.com**

DAHAB Hilton Resort ££££

Dahab Bay, Dahab **Tel** *(069) 3640 310* **Fax** *(069) 3640 424* **Rooms** *163* **Map** *F6*

The Hilton Resort is a landmark building in Dahab Bay. It is beautifully presented with lush gardens and whitewashed rooms that surround a lagoon-style swimming pool. It is situated on the beachside and offers some superb leisure amenities, including dive and windsurfing centres. **www.hilton.com**

NUWEIBA Basata £

Ras al-Burqa, Nuweiba **Tel** *(069) 3500 480/481* **Rooms** *26* **Map** *F5*

An extremely popular hotel and camp, Basata lies around 23 km (14 miles) north of Nuweiba. Guests live in mud and bamboo huts, a crucial part of the owner's policy on eco-friendliness. Basata has its own kitchen and bakery, and is known for its good snorkelling. Scuba diving, however, is not allowed.

NUWEIBA Habiba Village ££

Nuweiba City, South Sinai **Tel** *(069) 3500 770* **Rooms** *21* **Map** *F5*

Situated right on the beachside, this traditionally built hotel village has a beach restaurant to enjoy an evening under the stars, along with its Mataamak eaterie for more formal dining. Rooms are well presented, with most having air conditioning and a private bathroom. Wooden cabins have fans. **www.sinai4you.com**

NUWEIBA La Sirene £££

Beach Road, Nuweiba **Tel** *(069) 3500 701* **Fax** *(069) 3500 702* **Rooms** *45* **Map** *F5*

La Sirene Hotel is the centrepiece of a resort set right on the beach at Nuweiba, between the port and the city. It is a compact hotel that is pleasingly presented and well-located for local amenities. As such, it is extremely popular. The hotel's leisure amenities are few, but do include diving.

NUWEIBA Hilton Nuweiba Coral Resort ££££

Nuweiba City, South Sinai ***Tel*** *(069) 3520 320* ***Rooms*** *200* ***Map*** *F5*

This large resort is located amidst quiet, beachside gardens and is known for its tranquility and beauty. It has lots of amenities, including diving, windsurfing, tennis and squash courts. Other activities, like kayaking, windsurfing, disco dancing and even swimming with dolphins, are all within walking distance. **www.hilton.com**

SHARM EL-SHEIKH Shark's Bay UMBI Camp ££

Shark's Bay, Sharm el-Sheikh ***Tel*** *(069) 3600 942* ***Fax*** *(069) 3600 944* ***Rooms*** *64* ***Map*** *E7*

One of the most frequently revisited camps in Sharm el-Sheikh and popular with local families, Shark's Bay Camp sits right on the beach in an isolated location, and has its own reef and dive centre. Rooms are either chalets with air conditioning, or bamboo huts without air conditioning. Safari and desert activities available. **www.sharksbay.net**

SHARM EL-SHEIKH Amar Sina £££

Ras Um Sid, Sharm el-Sheikh ***Tel*** *(069) 3662 222* ***Fax*** *(069) 3662 233* ***Rooms*** *91* ***Map*** *E7*

The Amar Sina is designed and built to resemble a traditional whitewashed Egyptian village, with architectural features such as domes and arches. Facilities include a bar, shops, its own fitness centre and restaurants radiating from a central square. Rooms are pleasant and air-conditioned.

SHARM EL-SHEIKH Camel Hotel £££

Naama Bay, Sharm el-Sheikh ***Tel*** *(069) 3600 700* ***Fax*** *(069) 3600 601* ***Rooms*** *38* ***Map*** *E7*

Pretty and compact, the Camel Hotel has gained a reputation for not only providing top-class diving facilities but also offering great cuisine in its award-winning restaurants. The hotel, which is located in Naama Bay and minutes from Sharm el Sheikh, is known for its extensive facilities for disabled guests. **www.cameldive.com**

SHARM EL-SHEIKH Sanafir Hotel £££

Naama Bay, Sharm el-Sheikh ***Tel*** *(069) 3600 197* ***Fax*** *(069) 3600 196* ***Rooms*** *50* ***Map*** *E7*

The Sanafir Hotel is best known for being the venue for one of Naama Bay's most popular nightclubs – it comes alive after dark and is ideal for travellers looking for nightly entertainment. It has a good choice of bars and restaurants too. The hotel's air-conditioned rooms are well-presented. **www.sanafirhotel.com**

SHARM EL-SHEIKH Ritz-Carlton Resort ££££

Om El Seed Peninsula, Sharm el-Sheikh ***Tel*** *(069) 3661 919* ***Fax*** *(069) 3661 920* ***Rooms*** *321* ***Map*** *E7*

Oozing luxury, this top-class hotel sits in beautifully landscaped gardens where cascading waterfalls combine with subtly lit pools and shrubbery. Its rooms are equally well-presented, while on-site facilities include everything from fine international dining to superb golf, watersports and family fun. **www.ritzcarlton.com**

SHARM EL-SHEIKH Sofitel Sharm el-Sheikh ££££

Naama Bay, Sharm el-Sheikh ***Tel*** *(069) 3600 081* ***Fax*** *(069) 3600 085* ***Rooms*** *298* ***Map*** *E7*

Perched high on the coastline next to the beach, this hotel offers a wonderful panoramic view of Naama Bay. Amenities are in abundance and include a Turkish bath complex, archery, ice cream parlour and numerous restaurants, while rooms are attractive and most have sea views. **www.sofitel.com**

ST CATHERINE Morgenland Village £££

St Catherine City ***Tel*** *(02) 7956 856/ (069) 3470 331* ***Fax*** *(069) 3470 331* ***Rooms*** *230* ***Map*** *E6*

The Morgenland Village offers adequate rooms in the main building and a series of chalets in the grounds. Facilities include restaurants and a pool, plus a traditional-style shopping centre where one of the shops sells a selection of medicinal herbs from Sinai. The spectacular views are the best feature of the hotel.

ST CATHERINE St Catherine Guest House £££

St Catherine's Monastery, St Catherine ***Tel*** *(069) 3470 353* ***Fax*** *(069) 3470 543* ***Rooms*** *40* ***Map*** *E6*

Although a little lacking in luxuries, the auberge at St Catherine more than makes up for it in atmosphere. Set against a backdrop of countryside at the foot of Mount Sinai and right next to St Catherine's Monastery, it remains a firm favourite with travellers looking for a relaxing, "away from it all" location.

TABA Tobya Boutique Hotel £££

Taba International Road, Taba ***Tel*** *(069) 3530 274* ***Fax*** *(069) 3530 269* ***Rooms*** *100* ***Map*** *F5*

Outstanding architecture, lush greenery and distinctive decor using stone, wood and handwoven rugs, combined with excellent service make the Tobya a relaxing, luxurious retreat. Most rooms have a terrace and many have a view of the Red Sea. **www.tobyaboutiquehotel.com**

TABA Marriott Taba Heights Beach Resort ££££

Taba and Nuweiba Highway, Taba ***Tel*** *(069) 3580 100* ***Fax*** *(069) 3580 109* ***Rooms*** *394* ***Map*** *F5*

This is one of many hotels on the extensive Taba Heights development. The luxury hotel's amenities are complemented by the resort's marina, 18-hole golf course, a casino, safari programme and top-class spas, while restaurants offer everything from Japanese sushi and Indian dishes to European meals. **www.marriott.com**

TABA Three Corners El Wekala Golf Resort ££££

Taba Heights, Taba ***Tel*** *(069) 3580 150* ***Fax*** *(069) 3580 156* ***Rooms*** *215* ***Map*** *F5*

This luxury resort hotel occupies a prime location in the Taba Heights Resort, offering excellent facilities in a beautiful setting. The complex includes restaurants, bars and pools, and there's a daily programme of organized activities, such as exercise classes, water games and dancing. A shuttle bus takes guests to a private beach. **www.threecorners.com**

RESTAURANTS, CAFES AND BARS

Middle Eastern food is often overshadowed by other more glamorous world cuisine, and as such, the Holy Land has been seen by many as a gastronomic desert. Often simple and unpretentious, the food is, however, usually tasty and substantial *(see pp268–9)*. A constantly changing restaurant culture reflects the huge interest in food in the Holy Land, and many restaurants are of a very high standard, offering a wide range of Middle Eastern food sure to excite even the most sceptical palate.

Vendor selling the iced drink *tamahindi*

Aside from the native cuisine, there are many other restaurants offering more international food, reflecting the broad ethnic mix of the Holy Land. You can find South American, Chinese, Indonesian, Italian and French food, along with the ever popular American fast food. There are also many busy and informal cafés, which offer a cheaper, more limited menu. For a quick snack, street food revolves around the *shawarma* and *falafel* stalls, which can be found almost everywhere.

The restaurant in the Arabesque American Colony Hotel *(see p273)*

PRACTICALITIES

In most Israeli cities, especially Tel Aviv and in Jerusalem's Nakhalat Shiva neighbourhood, you will see people eating at all hours of the day, seated at cafés and restaurants or walking along the street with a pitta or *boureka (see p268)*. In the evening, people tend to eat late, and spend a long time over their meals. Eating is a big social event, with children accepted in many restaurants. Dining, when possible, is alfresco, and restaurants often stay open until after midnight, especially during the summer. However, restaurants are not always open all week, especially the Jewish ones. These always close for Shabbat (sundown on Friday until after sundown on Saturday), as well as for Yom Kippur, Shavuot, and the first and last day of Sukkoth and Passover. In addition, all Jewish-owned restaurants, whether kosher or not, are closed on Holocaust Day and Remembrance Day *(see p36)*.

Service is not generally included on the bill. You should expect to tip around 10–15 per cent, depending on the type of establishment. Most major credit cards are accepted in nearly all restaurant throughout Israel.

TYPES OF RESTAURANT

Food is a major part of Middle Eastern life, and there is a huge range of places to eat. With no fixed cuisine of its own, Israeli food is a melting pot of flavours, reflecting the cultural mix of the nation and adopting influences from the Middle East, the Mediterranean, and Eastern Europe. The main Israeli food is that of the Jews, largely the Oriental (Middle Eastern) and Ashkenazi (eastern European) communities. Their food is as different as their origins. Oriental dishes revolve mainly around grilled meats and fish, stuffed vegetables, and a range of *meze*. The Ashkenazi specialities are spicy stews, fish balls and stuffed pancakes, known as *blintzes*.

Other major ethnic groups have also brought their own unique and unusual dishes. Armenian favourites include spicy meat stews and sausages, while the Yemenites are famous for their *malawach* – large, flaky-pastry pancakes, stuffed with a variety of fillings.

Aside from Israeli fare, you can also find restaurants serving more international food, including French and Italian, which tend to be very expensive, and Chinese, Thai and Korean. There are also the usual fast-food chains. The selection of restaurants is far more limited if travelling in Jordan or Sinai, however, as most are found in the hotels.

Dining outside in the spectacular setting of Petra *(see pp220–31)*

Café culture in Israel is huge, and if you are after something cheaper and less substantial, then cafés offer salads, pizzas, club sandwiches and simple pasta dishes that will provide a tasty light meal. Cafés are also great places to sit and soak up the local atmosphere, and join in with Israeli life.

KOSHER RESTAURANTS

The Jewish dietary laws of *Kashrut* (literally, fitness), determine many of the eating habits in the Holy Land. To the outsider these can prove very confusing, especially as you will find that not all Jewish restaurants adhere to these strict rules.

Bourj al-Haman Intercontinental restaurant, Jordan ***(see p280)***

What these laws mean in practice is that meat considered impure (for example pork, rabbit and horse meat), as well as certain types of seafood (anything without scales and fins), cannot be eaten. Animals that are permitted for consumption have to be slaughtered according to Jewish religious practice and cleansed of all traces of blood before cooking. Furthermore, during Passover a kosher restaurant cannot serve any leavened food, such as bread or pastries.

The major complications of these laws revolve around the fact that meat and dairy produce can never be eaten together in the same meal. Dishes are consequently based on either one or the other, with many of the resulting problems deftly overcome through the use of dairy substitutes.

VEGETARIAN FOOD

As a vegetarian visiting the Holy Land, your dining options are surprisingly varied. Kosher restaurants serve all types of dairy-only food, such as creamy pasta and yogurt-based dishes, as well as many potato dishes and salads.

Secular restaurants also have a large number of vegetarian options. Much of the cuisine is based around pulses, which are found in anything from houmous to hearty bean stews. Roasted and stuffed vegetables also feature widely, along with a variety of savoury pastries. For a quick vegetarian snack, the *falafel* is hard to beat.

JORDAN AND SINAI

Jordanian food is a mix of the Lebanese-Syrian-Egyptian fare common throughout the Middle East, mixed in with local Bedouin cuisine. Expect lots of good, fresh *meze*, salads and grilled meats, plus traditional specialties such as *mansaf*: lamb on a bed of rice sprinkled with pine nuts. You may also be offered *maqlubbeh*, which is steamed rice pressed into a small bowl then turned out and topped with slices of grilled eggplant. Otherwise, places like Amman have plenty of international restaurants, cafés and takeaways.

Food in the Sinai resorts tends to cater to the tastes of package holidaymakers. Most restaurants are attached to hotels and favour Italian and other safe international dishes. Genuine Egyptian cuisine is rare, although the fish and seafood can be excellent.

Sidewalk restaurants in Nachlat Shiva, Jerusalem

SMOKING

There has been a smoking ban in all public places within Israel for many years. Restaurants are allowed to have a completely separate smoking area and smoking is allowed on terraces. In practice many people still light up despite the ban, however, visitors should not follow suit.

Egypt also does not allow smoking in all public places, but like Israel you will find that most locals do not pay much attention to the no-smoking signs.

In Jordan tourists are allowed to smoke in restaurants, cafés and bars, except during Ramadan when smoking is prohibited. Some restaurants do provide no-smoking areas.

Elvis American Diner ***(see p274)***

The Flavours of Jerusalem and the Holy Land

The cuisines of the Holy Land are as varied as its people. Over the centuries, the region has embraced rich culinary traditions from around the Mediterranean, Central and Eastern Europe, the Middle East, North Africa and South Asia. More recently, dishes brought by Jewish immigrants from Ethiopia have appeared, and a growth in travel to East Asia has resulted in the food from this region becoming hugely popular. The local dining scene has come a long way since the spartan communal dining halls of the early *kibbutzim*, and recently an increasingly sophisticated gastronomic culture has transformed the restaurant scene.

Pomegranates

Fish seller's stall at Jerusalem's Mahane Yehuda market

STREET FOOD

Stalls and storefront eateries offer a varied array of cheap, nutritious and relatively healthy "fast food". *Falafel* is an excellent option for vegetarians, as are houmous and *bourekas*, a filo pastry from the Balkans filled with salty *kashkaval* cheese, potatoes, spinach or mushrooms. Somewhat less well known is *sabih*, an Iraqi speciality that consists of potato chunks, fried aubergine (eggplant), a hard-boiled egg, salad, *tahina* (sesame paste), hot sauce and chopped parsley, served in a pitta. A carnivore favourite is *shwarma*, the local, often turkey-based, version of gyros or doner kebab. Griddled meats such as *me'urav yerushalmi* (a mixed grill of chicken livers, hearts and other offal) are served in, or with, a pitta.

MEZE OR SALATIM

A meal typically begins with a large selection of starters (*meze* in Arabic, *salatim* in Hebrew). Middle Eastern restaurants serve *meze* either as a starter or as a full meal. Dishes you are likely to encounter include houmous (chickpea/garbanzo paste with olive oil, lemon and garlic), tabouleh (cracked wheat with masses of chopped mint and parsley,

Some of the small dishes that make up a *meze* or *salatim*

DISHES AND SPECIALITIES OF THE HOLY LAND

The traditional dishes you'll find served throughout the Holy Land range from stuffed grape leaves and *mansaf* (rice and lamb with a sour yoghurt sauce), sometimes called the national dish of Jordan, to gefilte fish and chicken soup with matzo balls, favoured by Jews with roots in Eastern Europe. Popular Palestinian Arab specialities include *meze* salads and sumac-flavoured meat dishes such as *mussakhan*. About half of Israeli Jews have family roots in Asia and Africa, which is why the menus of ethnic restaurants often feature Moroccan couscous, fiery fish dishes from Libya, doughy *malawah* (pan-fried bread) and *jahnoun* (a heavy, slow-baked bread roll) from Yemen, and *kubbe* (or *kibbe*) from Iraq – also a Palestinian speciality.

Selection of sweets

Shashlik and kebab *are, respectively, pieces of meat and spiced ground meat grilled on a skewer.*

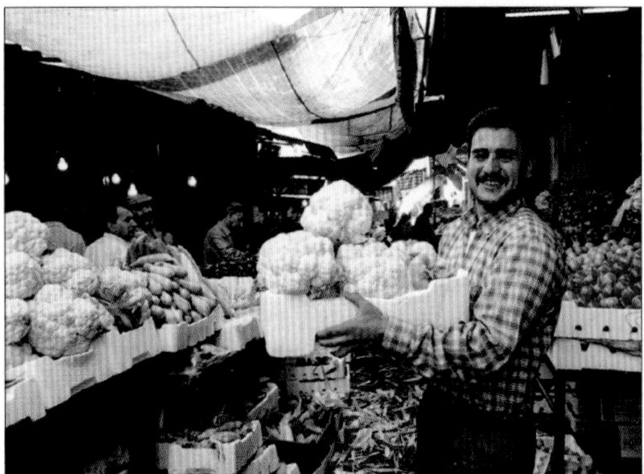

Market trader offers the superb fresh produce of the Holy Land

tomato, cucumber oil and lemon), *babaghanoush* (aubergine baked for a smoky flavour and then puréed), along with pickled vegetables and olives. *Kibbe* (cracked wheat and minced meat croquettes with onions and pine nuts) are among the few non-vegetarian dishes.

FRESH FRUIT

The Bible is filled with references to the produce of the land, and today the Holy Land grows not only ancient favourites such as grapes, pomegranates, figs and dates, but also citrus fruits, which have been exported to Europe under the Jaffa labels, since the 19th century. The largest type of citrus is the pomelo, a thick-skinned fruit that can grow to the size of a volleyball and is a speciality of the Jericho area. The fragrant Galia melon was developed in Israel. Watermelon is often eaten with chunks of salty Bulgarian cheese, similar to féta. Widely available exotic fruits include persimmon, kiwi and passionfruit.

Dried red peppers in the market at Shuk Levinsky, Tel Aviv

LOCAL PRODUCE

Israel has long been known for its excellent selection of cheeses, but in recent years a growing number of so-called "boutique" dairies has been setting ever-higher standards with their goats' and sheep's milk products. These go extremely well with classic Mediterranean specialities such as olives and extra virgin olive oil, produced with great pride by both Jews and Arabs. In both the Galilee and the Negev, travellers will often come across family-run roadside eateries where local farmers sell their own produce, such as delicious honey.

ON THE MENU

Baklava Honey-soaked chopped nut and filo pastries.

Cholent Sabbath lunch stew of beef, potatoes, carrots, barley, onions and beans.

Knafeh Palestinian pastry of cheese, crunchy wheat threads and very sweet syrup.

Kugel Egg noodle casserole, either sweet or savoury.

Labane Sharp, spreadable white "yoghurt" cheese, often preserved in olive oil

Za'atar Seasoning mix of hyssop, sesame seeds and salt.

Zchug fiery red or green Yemenite condiment.

Falafel *are deep-fried balls of mashed chickpeas (garbanzos) served stuffed into a pitta bread with salad.*

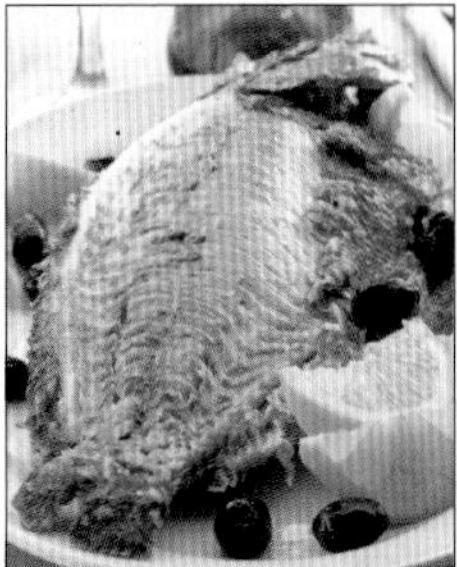

Tilapia, *or St Peter's Fish, is popular around the Sea of Galilee, simply grilled and served with lemon slices.*

Jerusalem salad *is a meal in itself with olives, féta, and sometimes pomegranate and* za'atar *sprinkled over.*

What to Drink in Jerusalem and the Holy Land

Tea with fresh mint leaves

Jews and Arabs alike adore coffee but have different ways of making it. It will be offered to you at any hour of the day or night. Teas of many kinds and herbal infusions are also popular. However, the hot, very dry climate makes water of the utmost importance. It is advisable to carry a bottle of it with you at all times and drink some before you feel thirsty to avoid dehydration. Israel now produces a lot of affordable medium to high quality wine. Beer is available in all the areas covered in this guide, but neither the Israelis nor the Arabs consume large quantities, preferring to go to cafés or coffeeshops for socializing.

Enjoying outdoor café life on traffic-free Lunz Street in Jerusalem

WATER AND SOFT DRINKS

Bottled water

In the entire area described in this guide, bottled mineral water is readily available everywhere. Although tap water throughout Israel is safe to drink, it is more advisable to drink bottled water because it tastes better, especially in the Red Sea area, where tap water is so heavily chlorinated that it is unpalatable. Always make sure that the bottle is sealed when you buy water.

Bottled fruit juice is also popular, but remember that even juices that are sold as "natural" are really long-life juices produced on an industrial scale. Fruit juices freshly squeezed in front of you, especially citrus and pomegranate, are very good. All non-alcoholic beverages except for freshly-squeezed juice are almost always served very cold and with a lot of ice (which may be made of heavily chlorinated water), so if you don't want your drinks this way, remember to say so when ordering.

BEERS AND SPIRITS

Many restaurants and cafés have draught beer, most of which is locally produced. The main Israeli beers are Maccabee, a slightly bitter, light lager, and Goldstar, which is reminiscent of British ale with a dash of malt. Taybeeh, similar to light, south German beer, is found in the Palestinian regions, East Jerusalem and some Israeli bars. Carlsberg is produced in Israel and Heineken in Jordan, both under licence, while most other major European brands are imported, especially into Israel.

Spirits are less widely available, but are always sold in hotel bars throughout the region. The commonest is arak, the typical Mediterranean distillate of anise.

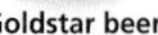
Goldstar beer

Arak

COFFEE AND TEA

In Jewish areas, coffee and tea are drunk in European- or American-style cafés. The most widely available type of coffee is filter coffee, which is always served for breakfast in hotels. Many places also offer espresso coffee, but it is almost always rather weak. For a real espresso, you must ask for a *katzar* (strong coffee). What is called cappuccino almost always has a huge amount of whipped cream added to it. Tea is almost invariably served in tea-bag form, and caffeine-free herbal tea *(zmachim)* is becoming increasingly popular.

Tea and coffee in Arab areas are drunk in coffeehouses *(qahwa)*, which serve nothing else – except sometimes traditional water pipes *(nargileh)* to accompany the drink. Arabic coffee (also called *qahwa*) is strong and aromatic because of the spices, in particular the cardamom, added to it. It is served in tiny cups holding only a few sips. If you do not specify little or no sugar, it always arrives heavily sweetened. To avoid a gritty texture, allow the sediment to settle in the cup first.

Elaborate Arabic coffee set

Arabic tea *(shai)* is more aromatic and stronger than Western-style tea and is also drunk without milk and with a lot of sugar. In restaurants it is often served after a meal with fresh mint leaves *(naana)*.

In Arab coffeehouses, if you want Western-style tea, ask for *shai-Libton*; in Arab or Jewish establishments, for Western-style coffee ask for *nes* (short for Nescafé).

WINE

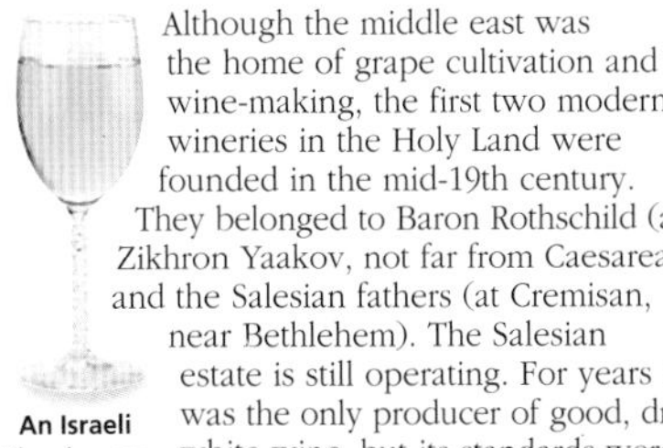

An Israeli Chardonnay

Although the middle east was the home of grape cultivation and wine-making, the first two modern wineries in the Holy Land were founded in the mid-19th century. They belonged to Baron Rothschild (at Zikhron Yaakov, not far from Caesarea) and the Salesian fathers (at Cremisan, near Bethlehem). The Salesian estate is still operating. For years it was the only producer of good, dry white wine, but its standards were later matched by the Latrun Trappist monks' winery, which has French vines and uses French wine-making techniques.

The number of vineyards then increased steadily and wine quality has improved dramatically since the early 1980s. The main wine areas are now: Golan and Upper Galilee at around 500 m (1,640 ft) above sea level, with ideal volcanic soil; Lower Galilee, the Jezreel Valley, the Mount Carmel region and Sharon, which are lower and more humid; Samson, the coastal plain south of Tel Aviv; and the hills of Judaea, which have poorer terrain and are very dry. A number of experimental vineyards in the Negev Desert are now in production.

The largest producers are the Carmel Winery, based in Zikhron Yaakov, whose Mizrachi "Private" series is especially good, and the Golan Heights Winery, based in Katsrin, whose main labels are Golan, Yarden, Gamla and Tishbi. Wines from small producers such as Kibbutz Tsora can be excellent.

Jordanian and Egyptian wines are very poor value for their price and, in both countries, imported wine is prohibitively expensive.

Israeli white wines, *especially the Chardonnays and Sauvignon Blancs, are generally very enjoyable: often aromatic, sometimes fruity, smooth and full-bodied. Many of the reasonably-priced whites are produced by the Golan Heights Winery.*

Yarden white

Gamla Chardonnay

Tishbi Muscat

WINE-GROWING REGIONS OF ISRAEL

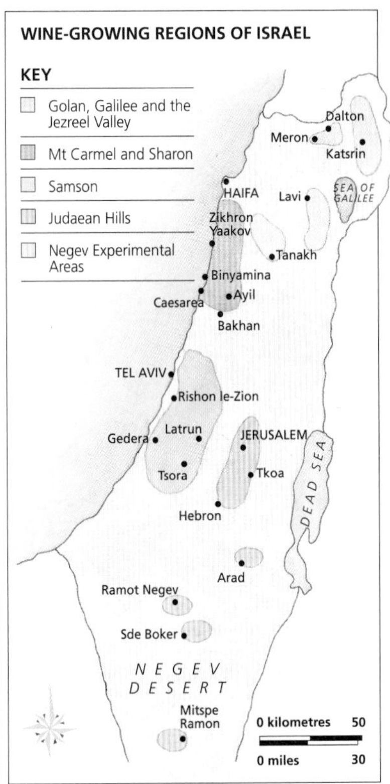

Israeli red wines *are also good, but, with some notable exceptions, tend either to lack body or to be slightly heavy. The grapes most commonly used are Carignan, Cabernet Sauvignon and Merlot, with many wines being a blend of the last two. Among the wines now produced by a growing number of small-scale, specialist wine makers are the fine Cabernets produced by Castel, and the Margalit reds.*

Carmel Mizrachi

Margalit red

Kibbutz Tsora

Choosing a Restaurant

The restaurants listed here have been selected for their good food, atmosphere and location within the Jerusalem area. The symbols cover some of the factors which may influence your choice (see back cover flap for key). For Jerusalem map references, see the Street Finder on pages 156–9; for restaurants further afield see back endpaper.

PRICE CATEGORIES
Prices are based on a three-course meal for one including half a bottle of wine, tax and service.
($) Under $15
($)($) $15–$30
($)($)($) $30–$60
($)($)($)($) Over $60

THE MUSLIM QUARTER

Abu Shukri ($)

63 El-Wad St, cnr Via Dolorosa, 97500 ***Tel*** *(02) 627 1538* **Map** *4 D2*

This small, simple restaurant, on the main street leading from Damascus Gate into the Muslim Quarter, is renowned for its plates of houmous, topped with whole chickpeas and freshly chopped parsley. It also serves up tahini, freshly baked bread and lots of other dishes. Excellent quality and very reasonable prices.

Quarter Café ($)($)

Tiferet Yisrael St, 97500 ***Tel*** *(02) 628 7770* **Map** *4 D4*

Known more for its view of the Temple Mount and the Mount of Olives than for its cuisine, this dairy café serves light meals and snacks. The food is not particularly exciting, but the location is a good stopping-off point in the Jewish Quarter as it is situated just up the hill from the Western Wall.

MODERN JERUSALEM

Big Apple ($)

13 Dorot Rishonim St, 94625 ***Tel*** *(02) 625 6252* **Map** *1 A3*

In the pedestrian zone just off Ben Yehuda St, this pizzeria serves New York-style thin-crust pizzas and is especially popular with Orthodox Jewish teenagers from New York who are in Jerusalem to take intensive religious studies courses. The restaurant is open until late and also offers a takeaway service.

Burgers Bar ($)

20 Emeq Refaim St, German Colony, 93105 ***Tel*** *(02) 561 2333*

Acclaimed by many Jerusalem old-timers as having the city's best burgers, this popular place also serves up other types of reasonably priced meat dishes. It is situated about 1 km (half a mile) south of the King David Hotel in the atmospheric German Colony, in the heart of a strip of popular cafés and restaurants.

Pinati ($)

13 King George St, 94229 ***Tel*** *(02) 625 4540* **Map** *1 A3*

Situated at the spot where Ben Hillel St meets King George St at an oblique angle (hence the name, which means "on the corner"), this popular eatery has long been regarded by many local connoisseurs as the source of the city's most delicious houmous – creamy, delicately seasoned, never too heavy.

Rahmo ($)

5 Eshkol St, Mahane Yehuda, 94322 ***Tel*** *(02) 623 4595*

A Jerusalem institution, Rahmo serves Israeli and Aleppo-style cuisine as well as authentic Jerusalem houmous prepared according to a secret recipe from the owner's mother. Situated on one of the tiny pedestrianized alleyways in the colourful Mahane Yehuda market, which is a feast for the eyes as well as the stomach.

Agas ve-Tapuach ba-Kikar ($)($)

6 Safra Square, 94141 ***Tel*** *(02) 623 0280* **Map** *1 B3*

Known in Italian as Pera e Mela in Piazza (The Pear and the Apple on the Square), this venerable Italian restaurant has been serving home-style Italian cuisine, made with recipes from the owner's grandparents, since 1978. Dishes come from both northern and southern Italy and include antipasti, foccacia, bruschetta and, of course, pasta.

Focaccia Bar ($)($)

4 Rabi Akiva St, 94582 ***Tel*** *(02) 624 2273* **Map** *1 A3*

Situated on a quaint courtyard, this romantic place, built entirely of stone, is evocative of the early 1900s. Specialities include meat dishes, pasta, seafood, salads and, naturally, focaccia. Situated in the heart of West Jerusalem, two blocks south of Ben Yehuda St and just off Hillel St.

Key to Symbols *see back cover flap*

HaShipudia $$

5 HaArmonim St, Mahane Yehuda, 94322 ***Tel*** *(02) 625 4036*

In the heart of the Mahane Yehuda market, this good-value eatery serves a range of excellent local dishes, including soups, stuffed vegetables, houmous and grilled meats. Situated just one block west of the main street of Jerusalem's largest fruit and vegetable market. For dessert, try one of the nearby pastry shops.

Kan Zaman $$

Nablus Rd, 97200 ***Tel*** *(02) 628 3282* ***Map*** *1 C2*

Situated in a 19th-century house just north of Damascus Gate, this is the restaurant of the Jerusalem Hotel. The vaulted ceilings, shaded terrace and Oriental decor create a typically Arab atmosphere. The Palestinian cuisine is carefully prepared. There are often live concerts of Arab music on Friday from 8pm.

Shanti $$

4 Nahalat Shiva St, 94240 ***Tel*** *(02) 624 3434* ***Map*** *1 A3*

On a tiny alleyway in the 19th-century Nahalat Yitzhak quarter, this pub-restaurant is popular with young and old alike. Served in a warm and authentic Jerusalem atmosphere, the salads are huge, as are the steaks. The chicken wings prepared with soy sauce, honey and ginger are delicious. Open only in the evening, from 7pm to 3am.

Tmol Shilshom $$

5 Yoel Moshe Salomon St, 91316 ***Tel*** *(02) 623 2758* ***Map*** *1 A3*

Hidden at the end of a Nahalat Shiva courtyard in a private house built in the 1870s is this mellow café-restaurant-bookshop. Dining options include superb whole trout, soups, quiches, creative salads, pasta and stuffed mushrooms. Great for a quiet conversation. Has a superb Friday morning buffet (9am to 1pm). Reservations are recommended.

Barood $$$

31 Jaffa St, 94221 ***Tel*** *(02) 625 9081* ***Map*** *1 A3*

Just off Jaffa St on Feingold Courtyard, this laid-back, stone-built place is known for its Spanioli (Sepharadi) cuisine, as well as its juicy steaks and heavenly chocolate soufflé. The bar is decorated with a surprising collection of bottle openers and offers a wide selection of alcoholic beverages. Hosts art exhibits and live concerts of mellow music.

Darna $$$

3 Horkanos St, 94235 ***Tel*** *(02) 624 5406* ***Map*** *1 A3*

Moorish-inspired decor, Moroccan ceramics and lots of cushions adorn this gourmet Moroccan restaurant, whose name means "our home" in Arabic. Specialities include meze, *harira marrakshia* (veal and lentil soup made with fresh coriander) and *mechoui* (roast lamb with almonds). Situated two blocks north of Jaffa St.

Dolphin Yam $$$

9 Shimon Ben Shetah St, 94147 ***Tel*** *(02) 623 2272* ***Map*** *1 A3*

A Jerusalem favourite for fresh fish and seafood, this place, on the edge of Nahalat Shiva, also serves meat dishes and pasta. The decor is understated and informal. Recommended dishes include shrimp in cream sauce and grilled whole calamari. Has a wide selection of fried or grilled fish.

Karma $$$

73 Ein Karem St, Ein Karem, 95744 ***Tel*** *(02) 641 7430*

In the pastoral neighbourhood of Ein Karem, on the far western edge of Jerusalem, this modern, informal restaurant has a great atmosphere and specializes in meat dishes. Try the entrecôte steak or the delicious focaccia, which is baked with a range of tasty toppings, and for dessert there is a Nutella and mascarpone "pizza".

Link $$$

3 HaMaalot St, 94263 ***Tel*** *(02) 625 3446*

Housed in a century-old Jerusalem-style building, this café-bistro is known for its superb spicy chicken wings, made with soy sauce and honey, its juicy steaks and, for vegetarians, the soy-and-honey tofu salad. Link has a generously shaded terrace and a congenial atmosphere. It is situated just off King George St, across from Independence Park.

Mona $$$

12 Shmuel HaNagid St, 94592 ***Tel*** *(02) 622 2283*

Housed in the historic, stone-built home of the century-old Bezalel Art School, with its high ceilings and fanciful crenellations, this café-restaurant combines great food with a magical, arty Jerusalem atmosphere and exhibits of contemporary and historic Israeli art. The cuisine is international and includes salads, soups, antipasti and meat dishes.

Philadelphia $$$

9 El-Zahra St, 97200 ***Tel*** *(02) 628 9770* ***Map*** *2 D2*

One of East Jerusalem's best-known Arab restaurants, Philadelphia is much appreciated for its Palestinian-style stuffed vegetables, spit-roasted meats, fish (including St Peter's fish) and seafood. The ambience is welcoming, if a little formal. Live music on Friday from 9pm. Three blocks north of the Old City's Herod's Gate, which leads to the Muslim Quarter.

Rooftop $$$

Mamilla Hotel, 11 King Solomon St, 94182 ***Tel*** *(02) 548 2222* ***Map*** *1 B5*

This relaxed, kosher restaurant with breathtaking views over the Old City serves good Italian food made with fresh ingredients. Dishes include grilled meat and fish, salads, pasta, focaccia bread and delicious desserts. The informal setting and natural decor make this the perfect place to relax after a day's sightseeing.

Sakura $$$

31 Jaffa St, 94221 ***Tel*** *(02) 623 5464* ***Map*** *1 A3*

Acclaimed as the city's best sushi bar and Japanese restaurant, this place has authentic Japanese furnishings. Sushi and sashimi, served on little wooden platters, are classic mainstays but you can also order dishes such as tempura with almonds and chicken yakitori. Drinks include sake and Japanese beers. Situated on the edge of Nahalat Shiva.

Te'enim $$$

12 Emile Botta St, 94109 ***Tel*** *(02) 625 1967* ***Map*** *1 B4*

Beautifully situated at the northern edge of Yemin Moshe, in an old stone building known as Beit HaKonfederatzia, this small place is one of Jerusalem's oldest, and best, vegetarian restaurants. The decor is modern, with Armenian ceramic highlights. Diners enjoy a superb panorama of the walls of the Old City and Mount Zion.

Terasa $$$

2 Naomi St, Abu Tor, 93552 ***Tel*** *(02) 671 9796*

A classy, contemporary restaurant that has rave reviews, Terasa is situated in elegant stone pavilion on the Sherover Promenade. It offers a superb panorama over southeast Jerusalem, towards the Dead Sea and the mountains of Jordan. Food is dairy and Mediterranean in style; desserts are particularly good. Situated off Hebron Road.

Village Green $$$

33 Jaffa St, 94221 ***Tel*** *(02) 625 3065* ***Map*** *1 A3*

In the low-rise,19th-century Nahalat Shiva quarter, this veteran vegetarian restaurant serves up everything from miso soup and Greek salad to quiches, ratatouille, lasagna and tofu dishes. Culinary inspiration comes from Europe, Africa, the Middle East and the Mediterranean Basin. Dessert options include fresh, home-made cakes.

Adom $$$$

31 Jaffa St, 94221 ***Tel*** *(02) 624 6242* ***Map*** *1 A3*

On the 19th-century Feingold Courtyard, this restaurant and wine bar serves meat, fish, seafood and vegetable dishes in the traditions of France and Belgium, with light Israeli touches. The daily specials are based on seasonal products fresh from the market. Great selection of wines and beers. Good value business lunch specials.

Arabesque $$$$

23 Nablus Rd, 97200 ***Tel*** *(02) 627 9777* ***Map*** *1 C1*

The elegant house restaurant of the legendary American Colony Hotel serves a fine selection of hearty, traditional Arab dishes, some based on lamb, as well as European cuisine and, often, a few off-beat surprises. The wine cellar is excellent and the Saturday lunch buffet is legendary. Turkish-style courtyard and lovely gardens.

Arcadia $$$$

10 Agripas St, 94301 ***Tel*** *(02) 624 9138*

One of Israel's most talked-about restaurants, Arcadia is next to Mahane Yehuda market and its super-fresh ingredients. French and Mediterranean traditions are skillfully brought together with dishes from the Jerusalem-Sepharadi tradition and the chef's family's native Iraq to produce cuisine that is uniquely Israeli. Reservations advisable.

Cavalier $$$$

1 Ben Sira St, 94181 ***Tel*** *(02) 624 2945* ***Map*** *1 A3*

This up-market French bistro and bar, in Nahalat Shiva, offers classic French cuisine as well as Mediterranean-influenced dishes, all made with only the freshest ingredients and presented with supreme elegance. Dishes include entrecôte in pepper and cream sauce and chocolate volcano dessert. Good deals between noon and 3:30pm.

Scala Chef Kitchen & Bar $$$$

7 King David St, 94101 ***Tel*** *(02) 621 2030* ***Map*** *1 B4*

Mouthwatering original recipes are served in an informal yet elegant setting at the Scala. All dishes are prepared using the freshest local ingredients. The menu changes regularly but favourites include onion layers stuffed with lamb and served with black lentils, root vegetables, tahini and date honey, and baked fish with chestnuts and artichokes.

FURTHER AFIELD

Abu Shukri $

4 Mahmoud Rashid St, Abu Ghosh, 90845 ***Tel*** *(02) 533 4963*

About 10 km (6 miles) west of Jerusalem (along the highway to Tel Aviv) in Abu Ghosh, near the top of the hill, this lively, informal restaurant is renowned for its houmous. The establishment has splendid views over the valley below and of its great rival, another houmous eatery run by a cousin, and also called Abu Shukri.

Elvis American Diner $$$

Neve Ilan, 90850 ***Tel*** *(02) 534 1275*

A 1950s-style American diner dedicated to worship of the King of Rock 'n Roll is not what you would expect to find in the Judean Hills, 12 km (7 miles) west of Jerusalem. This proudly kitsch place was founded in the 1970s by a dedicated local Elvis fan and serves both American and Middle Eastern food. The sound track, though, is pure Elvis.

Key to Price Guide *see p272* **Key to Symbols** *see back cover flap*

THE COAST AND GALILEE

AKKO Humous Sa'eid $

Market, Old City ***Tel*** *(04) 991 3945* ***Map*** *B2*

A perennial contender for the title of "Israel's best houmous restaurant", this small, unpretentious place has fast, efficient service, incredibly reasonable prices and houmous that melts in your mouth. Situated in the heart of the Old City market – just ask anyone for directions. Open only for breakfast and lunch, from 6am to 2:30pm.

AKKO Uri Buri $$$

Lighthouse Square, Old City, 24713 ***Tel*** *(04) 955 2212* ***Map*** *B2*

Considered to be one of the best places in Israel for fish and seafood, this restaurant has attentive, personal service and some unconventional menu items. The chef loves to serve meals based on lots of different dishes, with everyone at the table sharing them. Regulars say the daily special, whatever it is, is almost always a good bet.

BETH SHE'AN Herb Farm on Mount Gilboa $$$

Hwy 667, Mount Gilboa, 19122 ***Tel*** *(04) 653 1093* ***Map*** *C2*

The Mediterranean and Israeli dishes at this country-style restaurant receive consistently excellent reviews. The fresh mountain air and the panoramic views of Mount Gilboa add to the charm of the place. Specialities include salads, home-made bread and pumpkin soup with apples and sour cream. Situated 10 km (6 miles) southeast of Afula on Hwy 667.

CAESAREA Pundak HaTzalbanim $$$

Old City, 30889 ***Tel*** *(04) 636 1679* ***Map*** *B2*

Overlooking the Mediterranean and Caesarea's ancient port, this classy place serves Mediterranean-style fish and seafood, such as crab in white wine and garlic sauce. The menu also has chicken and meat dishes, and for dessert, *crème brulée*. The attentive service and breathtaking views make this the perfect place for a romantic meal.

GOLAN HEIGHTS Mis'edet HaShalom $

Southern entrance to Mas'adeh, 12435 ***Tel*** *(04) 687 0359* ***Map*** *C1*

In one of the four Druze villages on the Golan Heights, this restaurant serves excellent Druze cooking. Options include salads (such as cabbage seasoned with the spice sumac), sour *labaneh* cheese, houmous, sesame-coated *falafel*, soups, grilled meats and fish. Traditional desserts are available here or at the nearby Abu Zayd sweet shop.

HAIFA Falafel HaZekenim $

18 HaWadi St, Wadi Nisnas, 33044 ***Tel*** *(04) 851 4959* ***Map*** *B2*

This veteran establishment, on a lively street in the mainly Arab Wadi Nisnas neighbourhood (four blocks southeast of the German Colony), serves what some say is the best *falafel* in the country. Made according to a secret recipe, the fried green chickpea balls are crispy and always fresh. Guests are greeted with a *falafel* ball dipped in *tahina*.

HAIFA Shwarma Hazan $

140 Jaffa Rd, 35252 ***Tel*** *(04) 855 8075* ***Map*** *B2*

Confirming Haifa's position as a quality leader in the Israeli street food scene, this place is acclaimed by many as serving nothing less than the best *shwarma* in Israel. Situated on the main thoroughfare of the flat, sea-level part of the city, a few blocks south of Rambam Hospital and just a block from the Commonwealth Military Cemetery.

HAIFA Duzan $$$

35 Ben Gurion Ave, German Colony, 35021 ***Tel*** *(04) 852 5444* ***Map*** *B2*

East meets West in the form of delicious cuisine at this attractive restaurant and bar, ensconced in a German Templar house built in 1870. Menu items come from Lebanon (*kubbe*, *sambusak*, stuffed grape leaves), Italy and France. The interior mixes modern design with antique furnishings and a colourful tiled floor. There's a tree-shaded patio.

HAIFA Fattoush $$$

38 Ben Gurion Ave, German Colony, 35023 ***Tel*** *(04) 852 4930* ***Map*** *B2*

The specialities at this restaurant, a favourite of Haifa's Jewish and Arab elite, include Lebanese *fatoush* salad (fried pieces of pitta, cucumbers and tomatoes seasoned with olive oil, lemon and sumac) and, for dessert, *knafeh*. Diners can sit outside or on Damascene silk couches along the walls of a barrel-vaulted, Oriental-style chamber.

KIRYAT SHEMONA Focaccia Bar $$

Gan HaTzafon, Hwy 99, near Kibbutz HaGoshrim ***Tel*** *(04) 690 4474* ***Map*** *C1*

Very popular with locals, this eatery also attracts visitors from around the country, especially after they have spent the day exploring the Galilee Panhandle and the Golan. Focaccia Bar has a wide selection of tasty, reasonably-priced dishes of generous proportions, including salads, juicy steaks, fish, seafood and pizza. A good choice for families.

KIRYAT SHEMONA Dag Al HaDan ✡ $$$

Just north of Kibbutz HaGoshrim, 11016 ***Tel*** *(04) 695 0225* ***Map*** *C1*

Situated 5 km (3 miles) east of Kiryat Shemona at the confluence of two major tributaries of the Jordan, the Dan and the Hatzbani, this dairy restaurant has specialized in freshwater and saltwater fish since 1986. Other popular options include sandwiches, salads, pasta and cakes. Surrounded by a lush forest of willow, fig and plane trees.

KIRYAT SHEMONA Dagei Dafna $$$

Kibbutz Dafna, 12235 **Tel** *(04) 694 1154* **Map** *C1*

Right next to a trout farm so you know what you are eating is fresh, this rural and very informal fish restaurant is situated on the Dan River, a short walk from the Horshat Tal park. It has seating inside a rough-hewn, wooden structure and outside on shaded picnic benches. The menu also includes salads, chicken and steak.

KIRYAT SHEMONA HaTachana $$$

1 HaRishonim St, Metulla, 10292 **Tel** *(04) 694 4810* **Map** *C1*

Situated 8 km (5 miles) north of Kiryat Shemona in the charming border village of Metulla, this romantic, if somewhat pricey, restaurant is named after a nearby waterfall. It receives excellent reviews for its attentive service and succulent meat dishes, especially the steaks. Reservations are recommended.

KIRYAT SHEMONA Nechalim $$$$

Gan HaTzafon, Hwy 99, near Kibbutz HaGoshrim **Tel** *(04) 690 4875* **Map** *C1*

Revered by locals as one of the area's finest restaurants, this romantic, country-style venue delivers a truly first-rate dining experience on the banks of a tributary of the Jordan. The menu is Italian- and French-influenced and the speciality is fresh fish, but seafood and meat dishes are also served. Surrounded by rich vegetation and delightful views.

NAZARETH Diana $$$

51 Paul VI St, 16224 **Tel** *(04) 657 2919* **Map** *B2*

The most famous restaurant in Nazareth, this unpretentious, white-tablecloth place is known for its meat dishes, which range from lamb chops to Arab-style kebab and *shishlik*, and for its *meze* salads, including *tabouleh*, houmous and *fatoush*. Also on offer are steaks, fish, seafood and several dozen sorts of wine.

ROSH PINA Auberge Shulamit $$$

David Shuv St, 12000 **Tel** *(04) 693 1485* **Map** *C2*

A sophisticated, country-style restaurant in the charming Galilee village of Rosh Pina. Specialities include stuffed vine leaves, seasonal soups, home-smoked goose breast, shrimp in Roquefort sauce, buffalo wings, sautéed trout and *filet mignon*. Outstanding *tarte Tatin* is a good dessert choice. Perfect for a romantic dinner. Reservations recommended.

ROSH PINA Doris Katzavim $$$

Main road, Rosh Pina, 12000 **Tel** *(04) 680 1313* **Map** *C2*

After hard a day's Galilee or Golan hiking, this is a good place for a hearty, meaty meal. Specialities, many made with Golan Heights-grown beef, include steaks (New York, Porter House), lamb chops and hamburgers, all of generous proportions. Main courses come with a selection of *meze* salads.

ROSH PINA Pina BaRosh $$$

8 HeChalutzim St, 12000 **Tel** *(04) 693 7028* **Map** *C2*

This rustic, stone-built restaurant, just a short stroll from Rosh Pina's famous art galleries, affords panoramic views of the Hula Valley, the Golan Heights and often-snow-capped Mount Hermon. The French-inspired onion soup, hen-on-rice with lentils, fish dishes, entrecôte and pasta all get excellent reviews, as does the personalized service.

SEA OF GALILEE Ein Camonim $$

Hwy 85, 10 km west of Amiad Junction, 20109 **Tel** *(04) 698 9680* **Map** *C2*

Located on a family-run dairy farm in the rugged hills and olive trees northwest of Tabkha is this very rustic, vegetarian eatery. It is known for its fresh, farm-grown products, including goat's cheeses, olive oil and ice cream, and for its all-you-can-eat cheese meals, served with lemonade and red or white wine. Cosy fireplace in winter.

SEA OF GALILEE Ein Gev Fish Restaurant $$$

Kibbutz Ein Gev, 14940 **Tel** *(04) 665 8136* **Map** *C2*

On the eastern shore of the Sea of Galilee, this rather utilitarian restaurant is right on the water, near a swimming beach and a lakefront promenade. It has had a loyal following for decades thanks to the reasonable prices and excellent baked, fried and grilled fish, served with salad, French fries and pickles/gherkins. Also has pasta and quiche.

SEA OF GALILEE Vered HaGalil $$$

Hwy 90, Corazim Junction, 12928 **Tel** *(04) 693 5785* **Map** *C2*

On a family-run horse ranch that rents out rooms, this rustic, vaguely American-style restaurant, built of local boulders and wood beams, garners enthusiastic reviews. Specialities include juicy steaks, hamburgers, chicken dishes, salmon and aubergine lasagna. Children's meals available. Situated 5 km (3 miles) north of Capernaum.

SEA OF GALILEE Yarden $$$

Beit Gavriel, Tzemah, 15132 **Tel** *(04) 670 9302* **Map** *C2*

At the far southern tip of the Sea of Galilee, this non-meat restaurant, inside the stunning Beit Gavriel cultural centre, affords unsurpassed views of the sea and the Golan. Specializes in fresh fish and Italian dishes (pizza, pasta, lasagna) but also serves meal-sized salads, stuffed mushrooms, quiche and a good selection of classic desserts.

TEL AVIV Ashkara $

45 Yermiyahu St, 62594 **Tel** *(03) 546 4547* **Map** *B3*

Everyone has an opinion on where to find Tel Aviv's best houmous and houmous comparisons often arouse great passions, but Ashkara is certainly a contender. It is situated just a block from Park HaYarkon, Tel Aviv's "Central Park", where lawns, lakes and bike paths stretch along the Yarkon River eastwards from the old Tel Aviv Port.

Key to Price Guide *see p272* **Key to Symbols** *see back cover flap*

TEL AVIV Iceberg

$

108 Ben Yehuda St, 63401 ***Tel*** *(03) 522 5025* ***Map*** *B3*

This ice cream parlour serves the city's best ice cream, sorbet and frozen yogurt. Tel Aviv has some fine gelaterias, but Iceberg's products, made on the premises from all-natural ingredients, are in a class of their own. Photos of huge, natural chunks of ice adorn the walls.

TEL AVIV Aboulafia Bakery

$$

7 Yefet St, Jaffa, 68028 ***Tel*** *(03) 681 2334* ***Map*** *B3*

A visit to this bakery is a classic Jaffa experience that has been enjoyed for decades – queuing on the pavement at the high glass counters and choosing fresh pittas and sesame rolls topped, or filled, with *za'atar*, olives, a fried egg or cheese. Purchases are best enjoyed as a picnic in nearby Old Jaffa, overlooking the sea or in the hilltop park.

TEL AVIV Elimelech

$$

35 Wolfson St, Florentine, 66528 ***Tel*** *(03) 681 4545* ***Map*** *B3*

This quaint restaurant, in the rundown but lively Florentine district of south Tel Aviv, serves traditional Eastern European Jewish food, including chopped liver, chicken soup with *kneidelach* (matza balls), steamed cabbage and schnitzel. Traditional *cholent* is the big hit on Saturday. Also serves excellent on-tap beer.

TEL AVIV Frida Hecht

$$

20 Ben Yehuda St, 63802 ***Tel*** *(03) 620 1471* ***Map*** *B3*

Once a home delivery service for ethnic Jewish food, this is now an informal, cafeteria-style restaurant with good prices and some excellent home-style cooking. Regulars recommend the lovingly made *gefilte* fish, meatballs, *madjadra* (rice with lentils) and *mafrum* (Libyan-style potatoes stuffed with meat). Open until 7pm, closed Sunday.

TEL AVIV Lehem Erez

$$

52 Ibn Gabirol St, 64361 ***Tel*** *(03) 696 9381* ***Map*** *B3*

A Tel Aviv institution, this popular place is the original venue of what is now a growing chain. It specializes in gourmet sandwiches, some of them with an unexpected fusion of flavours, and excellent, fresh salads. Lehem Erez is also a good place for breakfast. It sits on a main avenue that has become Tel Aviv's hottest café strip.

TEL AVIV Pinxox Tapas

$$

57 Nahalat Binyamin St, 65163 ***Tel*** *(03) 566 5505* ***Map*** *B3*

In a 1930s Bauhaus-style building in the historic Nahalat Binyamin area, this very civilized restaurant has elegant table settings, a sleek wooden bar and some surprising artwork on the walls. Fish and seafood, tapas and a few meat dishes, are prepared with a distinct French accent and served with panache.

TEL AVIV Barbunia

$$$

192 Ben Yehuda St, 63471 ***Tel*** *(03) 524 0961* ***Map*** *B3*

A simple, immaculate restaurant, Barbunia is small but popular and serves excellent fresh fish at reasonable prices. The service is quick and professional. The restaurant is situated two blocks inland from the Hilton Hotel tower and from a cliff that overlooks the beach. Barbunia's bar is just across the street.

TEL AVIV Boya

$$$

3 HaTa'arukha St, Tel Aviv Port, 63509 ***Tel*** *(03) 544 6166* ***Map*** *B3*

Situated in the north-western corner of the old Tel Aviv Port, one of the city's major dining and nightlife districts, this chic restaurant has an outside bar so close to the Mediterranean that you could fish while eating your tapas, focaccia, linguini, steak or *Tarte Tatin*. Perfect for a romantic snack or drink, especially as the sun sets.

TEL AVIV Deca

$$$

10 HaTa'asiya St, 64739 ***Tel*** *(03) 562 9900* ***Map*** *B3*

Elegant decor and gourmet fish and dairy dishes make this delightful restaurant worth visiting. Expect Mediterranean flavours and fresh ingredients, such as grilled sea bass with quinoa ragout and beetroot salad, or grey mullet in a red wine sauce. It's set in the eastern industrial zone, close to many of Tel Aviv's biggest clubs.

TEL AVIV Il Pastaio

$$$

27 Ibn Gabirol St, 64078 ***Tel*** *(03) 525 1166* ***Map*** *B3*

Walk into this restaurant and you will feel almost like you are in Italy. The home-made pasta, lasagna and risotto with porcini mushrooms garner rave reviews. The tiramisu, too, is heavenly. Perfect for a long, slow, delicious meal. Open from noon to 3:30pm and 7 to 11pm; closed Sunday evenings and Saturdays.

TEL AVIV Kyoto

$$$

7 Shenkar St, Herzliya Pituach, 46725 ***Tel*** *(09) 958 7770* ***Map*** *B3*

On a stylish street lined with trendy restaurants, packed at lunchtime with Israel's high-tech elite, this Japanese restaurant and sushi bar is known for its modern, Japanese-inspired decor, attentive service and professionally prepared cuisine. Tuna blue laguna (tuna braised on the outside, raw on the inside) is highly recommended.

TEL AVIV Maganda

$$$

26 Rabi Meir St, Kerem ha-Teymanim, 65605 ***Tel*** *(03) 517 9990* ***Map*** *B3*

On a narrow street in the old Yemenite quarter, very near the bustling Carmel Market, this Middle Eastern meat restaurant is friendly and down-to-earth. It specializes in carnivorous treats such as grilled steak, *shishlik* and kebabs. Meals begin with a big selection of *meze*, stuffed vegetables and Moroccan-style, meat-filled "cigars".

TEL AVIV Margaret Tayar $$$

HaAliya HaSheniya Quay, Jaffa, 68128 **Tel** *(03) 682 4741* **Map** *B3*

Authentic Tunisian, Libyan and Mediterranean cuisine is what keeps bringing people back to this unpretentious, if somewhat pricey, place on the quay below Jaffa's Old City. The service is not quick but the specialities – fish, couscous with mutton and stuffed grape leaves – are very tasty indeed.

TEL AVIV Moon $$$

58 Bugrashov St, 63145 **Tel** *(03) 629 1155* **Map** *B3*

On a street that leads to the sea and is home to a number of relaxed cafés, this sushi bar has sleek, modern decor and a conveyor belt for the transport of raw fish delicacies. Prices are reasonable and quality is high, attracting a loyal following. *Yakitori* and *tempura* are also on offer. Good-value business lunches from noon to 6pm, except Saturday.

TEL AVIV Nanouchka $$$

28 Lilienblum St, 65133 **Tel** *(03) 516 2254* **Map** *B3*

Elegantly and very comfortably furnished, with one corner devoted to low seats with huge cushions, this restaurant is an excellent place to sample the little-known cuisine of the Caucasus nation of Georgia. Favourites include *badridjani* (aubergine stuffed with nuts), *lubio* (thick, sour bean soup) and *khachapuri* (cheese-filled pastries).

TEL AVIV Orna v'Ela $$$

33 Sheinkin St, 65232 **Tel** *(03) 620 4753* **Map** *B3*

This creative, Israeli-style café and restaurant, long a fixture on Tel Aviv's most Bohemian street, serves both home-style and gourmet dishes. Favourites range from goat kebab and pumpkin *kubbe* to pasta and gnocchi with mozzarella and parmesan. Breads are baked fresh every morning. Delicious desserts.

TEL AVIV Susannah $$$

9 Shabazi St, 65150 **Tel** *(03) 517 7580* **Map** *B3*

In the increasingly chic 19th-century Neve Tzedek district, this café-restaurant is across the street from the Susan Dallal Cultural Centre, the city's premier dance venue. Mediterranean and home-style Israeli specialities, served on a shaded balcony, include hearty soups, generous salads, *kubbe*, stuffed vegetables and grilled meats. Great breakfasts.

TEL AVIV Unami $$$

18 HaArba'a St, 64739 **Tel** *(03) 562 1172* **Map** *B3*

Situated on a street lined with trendy, excellent restaurants, this place is considered by local connoisseurs to be one of the city's finest purveyors of Japanese cuisine. Amid elegant surroundings, the outstanding dishes served here include a huge selection of sushi and sashimi. There's also a bar.

TEL AVIV Brasserie $$$$

70 Ibn Gabirol St, 64952 **Tel** *(03) 696 7111* **Map** *B3*

Dining at this café-bistro, facing Rabin Square, is like a quick trip to Paris. The excellent, traditional French cuisine includes oysters (a rare treat in Israel), *bouillabaisse*, juicy pepper steak, *coq au vin* and *cassoulet* (every Saturday). Reservations are recommended in the evening. Friday brunch is served from 7am to 5pm. Open 24 hours a day.

TEL AVIV Brew House $$$$

11 Rothschild Blvd, 66881 **Tel** *(03) 516 8666* **Map** *B3*

This micro-brewery, with bulbous copper brewing tanks as the centrepiece, has the sort of warm, beer-infused atmosphere and meaty menu selection that you would expect to find in Düsseldorf or Stuttgart. Main course options include steak, spare ribs, bratwurst, buffalo wings, chicken breast in BBQ sauce, fish and seafood.

TEL AVIV Mul Yam $$$$

Hangar 24, Tel Aviv Port, 63506 **Tel** *(03) 546 9920* **Map** *B3*

Israeli restaurants do not get any finer, more exclusive or pricier than this world-class seafood and fish place, in the midst of some of the city's most fashionable pubs, restaurants and nightspots. Specialities include Breton oysters, beef *carpaccio* with Jerusalem artichoke and asparagus, and various shrimp and lobster dishes. Incredible wine list.

TEL AVIV Yo'ezer Bar Yayin $$$$

2 Yo'ezer Ish HaBira St, Jaffa, 68027 **Tel** *(03) 683 9115* **Map** *B3*

Situated just outside Jaffa's picturesque Old City, around the corner from the Clock Tower, this very romantic wine bar and restaurant serves outstanding French-style delicacies such as oysters, beef *carpaccio*, salmon fillet and *boeuf bourguignon*. Exceptional selection of wines from Burgundy, Bordeaux, Tuscany and around the world.

THE DEAD SEA AND THE NEGEV DESERT

BETHLEHEM Al-Atlal $$$

Manger St, Paradise Hotel **Tel** *(02) 274 4542* **Map** *B3*

The sweet smell of *nargila* (water pipe) smoke often wafts through this Arab-style restaurant, decorated with Bedouin-inspired furnishings. Specialities include *gedra* (lamb and rice with yogurt sauce), 10 kinds of salad and grilled meats. Generally open only on weekends; reservations are recommended. Live music Saturday evening.

Key to Price Guide *see p272* **Key to Symbols** *see back cover flap*

EILAT Last Refuge $$$
Coral Beach, 88000 ***Tel*** *(08) 637 3627* ***Map*** *B7*

Several kilometres south of the city centre, this restaurant has some of the best fish and seafood in town. Seating is either inside, in a dining room decorated with old nautical equipment reminiscent of New England, or outside (except in winter). The calamari and *coquilles St Jacques* are especially good.

EILAT Chao-Phya $$$
Southern Beach, 88000 ***Tel*** *(08) 636 0360* ***Map*** *B7*

About 5 km (3 miles) south of town, inside the Orchid Hotel complex, this romantic Thai restaurant occupies a soaring, all-wood building brought over from Thailand, which is where the staff are from too. A great place to dine on delicious, spicy cuisine from another place and time, but with great views of the Gulf of Aqaba.

EILAT Pastory $$$
7 Tarshish St, 88000 ***Tel*** *(08) 634 5111* ***Map*** *B7*

A little north of the main beach, this well-regarded Italian restaurant has slightly overdone rustic Italian decor and a kitchen area that is visible to diners. Specialities include entrecôte, pasta with shrimp sauce and delicious desserts such as *tiramisu*. The pasta is fresh and home-made, as are the Tuscan-style sauces.

EILAT Brasserie $$$$
North Beach, 88000 ***Tel*** *(08) 636 3444* ***Map*** *B7*

Inside the King Solomon Hotel, this kosher establishment, with sparkling glasses on white tablecloths, is an ideal retreat for lovers of classic French cuisine, although influences from Italy and East Asia are also in evidence. Specialities range from beef Wellington and roasted goose with potato purée to fish and vegetable dishes.

JERICHO Al-Rawada $$
Ket f'il Wad neighbourhood ***Tel*** *(02) 232 2555* ***Map*** *C3*

This attractive, family-run garden restaurant, is hidden in a grove of citrus and palm trees located off the main road, a little south of the centre of town. The salads and other starters, grilled meat dishes and freshly-squeezed lemon drink are excellent, as is the verdant, tree-shaded setting. Attentive service. Open from 8am until late afternoon.

PETRA AND WESTERN JORDAN

AMMAN Hashem $
Opposite Cliff Hotel, Downtown ***Tel*** *(06) 463 6440* ***Map*** *C3*

Founded in the 1920s, this no-nonsense 24-hour budget restaurant is an Amman institution, packed with locals. Only two dishes are served – houmous and *fuul* (hot beans), both with flat bread – although you can pick up a bag of *falafel* balls from the stand next door. Wash it down with a glass of scalding hot, milkless, sweet tea.

AMMAN Tarweea $
Opposite KFC, Shmeisani ***Tel*** *(06) 569 1000* ***Map*** *C3*

A pleasant, quiet, budget-priced Arabic restaurant tucked away off the main street in this bustling West Amman neighbourhood, with no sign in English (it is attached to the Haya Cultural Centre). The dining area is open, airy and spacious – an unusual setting to try Arabic *meze* and grills, fresh-baked *manaqeesh* bread and stuffed *falafel*.

AMMAN Champions $$
At the Marriott Hotel, Issam al-Ajlouni St, Shmeisani ***Tel*** *(06) 560 7607* ***Map*** *C3*

Amman has the full range familiar Western fast-food outlets, but you would do better at Champions. Here, in a brisk and breezy US-style sports bar ambience, with TVs showing live sports events, you can tuck into high-quality burgers with fries, nachos, salads and other fast food offerings, in huge portions. Also at the Marriott on the Dead Sea.

AMMAN Reem al-Bawadi $$
Near Waha Circle, Tlaa al-Ali, West Amman ***Tel*** *(06) 551 5419* ***Map*** *C3*

An excellent choice for top-notch Arabic cuisine in an authentic, informal setting, much favoured by Jordanian families and business-people. Seating is either in the vast interior, or – in warmer months – outside in a gigantic Bedouin-style tent pitched in the gardens. Service is welcoming, accommodating and discreet.

AMMAN Blue Fig $$$
Prince Hashem bin al-Hussein St, Abdoun, 11844 ***Tel*** *(06) 592 8800* ***Map*** *C3*

One of Amman's hippest places to hang out, located on the fringes of the city proper. The interior is all subtle lighting and contemporary design, with chic, wealthy Ammanis enjoying the international fare. The wraps, salads and light bites are all done with panache. A fascinating glimpse of Jordan's "beautiful people".

AMMAN Noodasia $$$
Abdoun Circle ***Tel*** *(06) 593 6999* ***Map*** *C3*

Highly acclaimed Asian restaurant, in the heart of the Abdoun buzz. The building design is smart and contemporary, with exceptionally well-prepared and presented food to match, ranging from Szechuan staples to Thai dishes and sushi, all very authentic. Afterwards, roam the bars and cafés of Abdoun for something sweet or a nightcap.

AMMAN Wild Jordan $$$

Othman bin Affan St, off Rainbow St, below 1st Circle, Jabal Amman **Tel** *(06) 463 3542* **Map** *C3*

A wonderful wholefood café/restaurant attached to the offices of the Royal Society for the Conservation of Nature (RSCN). The building is perched on a hillside overlooking Downtown Amman, with spectacular views from its open terrace. All the food is organically produced and sourced locally – salads, wraps, smoothies and minty iced lemonade.

AMMAN Fakhr el-Din $$$$

40 Taha Hussein St, between 1st and 2nd circles, Jabal Amman **Tel** *(06) 465 2399* **Map** *C3*

Quite simply one of Jordan's loveliest restaurants, and one of its best. An elegant 1920s town house, on a quiet residential street, has been beautifully restored and converted into a formal Arabic restaurant of the highest quality. The *meze* and grills are impeccable, as is the service. Reservations are essential: in summer, book a table on the terrace.

AMMAN Romero $$$$

Off 3rd Circle, Jabal Amman **Tel** *(06) 464 4227* **Map** *C3*

Perhaps Amman's finest Italian restaurant, tucked away down a leafy side street opposite the InterContinental Hotel. The ambience is perfect, with bow-tied waiters gliding noiselessly around a cosy, tasteful dining room, and the food is exquisite, using the freshest of ingredients. Upstairs is the informal Living Room, for light bites and lounging.

AMMAN Tannoureen $$$$

Shatt al-Arab St, Umm Uthaina, West Amman **Tel** *(06) 551 5987* **Map** *C3*

Vying for the title of Jordan's best restaurant, this is an outstanding place to sample the finest of Lebanese cuisine in an elegant, formal setting. The *meze* are exceptionally good, the mains cover the range of grills and fish and the desserts, if you make it that far, are out of this world. Service is warm, smooth and courteous.

AQABA Ali Baba $$

Princess Haya Circle **Tel** *(03) 201 3901* **Map** *B7*

Situated in a great location, on a bustling corner overlooking Aqaba's main Princess Haya Circle, Ali Baba is a long-established, informal local restaurant, generally packed with both Aqabawis and tourists sampling Lebanese cuisine, chatting and watching the town go by. The fish and seafood are notably good and the service is genial.

AQABA Floka $$

Al Nahda St, 77110 **Tel** *(03) 203 0860* **Map** *B7*

In the centre of town, the street behind the Aqaba Gulf hotel is lined with interesting cafés and restaurants. Alongside the Alcazar Hotel stands Floka, a great little fish and seafood restaurant. Choose from the catch of the day or pick Arabic specialities from the extensive menu.

AQABA Bourj al-Hamam $$$$

At the InterContinental Hotel, King Hussein St, North Beach 77110 **Tel** *(03) 209 2222* **Map** *B7*

Aqaba's big hotels all have excellent restaurants, but the InterContinental's Bourj al-Hamam is exceptional, offering exquisite Lebanese specialities alongside the hotel's pool and palms, looking out to the beach and the Red Sea. The *meze* are superb, as are the fish specialities. The restaurant in the InterContinental in Amman is as good.

AQABA Royal Yacht Club $$$$

Off the main corniche, 77110 **Tel** *(03) 202 2404* **Map** *B7*

Off the main Princess Haya Circle, a side-road (with staffed gates) leads down to the marina, where you will find this wonderful restaurant. Catered by Romero of Amman, it offers a range of Mediterranean cuisine, from fish and wood-fired pizza to salads and Arabic *meze*, in a formal, airy space on the waterfront, with spectacular views.

DEAD SEA Mövenpick $$$

Dead Sea Rd, Sweimeh, 11180 **Tel** *(05) 356 1111* **Map** *C4*

Dining at the Dead Sea is a case of picking a hotel. The Marriott and the Kempinski both have excellent restaurants and terrace cafés, but the Mövenpick is perhaps the most atmospheric and offers exceptional quality. Try the lavish Mediterranean buffets at Saraya, in the main building, or Luigi's, a great little Italian on the resort's "village square".

MADABA Haret Jdoudna $$$

King Talal St, 11181 **Tel** *(05) 324 8650* **Map** *C4*

On a journey north or south through Jordan, or a trip out of Amman, it is worth the detour to Madaba to try this splendid, traditional Arabic restaurant, occupying a historic building in the town's old quarter. The setting is perfect, with tables dotted around an old courtyard home, and the food – *meze*, grills, fresh-baked bread – is exquisite.

PETRA Mövenpick $$$$

Wadi Musa, 218101 **Tel** *(03) 215 7111* **Map** *C5*

Dining in Wadi Musa is mostly quite ordinary. For something special, head to the Mövenpick: on one side of their beautiful internal atrium is the Saraya restaurant, offering extensive buffets; on another is the Maqaad bar; and opposite is the formal Liwan restaurant, Wadi Musa's best, with a high-priced menu of Mediterranean specialities.

UMM QAIS Resthouse $$

Umm Qais **Tel** *(02) 750 0555 or book through Romero in Amman (06) 464 4228* **Map** *C2*

Umm Qais – once the Roman city of Gadara – is located in the northernmost corner of Jordan. Within the ruins, an Ottoman school has been beautifully converted into a splendid restaurant. The menu is simple – salads, *meze*, grills, pasta – but the location is exceptional, on a high plateau overlooking the Sea of Galilee and Golan Heights.

Key to Price Guide *see p272* **Key to Symbols** *see back cover flap*

Choosing a Restaurant in the Red Sea and Sinai

The restaurants listed on this page have been selected for their value, good food, atmosphere and interesting location within the Red Sea and Sinai area. For key to symbols and map references, see back endpaper.

PRICE CATEGORIES
Prices are based on a three-course meal for one including coffee, tax and service.

£ Under LE 30
££ LE 30–LE 50
£££ LE 50–LE 100
££££ LE 100–LE 150
£££££ Over LE 150

THE RED SEA AND SINAI

DAHAB INMO Divers' Home Restaurant £££

Al-Mashraba, Dahab ***Tel*** *(069) 3640 370* ***Map*** *F6*

Located right on the beach at Dahab, this restaurant forms part of the INMO Divers' Resort and is built to the same architectural style as the main building, with lots of arches and domes. It serves good Oriental, vegetarian and international cuisine, with its speciality being Egyptian buffets and drinks such as *Sahlab*.

DAHAB Nirvana Indian Restaurant £££

Nirvana Dive Center, Lighthouse, Dahab ***Tel*** *(061) 046 061* ***Map*** *F6*

Open for breakfast, lunch and dinner, the Nirvana offers a mouth-watering selection of fresh Indian dishes, prepared by Indian chefs using only the finest imported spices and ingredients. Food and drinks are served on the beach or the patio. Hotel guests enjoy a discount on their meals and have the option to eat on the first-floor deck with a sea view.

DAHAB Nesima Restaurant ££££

Mashraba, Dahab ***Tel*** *(069) 3640 320* ***Map*** *F6*

The Nesima is a cosy and intimate restaurant within the Nesima Hotel, which is renowned for its excellent diving centre. The restaurant serves international cuisine, along with Egyptian dishes such as *koshari* followed by traditional desserts. There is a rooftop bar.

NAAMA BAY Kokai Grill Room ££££

Ghazala Hotel, Naama Bay, Sharm el-Sheikh ***Tel*** *(069) 3600 150* ***Map*** *E7*

The Kokai Grill Room at the Ghazala Hotel offers the finest in Polynesian and Chinese cuisine in an elegant setting. Each evening the restaurant's team of chefs theatrically prepares the grilled dishes at table grills. The duck, spring rolls and rice dishes are also recommended.

NUWEIBA Oasis Restaurant £££

Nuweiba Resort, Nuweiba City ***Tel*** *(069) 3500 402* ***Map*** *F5*

The Leserena has a great view over the Nuweiba Resort's beachside gardens and pool area. The cuisine is largely Egyptian and classic international, with dishes like pizzas and pasta, fresh fruits, salads, vegetarian dishes and desserts in abundance.

SHARM EL-SHEIKH Al-Fanar Restaurant £££££

Ras Un Sid Beach, Sharm el-Sheikh ***Tel*** *(069) 3662 218* ***Map*** *E7*

Like many of the restaurants in Sharm el-Sheikh, the Al-Fanar serves Italian cuisine, but what makes this one stand out from the rest is that fresh produce from Italy is regularly used and its location right on the beach, below the lighthouse, provides an intimate setting. Good wine list.

SHARM EL-SHEIKH La Luna Restaurant £££££

Ritz-Carlton Resort, Om El Seed, Sharm el-Sheikh ***Tel*** *(069) 3661 919* ***Map*** *E7*

With an experienced Italian chef who excels in specialities such as calamari, potato *gnocchi* and home-made pasta, a visit to La Luna will be a memorable experience. A list of fine Italian wines and grappa is served. The restaurant has a luxurious feel and is located within the Ritz-Carlton Resort.

SHARM-EL-SHEIKH Safsafa £££££

Asia Mall, Sharm-el Sheikh ***Tel*** *(069) 3660 474* ***Map*** *F5*

The Safsafa restaurant serves well presented fresh fish dishes, along with a good selection of vegetarian meals, such as *Fuul* and *Taamiyya*. It can usually be found full of discerning diners. Located right on the waterside at Naama Bay, in the shopping centre, the restaurant is bright and welcoming.

TABA Castle Zaman Restaurant £££££

Nuweiba–Taba Road, Taba ***Tel*** *(069) 3501 234* ***Map*** *F5*

This impressive monument commands a dramatic mountainous view of four countries. Castle Zaman's speciality is slow-cooked food – some dishes are cooked for up to three hours, leaving guests time to enjoy the pool, have a massage, explore the underground treasure room or relax with a fresh cocktail at the bar. Not suitable for children.

SHOPS AND MARKETS

When it comes to shopping, the main attraction in Jerusalem is undoubtedly the souks, or bazaars, of the Old City. In comparison with the great bazaars of Istanbul or Cairo, Jerusalem's souks can seem small and overly touristy, but they still deserve exploration *(see pp148–9)*. The streets of the Old City away from the souks are also dotted with interesting small shops, handicraft centres, workshops and boutiques. Most other towns and cities throughout the Holy Land also have souks, with particularly good ones in Akko, Amman, Hebron and Nazareth. Anybody intending shopping in the souks must become acquainted with the art of bargaining. In contrast to the traditional nature of the souk, bigger centres such as Jerusalem, Tel Aviv and Amman, all possess modern shopping districts, as well as large American-style malls, filled with familiar brand names from the West.

Armenian ceramic tile

In Jordan, the major tourist sites such as Petra and Jerash have small clusters of tourist-oriented shops where, sometimes, you can find local handicrafts and products of interest.

A typical fruit and vegetable stall

OPENING HOURS

Throughout the Holy Land there are often no strictly defined opening hours; it depends on the individual proprietor. In general, however, except for food shops, which open quite early, business activity begins at roughly 9am. Some shops close from 1 to 4pm, but most remain open all day until around 7pm. In Jerusalem's Old City and elsewhere, the souks don't really get going until perhaps 10am and they close around sunset. Many shops and stalls in the souks are closed all day Sunday, as many of the shop owners are Christian, although others are Muslim and they stay closed on Friday instead. During the holy month of Ramadan, Muslim-owned shops throughout the Holy Land close 30 minutes to one hour before sunset.

All Jewish-owned businesses in Jerusalem and throughout Israel close from Friday afternoon to sunset on Saturday for Shabbat (Sabbath). These shops are also closed during Jewish holidays *(see pp36–9)*.

HOW TO PAY

Major credit cards, such as Visa, American Express and MasterCard, are accepted in almost all shops throughout Israel; travellers' cheques are not. In Jordan and Sinai, credit cards are less widely accepted. Only in top-end and mid-range hotels and international restaurants are cards usually accepted; in most places, you will have to pay in cash. It is usual to pay in the local currency (in Jordan and Sinai use of any other currency is illegal), but in Israel, if you are making a large purchase, it is possible to get a discount by paying in US dollars. This is because transactions made in a foreign currency are not subject to Israeli VAT.

VAT EXEMPTIONS

A wide range of goods in Israel is subject to a Value Added Tax (*Mam* in Hebrew) of 17 per cent. Tourists are entitled to a refund on this for any purchases amounting to over 400 shekels (about US$100). Make sure the shop you buy from has a VAT (or tax) refund sign displayed. You need to ask the sales assistant for a special invoice showing the VAT paid in both dollars and shekels. This is then presented at the VAT counter at the airport at the time of your departure. You must have the purchases with you to cross-check against the invoice. Queues at this

Examining the wares at an Old City souvenir shop

Malcha Kanyon Mall in Malcha, Jerusalem

counter can be very long, so get there with time to spare.

DEPARTMENT STORES AND SHOPPING MALLS

Israel has a rapidly growing number of large shopping centres and US-style out-of-town malls. Both are filled with standard mall-type outlets that sell everything from greetings cards to electronics items, most of which are imported from Europe and the United States. Jerusalem has several large malls, including one of the biggest in the country, the **Malcha Kanyon Mall**, out in the Malcha suburb of West Jerusalem. In the centre of the city, **Mamilla Alrov Quarter** is a high-end shopping strip with international and local stores as well as many attractive restaurants and cafés. Tel Aviv's biggest mall is the **Azrieli Centre**, in the base of three modern towers on the northeastern edge of town. More centrally located malls in Tel Aviv include the **Dizengoff Centre** on Dizengoff Street and the **Gan ha-Ir Shopping Centre** just north of Rabin Square.

As well as the shopping opportunities, Israel's malls are typically full of good, moderately priced restaurants, snack bars and cafés. Given that they are air-conditioned, they can be great places for pedestrians to escape from the often stifling heat outside.

Jordan's capital, Amman, has also succumbed to the mall craze. The city's biggest is **Mecca Mall**, out in the north-western suburbs, which also contains a food court, cinema and bowling alley. There's also the smaller but more centrally located **Abdoun Mall**.

MARKETS

In addition to the souks of Jerusalem's Old City, there are lots of good buys at the **Makhane Yehuda** market in modern West Jerusalem *(see p131)*. Tel Aviv has **Carmel Market** *(see p172)*, which operates every day except Saturday, and, also in the same neighbourhood, the Nakhalat Binyamin **craft market** *(see p172)*, held every Tuesday and Friday. In Jordan, Downtown Amman has several streets filled with colourful market shopping *(see p212)*.

BUYING ANTIQUES

In Jordan and Sinai it is forbidden to export any antique or archaeological find unless you have obtained special permission in advance. The border authorities are extremely thorough in their checks in this regard. On the other hand, in Jerusalem and Israel you may buy objects from excavations. For more details and for the addresses of some reputable dealers, see pages 148–9.

HOW TO BARGAIN

Buying and selling in the Middle East is traditionally a highly ritualized affair, in which bargaining is far more than just haggling for a cheap price. The aim of the exercise is to establish a fair price that both vendor and buyer are happy with. As part of the process, a shop owner may well invite you to have a cup of tea or coffee and may literally turn the place upside down to show you something; you should not feel obliged to buy because of this, it is common sales practice and all part of the ritual.

Bargaining, by the way, is not socially acceptable in city-centre shops, but it is unavoidable in the souks if you don't wish to pay greatly over the odds.

The way to go about it is that once you identify an article that interests you, especially an expensive one, be brave enough to offer half the price quoted by the shop owner. Don't be put off by any feigned indignation on the part of the shopkeeper and only raise your next offer by a small amount. Through offer and counter-offer you should arrive at a mutually agreeable price. If you don't reach a price you think is fair then simply say thank you and leave. Making to walk away often has the effect of bringing the price plummeting down.

Haggling over the price – time-consuming but essential to avoid paying over the odds

In theory, no one gets cheated because you, the buyer, have set the price yourself; it follows that you are happy with what you have agreed to pay, and the shopkeeper will certainly never sell at a loss.

Where to Shop in the Holy Land

Jerusalem's souks are the first place to look for many of the items produced in this region (for shopping in Jerusalem, see pages 148–9), but there is also plenty of other good shopping in the Holy Land. Tel Aviv is probably Israel's finest shopping city, with several malls and markets, and lots of great boutique stores on and off Dizengoff Street. Amman, in Jordan, has lots of great arts and crafts items, a lot of which can also be found at stores in the more popular tourist destinations such as Madaba, Petra and Jerash.

Jewish menorahs for sale in the Old City of Jerusalem

RELIGIOUS ARTICLES

For Christian religious items there are any number of shops in Jerusalem's Old City *(see p149)*. However, prices are generally lower in Bethlehem, which is where many of these items are made. One place worth visiting here is the **Holy Land Arts Museum** on Milk Grotto Street, which specializes in wooden objects with mother-of-pearl inlay, and inlaid metalwork (damascene).

For Judaica, visit the Jewish Quarter in Jerusalem's Old City and along central Ben Yehuda Street in Tel Aviv. Visit **Pninat-ha'kesef** in Tel Aviv for a wide selection of candlesticks and paintings.

CERAMICS

Jerusalem is the place for beautifully coloured Armenian ceramics, but there are other styles produced elsewhere in the region. **Beit el-Badawi** in Amman sells the designs of local craftspeople who work in both traditional and modern styles. Pieces incorporate Arab calligraphy.

Also in Amman, **Silsal Ceramics** is another good sales studio specializing in modern pottery.

For something really chic, visit **Blue Bandana** in northern Tel Aviv, which stocks a fine array of beautiful tableware, much of which is designed specially for the store.

TEXTILES AND RUGS

The shops and market in the centre of Ramallah are a good place to look for densely embroidered Palestinian textiles. Cushions and bags made from Bedouin textiles are found in most souvenir shops in Israel. Prices vary little, but for Bedouin rugs, you would do better to buy in Jordan. Madaba *(see p216)*, in particular, is famous for its colourful rugs. These can be bought around town, but one recommended place is **Madaba Oriental Gifts**, which is opposite St George's Church. **Shtihei Carmel** in Rehovot, near Tel Aviv, specializes in Carmel rugs.

JEWELLERY

Some of the region's most distinctive jewellery is made by the Bedouin. It is sold at the street markets of Nakhalat Binyamin *(see p172)* in Tel Aviv, in many of the boutiques in Jaffa and at the Thursday market in Beersheva.

For more contemporary pieces **Agas and Tamar** is an upmarket boutique selling exquisite own-designed, one-off pieces. Even if your budget doesn't stretch this far, it's a beautiful shop in one of Tel Aviv's most interesting neighbourhoods.

HEBRON GLASSWARE

In Jerusalem, the first three shops on the left-hand side of David Street, going from Jaffa Gate, have the best selection of glassware. However, much lower prices are offered in the souk at Hebron. At Madaba in Jordan, **Madaba Oriental Gifts** has a good range of Hebron glassware, often at prices even lower than those in Hebron.

COSMETICS

The Arab town of Nablus is famed for its olive-oil soap, available at almost any East Jerusalem grocer's and in the Old City souks, especially on Khan el-Zeit Street. In Galilee the soap is sold in many souvenir shops, particularly in Nazareth, but at higher prices.

The reputed health-giving properties of the Dead Sea are exploited in the

Craftsman hand-knotting the fringe of a rug

cosmetic products made by the two companies, Ahava and Mineral. These are sold at all well-stocked pharmacies and at the Duty-Free Shop at Ben Gurion airport. When visiting the Dead Sea, you can buy directly from the **Ahava Factory**, north of Ein Gedi. It is open daily, but closes at 2pm on Fridays. There is also an **Ahava** at the Hilton Tel Aviv and a major Ahava outlet at the Ein Bokek spa resort on the Dead Sea shore.

A range of Dead Sea products is also sold at a shop called **Holy Treasures**, opposite St George Church in the town of Madaba, Jordan.

SOUVENIRS

Sandals, bags and belts are good articles to buy throughout the Holy Land. Copperware is also a good buy, notably coffeepots and trays, often etched with arabesque patterns. A more exotic souvenir is a *nargileh*, or Arab water pipe. All of these can be found in Jerusalem and also in Amman, where a particularly good one-stop shopping opportunity is offerd by the **El-Alaydi Jordan Craft Centre**, which has a vast selection of locally produced items, including Hebron glassware, Palestinian embroidery and Bedouin tent accessories.

In Madaba in Jordan, there is a complex of excellent **craft shops** just north of the Madaba Museum, offering everything from textiles to jewellery to mosaics. At Petra, look out for the **Made In Jordan** shop, which is near the entry gate to the site, and which has top-quality locally made items, including camel-hair shawls and olive oil.

Water pipes, or nargilehs

Decorative bottles filled with coloured sand are popular Jordanian souvenirs, especially at Wadi Rum and Petra.

For a very different sort of souvenir, an extensive range of recordings of modern and traditional Jewish music can be found at **Tower Records** in Tel Aviv. Alternatively, the **Bauhaus Centre** in Tel Aviv has a gift shop selling miniature models of some of the city's landmark 1930s architecture in the International Modern, or Bauhaus, style *(see p171)*, as well as books and prints.

Making sand-filled bottles, Jordan's most prevalent souvenir

DIRECTORY

SHOPPING MALLS

Abdoun Mall
El-Hashimi St, Abdoun, Amman, Jordan.
Tel *(06) 5920296.*

Azrieli Center
132 Petach-Tikva Hwy, Tel Aviv. ***Tel*** *(03) 6081199.*

Dizengoff Centre
50 Dizengoff St, Tel Aviv.
Tel *(03) 621 2416.*

Gan ha-Ir Shopping Centre
71 Ibn Gabirol St, Tel Aviv.
Tel *(03) 527 9111.*

Malcha Kanyon Mall
Malcha, West Jerusalem.
Tel *(02) 679 1333.*

Mamilla Alrov Quarter
Tel *(02) 636 0000.*
www.alrovmamilla.com

Mecca Mall
Mekka el-Mukkaramah Rd, Amman, Jordan.
Tel *(06) 552 7945.*

RELIGIOUS ARTICLES

Holy Land Arts Museum
Milk Grotto St, Bethlehem.
Tel *(02) 274 4819.*
www.holylandarts museum.com

Pninat-ha'kesef 1/86 Ha'kishor St, Tel Aviv
Tel *(03) 518 1406.*
www.pninat-hakesef.ybay.co.il

CERAMICS

Beit el-Bawadi
Fawzi el-Qawoaji St, Amman, Jordan.
Tel *(06) 593 0070.*

Blue Bandana
52 Hei Beyar, Kikar ha-Medina, Tel Aviv.
Tel *(03) 602 1686.*

Silsal Ceramics
Innabeh St, North Abdoun, Amman, Jordan. ***Tel*** *(06) 593 1128.* **www**.silsal.com

TEXTILES AND RUGS

Madaba Oriental Gifts
Madaba, Jordan.

Shtihei Carmel
Bilu Center, Rehovot.
Tel *(08) 935 5557.*

JEWELLERY

Agas and Tamar
43 Shabazi St, Neve Tzedek, Tel Aviv.
Tel *(03) 516 8421.*

COSMETICS

Ahava Factory
Kibbutz Mitspe Shalem, Route 90, Dead Sea.
Tel *(02) 994 5100.*

Ahava
Tel Aviv Hilton, Independence Park, Tel Aviv.
Tel *(03) 520 2222.*

Holy Treasures
Talal St, Madaba, Jordan.
Tel *(05) 324 8481.*

SOUVENIRS

El-Alaydi Jordan Craft Centre
El-Kulliyah el-Islamiyah St, Jebel Amman, Amman, Jordan.
Tel *(06) 464 4555.*

Bauhaus Centre
99 Dizengoff St, Tel Aviv.
Tel *(03) 522 02459.*
www.bauhaus-center.com

Craft shops
Haret Jdoudna Complex, Talal St, Madaba, Jordan.
Tel *(05) 324 8650.*

Made In Jordan
Petra, Jordan.
Tel *(03) 215 6665.*
www.petramoon.com

Tower Records
1 Allenby St, Tel Aviv.
Tel *(03) 517 4044.*

What to Buy in the Holy Land

Fish pendants

Visitors on the lookout for unusual souvenirs, or the products of different cultures and ages, will certainly find something to their liking in Jerusalem, either in the souks and alleyways of the Old City, or in particular districts of the modern city. Some artifacts, such as pottery, brass and silver objects, Bedouin textiles and Arab jewellery, are sold throughout the Holy Land. However, in Jerusalem you will find an especially wide range of Jewish religious articles (while other places concentrate on Christian or Muslim items), and Armenian pottery.

Copper goblets

Firjan with spirit stove

Blue Hebron Glass
Most of this attractive glass, in shades of light blue and turquoise, is made to imitate Roman and Phoenician vessels. Some modern designs and full dinner services are also produced.

Copper- and Brassware
Copper plates, jugs, pots, trays and goblets, all usually engraved, are found everywhere. So, too, are traditional firjan (coffee pots) and large platters made of beaten brass.

Armenian Ceramics
The best-known decorative pottery is produced by the Armenian community, which has had a presence in Jerusalem since the 4th century (see pp106–7). *It is characterized by the abundant use of blue and yellow, and of floral motifs. The designs are usually intricate and painted on a white ground.*

Silver and Pewter Jewellery
The Yemenite tradition of silver filigree work has been extensively adopted by religious and secular jewellers in the Holy Land. Look out also for attractive, modern pewter jewellery set with semi-precious stones, as well as traditional blue glass-eye and khamsa (hand-shaped) lucky charms, popular with Arabs and Jews alike.

Olive-wood Objects
Crucifixes, rosaries, Nativity scenes and figures of Christ, the Virgin Mary and the saints carved in hard, light-coloured and attractively-grained olive wood make evocative souvenirs. The best come from the Bethlehem area.

Blue glass-eye pendants

Modern brooch

Silver *khamsa*

Olive-wood sculpture

Jewish Liturgical Articles

These often beautifully-made objects include the kippah (male skullcap), tallit (pure wool prayer shawl), kiddush (blessing) cup, besamim (spice-holder), mezuzah (prayer container hung at front doors) and shofar (ram's horn blown for Yom Kippur).

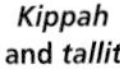

Kippah and *tallit*

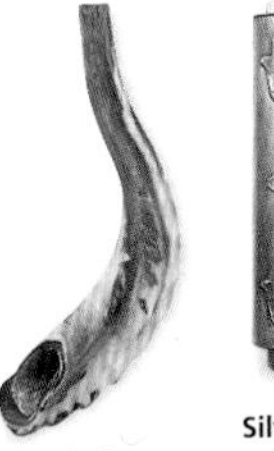

Shofar

Silver *mezuzah*

Silver *besamim*

Rugs and Fabrics

Robust and vividly coloured Bedouin rugs, cushions and bags made from the cloth formerly used as Bedouin saddle covers, and traditional, finely embroidered Palestinian dresses are popular buys.

Bedouin cushion covers

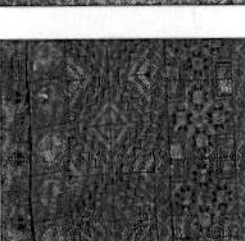

Bedouin fabrics

Palestinian fabrics

Ancient Household Articles and Coins

Reputable dealers in finds from archaeological sites will often have attractive basalt, earthenware and stone kitchen vessels, small terracotta amphorae and Roman and Phoenician glassware. Coins from many historical periods are fairly plentiful, but beware of fakes.

Beauty Products from the Dead Sea and Nablus

A vast range of creams, soap, salts and Dead Sea mud, using the mineral properties of the unique Dead Sea salt, is sold to alleviate skin conditions. Nablus soap, which has olive oil and less than two per cent caustic soda as its only ingredients, is cheap, fragrant and long-lasting, and is good for use in dry climates.

Nablus soap

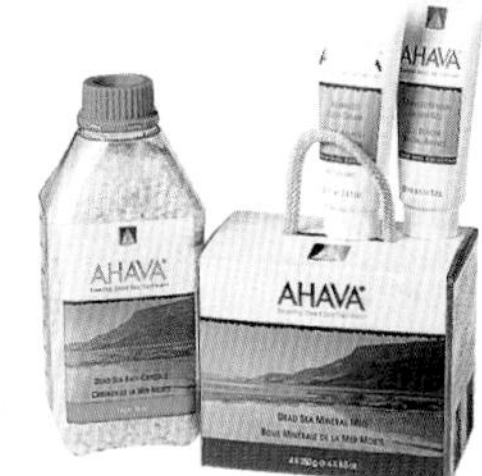

Dead Sea lotions

Local Delicacies

Specialist shops stocked with large sacks of nuts, dried fruits, pulses and dried vegetables are fascinating places to explore. They often sell spices, too. All these products make good buys as they are easy to carry, and keep well at home.

Dried apricots

Dried red peppers and aubergines

Chickpeas

Mulberries

Almonds

Pistachio nuts

ENTERTAINMENT IN THE HOLY LAND

While Jerusalem has its theatres, concert halls and cinematheque *(see pp150–51)*, Israel's real centre of entertainment lies some 60 km (37 miles) west in Tel Aviv. If Jerusalem is, as Israelis often say, the city where they pray and Haifa is the city where they work, then Tel Aviv is definitely where they play. High culture is catered for by a fine modern opera house, several theatres, and a busy dance and performing arts centre. Popular culture is supported by myriad bars, clubs and live music venues. Elsewhere, there is far less going on, although Jordan's capital Amman boasts several busy cultural centres and cinema complexes. Down on the Red Sea coast and in Sinai, entertainment is largely limited to bars and nightclubs.

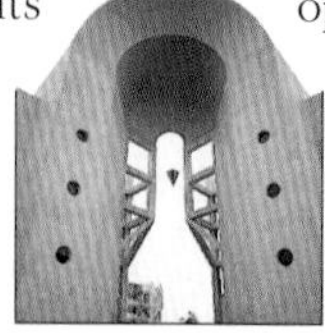

Entrance, Performing Arts Centre, Tel Aviv

The Israel Philharmonic Orchestra

INFORMATION

The *Jerusalem Post* and the English-language edition of the newspaper *Ha-Aretz*, both of which are available throughout Israel, carry daily entertainment listings. Both also have extensive cultural supplements on Fridays with detailed listings of events for the week to come. There is also an English-language listings magazine *Time Out Tel Aviv* published every two months and available free at certain bars and hotels. Tourist offices *(see p299)* also have abundant events magazines.

In Jordan, look out for the *Jordan Times* and the weekly *The Star*, or, when in Amman, visit **books@cafe**, an internet café-cum-bookshop, whose notice boards provide the best way of finding out what's on in the capital.

In Sinai, look out for the monthly *Egypt Today*, which carries what's on information.

CLASSICAL MUSIC

The Israel Philharmonic, one of the world's most prestigious orchestras, is based in Tel Aviv at the **Performing Arts Centre**. The neighbouring **Tel Aviv Museum of Art** also hosts regular chamber music concerts and other classical events in its Recanati Hall. Smaller venues include the **Felicia Blumenthal Centre** and **Einav Cultural Centre**, both of which host local and international classical muscians.

In the village of Ein Kerem *(see p138)* near Jerusalem, young musicians give free recitals of chamber music every Friday at noon from October to May at the Fountain of the Virgin in the **Targ Centre**.

Classical street musician

In Amman, there's large modern **Royal Cultural Centre**, which presents a varied programme of traditional Arabic music, theatre and dance. The **King Hussein Cultural Centre** also hosts occasional classical and Arab music performances.

OPERA

Tel Aviv's Performing Arts Centre is home to the **New Israeli Opera**, a world-class company, which puts on four or five new productions a year. The centre also frequently hosts visiting productions from Europe and America.

ROCK, JAZZ AND BLUES

Even in partying Tel Aviv, the live music scene is surprisingly disppointing. Local rock bands of variable quality perform most nights at **Goldstar Zappa**, **Ha-Bima Club**, and also at **Mike's Place**, which is a foreigner-friendly bar down on the seafront. **Benchmark** is a relative newcomer featuring live music in the bar-saturated area of Nakhalat Binyamin. For hardcore fans who are prepared to travel, **Barbie** mixes Israeli rock with Russian hard rock.

Cafe Henrietta, **Shablul Jaz Club** and **Green Racoon**, all of which are in central Tel Aviv, all feature jazz at least one night a week. Call or see the local press for details.

In Amman, weekly concerts of very varied music are put on by **books@cafe**.

Tel Aviv's Suzanne Dellal Centre, renowned for excellence in modern dance

BARS AND CLUBS

In Tel Aviv, the main cluster of bars is in the Nakhalat Binyamin district, particularly around the southern end of Rothschild Avenue and Lillenblum. The venue that has young hipsters queuing outside every night is **Nanuchka**, a rowdy but classy bar-restaurant with surreal decor and a permanent party vibe. Around the corner, the splendidly named **Betty Ford** is a New York-style bar with a SoHo-style buzz. There are also plenty of good late-night spots around the Cinematheque on Ha-Arbaa Street and up in the Old Port *(see p168)* area, which is where you'll find **TLV**, one of the best and biggest open-air clubs in the city.

Also up at the Old Port are a couple of super clubs, including long-time favourite **Whisky a Go-Go**. However, perhaps the most fascinating and singular club is **Ha-Hamman**, a strikingly beautiful, converted Turkish bathhouse in Jaffa.

For something more casual and laidback, there's **Mike's Place** down near the seafront in central Tel Aviv or the **Gordon Inn**, a local pub with a pool table and a reliably friendly crowd.

In Amman, there are plenty of bars and clubs in the uptown neighbourhoods such as Abdoun and Shmeisani. One of the most popular places is the **Big Fellow Irish Pub**, which is run by the Sherton group. Drinks include, of course, Guinness and there's Guinness pie to eat. **Champions** is an American-style sports bar at the Marriott, while the **Living Room** is an attractive lounge bar with a good, American-influenced food menu.

In Petra, do not miss the chance to have a drink at the **Cave Bar**, which occupies a geniune 2,000-year-old Nabataean rock tomb. There's live bedouin music most nights too.

In Sinai, bars and clubs are generally found in the many resort hotels.

Dining, drinking and dancing al fresco in Atarim Square, Tel Aviv

DANCE

The internationally-known Bat Sheva company is the mainstay of modern Israeli dance. There are no classical ballet companies in Israel, but contemporary dance is very much alive here. The focal point of dance activity is the

Spontaneous outdoor dancing on the beach

Suzanne Dellal Centre, a superb, old Ottoman building at the heart of the historic, southern Tel Aviv district of Neve Tzedek, which has benefited from extensive architectural renewal. In Jerusalem, dance can be seen at the Centre for Performing Arts in the Jerusalem Sherover Theatre complex, while Jewish and Arabic folk dancing performances take place on Monday, Thursday and Saturday evenings in the **YMCA** auditorium.

In Jordan, there are two well-established national folkloric groups. Both dance at the **Royal Cultural Centre** and, occasionally, at the Roman Theatre, both in Amman *(see p212)*. Folkloric dance also features quite heavily at the **Jordan Festival** *(see p37)*.

Tel Aviv's Ha-Bima Theatre

THEATRE

Plays in Israel are almost always performed in Hebrew (or, less commonly, Arabic), although some of the bigger theatres such as Tel Aviv's **Ha-Bima Theatre and New Cameri Theatre** (and Jerusalem's Sherover Theatre, *see p151*) have headphones providing English-language translation for some performances, though there are a lot of performances in English as well. Productions, in all cases, range from revivals of the classics of world drama (both old and modern) to first-run stagings of new Israeli plays.

There are several theatre festivals throughout the year in Israel *(see pp36–9)*, the most exciting of which is the **Acre Fringe Theatre Festival**, which stages some performances in the city's subterranean Crusader halls.

In Amman, theatre takes place at the **Royal Cultural Centre** and **King Hussein Cultural Centre**. However, the premier theatrical event is the **Jordan Festival** *(see p37)*, which brings together performers from all over the world to present their work amid the ancient ruins.

CINEMA

Foreign films shown in Israel are not dubbed, but carry Hebrew subtitles. Cinemas are plentiful, especially in Tel Aviv, where complexes such as the **Rav-Chen 1–5** are modern, comfortable and airconditioned. They tend to screen first-run Hollywood fare. The **Cinematheques**, of which there is one in Jerusalem *(see p150)* and one in Tel Aviv, specialize in art-house and independent films, as well as holding themed seasons and retrospectives. Israel's biggest movie theatre complex is **Cinema City**, which has 21 auditoriums and three 3D screens.

There are several modern cinema-plexes in Amman offering recent releases, including the **Century Cinemas** in the Zara Centre behind the Grand Hyatt and **Galleria**. Films are shown in their original language with Arabic subtitles.

SPECTATOR SPORTS

Football is by far the most popular sport throughout the Holy Land. Two teams from Jerusalem play in Israel's premier league, Beitar and Ha-Poel. Matches take place in the **Teddy Stadium** at Malcha in West Jerusalem, which was opened in 1992.

A basketball match at the Yad Eliahu Arena

Basketball is the next most popular sport. The Jerusalem team, Ha-Poel, plays in the Sports Arena near the Teddy Stadium, while the Maccabee Tel Aviv plays at the **Yad Eliahu Arena** just off the Ayalon highway.

Football is also followed religiously in Jordan. The two main teams in Amman are Wahadat and Faisaly. Games are mostly played at the **Jordan International Stadium**, in the Shmeisani district.

SWIMMING

Almost all the large hotels have outdoor swimming pools; the YMCA in Jerusalem also has an indoor pool. You can also swim all year round at the **Jerusalem Swimming Pool**, in the German Colony district, south of the centre.

The Red Sea is warm enough for year-round swimming, although most resort hotels also have swimming pools. Swimming in the Mediterranean is fine in summer but it's too cold from around October to April.

CHILDREN

For information on Jerusalem for children, see page 150. In northern Tel Aviv, the **Ramat Gan Safari Zoo** makes a good outing for children. You can drive through and observe the wildlife in its natural habitat. The **Children's Museum** in Holon, a short drive from Tel Aviv, has lots of fun, interactive exhibits. **Mini Israel**, which is just off the main highway that runs between Jerusalem and Tel Aviv, has over 350 miniature models of the Holy Land's most important landmarks.

On the shores of the Dead Sea, just south of Jericho, **Attraktsion** is a large aquatic amusement park with water slides and splash pools. However, it is only open from April to October.

Tel Aviv's beach, starting to attract swimmers in spring

DIRECTORY

INFORMATION

books@cafe
Mango St, Jebel Amman, Amman, Jordan.
Tel *(06) 465 0457.*

CLASSICAL MUSIC

Einav Cultural Centre
71 Ibn Gvirol St, Tel Aviv.
Tel *(03) 521 7763.*

Felicia Blumenthal Centre
26 Bialik St, Tel Aviv.
Tel *(03) 620 1185.*
www.fbmc.co.il

King Hussein Cultural Centre
Omar Matar St, El-Muhajareen, Amman.
Tel *(06) 473 9953.*

Performing Arts Centre
19 Ha-Melekh Shaul Ave, Tel Aviv.
Tel *(03) 692 7777.*

Royal Cultural Centre
Al-Malekah Alia St, Shmeisani, Amman.
Tel *(06) 566 1026.*

Targ Centre
Ein Kerem, near Jerusalem.
Tel *(02) 641 4250.*
www.klassi.net/targ

Tel Aviv Museum of Art
27 Ha-Melekh Shaul Ave, Tel Aviv.
Tel *(03) 696 1297.*
www.tamuseum.com

OPERA

New Israeli Opera
Performing Arts Centre, 19 Ha-Melekh Shaul Ave, Tel Aviv.
Tel *(03) 692 7777.*
www.israel-opera.co.il

ROCK, JAZZ AND BLUES

Barbie
52 Kibbutz Gayulot St, Tel Aviv.
Tel *(03) 518 8123.*

Benchmark
37 Nakhalat Binyamin St, Tel Aviv.

books@cafe
See Information.

Cafe Henrietta
186 Arlozorov St, Tel Aviv.
Tel *(03) 691 1715.*

Goldstar Zappa
24 Raul Wallenberg St, Ramat HaChayal, Tel Aviv.
Tel *(03) 767 4646.*
www.zappa-club.co.il

Green Racoon
186 Ben Yehuda St, Tel Aviv.
Tel *(03) 529 8513.*

Ha-Bima Club
Basement, 2 Tarsat St, Tel Aviv.
Tel *(03) 528 2174.*

Mike's Place
86 Herbert Samuel, Tel Aviv.
Tel *(054) 819 2089.*

Shablul Jazz Club
Hangar 13, Tel Aviv Port.
Tel *(03) 546 1891.*

BARS AND CLUBS

Betty Ford
48 Nakhalat Binyamin St, Tel Aviv.
Tel *(03) 510 0650.*

Big Fellow Irish Pub
Abdoun Circle, Amman.
Tel *(06) 593 4766.*

Cave Bar
Behind the Visitors' Centre, Petra.
Tel *(03) 215 6266.*

Champions
Amman Marriott, Isam el-Ajlouni St, Shmeisani, Amman.
Tel *(06) 560 7607.*

Gordon Inn
17 Gordon St, Tel Aviv.
Tel *(03) 523 8239.*

Ha-Hammam
10 Mifraz Shlomo St, Jaffa.
Tel *(03) 681 3261.*

Living Room
Mohammed Hussein Heikal St, Amman.
Tel *(06) 465 5988.*

Mike's Place
See Rock, Jazz and Blues.

Nanuchka
30 Lilenblum St, Tel Aviv.
Tel *(03) 516 2254.*

TLV
Tel Aviv Port, Hayarkon Estuary Compound.
Tel *(03) 544 4194.*

Whisky a Go-Go
Tel Aviv Port, Hayarkon Estuary Compound.
Tel *(03) 544 0633.*

DANCE

Jordan Festival
Jerash Festival Office, Amman, Jordan.
Tel *(06) 566 0156.*

Royal Cultural Centre
See Classical Music.

Suzanne Dellal Centre
5 Yehieli St, Neve Tzedek, Tel Aviv.
Tel *(03) 510 5656.*

YMCA
King David St, Jerusalem.
Tel *(02) 569 2692.*

THEATRE

Acre Fringe Theatre Festival
Tel *(04) 955 6706.*

Ha-Bima Theatre
Habima Square, Tel Aviv.
Tel *(03) 629 5555.*
www.habima.org.il

Jordan Festival
See Dance.

King Hussein Cultural Centre
See Classical Music.

New Cameri Theatre
30 Leonardo Da Vinci St, Tel Aviv.
Tel *(03) 606 0960.*
www.cameri.co.il

Royal Cultural Centre
See Classical Music.

CINEMA

Century Cinemas
3rd Circle, Jebel Amman, Amman, Jordan.
Tel *(06) 461 3200.*

Cinema City
Gilot Junction, Tel Aviv.
Tel *(1-700) 702 255.*

Galleria
Abdoun Circle, Amman, Jordan.
Tel *(06) 593 4793.*

Rav-Chen 1-5
Opera Towers, 1 Allenby St, Tel Aviv.
Tel *(03) 510 2674.*

Tel Aviv Cinematheque
2 Sprinzhak St, Tel Aviv.
Tel *(03) 606 0800.*

SPECTATOR SPORTS

Jordan International Stadium
Shmeisani, Amman.

Teddy Stadium
Agudat Sport Beitar, Malkha, West Jerusalem.
Tel *(02) 678 8320.*

Yad Eliahu Arena
51 Yigal Allon St, Tel Aviv.
Tel *(03) 537 6376.*

SWIMMING

Jerusalem Swimming Pool
43 Emek Refaim St.
Tel *(02) 563 2092.*

CHILDREN

Attraktsion
Kalia Beach, Dead Sea, Israel.
Tel *(02) 994 2391.*

Children's Museum
1 Mifratz Shlomo St, Holon, Israel.
Tel *(03) 650 3000.*

Mini Israel
Kibbutz Nacsho, Latrun, Israel.
Tel *1 700 559 559.*

Ramat Gan Safari Zoo
Ramat Gan, Tel Aviv.
Tel *(03) 631 2181.*

Sporting and Specialist Holidays in the Holy Land

With terrain that runs from reefs rich in marine life to sometimes snow-capped peaks, and from coniferous forests to stony desert, the region offers a wide assortment of outdoor activities. Added to this, Israel is very much an "outdoors" society. As a consequence, the region is criss-crossed with hiking trails and treks, rivers are busy with rafts and canoes, parks offer opportunities for horse riding, and deserts for exploration by camel. All this is primarily for the locals but visitors can enjoy these facilities too.

A clown fish swims by brightly coloured soft corals

DIVING

Experienced divers claim that the Red Sea offers some of the world's best diving. The various scuba diving centres in Eilat, Aqaba and, especially, Sinai organize courses for beginners, as well as for more experienced divers who wish to qualify for the various international licences. Most centres hire out all the diving equipment you need (the daily rate is about $35–50), including, if desired, underwater photographic equipment.

Although the entire Red Sea teems with marine life, some of the richest dive sites are undoubtedly those within the Ras Muhammad National Park *(see p243)*, which is close to Sharm el-Sheikh at the tip of the Sinai peninsula. Dives in the park must be organized through a dive club.

While it is possible to sort out your own diving arrangements with a local company once you arrive, there are also many international agencies specializing in Red Sea diving holidays.

In Eilat, reputable diving centres include **Aqua Sport**, which organizes daily boat excursions along the Sinai coast to less-dived locations, **Divers' Village** and **Marina Divers**.

In Sinai, some of the better outfits include **INMO** and the **Nesima Dive Centre** in Dahab, and the **Camel Dive Club**, **Emperor Divers**, **Oonas Dive Centre** and **Sinai Divers** in Sharm el-Sheikh. You can visit their websites *(see p297)* for more information.

A diver enters the Red Sea just off Aqaba in Jordan

Windsurfing between Eilat and Taba in the Gulf of Aqaba

For a different kind of diving experience, **Caesarea Diving** at the Caesarea National Park *(see p176)* on Israel's Mediterranean coast offers scuba trips that allow you to explore the submerged ruins of Herod's ancient harbour.

SNORKELLING

Another way of viewing the rich marine life and beauty of the reefs is to snorkel. This has the advantage of being cheap and of not requiring any complicated equipment or specialised training. Dahab and Sharm el-Sheikh in Egypt *(see pp242–3)* are the best locations, and each has plenty of snorkel-hire shops. It is also possible to snorkel in Israel at Eilat *(see p205)* and in Jordan at Aqaba *(see p235)*.

WATER SPORTS

The windsurfing is good in the Gulf of Aqaba, particularly on the coast between Eilat and the border at Taba; there are plenty of places to rent boards, many of them near the small marina by the Club Med hotel. The region's centre for water sports is Eilat *(see p205)*, with everything from snorkels to jetskis for hire, plus a multitude of other activities, including paragliding and glass-bottomed boats. Israel's Mediterranean coast is more exposed, with dangerous currents, but there are water sports activities at Tel Aviv and a few other coastal towns, such as Netanya.

In Egypt, all the larger Sinai resorts, including Taba Heights, Dahab and Sharm el-Sheikh offer extensive water sports facilities.

RAFTING AND CANOEING

Possibilities exist for rafting and canoeing on the Jordan River in the Golan Heights *(see 181)*; these activities are supervised by **Abu Kayak** in the Jordan River Park, at Tel Bethsaida.

DESERT HIKING

A large number of specialist organisations lead hikes throughout Israel. A good starting point for finding out about such trips is to visit the **Society for the Protection of Nature in Israel (SPNI)**. Its offices/bookshops in Tel Aviv and Jerusalem carry a wide range of specialized maps and useful publications. The SPNI also runs plenty of hikes itself. Some of the best routes are around Maktesh Ramon *(see p204)* and Ein Gedi *(see p196)*, and up in the Golan Heights *(see p181)*.

The best hiking in Jordan is, without doubt, in and around Wadi Rum *(see p232–4)*. Here you'll find trails that last anything from a couple of hours to several days, all of which are described in the essential *Treks and Climbs in Wadi Rum* by Tony Howard. There are numerous guide agencies based in the area; some of the better ones include **Bedouin Roads**, **Terhaal**, **Sunset Camp** and **Wadi Rum Adventures**. There is also some excellent hiking around Petra *(see p220–31)* and at Wadi Mujib *(see p197)*. For more information on treks and hikes visit the **Wild Jordan Centre** in Amman.

Trekking in one of the canyons of the Judaean Desert

While not as magnificent as Wadi Rum or Petra, Egypt's Sinai peninsula has an interior that is starkly beautiful and well worth exploring; this can be arranged at most hotels in Nuweiba, Dahab or Sharm el-Sheikh. Some of the most rewarding trekking is around the St Catherine's Monastery region *(see p246–8)*. All treks must be done with a Bedouin guide, and this can be arranged through the services of **Sheikh Musa**, a local Bedouin leader.

CAMEL TREKKING

One of the best ways to explore the vast sandy expanses of Wadi Rum *(see p232–4)* is on the back of a camel. A wide variety of treks are available, ranging from half-hour explorations to overnight expeditions. It is also possible to arrange longer camel excursions from Wadi Rum – or Petra – down to Aqaba. These take from three to six nights, depending on the route. For more details contact an agency such as **Bait Ali**, **Bedouin Roads**, **Petra Moon Tourism**, **Sunset Camp** or **Wadi Rum Adventures** *(see p297)*.

In Israel, the **Mamshit Camel Ranch**, near Dimona on Route 25 between Beersheva and Sodom, offers desert trips on camels. In Egypt's Sinai, camel trekking can also be arranged by most hotels in Nuweiba, Dahab and Sharm el-Sheikh.

Tourists enjoying a camel trek along the rugged shoreline of Egypt's southern Sinai

Horse riders passing the Bab el-Siq Triclinium en route to the entrance at Petra

CLIMBING

Wadi Rum *(see p232–4)* offers some of the Holy Land's best rock climbing, with the ascent of Jebel Rum high on most climbers' lists. For information on route options see the book *Treks & Climbs in Wadi Rum, Jordan* by Tony Howard and Di Taylor (easily available in Jordan) or try the website www.wadirum.net. Several guides offer instruction in basic climbing techniques, including **Wadi Rum Mountain Guides**, which is run by Attayak Aouda, one of Rum's best climbing guides. Experienced climbers should bring their own equipment.

Jebel Umm Adaami, near the border with Saudi Arabia, is Jordan's highest peak at 1,832 m (or 6,045 ft). It's a fairly easy hike to the summit, plus an hour-long jeep drive each way, and you can stop off at some interesting petroglyphs and lovely scenery en route.

Rope-assisted descents of spectacular gorges in Israel's Judaean Desert can be organized by the **Metzoke Dragot Centre**. The same company also offers climbing, hiking, and jeep or truck excursions into the desert.

HORSE RIDING

Stables and riding schools are located throughout Israel, particularly in Upper Galilee, the Golan region and on the coast between Tel Aviv and Haifa. **Vered ha-Galil**, just north of the Sea of Galilee, is the largest riding school in the country, while the **Haela Ranch** is conveniently close to Jerusalem, up in the hills east of the city.

In Jordan it is possible to explore the desert landscapes of Wadi Rum on horseback. Among the agencies who can organize this are **Bait Ali** and **Rum Horses**. It is also possible to ride at Petra, although this is limited to a one km (half a mile) canter to the site entrance.

In Sinai, several resort hotels offer horse riding by the hour, while in Dahab, Bedouin rent horses on the beach.

GOLF

Israel has precisely two golf courses and, of these, the **Caesarea Golf Club** is the only one that meets international 18-hole standards. The course, designed in 1961, passes through ancient Roman and Byzantine ruins. In recent years Egypt has sought to market itself as a golfing destination and it has several new courses. Two of these are in Sinai: the **Jolie Ville Golf Resort** at Sharm el-Sheikh, opened in 1998, and, further north, the **Taba Heights Golf Resort** with its views across the Red Sea to Saudia Arabia and Jordan, which opened in 2006.

BIRDWATCHING

Israel and Sinai lie on one of the principal bird migration routes between Europe and Africa and so are something of a birdwatcher's paradise. In Israel, interested parties should visit the **International Birding and Research Centre**, which is in Eilat, near the Arava border crossing with Jordan, a short distance north-east of the town centre.

In Jordan, the Royal Society for the Conservation of Nature organizes birding trips (visit them at the **Wild Jordan Centre** in Amman), typically out to the Azraq Wetland Reserve, which is about 80 km (50 miles) east of Amman. For information on birding in Sinai, and throughout Egypt, see www.birdingegypt.com.

WORKING ON A KIBBUTZ

Not as popular as it once was, Israel's pioneering, socialist-style kibbutz movement continues to employ young volunteers (who must be aged between 18 and 32) from abroad to carry out manual work. Typical work involves picking fruit out in the fields, working on a factory production line, or being attached to a dining room, kitchen or laundry. The kibbutz will normally expect a minimum commitment of two months, during which time volunteers work for their accommodation, meals and a small personal allowance, plus one day a week holiday. The kibbutz facilities are

Volunteers working on a kibbutz in northern Israel

available to volunteers; these may include such things as a swimming pool or gym.

Volunteers usually apply through a special kibbutz office in their home country, although there is also a kibbutz office in Tel Aviv, through which online applications can be made (see the directory, below).

HAMMAMS

Hammams are what are known elsewhere as Turkish baths. At one time, every Arab town would have had several such institutions. They were as much social centres as places to get clean. The advent of domestic plumbing has rendered them largely obsolete but a handful remain. In Amman, Jordan, is the grand **Hammam el-Pasha**, which has separate areas for men and woman. In Aqaba *(see p235)*, the **Aqaba Turkish Baths** are men only – although women may visit by special appointment, in which instance they get the whole place to themselves.

DIRECTORY

DIVING AND SNORKELLING

Aqua Sport
Coral Beach, Eilat, Israel.
Tel *(08) 633 4404.*
www.aqua-sport.com

Caesarea Diving
Caesarea National Park, Israel.
Tel *(04) 626 5898.*
www.caesarea-diving.com/eng

Camel Dive Club
Sharm el-Sheikh, Egypt.
Tel *(069) 360 0700.*
www.cameldive.com

Divers' Village
Coral Beach, Eilat, Israel.
Tel *(08) 637 2268.*

Emperor Divers
Dahab, Nuweiba and Sharm el-Sheikh, Egypt.
Tel *(012) 350 2433.*
www.emperordivers.com

INMO
Dahab, Egypt.
Tel *(069) 364 0370.*
www.inmodivers.de

Marina Divers
Coral Beach, Eilat, Israel.
Tel *(08) 637 6787.*
www.marinadivers.co.il

Nesima Dive Centre
Dahab, Egypt.
Tel *(069) 364 0320.*
www.nesima-resort.com

Oonas Dive Centre
Dahab and Sharm el-Sheikh, Egypt.
Tel *UK (01323) 648 924.*
www.oonasdivers.com

Sinai Divers
Dahab, Sharm el-Sheikh and Taba, Egypt.
Tel *(069) 360 0697.*
www.sinaidivers.com

RAFTING AND CANOEING

Abu Kayak
Jordan River Park, Beth Saida, Israel.
Tel *(04) 692 1078.*

DESERT HIKING

Bedouin Roads
Wadi Rum, Jordan.
Tel *(079) 589 9723.*
www.bedouinroads.com

Sheikh Musa
St Catherine's, Egypt.
Tel *(010) 641 3575.*

Society for the Protection of Nature in Israel (SPNI)
13 Heleni ha-Malka Street, West Jerusalem. **Tel** *(02) 624 4605.* 4 Ha-Shfela Street, Tel Aviv. **Tel** *(03) 638 8674.* **www**.teva.org.il

Sunset Camp
Wadi Rum, Jordan.
Tel *(077) 731 4688.*
www.mohammedwadirum.8m.com

Terhaal
48 Ali Nasuh Al Tahir St. Amman, Jordan.
Tel *(06) 581 3061.*
www.terhaal.com

Wadi Rum Adventures
Wadi Rum, Jordan. **Tel** *(077) 747 2074.* **www**.wadirumadventures. com

Wild Jordan Centre
Amman, Jordan.
Tel *(06) 461 6523.*
www.rscn.org.jo

CAMEL TREKKING

Bait Ali
Wadi Rum, Jordan.
Tel *(079) 554 8133.*
www.baitali.com

Bedouin Roads
See Desert Hiking.

Mamshit Camel Ranch
Mamshit, Western Negev, Israel.
Tel *(08) 943 6882.*

Petra Moon Tourism
Petra, Jordan.
Tel *(03) 215 6665.*
www.petramoon.com

Sunset Camp
See Desert Hiking.

Wadi Rum Adventures
See Desert Hiking.

CLIMBING

Metzoke Dragot Centre
Metzoke Dragot, Dead Sea, Israel.
Tel *(02) 994 4222.*

Wadi Rum Mountain Guides
Wadi Rum, Jordan.
Tel *(079) 583 4736.*
www.rumguides.com

HORSE RIDING

Bait Ali
See Camel Trekking.

Haela Ranch
Nes Harim, Israel.
Tel *(050) 444 3902.*
www.haelaranch.com

Rum Horses
Wadi Rum, Jordan.
Tel *(03) 203 3508.*
www.desertguides.com

Vered ha-Galil
Korazim, 20 km north of Tiberias, Galilee, Israel.
Tel *(04) 693 5785.*
www.veredhagalil.co.il

GOLF

Caesarea Golf Course
Caesarea, Israel.
Tel *(04) 610 9600.*
www.caesarea.co.il

Jolie Ville Golf Resort
Mövenpick Resort, Sharm el-Sheikh, Egypt.
Tel *(069) 360 0100.*
www.jolieville-hotels.com

Taba Heights Golf Resort
Taba Heights, Egypt.
Tel *(069) 358 0073.*
www.tabaheights.com

BIRDWATCHING

International Birding & Research Centre
Near Arava Crossing, Eilat, Israel.
Tel *(08) 633 5339.*
www.eilat-birds.org

Wild Jordan Centre
See Desert Hiking.

WORKING ON A KIBBUTZ

Kibbutz Programme Centre
6 Frishmann Street, Tel Aviv, Israel.
Tel *(03) 524 6154.*
www.kibbutz.org.il

HAMMAMS

Aqaba Turkish Baths
King Hussein Street, Aqaba, Jordan.
Tel *(03) 203 1605.*

Hammam el-Pasha
El-Mahmoud Taha Street, Jebel Amman, Amman, Jordan.
www.pashaturkishbath.com

SURVIVAL GUIDE

PRACTICAL INFORMATION

The area covered by this guide is not very large, but because it includes the territory of three nations (Israel, Jordan and Egypt), as well as the Autonomous Palestinian Territories, getting about from one place to another may not always be straightforward. The political situation in this part of the world changes frequently, and before embarking on a trip that involves any crossing of borders, you should make sure that there have been no significant changes to the international agreements between these countries. Israel, Jordan and Egypt all have their own tourist organizations, which have offices abroad *(see p301* for a directory of contact details for these).

Israeli tourist board logo

CROSSING BORDERS

Peace agreements of recent years have made it possible to travel overland between Israel and Egypt, and between Israel and Jordan. There are three commonly used crossings between Jordan and Israel.

The King Hussein Bridge (also known as the Allenby Bridge) is 16 km (10 miles) east of Jericho. From East Jerusalem (opposite Damascus Gate) you can take a taxi or minibus to the border then, once across, pick up transportation on to Amman. There are hefty Israeli exit and Jordanian entry taxes to pay. The crossing is open 8am–8pm Sunday–Thursday and 8am–1pm Friday and Saturday. The border crossing point at Wadi Arava (also known as Yitzhak Rabin Terminal) is 4 km (2 miles) from Eilat and 10 km (6 miles) from Aqaba. It is open 6am–8pm Sunday–Thursday and 8am–8pm Friday–Saturday. The third crossing, the Jordan River Border Terminal, is open 6:30am–9pm Sunday–Thursday and 8am–8pm Friday.

To enter Sinai you can take the ferry or catamaran from Aqaba in Jordan to Nuweiba. Both depart once a day, and you can get your Sinai Permit on board. You can also cross overland using public transport from Eilat in Israel to Taba. Allow up to three hours for crossing any of these borders, as there are strict security measures in place.

Israeli soldiers checking cars coming from the Palestinian Autonomous Territories

All borders are closed on Yom Kippur and the Muslim Feast of Sacrifice, apart from Wadi Arava and the Jordan River Border Terminal, which are closed on the Muslim New Year instead.

VISAS FOR ISRAEL

You must have a passport that is valid for at least six months to enter Israel. Citizens of European nations, as well as those from North America, Australia and New Zealand do not need a visa. Citizens of most Arab, Asian, African and South American countries do need visas, and must obtain them in advance from an Israeli consulate in their home country. The visa is usually valid for up to a three-month stay, but can be extended. You can also obtain a "volunteer visa" (valid for 6–12 months) that allows you to work temporarily in a kibbutz *(see pp294–5)*.

An Israeli visa in your passport will bar you from entering some Arab countries, notably Syria and Lebanon, but not Egypt or Jordan. To avoid this ask at the airport for the visa to be stamped on a separate piece of paper. Other than at the Allenby Bridge crossing, this cannot be done at the land borders. Bear in mind that it is better to have the visa in your passport if visiting the Palestinian Territories.

Entry card for Israel, and visa required to enter Jordan

At checkpoints between Israel and the Palestinian territories, Israeli or Palestinian police will ask to see your passport and may carry out security checks.

VISAS FOR JORDAN

Tourists arriving in Jordan must have a passport valid for at least six months, and also a visa. If you are arriving at Queen Alia airport you can obtain a two-week tourist visa upon arrival, which is easily extended at any police station. The price of this can vary depending on your nationality.

If entering Jordan by land at Allenby Bridge you must obtain your visa in advance. Visas can be issued by the Jordanian consulate or embassy in your home country, or by

◁ Bedouin tents in the mountainous wilderness of Wadi Rum

those in Tel Aviv or Cairo. At the Wadi Arava and Jordan River crossings visas can be issued at the border post.

VISAS FOR EGYPT

If you are entering Sinai from Israel, you can get a Sinai Permit that allows you to stay for up to 14 days; this is obtained at the border and is free. Bear in mind, however, that the Sinai Permit cannot be changed into a full visa. Neither can a full visa for Egypt be obtained at the border. If you plan to visit other parts of Egypt beyond Sinai, you must obtain a visa in advance from an Egyptian consulate or embassy in your country, or else in Amman, Aqaba, Tel Aviv or Eilat.

DUTY-FREE ARTICLES AND CUSTOMS

The duty-free allowance in all three countries is 200 cigarettes or 200 grams of tobacco, a litre of spirits and two bottles of wine. Valuable electrical objects such as computers and video cameras will be entered in passports by customs officers to prevent their resale in the country.

LANGUAGE

English is a second language in Israel, where many immigrants do not speak Hebrew. All signs are bilingual and most people speak some English. This is not the case in Palestinian areas and in Jordan, however, but Arabs will make every effort to communicate with foreigners, even if it means resorting to sign language. In areas frequented by tourists it is easier to find English speakers, though attempts to speak Arabic will always be welcomed. Away from the main tourist areas it can be much harder to get your message across without a rudimentary grasp of the language.

Israeli road signs

ETIQUETTE

Israeli society, on the whole, is not that different from the West. There are exceptions such as ultra-Orthodox areas where your behaviour and dress should err on the side of conservatism. This is also the case in Arab areas, both in the Palestinian Autonomous Territories and in Jordan. Arab women usually cover their arms, legs and sometimes their heads in public, and men do not wear shorts. Visitors are not always expected to cover up in the same way, but you must be suitably clothed when visiting certain public places and any of the holy sites *(see p300)*.

Intimate physical contact is also taboo; Arabic couples are rarely seen kissing, embracing or even holding hands. In Egypt, photography at certain places such as bridges or military installations is prohibited.

Arab women in customary dress, outside the Dome of the Rock, Jerusalem

DIRECTORY

EMBASSIES AND CONSULATES

In Israel

UK Embassy
192 Ha-Yarkon Street, Tel Aviv.
Tel *(03) 725 1222.*
www.britemb.org.il

UK Consulates
19 Nashashibi Street,
Sheikh Jarah, East Jerusalem.
Tel *(02) 541 4100.*

US Embassy
71 Ha-Yarkon Street, Tel Aviv.
Tel *(03) 519 7575.*
http://telaviv.usembassy.gov

US Consulates
18 Agron Street, West Jerusalem.
Tel *(02) 622 7230.*

In Jordan

UK Embassy
Damascus Street, Abdoun, Amman.
Tel *(06) 590 9200.*

US Embassy
Damascus St, Abdoun, Amman.
Tel *(06) 590 6950.*

In Egypt

UK Embassy
7 Ahmed Ragheb Street,
Garden City, Cairo.
Tel *(02) 2794 0852.*

US Embassy
5 Latin America Street,
Garden City, Cairo.
Tel *(02) 2795 7371.*

In the UK

Egyptian Consulate
2 Lowndes Street, London SW1.
Tel *(020) 7235 9777.*

Israeli Embassy
2 Palace Green, London W8.
Tel *(020) 7957 9500.*

Jordanian Embassy
6 Upper Phillimore Gardens,
London W8.
Tel *(020) 7937 3685.*

In the US

Egyptian Consulate
1110 2nd Avenue, New York.
Tel *(212) 759 7120.*

Israeli Embassy
3514 International Drive NW,
Washington DC.
Tel *(202) 364 5500.*

Jordanian Embassy
3504 International Drive NW,
Washington DC.
Tel *(202) 966 2664.*

Tips for Tourists

Israeli tourist office sign

Tourism in Jerusalem and the Holy Land is considerable, given the region's major historical and religious importance, as well as its great natural beauty. As such, most towns are well adapted for visitors, with good public facilities and helpful tourist offices. Major sites are open long hours for much of the week, and also have good facilities as well as useful educational material. Some sites, however, are well off the beaten track, and difficult to reach using public transport. If visiting desert areas, make sure you arrive early, to avoid the extreme afternoon heat.

Roman ruins, Caesarea National Park *(see p176)*, free with a Green Card

Visit Jordan logo

TOURIST INFORMATION

As well as providing useful information in the form of free brochures and maps, Israeli tourist offices are usually able to help with other matters, such as finding accommodation. In smaller towns, or at archaeological sites, the tourist offices are of more limited use, and information is usually confined to the immediate area. The Autonomous Palestinian Territories are also in the process of organizing a network of information bureaux, but for the present, their sole office is in Bethlehem.

In Jordan the only tourist information offices are in the main tourist destinations such as Amman, Petra and Jerash, while in Sinai there are no tourist information offices at all. All three countries have international tourist bureaux, however, which you can use before you leave. The national airline offices can also often help with travel information.

ENTRANCE FEES

Most of the historic and archaeological sites in Jerusalem and the Holy Land have some kind of admission charge, although some smaller churches and mosques have no fixed fee at all. In these cases a small donation is customary. Prices are generally very reasonable, with most minor sites in Israel charging only a few shekels. Larger places may charge slightly more, with the most expensive site to visit by far being Petra.

In Israel you can purchase a 14–day Green Card for around NIS 130, that gives free access to all sites under the control of the Nature and National Parks Protection Authority. These are mainly natural and more minor archaeological sites, but if you are planning to spend some time sightseeing in Israel, this may be a good investment.

OPENING HOURS

Because of the many religious holidays *(see pp36–9)* celebrated in the region (Jewish, Muslim and Christian), opening hours for the many tourist sites and historic monuments can vary greatly. As a general rule, however, sites in Israel are usually open daily, except for Friday, when they keep more restricted hours, and Saturday, when they are closed altogether. Christian sites, other than the churches, are open on Saturdays but closed on Sundays, while Muslim sites are closed on Fridays. In general, last entry to a site will be one hour before the stated closing time.

In Jordan the main sites (including Petra and Jerash) are open daily, but other, smaller sites, including many of the museums, are closed on Tuesdays. Most shops are closed on Fridays and Saturdays. Friday is the usual closing day in Egypt. From around October to March (considered the winter season), most sites in the Holy Land close an hour earlier than usual.

WHAT TO WEAR AT SACRED SITES

When visiting holy sites such as churches, synagogues and mosques, it is essential that you dress appropriately. This means that your arms and

Jewish worshippers praying at the Western Wall *(see p85)*

legs must be fully covered; shorts or short skirts and sleeveless tops are not acceptable. At certain places cloaks are provided to cover up visitors who are deemed to be immodestly dressed. Shoes must be removed before entering a mosque, and at some Jewish holy sites, such as the Western Wall, heads must also be covered. In such cases a *kippah* (skullcap) or headscarf will be provided.

TIME

The time in Israel, Jordan and Egypt is two hours ahead of Greenwich Mean Time (GMT), and seven hours ahead of Eastern Standard Time (EST). All three countries have daylight saving time which lasts from approximately March to September.

DISABLED VISITORS

In Israel many hotels and modern museums are adapted for disabled use. **MILBAT** is a useful advisory centre on such matters, while **JDC-Israel** is also able to advise on suitable hotel accommodation and site accessibility. The **Yad Sarah Organization** lends wheelchairs and other useful aids.

Jordan and Sinai make no real provision for the disabled, and as most sites are in rough terrain, visiting these areas can be very problematic.

Sign for public toilets

Student ISIC identity card

STUDENT INFORMATION

In Israel, the presentation of a recognized student card, such as an International Student Identity Card (ISIC), will get the holder a 10 per cent discount on bus fares, as well as discounts on most museum and site admissions. The **Israel Student Tourist Association (ISSTA)** can arrange cheap flights and accommodation, and provide information on student discounts, as well as arranging its own package holidays. There are no student discounts offered in Jordan, but Egypt offers a 50 per cent concession on most site admissions.

WCS

Public toilets are easily found throughout Israel, and are of the standard type found in the West. In Jordan they are much less common and a lot more rudimentary, but still usually clean, as they are tended by caretakers. In Sinai public toilets do not exist at all. It is always wise to have a supply of paper with you, as this is often not provided. All paper should be disposed of using the bins provided, and not put down the toilet, as the local plumbing cannot cope.

ELECTRICAL ADAPTORS

The electric current in Israel, Jordan and Sinai is 220V. Plugs in Israel are round-pronged and three-pinned, whereas in Jordan and Sinai they are round-pronged and two-pinned. Adaptors should be bought prior to departure.

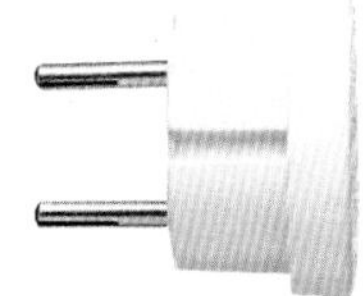

Two-pin plug adaptor for use in Jordan and Sinai

CONVERSION CHART

Imperial to Metric
1 inch = 2.54 centimetres
1 foot = 30 centimetres
1 mile = 1.6 kilometres
1 ounce = 28 grams
1 pound = 454 grams
1 pint = 0.6 litres
1 gallon = 4.6 litres

Metric to Imperial
1 centimetre = 0.4 inches
1 metre = 3 feet, 3 inches
1 kilometre = 0.6 miles
1 gram = 0.04 ounces
1 kilogram = 2.2 pounds
1 litre = 1.8 pints

DIRECTORY

TOURIST INFORMATION

Israel Ministry of Tourism
www.goisrael.com
UK: 180 Oxford St, London W1N 0EL. ***Tel*** *(020) 7299 1111.*
US: 800 Second Ave, New York 10017. ***Tel*** *(212) 499 5660.*

Israeli Tourphone
Tel **3888 (24-hour info for tourists).*

Jordan Tourist Board
www.visitjordan.com
UK: 2nd floor Masters House, 107 Hammersmith Rd, London W14 0QH. ***Tel*** *(020) 7371 6496.*
US: 535 Fifth Ave, New York. ***Tel*** *(212) 949 0060.*

Egyptian Tourist Authority
Tel *(02) 285 4509.* **http://**touregypt.net/tourism/
UK: 170 Piccadilly, London W1. ***Tel*** *(020) 7493 5282.*
US: Suite 1706, 630 Fifth Ave, New York. ***Tel*** *(212) 332 2570.*

Palestinian Authority
Tel *(02) 274 1581 / 2 / 3.*
www.travelpalestine.ps

National Parks Authority
Tel **3639.*

DISABLED VISITORS

MILBAT
Sheba Medical Centre, Tel ha-Shomer, Ramat Gan, Tel Aviv, Israel. ***Tel*** *(03) 530 3739.*
www.milbat.org.il

JDC-Israel
www.jdc.org

Yad Sarah Organization
124 Herzl Blvd, Jerusalem. ***Tel*** **6444.*

STUDENT INFORMATION

ISSTA
31 Ha-Neviim St, Jerusalem. ***Tel*** *(02) 621 3600.*

Security and Health

Israel and the Middle East suffer from a bad press when it comes to security. However, despite the occasional alarming headline, Israel and its neighbouring territories of Jordan and Sinai are perfectly safe for tourists. Visitors rarely encounter crime, and there are next to no hazards in the form of dangerous animals, or endemic diseases. Political unrest does from time to time result in acts of terrorism or rioting, but this hardly ever affects visitors. With the present ongoing attempts to reach peace between Israel and the Palestinians, even these infrequent incidents of violence may, hopefully, soon be a thing of the past.

Israeli Defence Force soldiers at Damascus Gate

LAW AND ORDER

Israel, Jordan and Sinai all have special tourist police to deal with any complaints or problems visitors may encounter. These police mostly speak English, and are posted at most major sites and at tourist resorts. They wear identifying armbands. The Jordanians have a special form of tourist police, active in the Wadi Rum area, known as the Desert Patrol. These officials are easily identified by their smart khaki uniforms, their distinctive red-and-white checked headdress and by the fact that they often ride camels.

Normal Israeli police wear navy blue uniforms and peaked caps. Also part of the police force are the border guards, who wear a military style uniform and a green beret. They operate mainly in the Israeli-controlled areas of the West Bank. The Palestinians also have their own security forces, who come in a multitude of guises.

Visitors will notice a preponderance of military personnel on the streets in Jerusalem and Israel. Every citizen must perform military service in the Israeli Defence Force (IDF) as soon as they reach the age of 18. The term of service is three years for men and two years for women. Men serve for an additional 30 days a year until the age of 35. Consequently, you will see armed soldiers around all the time, particularly at bus stations, as they are usually on the way to or from their bases.

PERSONAL SAFETY

On arrival at Ben Gurion Airport, you will almost immediately experience just how tight security is in Israel. During your stay, you may be subject to security checks on entering hotels, restaurants, bars, cinemas and shopping complexes, so it is wise always to carry some identification, preferably your passport. But as far as the visitor is concerned, terrorism is not a major worry. Tourists have never been the target of terrorists and most attacks have occurred well away from all tourist sites. Naturally, you have to be alert when in the streets, and also keep an eye on the local news. Among the "sensitive" areas are East Jerusalem and West Bank towns such as Hebron and Ramallah. In times of unrest you should definitely give such places a wide berth. Should you be unlucky enough to encounter a disturbance in the streets, move away from the scene quickly, and make it completely clear that you are a foreign tourist.

A member of the Desert Patrol, Wadi Rum, Jordan

Israeli policeman

Stories of theft, mugging and other similar opportunistic crimes are rare in the Holy Land. Crime is not the problem here that it is in many other parts of the world. As a rule, all areas are considered safe for visitors, unless the visitor is an unaccompanied woman. Lone females are frequently subjected to unwanted verbal pestering and harrassment from local males. This problem is particularly acute in Jerusalem's Old City and its surrounding areas, such as the Mount of Olives and Mount Zion. Incidences of rape have even been reported, and so our advice to women is that they should not walk alone in unpeopled areas or in secluded areas of the Old City after dark.

PERSONAL PROPERTY

On the whole Israelis and Arabs are very honest people. If you lose anything it is always worth going to the last place the item was seen, or going to the tourist police. On occasion, unpleasant experiences do happen. To minimize the risk of this, do not leave valuable objects inside a car or in full view in your hotel room. Leave your valuables in the hotel safe or at the reception desk. The fact that credit cards are accepted almost everywhere is a good reason not to carry a lot of cash with you. In case of theft, remember to make a report to the police and to ask for a copy of the report, which you will then have to present to your insurance company when you make your claim.

Security considerations mean that you should not leave luggage unattended (especially in airports and bus stations), as they might cause alarm or trigger a reaction on the part of the security forces. Don't accept packages from anyone asking you to carry something for them.

HEALTH PRECAUTIONS

Medical care in Jerusalem and the Holy Land is costly, making it inadvisable to travel without some form of medical insurance. The policy should at least cover the cost of a flight home.

No specific vaccinations are legally required before entering Israel, Jordan or Sinai, but doctors may advise inoculation against hepatitis A (spread through contaminated food or water), hepatitis B, tetanus and also typhoid.

There are no particular endemic diseases in the Middle East, but the hot climate necessitates that you take certain precautions, at least until you are used to the change in diet. It is advisable to drink mineral water (which is sold everywhere) and not use ice in your drinks. Avoid raw vegetables or food that has obviously been left standing for some time since it was cooked, and peel fruit. Continually

A small pharmacy in Jerusalem

drinking large quantities of liquids is essential: the lack of humidity in the air causes rapid dehydration, even though you may not be aware of it. Other than this, the most frequent problems are intestinal. A change of diet often upsets the stomach. It is recommended that you should always carry diarrhoea pills. If the upset continues then consult a doctor or pharmacist for more powerful medication.

Mosquitoes can sometimes be a nuisance, but there is no threat of malaria. Bring repellent lotion or spray from your own country – although, if you forget, it is easy to find in any pharmacy. If you go diving in the Red Sea, you need to be careful of sharp corals and be aware of which species of fish are poisonous and are to be avoided.

Pharmacy sign in Israel

PHARMACIES

Good pharmacies are easy to find throughout both Israel and Jordan. However, if you need a particular medicine, it is still advisable to travel with your own supplies and keep a note of the product and its composition so that, if worst comes to worst, a pharmacist will be able to find a local equivalent. In Israel, the *Jerusalem Post* lists the names and addresses of pharmacies that stay open late and during Shabbat and holidays.

Magen David Adom ambulance

MEDICAL TREATMENT

In an emergency in Israel, you can call 101 to request an ambulance or to ask about the nearest casualty department. Alternatively, contact the local branch of the **Magen David Adom** (Israel's equivalent of the Red Cross), or call its countrywide toll-free number.

In Jordan, if you need a doctor, call into a pharmacy and ask for a recommendation or call your embassy. In Sinai, most large hotels have a resident doctor. For divers, there is a special **Hyperbaric Medical Centre** in Sharm el-Sheikh equipped with a recompression chamber.

DIRECTORY

EMERGENCY NUMBERS

In Israel

Ambulance
Tel *101.*
Private Ambulances (Natali):
Tel *1-700-700-180.*
Police and General Enquiries
Tel *100.*
Fire Brigade
Tel *102.*
Directory Assistance
Tel *144.*

In Jordan

Ambulance/ Fire Brigade
Tel *199.*
Police
Tel *191.*
Local Directory Assistance
Tel *131.*

In Sinai

Ambulance
Tel *123.*
Hyperbaric Medical Centre
Tel *(069) 366 0922/3 (24 hr).*
Police
Tel *122.*
Tourist Police
Tel *126.*

Banking and Currency

Leumi Bank logo

Exchanging and obtaining money pose no problems in Israel, Jordan and Egypt. Cash and traveller's cheques can be exchanged at banks, exchange offices and in many hotels. Credit cards are widely accepted and can be used to obtain funds. The only issues to be aware of are the greatly varying levels of commission charged on transactions, and the limited opening hours of banks.

BANKS

Banks in Israel, Jordan and Sinai will exchange all major European currencies, but the most welcome currency of all is the US dollar. ATMs (automatic cash dispensers) linked into international banking networks, such as Cirrus or Plus, are widespread in Israel. You will find them in the foyers of most banks. These machines are less common in Jordan and Sinai, and found only in Amman, Petra and Sharm el-Sheikh. Some banks in Israel also have automatic currency exchange machines, which are accessible 24 hours a day. The drawback is that these machines usually charge a high transaction fee combined with a very poor rate of exchange.

Automatic currency exchange machine

Jerusalem's banking district is centred on Zion Square, at the bottom of Ben Yehuda Street in the New City. Banks are generally open from 8:30am to 12:30pm, reopening for another hour or two from around 4pm (but not on Wednesdays). They are shut on Fridays and Saturdays, as are banks in Jordan. In Sinai banking hours are similar to Israel, except that they are closed only on Fridays.

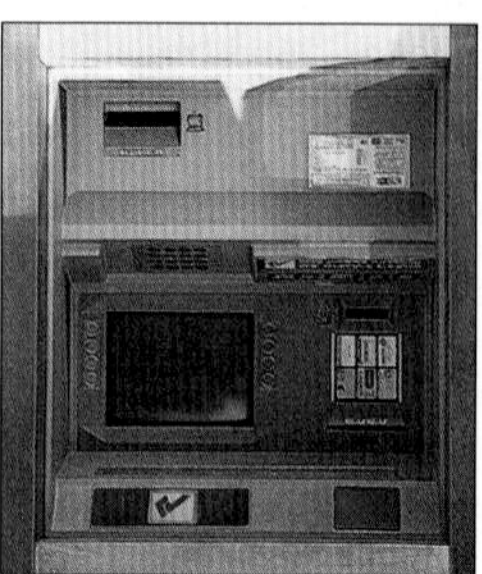
ATM machine at an Israeli bank

EXCHANGE OFFICES

The banks often charge a considerable commission on currency exchanges; one way to avoid this is to use an official exchange office such as the **post office**, **Western Union** or **Change Spot**. These places charge no commission. Change Spot also tends to be open much longer hours than the banks (from 9am to 9pm in some cases). Such exchange offices in Jerusalem can be found mainly on Jaffa Road and Ben Yehuda Street. There are also several small Arab exchange offices just inside Jaffa and Damascus gates in the Old City.

In Jordan, central Amman is full of small exchange offices, but there are not so many outside the capital. The exchange rate is set daily by the Jordanian Central Bank.

TRAVELLER'S CHEQUES AND CREDIT CARDS

Traveller's cheques can be exchanged at banks but commission is charged per cheque. It is better to cash them at exchange offices, where no commission is charged at all.

Major credit cards, such as VISA, MasterCard, Diners Club and American Express are widely accepted throughout Israel, Jordan and Sinai in shops, restaurants and hotels. If you have your PIN number you can draw cash from ATMs.

Official money exchange office

CURRENCY

Israel's national currency is the new Israeli shekel (NIS), referred to simply as the shekel. It is also the currency in the Palestinian Autonomous Territories, although a Palestinian national currency may be introduced in the future. Jordan has dinars (JD), while the currency in Sinai is the Egyptian pound (LE). These currencies are only valid in their home countries so, for example, you cannot spend excess Israeli shekels in Jordan. Exchange rates between the three tend to be extremely bad. This means, for example, that it is wise to use up all your shekels before leaving Israel and then to exchange dollars for dinars or pounds on arriving in Jordan or Egypt.

DIRECTORY

EXCHANGE OFFICES

Change Spot
5 Nordau Street,
Haifa.
Tel *(04) 864 4111.*
2 Ben Yehuda St,
Jerusalem.
Tel *(02) 624 0011.*
32 Jerusalem St,
Safed.
Tel *(04) 682 2777.*
13 Ben Yehuda St,
Tel Aviv.
Tel *(03) 510 0573.*
140 Dizengoff St,
Tel Aviv.
Tel *(03) 524 3393.*
www.changespot.co.il

Israel Post
www.israelpost.co.il

Western Union
www.westernunion.com

Israeli Banknotes

Israeli banknotes come in four different denominations: 200, 100, 50 and 20 NIS. There are also plans to release a 500 NIS banknote.

Two hundred shekels (200 NIS)

One hundred shekels (100 NIS)

Twenty shekels (20 NIS)

Israeli Coins

The shekel is divided into 100 agorot. There are coins to the value of 10, 5 and 1 shekels, as well as 50 and 10 agorot.

Ten shekels

Five shekels

One shekel

Fifty agorot

Ten agorot

Jordanian Currency

The Jordanian dinar is divided into 1,000 fils and, confusingly, also 100 piastres (100 fils therefore equals 10 piastres). Notes come in denominations of 20, 10, 5, 1 and ½ dinars. Coins exist to the value of 500, 250, 100, 50, 25, 10 and 5 fils, and 10, 5 and 2½ piastres.

20 dinars

10 dinars

5 dinars

Egyptian Currency

The currency in Egypt is the Egyptian pound (abbreviated to LE). The pound is divided into 100 piastres. Notes come in denominations of LE 200, 100, 50, 20, 10, 5, 1 and 50 and 25 piastres. Coins exist to the value of one Egyptian pound (LE 1), and 20, 10 and 5 piastres.

Twenty Egyptian pounds (LE 20)

Five Egyptian pounds (LE 5)

Communications and Media

Israeli post office logo

Israel's postal service is generally efficient, but letters to Europe and North America can still take a week or more to arrive. This, however, is quicker than the Jordanian or Egyptian postal systems, which are highly unpredictable. Calling overseas is very straightforward from Israel, and it is similarly easy to call overseas in Sinai, but telephone communications from Jordan are considerably more complicated, and expensive.

PUBLIC TELEPHONES IN ISRAEL

Israel's public telephones are almost all operated by the national phone company, Bezek. They take prepaid phonecards, which are sold at post offices, shops and lottery kiosks. They are available in denominations of 20 units (13 NIS), 50 units (29 NIS) or 120 units (60 NIS). Calls made from 10pm to 1am and all day Saturday and Sunday are 25 per cent cheaper than the standard rate. Calls made between 1am and 8am are 50 per cent cheaper. To dial abroad using Bezek, the international access code is 014.

Bezek competes for custom with other telephone companies, including Golden Lines (012 to dial abroad) and Barak (013 to dial abroad). These rival services are often cheaper than Bezek, although it does depend on the country you are calling. You can also make discounted calls from Solan Telecom, whose offices are found throughout Israel.

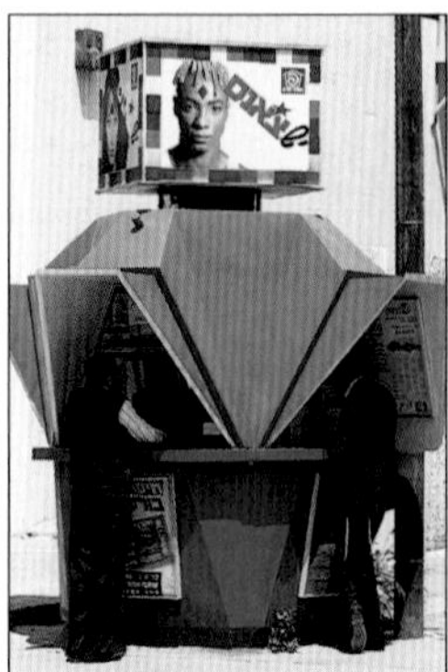
An Israeli lottery kiosk, where phonecards can also be bought

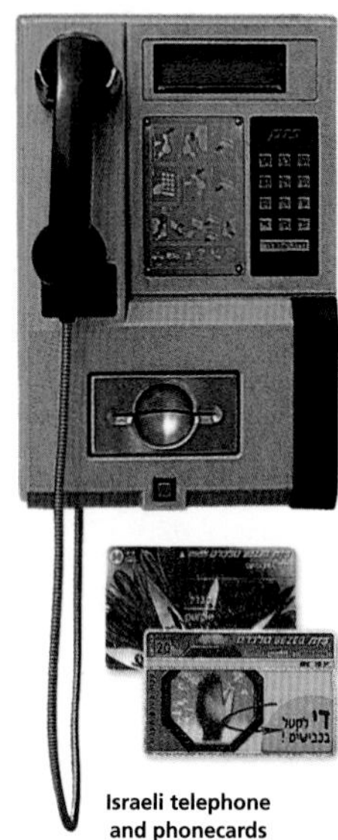
Israeli telephone and phonecards

Visitors can rent mobile phones on arrival at Ben Gurion Airport. Rental rates start at about US$1 per day. Israel's mobile network does not have reciprocal roaming arrangements with many countries. Anyone who plans to take their mobile with them should check with their home service whether it can be used in Israel.

PUBLIC TELEPHONES IN THE PALESTINIAN TERRITORIES

In the West Bank and Gaza Strip, the Palestinians have their own telephone network with their own phonecards. These Palestinian phonecards can be purchased in Arab post offices and some shops. They cannot, at present, be used in Israeli phones.

PUBLIC TELEPHONES IN JORDAN AND SINAI

Jordan's telephone network is creaky, but it is in the process of being upgraded. International calls can be made from public cardphones, for which the cards are purchased from nearby shops. However, phonecards for international calls only come in the denomination of JD 15. A better option is to use one of the many unofficial telephone bureaus, where you write the number you want on a piece of paper and the desk clerk makes the call. These calls are charged by the minute and, with a great many offices competing for custom, rates are reasonable.

The Egyptian telephone network in Sinai also uses phonecards. These come in denominations of LE 15, 20 or 30 and they can be bought at post offices.

POSTAL SERVICES

Using Israeli post offices is straightforward. The exception is if you are sending parcels or bulky items; this entails a series of security inspections. When posting letters, the yellow post boxes are for local correspondence and the red are for the rest of the country and abroad. Post office opening hours in Israel vary depending on the branch but all are closed on Tuesdays. Postal rates vary according to the type of post and its weight, but a standard airmail letter to Europe or the US costs the equivalent of half a US dollar. For postal information in Israel call 177 022 2121.

Red Israeli post box

The Palestinian Authority also has its own postal service, and issues its own stamps, but it is not as efficient as the Israeli service.

A letter posted in Jordan can take anything up to two weeks to reach Europe and a month to the US. It can help to speed things up if you post your letters at a five-star hotel or a main post office, rather than a post box on the street. Jordanian post offices are closed on Fridays.

NEWSPAPERS AND MAGAZINES

English-language readers are well catered for in Israel. The leading English-language publication is the daily *Jerusalem Post* (no Saturday edition). This is worth picking up on Fridays for its extensive cultural supplements and entertainment listings. *Time Out Israel* is a free weekly magazine available at the airports and in some hotels, and has good insider information – particularly on Tel Aviv. *Haaretz* is the oldest national daily newspaper in Israel, with an English insert distributed inside the *International Herald Tribune*. The weekly *Jerusalem Times* is a Palestinian publication, which is usually available only in East Jerusalem and Arab areas of the Old City. In Jordan, look out for the *Jordan Times*, published daily except for Fridays, and the weekly English-language paper *The Star*, published on Thursdays. Foreign newpapers and magazines, such as *The Times, The Washington Post and Newsweek*, are widely available, and are usually just one or two days old.

Local English-language press

TELEVISION AND RADIO

Israeli TV has two state channels, both of which show a large number of subtitled English-language programmes. Most hotels also offer satellite channels such as BBC, Sky News and CNN. In Jordan, Channel 2 devotes plenty of screen time to US programmes, and has English-language news nightly at 10pm. Most hotels have satellite TV, offering a wide choice of programmes.

Israel Radio is the national radio station. It broadcasts news in English each weekday evening at 6:30am, 12:30pm and 8:30pm. In addition, there are various independent and army radio stations.

INTERNET CAFÉS

Despite being an extremely computer literate society, there are few Internet cafés in Israel. This is possibly because most Israeli families have Internet access in their own at home, and because of the wide availability of free Wi-Fi hook-up in many places in Israel, including at Terminal 3 of Ben Gurion Airport *(see p308)*.

There are only a few Internet cafés in Jerusalem, and a handful dotted around the country.

Jordan has an excellent Internet venue, **books@cafe**, located in central Amman. There are further Internet cafés in Jordan at Wadi Musa (Petra), Madaba and Aqaba. In Sinai the Internet is available at many hotels and there is an Internet café near the Fayrouz Hilton in Sharm el-Sheikh. Internet services are also offered at various shops in Dahab. Online time is usually charged by the half hour.

Newspaper seller in Tiberias, Israel

DIRECTORY

TELEPHONE PREFIXES IN ISRAEL

Country code: *972*
Jerusalem: *02*
Tel Aviv: *03*
Haifa and the northern coast: *04*
Galilee and the Golan Heights: *04*
Negev and the Dead Sea: *08*
Coast south of Tel Aviv: *08*
Coast north of Tel Aviv: *09*

TELEPHONE PREFIXES IN THE PALESTINIAN TERRITORIES

Country code: *972*
Bethlehem, Jericho: *02*
Ramallah: *02*

TELEPHONE PREFIXES IN JORDAN

Country code: *962*
Amman: *06*
Jerash: *02*
Kerak, Petra, Aqaba: *03*

TELEPHONE PREFIXES IN EGYPT

Country code: *20*
Sharm el-Sheikh: *069*

INTERNET CAFÉS IN ISRAEL

Interfun
20 Allenby Street, Tel Aviv.
Tel *(03) 517 1448.*

Internet Café
31 Jaffa St, Jerusalem.
Tel *(02) 622 3377.*

Webcafé
7 Sha'ar Hagail St, Netanya.
Tel *(09) 832 1804.*

INTERNET CAFÉS IN JORDAN

books@cafe
Mango St, Jebel Amman, Amman.
Tel *(06) 465 0457.*

Let's Go Internet Café
Off El-Yarmouk St, Madaba.

INTERNET CAFÉS IN EGYPT

Naama Bay Internet Café
Naama Bay Hotel, Naama Bay, Sharm El Sheikh.
Tel *(012) 104 0761.*

Mina Com
Hilton Sharm El Sheikh Fayrouz Resort, Naama Bay, Sharm El Sheikh. ***Tel*** *(069) 600 136.*

TRAVEL INFORMATION

The logo of leading Israeli airline El Al

The easiest way to get to Jerusalem and the Holy Land is to fly direct. Jerusalem is served by Ben Gurion Airport, and there are also international airports at Eilat, Amman in Jordan and Sharm el-Sheikh in Sinai. There are frequent flights to Ben Gurion and, being a busy tourist destination, it is possible to get cheap deals, especially if you are prepared to travel with a smaller, lesser-known airline, or take advantage of a charter package. There are no direct sailings to Israel from mainland Europe; the only sea route is from Athens via Cyprus. Travelling overland is an arduous business as all European trains terminate at Istanbul.

Arrival hall of Terminal 3 at Ben Gurion, Israel's main international airport

FLYING TO ISRAEL

The Israeli national airline is **El Al**. It has direct flights to Ben Gurion Airport from most major European cities, as well as from New York, Los Angeles, Chicago, Miami, Baltimore and Orlando in the United States. Ben Gurion is also served by foreign airlines, including **Air France**, **Alitalia**, **British Airways**, **KLM**, **Lufthansa** and **Swissair**, and **American Airlines**, and **Delta**, as well as some low-cost airlines.

Fares are seasonal. The high season is during the Jewish and Christian holiday periods, in particular Passover, Easter and Rosh ha-Shanah *(see pp36–9)*. At such times fares are at a premium and it can often be hard to find seats.

It is always worth looking into flights to Eilat's Ovda airport. This largely caters for charter traffic, and it is on these flights that the cheapest fares are to be found. The drawbacks are that there are often restrictions on the dates you may travel and you have to make your own way up to Jerusalem and back, a bus journey of between four and five hours each way.

BEN GURION AIRPORT

Named after the first prime minister of Israel, Ben Gurion Airport lies southeast of Tel Aviv, just off the road to Jerusalem. All international flights arrive at and depart from the ultra-modern Terminal 3, which opened in 2004. Services at the airport include duty-free shops, a telecommunications office, foreign currency exchange offices, car-hire outlets and tourist information and hotel reservation desks. There is a domestic terminal for flights to Eilat; Jerusalem and Tel Aviv both have small city airports for internal flights.

Ben Gurion reputedly has the tightest security of any airport in the world. The time taken to inspect every item of baggage means that passengers must check in three hours before departure. However, anyone flying with El Al can check in luggage the day before at special offices in Jerusalem, Tel Aviv and Haifa. Those who do this need only turn up at the airport an hour and a quarter before departure.

El Al aeroplane on the runway at Ben Gurion airport

GETTING TO AND FROM BEN GURION AIRPORT

Ben Gurion Airport is at Lod, about 22 km (14 miles) from Tel Aviv and some 45 km (28 miles) from Jerusalem. Private taxis take about 45 minutes to Jerusalem, or you can take a shared taxi, or *sherut (see p310)*, which is much cheaper. These leave from just outside the arrivals hall but be aware that they do not set off until they are full. The *sheruts* run through the night and will drop passengers anywhere in the city.

Egged buses Nos. 945 and 947 depart every half hour from around 5:30am until 9pm for Jerusalem's Central Bus Station on Jaffa Road. While this is the cheapest method of getting from the airport into the city, the bus station is more than a kilometre from the centre of the New City, and most people will then have to catch a further bus or taxi on to their hotel. The buses do not run on Shabbat – sundown Friday to sundown Saturday.

To get to the airport from Jerusalem, book a taxi the day before departure. Most hotels can usually organize this or call the number for licensed taxis at Ben Gurion Airport.

FLYING TO JORDAN AND SINAI

Jordan's principal airport, and the home base for the national carrier **Royal Jordanian Airlines**, is Queen Alia International. Royal Jordanian has direct services between Amman and most major European capitals. It also flies, via Amsterdam, to New York and Chicago.

Other major carriers flying into Amman include Air France, Alitalia, **BMI** and **Emirates**. There are no non-stop flights from the US – instead you have to fly via a European hub. There is a second airport, known as Marka, about 5 km (3 miles) east of central Amman, but this handles only short-hop flights to Israel and Egypt. There is also a further airport about 10 km (6 miles) north of Aqaba, but

Compact Queen Alia International Airport, Jordan's main air transport hub

it receives very few international flights.

Flights to Amman are not cheap. In general, it is much more economical to fly into Ben Gurion or Eilat in Israel and take a bus across the border. The airport at Sharm el-Sheikh in Sinai lies about 17 km (11 miles) north of town. It is served by Air Sinai and Egypt Air, but these are not direct flights; they involve a change of plane in Cairo.

GETTING TO AND FROM QUEEN ALIA AIRPORT

Queen Alia Airport is about 30 km (19 miles) south of Amman. Comfortable Airport Express buses depart hourly between 7:15am and 9:15pm for Downtown from just outside the arrivals terminal.

Other buses head for the northern parts of town. Be sure to check the destination before boarding to make sure you have the correct bus. Baggage is charged extra on all buses. Alternatively, you can catch a private taxi, but bear in mind that the official going rate is some 15 times the fare on the bus.

FLIGHTS WITHIN THE HOLY LAND

Within Israel domestic flights are operated by **Arkia**. In Jerusalem these flights use Atarot Airport, 7 km (4 miles) north of the city centre. They connect to Tel Aviv (Sde Dov Airport), Eilat and Haifa. With distances in Israel being so short, it only makes sense to fly internally to or from Eilat.

El Al and Royal Jordanian both fly between Ben Gurion and Amman, while El Al and Air Sinai connect Ben Gurion with Sharm el-Sheikh and Cairo. Fares are not cheap, but you can, of course, save a lot of time by flying.

DIRECTORY

AIRPORTS

Ben Gurion
Tel *(03) 975 5555.*
www.iaa.gov.il

Eilat (Ovda)
Tel *(08) 638 4848.*

Queen Alia International
Tel *(06) 445 1739.*
www.qaia.gov.jo

NATIONAL AIRLINES

Arkia
11 Frishman St, Tel Aviv.
Tel *(03) 690 2222.*
www.arkia.co.il

El Al
12 Hillel St, Jerusalem.
Tel *(02) 677 0200.*
32 Ben Yehuda St, Tel Aviv.
Tel *(03) 526 1222.*
Eilat. **Tel** *(08) 632 6504.*
Amman.
Tel *(06) 562 2526.*
Cairo.
Tel *(20-2) 736 1795.*
www.elal.co.il

Royal Jordanian Airlines
Seventh Circle, Amman.
Tel *(06) 510 0000.*
www.rja.com.jo

OTHER AIRLINES

Air France
Tel Aviv.
Tel *(03) 755 5050.*
Amman.
Tel *(06) 566 6055.*
www.airfrance.com

Alitalia
Tel Aviv.
Tel *(03) 796 0700.*
Amman.
Tel *(06) 463 6038.*
www.alitalia.com

American Airlines
Tel Aviv.
Tel *(03) 795 2122.*
www.aa.com

BMI
Amman.
Tel *(06) 554 8951.*
Cairo.
Tel *(20-2) 2395 4888.*
www.flybmi.com

British Airways
Tel Aviv.
Tel *(03) 606 1555.*
www.britishairways.com

Delta Air Lines
Tel Aviv.
Tel *(03) 513 8000.*
www.delta.com

Emirates
Amman.
Tel *(06) 461 5222.*
Cairo.
Tel *(20-2) 1 9899.*
www.emirates.com

KLM
Tel Aviv.
Tel *(03) 796 7999.*
Amman.
Tel *(06) 510 0760.*
Cairo.
Tel *(20-2) 2770 6251.*
www.klm.com

Lufthansa
Tel Aviv.
Tel *(03) 513 5353.*
www.lufthansa.com

Swissair
Tel Aviv.
Tel *(03) 513 9000.*
www.swiss.com

AIRPORT TAXIS

Nesheri
(Shared taxis to and from airport)
Tel *(02) 623 1231.*

Getting Around Jerusalem

Street sign

Most of Jerusalem's major historical and religious sites are concentrated in the Old City, which has to be explored on foot, as it is almost a completely vehicle-free zone. Elsewhere, the city bus network functions efficiently and will get visitors to more or less everywhere they might want to go. This is just as well, as taxis tend to be very expensive for frequent use. The one time when visitors might have to use taxis is on Shabbat, when public transport stops running from sundown on Friday to sundown on Saturday.

JERUSALEM ON FOOT

The old city is very much a pedestrian zone. Its narrow streets and alleys do not allow for vehicles. Flat-soled footwear is essential, as many of the ancient streets are either cobbled or unevenly paved. There are some areas of the New City that are also easy to get around on foot, notably Yemin Moshe and Nakhalat Shiva, but elsewhere wide roads and aggressive traffic can make walking very unpleasant.

Finding your way around poses little problem as street signs are in at least two languages (either Hebrew and English, or Arabic and English). In the Old City, they are in the scripts of all three.

TAXIS

It is easy to find a taxi in Jerusalem. You can either book one by phone, hail one on the street, or find one at an official rank. Restaurant and hotel staff will always phone a cab for you.

Taxis are white if they are Israeli and yellow if they are Arab. Occasionally an Israeli driver may refuse to drive to an address in Arab East Jerusalem, while an Arab driver may balk at venturing into parts of West Jerusalem. All Jerusalem taxi drivers have a bad reputation for overcharging. Although the taxis have modern meters (which can print out a receipt on request), drivers are not in the habit of using them. However, you should insist that it is used. If it is not, you will pay a variable fare, which will be dependent on your haggling skills, but which will certainly be much more than the meter would have indicated. Note that taxi fares are officially higher 9pm–5:30am, on Shabbat and holidays.

White Israeli taxi

SHARED TAXIS

The shared taxi is popular in Jerusalem and throughout the Holy Land region. Known to the Israelis as a *sherut* and to the Arabs as a "service" (pronounced "servees"), shared taxis are a cross between a bus and a taxi. They operate fixed routes like a bus, but they run more frequently and, like a taxi, they can be hailed on the street. At the start of the route drivers wait until every seat is taken before setting off. Points of origin and final destinations are displayed in the front window (although in the case of "services", this will be in Arabic). There are no set stops; passengers indicate to the driver when they wish to be let off. Fares are of a similar rate to the equivalent bus ride and much cheaper than a taxi.

Israeli shared taxi, or *sherut*

Yellow Palestinian taxi

BUSES

Jerusalem's city bus system is run by Egged the national carrier. Tickets are bought from the driver on boarding. The fare is the same for all destinations – the equivalent of just over one US dollar.

Buses are identified by a number displayed in the front window. Major bus routes include: bus No. 1 from **Egged Central Bus Station** to Jaffa Gate and on to Mount Zion and the Western Wall bus station in the Old City's Jewish Quarter; bus No. 20, which runs between Jaffa Gate and Yad Vashem, via Jaffa Road; and bus No. 27, which runs from Hadassah Hospital, along Jaffa Road past the central bus station, terminating at Nablus Road Bus Station in East Jerusalem near Damascus Gate.

Most buses run between about 5:30am and midnight. Night buses run from midnight–3 or 4am Sun–Thu in July and August (Nos 101–107).

East Jerusalem is served by Arab-run buses. It is unlikely that many visitors to the city will find it necessary to use these buses.

Taxi rank on Omar ibn al-Khattab Square inside Jaffa Gate, the Old City

Jerusalem's Central Bus Station

THE NO. 99 BUS

A ride on the No. 99 bus is the best way to discover the city. This bus follows a circular route that in just under two hours takes in most of the important sites outside the Old City. It departs four times per day from Egged Central Bus Station on Jaffa Road: 9am, 11am, 1:30pm and 3:45pm (the last bus doesn't run on Friday, and there are no buses on Saturday). Tickets can be bought on the bus, but it is wise to book in advance as it is often full. Bookings can be made at Central Bus Station, or at the city tourist information office.

There are either one- or two-day tickets, and you can hop on and off wherever you like. A guided tour is available on a personal listening device available in eight languages.

JERUSALEM MASS TRANSIT SYSTEM

The Mass Transit System is the city's answer to the problem of the heavily congested streets. The centrepiece of the project are the two Light Rail lines. The first line connects Pisgat Ze'ev in the north via Jaffa Road to Mount Herzl in the south. The second line, which is due to open at the end of 2010, also runs north to south but with stops near Jabotinsky, Agron and King George V streets.

The BRT (Bus Rapid Transit) is a fleet of "new generation" buses with a dedicated bus lane that runs from Talpiot to Har Hozvim.

For more information on tickets and timetables go to the CityPass website.

USEFUL INFORMATION

CityPass
www.citypass.co.il.

Egged Central Bus Station
224 Jaffa Rd.
Tel **2800*

El-Ittihad Taxis
East Jerusalem.
Tel *(02) 628 4641.*

Ha-Palmakh Taxis
20 Shay Agnon Ave.
Tel *(02) 679 2333.*

Rehavia Taxis
3 Agron St. ***Tel*** *(02) 625 4444.*

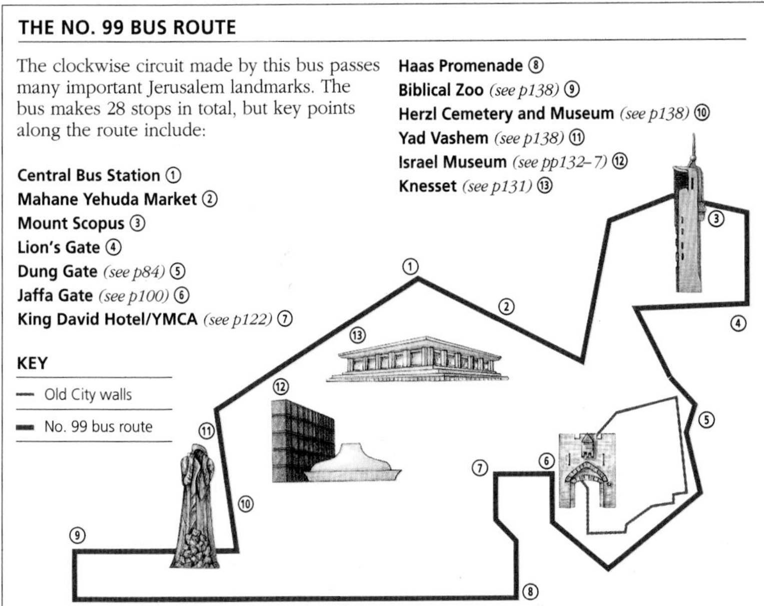

THE NO. 99 BUS ROUTE

The clockwise circuit made by this bus passes many important Jerusalem landmarks. The bus makes 28 stops in total, but key points along the route include:

Central Bus Station ①
Mahane Yehuda Market ②
Mount Scopus ③
Lion's Gate ④
Dung Gate *(see p84)* ⑤
Jaffa Gate *(see p100)* ⑥
King David Hotel/YMCA *(see p122)* ⑦
Haas Promenade ⑧
Biblical Zoo *(see p138)* ⑨
Herzl Cemetery and Museum *(see p138)* ⑩
Yad Vashem *(see p138)* ⑪
Israel Museum *(see pp132–7)* ⑫
Knesset *(see p131)* ⑬

Public Transport in the Holy Land

By far the best and most popular way of getting around Israel and the Holy Land is by bus. Every town and city has a bus station, and inter-urban services tend to be frequent and very affordable. In comparison, rail networks in this part of the world are extremely limited: Israel has just two lines, and Jordan one, which is of little use, running, as it does, north to Damascus. There are no railways at all in Sinai. Sea transport is limited to just one route, across the Red Sea between Jordan and Sinai.

Modern Egged bus, a popular and convenient way to travel in Israel

LONG-DISTANCE BUSES

Nearly all long-distance bus routes in Israel are operated by the Egged company. This virtual monopoly at least has the advantage of making bus travel straightfoward and simple. Except to the Dead Sea region, services are frequent. For example buses depart from Jerusalem to Tel Aviv every 15 minutes, to Haifa every 45 minutes, and to Tiberias every hour. There is rarely any need to book in advance; you can simply turn up at the city bus station and get a ticket for the next service out. The only time that you might need to book in advance is if you are travelling to Eilat, as there are only about four buses a day that head in this direction.

Given the small size of the country, journeys are never very long (the longest one is from Jerusalem to Eilat, which lasts around five hours). Egged buses are comfortable and air-conditioned, with plenty of space in the baggage holds.

There are passes for an unlimited number of journeys, which are valid for one or more weeks. These are called Israbus cards. For information on these passes and reduced fares for students, contact the central bus stations or **Egged Tours** website.

The one drawback to Israeli buses is that there are no services on Shabbat (Sabbath). This means that you should not plan to travel any time from late Friday afternoon to early evening Saturday. There are no buses either on Jewish holidays *(see pp36–9)*. This can prove highly disruptive for any visitors who may be caught unawares.

TRANSPORT IN THE PALESTINIAN TERRITORIES

With the constant new developments in the administrative situation, public transport in the Palestinian territories is forever changing. In general, there are two options: Arab buses or shared taxis *(see p310)*. Arab buses depart from two stations in East Jerusalem, one on Nablus Road (which is used mainly for city services), the other on Suleyman Street, opposite the Old City walls. From one of these two, visitors can catch services for West Bank Palestinian towns such as Bethlehem, Hebron, Jericho and Ramallah.

Arab shared taxis depart from a parking lot just outside the walls opposite Damascus Gate. They serve all the same destinations as the buses, but they are faster and depart far more frequently.

In general Arab buses do not go to Israeli towns, and vice versa. It is possible to catch an Israeli Egged bus to Bethlehem, but it drops you off on the highway outside town necessitating a 20-minute walk into the centre.

TRAVELLING BY TRAIN

Israel's very limited coastal railway system comprises just two lines: one from Tel Aviv to Jerusalem and a second from Tel Aviv to Nahariya. The latter runs up the northern Mediterranean coast to near the border with Lebanon. Although the line serves several important destinations, including Haifa and Acre, the drawback is that there are few services each day and on Jewish holidays trains are very crowded. Stations also tend to be some distance from the town centre, often requiring a taxi ride to reach them.

The other line, between Jerusalem and Tel Aviv (following on to Haifa), was upgraded in 2005 and it

The line up the north coast of Israel, slow but scenic

Bright yellow taxis amid the busy traffic of central Amman

passes through some particularly lovely scenery. However, the train is much slower than the bus – allow an extra 40 minutes. There are 6–10 trains per day.

TRANSPORT IN JORDAN

There are several national bus companies in Jordan. The main one is **JETT**, which runs blue-and-white air-conditioned buses between Amman and Aqaba, the King Hussein (Allenby) Bridge and Petra. Booking your seat in advance is advisable. The JETT bus station in Amman is on King Hussein Road. Ten minutes' walk downhill on King Hussein is the Abdali bus station, which is where all the other Jordanian bus companies depart from for routes north and west, including services to Ajlun, Jerash and the King Hussein Bridge. All non-JETT buses to the south (including services to Kerak, Petra and Aqaba) leave from the Wahdat station, some 5 km (3 miles) south of the city centre.

The one destination that is hard to reach from Amman is the Dead Sea. There are no scheduled bus services. The only way to get here is by minibus or shared taxi.

Shared taxis are common in Jordan and far more frequent and convenient than buses. A shared taxi ride from Amman to Aqaba takes about five hours and one from Amman to Petra about three.

The only regular rail service in Jordan is the three times a week train up to Damascus. It runs on the Hejaz Railway, built at the turn of the 20th century by the Turks but more famous for being repeatedly blown up by Lawrence of Arabia and his Arab fighters *(see p233)*. The trip takes about nine hours but you must have a visa in advance to enter Syria.

To get about in Amman there are city buses, but the destination is indicated only in Arabic. Taxi drivers tend to be honest and use the meter, making this an acceptable way of getting around. Only late in the evening or for longer journeys (such as to and from the airport) will you have to agree upon the price beforehand.

TRANSPORT IN SINAI

The resorts of the east coast of the Sinai peninsula are served by the buses of Egypt's East Delta Bus Company. Services are not particularly frequent with no more than about half a dozen buses a day. All of these buses are either coming from or heading to Cairo (which is between seven and nine hours away). Only one or two of these buses pass by St Catherine's Monastery, so you need to check timetables carefully.

A very informal shared taxi service also operates in Sinai, but it can take time for the cars to fill up and the drivers can be alarmingly reckless so this method of transport is not really recommended.

RED SEA FERRIES

Aqaba in Jordan and Nuweiba in Sinai are linked by a ferry and a catamaran. Both of these make one sailing each way, once a day. The ferry, which also carries cars, takes three hours, while the catamaran completes the trip in around one hour. Booking in advance is not necessary unless you are travelling with a car. It is possible for passengers to obtain a Sinai Permit *(see p299)* on board both vessels – this allows you to stay in the region for up to 14 days.

DIRECTORY

BUS INFORMATION IN ISRAEL

Dan Buses
Tel *(03) 639 4444.*

Egged Information
Tel **2800 or (03) 694 8888.*
www.egged.co.il

Egged Tours
Tel *(03) 527 1212 or 1 700 70 75 77 (within Israel only).*
www.eggedtours.com

Eilat Bus Station
Ha-Temarim St.
Tel *(08) 636 5120.*

Haifa Bus Station
Ha-Mifratz and Hof Ha-Karmel Stations.
Tel *(04) 847 3555.*

Jerusalem Bus Station
224 Jaffa Rd.
Tel *(02) 530 4704.*

Metrodan
Tel **5100.*

Tel Aviv Bus Station
Levinsky St.
Tel *(03) 694 8888.*

BUS INFORMATION IN AMMAN

JETT Bus Station
King Hussein Rd.
Tel *(06) 566 4146.*
www.jett.com.jo

TRAIN INFORMATION

Israel Railways
Tel **5700.*
www.rail.co.il

Travelling Around by Car

Road sign in three languages

With well-maintained roads, light traffic away from the big cities and Israel's coastal highway, short distances between towns and some enchanting scenery, the Holy Land should be a pleasure to drive around. The one black spot is other road users. Both Israelis and Arabs can be reckless behind the wheel, and road fatalities are high. While this should not put you off driving, you do need to be cautious. On the positive side again, Israel is full of small places of beauty and interest, located well off any bus route, and having a car at your disposal can really open up the country.

CAR HIRE

Most international car hire companies are represented in Israel, with offices (or counters) at Ben Gurion Airport, in Tel Aviv and in Jerusalem. For the sake of convenience, it is better to use one that has a representative at the airport. To rent a car, you must have a full, clean driving licence (an international driving licence is not necessary). Cars are rented only to those over 21 years old, although some companies require that you be 23. Prices vary dramatically and it is recommended that you shop around before settling on a deal. Local companies, such as **Eldan**, frequently offer the best rates. Be aware that rental charges are usually quoted exclusive of insurance and collision waivers.

Budget

Sign for a car rental company

Note that it is not allowed to take cars hired in Israel over into Jordan or Sinai.

Car hire is not very popular in Jordan and Sinai because there are so few roads to explore. It also works out as very expensive when compared with getting around by other forms of transport, such as the bus or hiring a taxi for a day or two.

Petrol stations in Jordan, Sinai and even certain parts of Israel, particularly the Negev and Dead Sea areas, are few and far between. You are strongly advised to fill up your tank before setting off on any long journeys.

Petrol station in Israel

THE RULES OF THE ROAD

Driving in Israel is on the right-hand side of the road. At unmarked junctions drivers give way to traffic on the right, and overtaking is done on the left. The speed limit in towns is 50 km/h (30 mph) and 90 km/h (55 mph) on out-of-town roads. On some motorways the speed limit is 100 km/h (60 mph). Seat belts must be worn. Children under 15 must sit in the back and children under four must be restrained in a suitable child's seat.

ROAD SIGNS IN ISRAEL

Although there is a lack of cautionary and warning signs on Israel's roads, all places of interest are well indicated. Signs are in both Hebrew and English (and sometimes in Arabic too). A problem arises, however, with the lack of consistency in the transliteration of place names from Hebrew and Arabic into English. You could be following directions for Beersheba one minute and for Be'er Sheva the next. These are, of course, the same place. In this book we have tried to present place names as you will see them spelled on Israeli road signs but local inconsistencies mean that this is not always the case.

No entry sign

School sign

Two-way sign

Right-hand bend

Tourist site sign

Parking sign

DRIVING IN THE PALESTINIAN TERRITORIES

Cars in Israel and the Palestinian Autonomous Territories have licence plates of different colours. Israeli cars have yellow plates, while Palestinian cars' plates are green. It is inadvisable to drive a car with yellow, Israeli plates into Palestinian areas, particularly frequent trouble-spots such as Hebron and Ramallah. Cars hired in Israel are usually not insured for the Palestinian Territories. Conversely, driving a car with Palestinian plates in Israel will make you the object of a great deal of unwelcome attention from the security forces.

DRIVING IN JORDAN

While driving is on the right, Jordanians seem to consider most other road rules open to interpretation. Overtaking takes place on both sides of the road and right of way goes to he or she

Typically heavy traffic on the seafront promenade in Tel Aviv

who hesitates least. Roads are often in a poor state of repair. Many are badly surfaced, and road markings are often absent.

Speed limits are generally 100 km/h (60 mph) on open roads and 40 km/h (25 mph) in built-up areas. Care is needed on desert roads, where drifting sand can put the car into a spin if hit at speed.

Direction signs are frequently positioned right at the junction, offering no advance warning and making it all too easy to drive past your turn-off.

DRIVING IN SINAI

There are very few roads in Sinai, so routes to drive are limited. They do, however, pass through some stunning scenery. Traffic is light but what traffic there is, is mainly composed of buses and large shared taxis; these generally travel at high speed, paying little heed to other road users. Car drivers must constantly be on the lookout and be prepared to take evasive action.

Other than on recognized trails, off-road driving is not encouraged as it can damage the fragile desert environment. Several such trails begin in the region of Nuweiba *(see p242)*.

DRIVING IN CITIES

Traffic in and around Tel Aviv and, to a lesser extent, Jerusalem is nightmarish. You should aim to avoid rush hour, which is roughly 7–9am and 4–6pm. That said, it is not unknown to encounter traffic jams in Tel Aviv at 1am.

HITCH-HIKING

Known in Israel as *tremping*, hitch-hiking used to be a common way of getting about the country. It was particularly popular with soldiers heading home or returning from leave. But recently hitch-hiking has become increasingly unsafe. Women soldiers are now banned from hitching and we recommend visitors do not hitch-hike either.

CYCLING

Parts of Israel are excellent places for cycle touring. The best regions are Galilee and the Golan Heights, where the scenery is at its most varied and the altitude serves to moderate the extreme summer temperatures. Even so, from June to August it is best to plan to cycle only in the mornings, to avoid the afternoon heat.

In Tiberias, it is possible to hire bicycles by the day to explore the shores of the Sea of Galilee *(see pp182–4)*. In Jerusalem you can rent bicycles by the day from **Walk Ways**, who will deliver to your hotel. For general cycling advice and to enquire about joining organized rides, enthusiasts could also try contacting the **Jerusalem Cycle Club**.

Cycling in Jaffa

DIRECTORY

CAR HIRE IN ISRAEL

Autoeurope
Tel Aviv.
Tel *(03) 524 4244.*
www.autoeurope.co.il

Avanti
www.avanti.co.il

Avis
Tel **2722 or 1-700-700-222.*
www.avis.co.il

Budget
Tel **2200 or 1-700-70-41-41.*
www.budget.co.il

Eldan
Tel *1-700-700-740.*
www.eldan.co.il

Europcar
www.europcar.com

Hertz
Ben Gurion Airport.
Tel *(03) 975 4505.*
Jerusalem.
Tel *(02) 623 1351.*
www.hertz.co.il

Sixt
Tel *(03) 975 4167.*
www.sixt.com/car-rental/israel

CAR HIRE IN JORDAN & SINAI

Avis
Amman.
Tel *(06) 569 9420/30.*
www.avis.com.jo

EuroDollar
Amman.
Tel *(06) 569 3399.* **www**.1stjordan.net/eurodollar

Europcar
Amman.
Tel *(06) 445 2012.*
www.europcar.jo

Hertz
Sharm el-Sheikh/Cairo.
www.hertzegypt.com
Amman.
www.hertzjordan.com

Oscar Car Rental
Amman.
Tel *(06) 553 5635.*
www.1stjordan.net/oscar

Rent a Reliable Car
Amman.
Tel *(06) 592 9676.*
www.rentareliablecar.com

CYCLING

Walk Ways
Tel *(02) 534 4452.*
www.walk-ways.com

General Index

Page numbers in **bold** refer to main entries

A

N

Acknowledgments

Dorling Kindersley would like to thank the following people whose invaluable contributions and assistance have made the preparation of this book possible.

Senior Managing Editor
Louise Bostock Lang.

Managing Art Editor
Jane Ewart.

Editorial Director
Vivian Crump.

Publishing Manager
Scarlett O'Hara.

Revisions Coordinator/Editor
Anna Freiberger, Rose Hudson.

Art Director
Gillian Allan.

Publisher
Douglas Amrine.

Main Consultants
Felicity Cobbing, Andrew Humphreys, Jonathan Tubb.

Translator
Richard Pierce.

Maps
Rob Clynes, James Macdonald (Colourmap Scanning Ltd).

Production
Imogen Boase, Marie Ingledew.

Additional Contributors and Consultants
Jonathan Elphick, Professor Jonathan Magonet, Peter Parr, Amir Reuveni, Matthew Teller, Wolfgang Tins.

Visualizer
Joy FitzSimmons.

Additional Illustrations
Richard Bonson.

Additional Photography
Steve Gorton, Noam Knoller, Ian O'Leary.

Design and Editorial Assistance
Gillian Andrews, Sam Borland, Camilla Gersh, Priya Kukadia, Esther Labi, Nicola Malone, Loren Minsky, Sonal Modha, Helen Partington, Pollyanna Poulter, Lee Redmond, Marisa Renzullo, Ellen Root.

Editor
Jude Ledger.

Factchecker
Noam Knoller.

Proof Reader
Stewart J Wild.

Indexer
Hilary Bird.

Special Assistance
Sheila Brull, Egyptian Tourist Authority, Giovanni Francesio and Mattia Goffetti at Fabio Ratti Editoria, Efrat Goller at Keter Publishing, Tony Howard and Di Taylor at N.O.M.A.D.S. (New Opportunities for Mountaineering and Desert Sports), Israel Ministry of Tourism, Jordan Tourism Board, Amalyah Keshet and Tal Sher at the Israel Museum, Deborah Lipson at the Tower of David Museum of the History of Jerusalem, Hila Reuveni, Shelly Shemer at the Israel Wine and Gourmet Magazine. Special thanks to Massimo Acanfora Torrefranca.

Additional Picture Research
Julia Harris-Voss.

Photographic and Artwork Reference
Dale Harris, Ben Johnson, Albatros, Jerusalem.

Photography Permissions
The publisher would like to thank all the churches, museums, hotels, restaurants, shops, galleries and sights too numerous to thank individually, for their co-operation and contribution to this publication.

Picture Credits
t = top; tl = top left; tlc = top left centre; tc = top centre; tr = top right; cla = centre left above; ca = centre above; cra = centre right above; cl = centre left; c = centre; cr = centre right; clb = centre left below; cb = centre below; crb = centre right below; bl = bottom left; b = bottom; bc = bottom centre; bcl = bottom centre left; br = bottom right; (d) = detail.

Works of art have been reproduced with the permission of the following copyright holders: *Reclining Figure* (1969–70) Henry

Moore, Gift of Maurice and Bella Wingrave, London. Through the British Friends of the Art Museums in Israel 170bc.

The publisher would like to thank the following individuals, companies and picture libraries for permission to reproduce their photographs:

ALAMY IMAGES: Tibor Bognar 214cla; Charles Bowman 204tr; Paul Doyle 214br; Eddie Gerald 192; Nick Hanna 243tl; 295br; ImageState/David South 11tr; Israel images/ Hanan Isachar 138tr; Shein Audio Visual 288cl; Jochen Tack 212bl; Danny Yanai 269c; AMERICAN COLONY HOTEL: 266cl; ANCIENT ART & ARCHITECTURE COLLECTION: 28cb, 29ca, 30cr, 41c, 41br, 42t, 44cl, 47ca; R Sheridan 24t, 41bl, 44tl, 46b, 50cb, 107cla, 107cl; G Tortoli 27cra; ANDES PRESS: Carlos Reyes-Manzo 23cr; AKG, LONDON: 42crb, 46t, 50t, 54bl, 233br; Erich Lessing 20cl, 21ca, 22br, 29b, 30b, 45cb, 45b, 46cb, 92ca, 189c, 216t, 216ca, 216br; Jean Louis Mou 59t; FABRIZIO ARDITO 20b, 26t, 35cra, 78tl, 103b, 110tl, 179t, 217b, 222t, 224t, 232t, 233t, 233cb, 235t, 235c, 298t, 299t, 300tl, 300tcr, 300tr, 300c, 304c, 304b, 306c, 306c, 306br, 307c, 310tbr, 310c, 314cl, 314cb; ASAP, JERUSALEM: 120ca; Eyal Bartov 34tl, 34cl, 35t, 35ca, 33bl; Lev Borodulin 255tr; Bridgeman Art Library 24cl, 27cr, 51bl, 71c; C.Z.A. 52ca; Shai Ginott 34clb, 58bl; Avi Hirschfield 79c, 79cb; Hanan Isachar 132tr, 137cra, 312b; Itsik Marom 34bl, 35cl, 35clb, 35cb; Garo Nalbandian 3 (inset), 24bl, 24br, 107cr, 110ca, 110cb, 246b, 247b; Richard Nowitz 110tr; Nitsan Shorer 308bc; Vivian Silver 53t; Israel Talby 34cr; AVIS BUDGET GROUP: 314cl.

BANK LEUMI (UK) PLC: 304tl; BRIDGEMAN ART LIBRARY: *Christ Carrying the Cross* Eustache Le Sueur (1651) 31b, 32b, *Jerusalem from the Mount of the Olives* Edward Lear (1859) 33t, *The Finding of the Saviour in the Temple* William Holman Hunt (1854–60) 33b, 40, 248b; Archeological Museum, Amman, Jordan 212c; Bibliothèque Municipale de Lyon 48clb; British Library 20cr, 21t; Galleria Borghese *St Jerome Writing* Caravaggio (1604) 195b; Giraudon 26cl; Musée Condé, Chantilly 28b; BRITISH LIBRARY: 22t.

CAMERA PRESS: Fred Adler 309t; CORBIS: Aaron Horowitz 268cl; Hanan Isachar 296br; Reuters/Ali Jarekji 213tc; Peter Turnley 55br.

JO DORAN: 90c.

EGGED – ISRAEL TRANSPORT COOPERATIVE SOCIETY LTD.: 312cl; EL AL ISRAEL AIRLINES: 308tc; E.T.ARCHIVE: 49t, 49br; MARY EVANS PICTURE LIBRARY: 9 (inset), 32t, 32c, 33c, 48bl, 48br, 57 (inset), 87tc, 251c.

ffOTOGRAFF: Patricia Aithie 17t, 50b, 60, 67t, 98b, 99t, 104ca, 115b, 122t, 122c, 125br, 129t, 138c, 138b, 143c, 197c, 253c, 284t, 287cra, 288b, 306bl, 310tl, 314t; Charles Aithie 22–23c, 31clb, 59tr, 114b, 121ca; GINO FRONGIA 16, 24–25c, 34cla, 68c, 70tr, 102tr, 104tr, 105ca, 174tl, 187b, 190b, 254c, 300b, 302tl.

CRISTINA GAMBARO: 5ca, 62ca, 62cb, 65t, 67c, 93b, 99b, 109t, 179c, 182b, 197t, 202t, 302tr, 304l; EDDIE GERALD: 11br, 30cl, 62b, 68tl, 78cla, 78c, 79ca, 83t, 106b, 107t, 107bl, 111cb, 120cb, 124t, 125bl, 164, 174tr, 174ca, 174bl, 182tl, 183t, 183c, 195cra, 207, 228b, 238b, 242c, 246cb, 247ca, 285c, 289t, 289c, 289b, 290b, 303t, 303c, 303b, 310tr, 312c, all 314cra, all 314cr, 314crb, 314b. GETTY IMAGES: AFP/Andre Brutmann 55crb.

SONIA HALLIDAY: Laura Lushington 137bl; ROBERT HARDING PICTURE LIBRARY: Adrian Neville 300br; 25cr; ASAP/Nalbandian 58t, 100br; Gascoigne 28c; HEMISPHERES IMAGES: Franck Guiziou 269tl; HOLMES PHOTOGRAPHY: 249, Jean Holmes 49crb, Reed Holmes 232b, TONY HOWARD:232ca.

IMAGES COLOUR LIBRARY: 244–245; INTER-CONTINENTAL HOTEL, AQUABA: 267cl; HANAN ISACHAR: 1c, 4b, 5t, 5clb, 11clb, 19b, 25t, 25cra, 25crb, 31t, 35crb, 36t, 36c, 36b, 37ca, 37br, 38cra, 38b, 39c, 39b, 61t, 88, 91t, 92cb, 93t, 93cra, 94cb, 100bc, 110b, 162ca, 166c, 167b, 175ca, 175cb, 175b, 176t, 176b, 178t, 182cb, 189t, 194ca, 194cb, 197clb, 200t, 205bl, 207, 216cl, 218–219, 221b, 226c, 249t, 282b, 284b, 311tl, 315b; WWW.ISRAELIMAGES.COM: Raffi Rondel 290tl; Israel Talby 267br, 267tr, 308cl; ISRAEL MINISTRY OF TOURISM: 300tr; ISRAEL MUSEUM: 44cbl, 133crb, *Destruction and Sack of the Temple of Jerusalem* Nicolas Poussin (1625–6) 45t, 45cr, 49cra, 53b, *Apple Core* Claes Oldenburg (1929), 132tl, 132cla, 132cra, *Red-Blue Chair* GT Rietveld (1918) © DACS, London 2006, 133t, *The Rabbi* Marc Chagall (1912–13) ©ADAGP, Paris and DACS, London 2006, 133ca, 134tl, *Jeanne Hebuterne seated* Modigliani (1918) 134tr, 134b, 135t, 135c, 135b, 136c, 137t, 137cl, 137c, 137cr, 201c; *Woman Combing Her Hair* Alexander Archipenko (1914) © ARS, NY and DACS, London 2006, 132bl, 136b;

David Harris 133cb, 133bl, 134c, 137br; Ann Levin 136t.

Paul Jackson: 174b, 175t, 225bl, 227br; www.Jerusalemshots.com: 94crb; Jordan Tourist Board: 197crb, 234bl. 234cla, 234tr, 294bl, 294cla, 296tl.

Magnum Photos:54t.

NHPA: Henry Ausloos 35br; Richard Nowitz: 2–3, 18t, 74–75, 92b, 93c, 95c, 162b, 163t, 163cb, 165b, 178c, 178b, 186, 188ca, 189b, 193tl, 196b, 197b, 200br, 201bl, 203br, 206, 217t, 223b, 225br, 230b, 236, 239t, 242t, 247t, 247bl, 249b, 250–251, 252t, 285t, 287ca, 299b, 307b; Air Photos, Israel 186.

Cristine Osborne Pictures: 163b, 229c, 248t, 248c.

Pa Photos: Ariel Schalit 290c; Planet Earth Pictures: Kurt Amsler 239b; Popperfoto: 51br, 52c, 52bl, 52br, 53c, 54br. Zev Radovan: 4–5t, 20tr, 21cb, 22cl, 22bl, 23t, 23tr, 28t, 29cb, 42bl, 42br, 43cb, 43b, 51t, 67b, 71b, 76, 77t, 130c; Fabio Ratti: 72c, 103ca, 117b, 212t, 212b, 213t, 213b, 301b; Retrograph Archive: 54c; Reuters: Ronen Zvulun 55tl; Rex Features: 19c.

Peter Sanders Photography: 26bl, 26br, 27t, 70tl, 70c; Science Photo Library: CNES, 1990 Distribution Spot Image 12ca; THE Original Shakespeare Company 37bl; Eitan Simanor: 39cra, 44b, 56–57, 182tr, 254b, 290t; Jon Spaull: 163ca, 208t, 208b, 209t, 209b, 296–297, 313c; STA Travel Group: 301tc.

Visions of the Land: American Colony Hotel 51cb; Tony Malmqvist 237b, 240t, 240c, 240cb, 240bl, 240bc, 240br, 241tl, 241tr, 241cra, 241cb, 241bl, 241br, 243c, 243b, 292t; Beni Mor 29t, 63t, 64b, 90tl, 90tr, 91b, 101c, 114c, 115t, 116t, 116b, 177b, 180t, 180b, 181t, 181b, 184t, 184b, 190t, 191t, 191b, 192t, 203b, 222cl, 233ca; Garo Nalbandian 26–27c, 35cla, 59crb, 63cb, 68b, 69cr, 69b, 72b, 73cr, 100t, 100clb, 106t, 106c, 107cb, 107br, 112tl, 112tr, 113t, 113b, 124b, 125t, 185t, 193t, 198–199, 211b, 223tl, 223tr, 223c, 224c, 224b, 226t, 226b, 227tr, 227tl, 227cla, 228t, 229cb, 229b, 230t, 230c, 231t, 231cl, 231cr, 231b, 235b, 205t, 246t, 246cla, 246b, 255t, 266t, 266c; Basilio Rodella 43ca, 62t, 63ca, 64t, 64c, 66t, 66c, 66b, 69t, 72tr, 73tr, 73b, 78b, 79b, 82t, 82c, 82b, 84t, 84c, 84b, 85t, 85c, 85b, 90ca, 92t, 104tl, 114t, 117t, 121cb, 122b, 123b, 126t, 126bl, 126br, 127c, 127b, 130t, 130b, 131c, 131b, 138t, 177c, 184cr,
185b, 190c, 192b, 196t, 200c, 200bl, 201t, 201ca, 202b, 203t, 210t, 210b, 253t, 288t, 293t; SPNI Collection/Yossi Eshbol 203c, 243t; Studium Biblicum Franciscanum Archive 216–217c, 217cr; Ilan Sztulman 5crb, 252c, 268tr, 268tl, 268tc, 268cla, 268ca, 268cra, 268cl, 268cr, 268clb, 268bl, 268bc, 268br, 269t, 269tl, 269tr, 269cla, 269cra, 269c, 269cb, 269bl, 269bra, 269br, 270tl, 270cl, 270cr, 270cr, 270b, 271bl, 271clb, 271bc, 271cbc, 271brc, 271br, 283t, all 286cla, 286crc, 286clb, 286cb, 286bcl, 286bc, 287tl, 287tc, 287tr, 287trr, 287cla, 287c, all 287crb, 298b, 312t; www.VisitJordan.com: 300clb.

Werner Forman Archive: British Museum 21b; PETER WILSON: 17b, 34tr, 34br, 160–161, 167t, 183b, 188b, 211t, 222–223, 227b, 228cb, 229t, 232cb, 233bl, 238t, 302b, 315t.

Front Endpaper: All commissioned photography except ffotograff: Patricia Aithie tr; Eddie Gerald tcl; Hanan Isachar c; Richard Nowitz tl, bcl, bl; ZEV Radovan brc.

COVER:
Front Cover: Corbis: JAI/Jon Arnold; main image; DK images: Eddie Gerald clb.
Back Cover: Kurt Amsler: cla; DK images: Alistair Cuncan clb, tl; Richard T Nowitz: bl.
Spine: Corbis: JAI/Jon Arnold t; DK Images: Magnus Rew b.

Hebrew Phrase Book

Hebrew has an alphabet of 22 letters. As in Arabic, the vowels do not appear in the written language and there are several systems of transliteration. In this phrasebook we have given a simple phonetic transcription only. Bold type indicates the syllable on which the stress falls. An apostrophe between two letters means that there is a break in the pronunciation. The letters "kh" represent the sound "ch" as in Scottish "loch", and "g" is hard as in "gate". Where necessary, the masculine form is given first, followed by the feminine.

In Emergency

Help!	Hatzi**lu**!
Stop!	At**zor**!
Call a doctor!	**Az**minu ro**fe**!
Call an ambulance!	**Az**minu **am**bulans!
Call the police!	Tzaltze**lu** lamishta**ra**!
Call the fire brigade!	Tzaltze**lu** lemekha**bei** esh!
Where is the nearest telephone?	**E**fo ha**te**lefon hatzibu**ri** ha**khi** ka**rov**?
Where is the nearest hospital?	**E**fo bet hakho**lim** ha**khi** ka**rov**?

Communication Essentials

Yes	Ken
No	Lo
Please	Bevaka**sha**
Thank you	To**da**
Many thanks	To**da** ra**ba**
Excuse me	Sli**kha**
Hello	Sha**lom**
Good day	**Bo**ker tov
Good evening	**E**rev tov
Good night	**Lai**la tov
Greetings (on the Sabbath)	Sha**bat** Sha**lom**
Have a good week (after the Sabbath)	Sha**vu**'a tov
morning	**bo**ker
afternoon	a**khar** hatzoho**ry**im
evening	**e**rev
night	**ly**la
today	ha**yom**
tomorrow	ma**khar**
here	po
there	sham
what?	ma?
which?	**ei**zeh?
when?	ma**tai**?
who?	mi?
where?	**e**fo?

Useful Phrases

How are you?	Ma shlom**kha**/shlo**mekh**?
Very well, thank you	Be**se**der, to**da**
Pleased to meet you	Na'**im**me**od**
Goodbye	Lehitra**ot**
(I'm) fine!	Be**se**der ga**mur**
Where is/Where are?	**Efo**...?
How many kilometres is it to...?	**Ka**ma kilome**trim** mi**po** le...?
What is the way to...?	Ekh megi'**im** le...?
Do you speak English?	A**ta**/at meda**ber**/meda**be**ret ang**lit**?
I don't understand	A**ni** lo me**vin**/me**vi**na
Could you speak more slowly, please?	Tu**khal**/tukh**li** leda**ber** yo**ter** le'**at**, bevaka**sha**?

Useful Words

large	ga**dol**
small	ka**tan**
hot	kham
cold	kar
bad	lo tov
enough	mas**pik**
well	be**se**der
open	pa**tu**akh
closed	sa**gur**
left	smol
right	ya**min**
straight	ya**shar**
near	ka**rov**
far	ra**khok**
up	le**ma**la
down	le**ma**ta
soon	muk**dam**
late	meu**khar**
entrance	kni**sa**
exit	yet**zia**
toilet	sheru**tim**
free, unoccupied	pa**nu**i
free, no charge	khi**nam**

Making a Telephone Call

I'd like to make a long-distance call	Haiti rot**ze**/rot**za** lehitka**sher** le**khutz** la**ir**
I'd like to make a reversed-charge call	Haiti rot**ze**/rot**za** lehitka**sher** gov**ai**na
I'll call back later	Etka**sher** meu**khar** yo**ter**
Can I leave a message?	Ef**shar** leha**shir** hoda'**a**?
Hold on	Ham**tin**/hamt**i**ni (Tam**tin**/tamt**i**ni)
Could you speak up a little, please?	Tu**khal**/tukh**li** leda**ber** be**kol** ram yo**ter**?
local call	si**kha** iro**nit**
international call	si**kha** benleu**mit**

Shopping

How much does it cost?	**Ka**ma zeh o**leh**?
I would like...	Haiti rot**zeh**/rot**za**...
Do you have...?	Yesh la**khem**...?
I'm just looking.	A**ni** rak mista**kel**/mista**ke**let.
Do you take credit cards?	Atem mekab**lim** kar**ti**sei ash**rai**?
Do you take traveller's cheques?	A**tem** mekab**lim** traveller's cheques?
What time do you open?	Ma**tai** pot**khim**?
What time do you close?	Ma**tai** so**grim**?
this one	zeh
that one	ha**hu**
expensive	Ya**kar**
inexpensive/cheap	lo ya**kar**/zol
size	mi**da**
shoe size	mi**da** (mi**dat** **na**'alyim)
white	la**van**
black	sha**khor**
red	a**dom**
yellow	tza**hov**
green	ya**rok**
blue	ka**khol**

Types of Shop

antiques shop	kha**nut** ati**kot**
bakery	ma'a**fia**
bank	bank
barber's	maspe**ra**
bookshop/newsagent	kha**nut** sfa**rim**/ve'ito**nim**
butcher's	it**liz**
cake shop	ma'ada**nia**
chemist's	bet mer**ka**khat
clothes shop	kha**nut** b'ga**dim**
greengrocer's	yar**kan**
grocer's	ma**ko**let
hairdresser's	maspe**ra**
jeweller's	kha**nut** takhshi**tim**
market	shuk
post office	snif ha**do**ar
shoe shop	kha**nut** na'**al**yim
supermarket	**su**permarket
travel agency	sokh**nut** nesi**yot**

Sightseeing

bus station	takha**na** merka**zit**
bus stop	takha**nat** **o**tobus
church	knesia
closed	sa**gur**
library	sif**ria**
mosque	mis**gad**
park	park
synagogue	bet hak**nes**set
taxi	mo**nit**
tourist information office	merkaz hamei**da** leta**yar**
town hall	bet ha'i**ria**
train station	takha**nat** ra**ke**vet

Staying in a Hotel

I have a reservation	Yesh li azma**na**
Do you have a free room?	Yesh la**khem** **khe**der pa**nu**i?
double room	**khe**der zugi
room with two beds	**khe**der im sh**tei** mi**tot**
room with a bath or a shower	**khe**der im sheru**tim** ve am**ba**tia o mik**la**khat
single room	**khe**der ya**khid**

key	maf**teakh**
lift	ma'a**lit**
Can someone help me with my luggage?	**Mi**shehu ya**khol** la'a**zor** li im hamisva**dot**?

Eating Out

Have you got a table free?	Yesh la**khem** shul**khan** pa**nui**?
I would like to book a table	Haiti rot**ze**/rot**za** lehaz**min** shul**khan**
The bill please	Khesh**bon**, bevaka**sha**
I am vegetarian	**A**ni tzimkho**ni**/tzimkho**nit**
menu	ta**frit**
fixed-price menu	ta**frit** is**kit**
wine list	ta**frit** hayei**not**
glass	kos
bottle	bak**buk**
knife	sa**kin**
spoon	kaf
fork	mas**leg**
breakfast	aru**khat** **bo**ker
lunch	aru**khat** tzoho**ry**im
dinner	aru**khat** **e**rev
starter	ma**na** risho**na**
main course	ma**na** ika**rit**
portion	ma**na**
rare	mevu**shal** me'**at**
well done	mevu**shal** he**tev**

Food and Drink

almonds	shke**dim**
apples	tapua**khei** etz
apricot	mish mish
aubergine/eggplant	khatzi**lim**
beans	shu'**it**
beef	**ba**kar
beer	**bi**ra
bread	**le**khem
broad beans	ful
broccoli	**bro**koli
butter	khem'**a**
cabbage	kruv
cake	ug**ha**
carrot	**ge**zer
cauliflower	kru**vit**
cheese	gvi**na**
cherries	dudva**nim**
chicken	off
chickpeas	**khu**mus
chips/fries	chips
chocolate	**sho**kolat
coffee	ka**fe**
cold cuts	pas**tra**ma
coriander	**kuz**bera
courgettes/zucchini	kishu**im**
crabs	sarta**nim**
cucumbers	melafefo**nim**
dessert	ki**nu**akh
draught beer	**bi**ra mihakha**vit**
dry	ya**vesh**
eggs	bet**za**
figs	te'e**nim**
fish	dag
French beans	shu'**it** yero**kha**
fried	metu**gan**
fruit	pei**rot**
garlic	shum
grapes	ana**vim**
grey mullet	**bu**ri
grilled	al ha**esh**
grouper	**lo**kus
hard-boiled eggs	bet**za** ka**sha**
herbal tea	**tei** tzma**khim**
hot (spicy)	kha**rif**
ice	**ke**rakh
icecream	**gli**da
kebab	shi**pud**
lamb, mutton	**ke**ves
lemon	li**mon**
liver	ka**ved**
meat	ba**sar**
milk	kha**lav**
mineral water	**my**im mine**ra**lim
nuts	ego**zim**
olive oil	**she**men **zy**it
omelette	khavi**ta**
onion	bat**zal**
orange juice (freshly squeezed)	mitz tapu**zim** (tiv'**i** sa**khut**)
oranges	tapu**zim**
peaches	afarse**kim**
pepper (condiment)	**pil**pel
peppers (capsicums)	pilpe**lim**
pickles	khamut**zim**
plums	shezi**fim**
potatoes	tapu**khei** ada**ma**
prawns/shrimps	shrimps
red snapper	**de**nis
red wine	**ya**in a**dom**
rice	**o**rez
roast	beta**nur**
salad	sa**lat** yera**kot**
salmon	sal**mon**
salt	**me**lakh
sandwich/filled roll	lakhmani**a**
sauce	**ro**tev
seafood	pei**rot** yam
smoked	me'u**shan**
soup	ma**rak**
spinach	**te**red
spinach beet (Swiss chard)	**alei se**lek
squid	kala**ma**ri
steak	steik
strawberries	tut sa**de** (tu**tim**)
stuffed vegetables	memula**im**
sugar	su**kar**
tea	**tei**
tomatoes	agvani**ot**
trout	fo**rel**
turkey	**ho**du
vegetables	yera**kot**
vinegar	**kho**metz **ya**in
water	**my**im
white wine	**ya**in la**van**

Numbers

0	**e**fes
1	a**khad**
2	**shta**im
3	sha**losh**
4	**ar**ba
5	kha**mesh**
6	shesh
7	**she**va
8	**shmo**ne
9	**tei**sha
10	**e**ser
11	a**ha**des**reh**
12	**shte**mes**reh**
13	**shlo**shes**reh**
14	ar**ba**es**re**
15	kha**me**shes**reh**
16	sheshes**reh**
17	**shva**es**reh**
18	shmona'es**reh**
19	**tsha**es**reh**
20	es**rim**
21	esrim vea**khad**
30	shlo**shim**
40	arba'**im**
50	khami**shim**
60	shi**shim**
70	shiv'**im**
80	shmo**nim**
90	tish'**im**
100	me**a**
200	ma**ty**im
300	shlosh me**ot**
1,000	**e**lef
2,000	al**py**im
3,000	shlo**sha e**lef
4,000	ar**ba e**lef
10,000	asa**ra e**lef

Time

one minute	da**ka**
one hour	sha'**a**
half an hour	khet**zi** sha'**a**
Sunday	yom ri**shon**
Monday	yom she**ni**
Tuesday	yom shli**shi**
Wednesday	yom revi'**i**
Thursday	yom khami**shi**
Friday	yom shi**shi**
Saturday	sha**bat**
week	sha**vu**'a
month	**kho**desh
year	sha**na**